T0230579

CRC HANDBOOK OF

THERMODYNAMIC DATA *of* COPOLYMER SOLUTIONS

CRC HANDBOOK OF

THERMODYNAMIC DATA *of* COPOLYMER SOLUTIONS

Christian Wohlfarth

CRC Press
Taylor & Francis Group
Boca Raton London New York

CRC Press is an imprint of the
Taylor & Francis Group, an **informa** business

CRC Press
Taylor & Francis Group
6000 Broken Sound Parkway NW, Suite 300
Boca Raton, FL 33487-2742

First issued in paperback 2021

ISBN 13: 978-1-03-210007-4 (pbk)
ISBN 13: 978-0-8493-1074-4 (hbk)

Publisher's Note
The publisher has gone to great lengths to ensure the quality of this reprint but points out that some imperfections in the original copies may be apparent.

Visit the Taylor & Francis Web site at
http://www.taylorandfrancis.com

and the CRC Press Web site at
http://www.crcpress.com

Library of Congress Cataloging-in-Publication Data

Catalog record is available from the Library of Congress

Foreword

Practical applications of thermodynamics as well as theoretical calculations of thermodynamic properties generally require data of real systems. In most cases these data rest on laboratory measurements of various physical properties. The study of phase behavior and thermodynamic properties of polymers and mixtures containing polymers has been subject of interest during the last 50 years. Investigations on such properties for copolymer systems were emphasized during the last decade when copolymers gained an increasing commercial interest because of their unique physical properties.

Much effort has been devoted over the years to compiling thermodynamic data for types of systems from literature and preparing compilations and databases for both scientific and industrial use. However, scarcely anything is found when one looks for compilations or databases that provide thermodynamic properties of polymer, or even more specially, copolymer solutions. Experimental information is spread over many articles and journals. There are only a small number of data books that cover this field. The author of this handbook wrote one of them on vapor-liquid equilibria of binary polymer solutions in 1994. He is known for his experience and his own experimental investigations on polymer and copolymer solutions for more than 20 years. With his new *Handbook of Thermodynamic Data of Copolymer Solutions* for the first time a compilation of thermodynamic data for copolymer solutions from the original literature is available.

Taking into account vapor-liquid equilibrium (VLE) data, liquid-liquid equilibrium (LLE) data, high-pressure phase equilibrium (HPPE) data of copolymer solutions in supercritical fluids, volumetric property (PVT) data of copolymer melts, enthalpy data, and second osmotic virial coefficients of copolymer solutions, the book covers all the necessary areas for researchers and engineers who work in this field.

When dealing with copolymer systems, one encounters the special problem of copolymer characterization since a copolymer is far from well-defined only by its chemical formula. Copolymers vary by a number of characterization variables. Molar mass, chemical composition, and distribution functions, tacticity, sequence distribution, branching, and end groups determine their thermodynamic behavior in solution. It is far from clear how these parameters influence the thermodynamic properties in detail. Unfortunately, there usually is not much information in the original papers; the available ones are added to each system in this book.

In comparison to low-molecular systems, the amount of data for copolymer solutions is still rather small. About 300 literature sources were perused for the purpose of this handbook, including some dissertations and diploma papers. Several hundred vapor-pressure isotherms, Henry's constants, LLE and HPPE data sets, a number of PVT data and some second osmotic virial coefficients are reported.

I am sure that readers interested in the field of thermodynamic properties of polymer solutions will benefit from this handbook and will identify the work that has to be done in the future.

Henry V. Kehiaian

Chairman
IUPAC-CODATA Task Group on Standard Physico-Chemical Data Formats

CRC Press
Taylor & Francis Group
6000 Broken Sound Parkway NW, Suite 300
Boca Raton, FL 33487-2742

First issued in paperback 2021

ISBN 13: 978-1-03-210007-4 (pbk)
ISBN 13: 978-0-8493-1074-4 (hbk)

Publisher's Note
The publisher has gone to great lengths to ensure the quality of this reprint but points out that some imperfections in the original copies may be apparent.

Visit the Taylor & Francis Web site at
http://www.taylorandfrancis.com

and the CRC Press Web site at
http://www.crcpress.com

Library of Congress Cataloging-in-Publication Data

Catalog record is available from the Library of Congress

PREFACE

Knowledge of thermodynamic data of copolymer solutions is a necessity for industrial and laboratory processes. Such data serve as essential tools for understanding the physical behavior of copolymer solutions, for studying intermolecular interactions, and for gaining insights into the molecular nature of mixtures. They also provide the necessary basis for any developments of theoretical thermodynamic models. Scientists and engineers in academic and industrial research need such data and will benefit from a careful collection of existing data. The *CRC Handbook of Thermodynamic Data of Copolymer Solutions* provides a reliable collection of such data for copolymer solutions from the original literature.

The *Handbook* is divided into seven chapters: (1) Introduction, (2) Vapor-Liquid Equilibrium (VLE) Data of Binary Copolymer Solutions, (3) Liquid-Liquid Equilibrium (LLE) Data of Quasibinary or Quasiternary Copolymer Solutions, (4) High-Pressure Phase Equilibrium (HPPE) Data of Quasibinary or Quasiternary Copolymer Solutions in Supercritical Fluids, (5) Enthalpy Changes for Binary Copolymer Solutions, (6) PVT Data of Molten Copolymers, and (7) Second Virial Coefficients (A_2) of Copolymer Solutions. Finally, four appendices quickly route the user to the desired data sets.

Original data have been gathered from approximately 300 literature sources, including also a number of dissertations and diploma papers. The *Handbook* provides about 250 vapor-pressure isotherms, 75 tables of Henry's constants, 50 LLE data sets, 175 HPPE data sets, and 70 PVT data tables for more than 165 copolymers and 165 solvents. Data are included only if numerical values were published or authors provided their numerical results by personal communication (and I wish to thank all those who did so). *No digitized data* have been included in this data collection, but some tables include systems data published in graphical form. The *Handbook* is the first complete overview about this subject in the world's literature. The closing day for the data collection was October 1, 2000. The *Handbook* results from parts of a more general database, *Thermodynamic Properties of Polymer Systems*, which is continuously updated by the author. Thus, the user who is in need for new additional data sets is kindly invited to ask for new information beyond this book via e-mail at wohlfarth@chemie.uni-halle.de. Additionally, the author will be grateful to users who call his attention to mistakes and make suggestions for improvements.

The *Handbook* also highlights the work still to be done – obvious, when one compares the relatively small number of copolymer solutions for which data exist with the number of copolymers in use today. Additionally, only a small minority of possible solutions of the copolymers covered by this book were properly investigated (in relation to the combinatorial number of copolymer/solvent pairs, although it is appreciated that not all make thermodynamic sense or are of practical use).

The *CRC Handbook of Thermodynamic Data of Copolymer Solutions* will be useful to researchers, specialists, and engineers working in the fields of polymer science, physical chemistry, chemical engineering, material science, and those developing computerized predictive packages. The *Handbook* should also be of use as a data source to Ph.D. students and faculty in Chemistry, Physics, Chemical Engineering, and Materials Science Departments at universities.

Merseburg, January 2001

Christian Wohlfarth

About the Author

Christian Wohlfarth is Associate Professor for Physical Chemistry at Martin Luther University Halle-Wittenberg, Germany. He earned his degree in Chemistry in 1974 and wrote his Ph.D. thesis on investigations on the second dielectric virial coefficient and the intermolecular pair potential in 1977, both at Carl Schorlemmer Technical University Merseburg. In 1985, he wrote his habilitation thesis, *Phase Equilibria in Systems with Polymers and Copolymers*, at Technical University Merseburg.

Since then, his main research is related to polymer systems. Currently, his research topics are molecular thermodynamics, continuous thermodynamics, phase equilibria in (co)polymer mixtures and solutions, (co)polymers in supercritical fluids, PVT-behavior and equations of state, sorption properties of (co)polymers, about which he has published approximately 90 original papers. He has also built a database, *Thermodynamic Properties of Polymer Systems,* and has written the book *Vapor-Liquid Equilibria of Binary Polymer Solutions.*

He is working on the evaluation, correlation, and calculation of thermophysical properties of pure compounds and mixtures resulting in 6 volumes of *Landolt-Börnstein New Series*. He is a member of the Editorial Board of *ELDATA: The International Electronic Journal of Physico-Chemical Data.*

CONTENTS

1. INTRODUCTION

1.1. Objectives of the handbook 1
1.2. Experimental methods involved 2
1.3. Guide to the data tables 21
1.4. List of symbols 25
1.5. References 27

2. VAPOR-LIQUID EQUILIBRIUM (VLE) DATA OF BINARY COPOLYMER SOLUTIONS

2.1. Partial solvent vapor pressures or solvent activities for copolymer solutions 29
2.2. Classical mass-fraction Henry's constants of solvent vapors in molten copolymers 88
2.3. References 142

3. LIQUID-LIQUID EQUILIBRIUM (LLE) DATA OF QUASIBINARY OR QUASITERNARY COPOLYMER SOLUTIONS

3.1. Cloud-point and/or coexistence curves of quasibinary solutions 145
3.2. Table of systems where binary LLE data were published only in graphical form as phase diagrams or related figures 160
3.3. Cloud-point and/or coexistence curves of quasiternary solutions containing at least one copolymer 163
3.4. Table of systems where ternary LLE data were published only in graphical form as phase diagrams or related figures 172
3.5. Lower critical (LCST) and/or upper (UCST) critical solution temperatures of copolymer solutions 175
3.6. References 183

4. HIGH-PRESSURE PHASE EQUILIBRIUM (HPPE) DATA OF COPOLYMER SOLUTIONS IN SUPERCRITICAL FLUIDS

4.1. Experimental data of quasibinary copolymer solutions 187
4.2. Table of systems where binary HPPE data were published only in graphical form as phase diagrams or related figures 225
4.3. Experimental data of quasiternary solutions containing at least one copolymer 229
4.4. Table of systems where ternary HPPE data were published only in graphical form as phase diagrams or related figures 264
4.5. References 266

5. ENTHALPY CHANGES FOR BINARY COPOLYMER SOLUTIONS

5.1. Enthalpies of mixing or intermediary enthalpies of dilution, and copolymer partial enthalpies of mixing (at infinite dilution), or copolymer (first) integral enthalpies of solution 269
5.2. Partial molar enthalpies of mixing at infinite dilution of solvents and enthalpies of solution of gases/vapors of solvents in molten copolymers from inverse gas-liquid chromatography (IGC)....307
5.3. Table of systems where additional information on enthalpy effects in copolymer solutions can be found....333
5.4. References....337

6. PVT DATA OF MOLTEN COPOLYMERS

6.1. Experimental data and/or Tait equation parameters....341
6.2. References....430

7. SECOND VIRIAL COEFFICIENTS (A_2) OF COPOLYMER SOLUTIONS

7.1. Experimental A_2 data....431
7.2. References....445

8. APPENDICES

8.1. List of copolymer acronyms....447
8.2. List of systems and properties in order of the copolymers....451
8.3. List of solvents in alphabetical order....494
8.4. List of solvents in order of their molecular formulas....499

1. INTRODUCTION

1.1. Objectives of the handbook

Knowledge of thermodynamic data of copolymer solutions is a necessity for industrial and laboratory processes. Furthermore, such data serve as essential tools for understanding the physical behavior of copolymer solutions, for studying intermolecular interactions, and for gaining insights into the molecular nature of mixtures. They also provide the necessary basis for any developments of theoretical thermodynamic models. Scientists and engineers in academic and industrial research need such data and will benefit from a careful collection of existing data. However, the database for polymer solutions is still modest in comparison with the enormous amount of data for low-molecular mixtures, and the specialized database for copolymer solutions is even smaller. On the other hand, copolymers are gaining increasing commercial interest because of their unique physical properties, and thermodynamic data are needed for optimizing their synthesis, production, and application.

Basic information on polymers as well as copolymers can be found in the *Polymer Handbook* (99BRA). Some data books on polymer solutions appeared in the early 1990s (90BAR, 92WEN, 93DAN, and 94WOH), but most data for copolymer solutions have been compiled during the last decade. No books or databases dedicated to copolymer solutions presently exist. Thus, the intention of the *Handbook* is to fill this gap and to provide scientists and engineers with an up-to-date compilation from the literature of the available thermodynamic data on copolymer solutions. The *Handbook* does not present theories and models for (co)polymer solution thermodynamics. Other publications (71YAM, 90FUJ, 90KAM, and 99PRA) can serve as starting points for investigating those issues.

The data within this book are divided into six chapters:

- Vapor-liquid equilibrium (VLE) data of binary copolymer solutions
- Liquid-liquid equilibrium (LLE) data of quasibinary or quasiternary copolymer solutions
- High-pressure phase equilibrium (HPPE) data of copolymer solutions in supercritical fluids
- Enthalpy changes for binary copolymer solutions
- PVT data of molten copolymers
- Second virial coefficients (A_2) of copolymer solutions

Data from investigations applying to more than one chapter are divided and appear in the relevant chapters. Data are included only if numerical values were published or authors provided their numerical results by personal communication (and I wish to thank all those who did so). No digitized data have been included in this data collection, but some tables include systems data published in graphical form.

This volume also highlights the work still to be done – obvious, when one compares the relatively small number of copolymer solutions for which data exist with the number of copolymers in use today. Additionally, only a small minority of possible solutions of the copolymers covered by this book were properly investigated (in relation to the combinatorial number of copolymer/solvent pairs, although it is appreciated that not all make thermodynamic sense or are of practical use).

Very few investigations involved thermodynamic data for particular copolymer solutions, and the temperature (and/or pressure) ranges usually investigated are rather limited. The *Handbook* provides the results of recent research, and clearly identifies areas that require further exploration in the future.

1.2. Experimental methods involved

Vapor-liquid equilibrium (VLE) measurements

Investigations on vapor-liquid equilibrium of polymer solutions can be made by various methods:

1. Absolute vapor pressure measurement
2. Differential vapor pressure measurement
3. Isopiestic sorption/desorption methods, i.e., gravimetric sorption, piezoelectric sorption, or isothermal distillation
4. Inverse gas-liquid chromatography (IGC) at infinite dilution, IGC at finite concentrations, and headspace gas chromatography (HSGC)
5. Steady-state vapor-pressure osmometry (VPO)

Experimental techniques for vapor pressure measurements were reviewed in 75BON and 2000WOH. Methods and results of the application of IGC to polymers and polymer solutions were more often reviewed (76NES, 88NES, 89LLO, 89VIL, and 91MU1). Reviews on ebulliometry and/or vapor-pressure osmometry can be found in 74TOM, 75GLO, 87COO, 91MAY, and 99PET.

The measurement of vapor pressures for polymer solutions is generally more difficult and more time-consuming than that of low-molecular mixtures. The main difficulties can be summarized as follows: Polymer solutions exhibit strong negative deviations from Raoult's law. These are mainly due to the large entropic contributions caused by the difference between the molar volumes of solvents and polymers as was explained by the classical Flory-Huggins theory about 60 years ago. However, because of this large difference in molar mass, vapor pressures of dilute solutions do not differ markedly from the vapor pressure of the pure solvent at the same temperature, even at polymer concentrations of 10-20 wt%. This requires special techniques to measure very small differences in solvent activities. Concentrated polymer solutions are characterized by rapidly increasing viscosities with increasing polymer concentration. This leads to a strong increase in time required to obtain real thermodynamic equilibrium caused by slow solvent diffusion effects (in or out of a non-equilibrium-state polymer solution). Furthermore, only the solvent coexists in both phases because polymers do not evaporate. The experimental techniques used for the measurement of vapor pressures of polymer solutions have to take into account all these effects.

Vapor pressures of polymer solutions are usually measured in the isothermal mode by static methods. Dynamic methods are seldom applied, e.g., ebulliometry (75GLO and 87COO). At least, one can consider measurements by VPO to be dynamic ones, where a dynamic (steady-state) balance is obtained. Limits for the applicable ranges of polymer concentration and polymer molar mass, limits for the solvent vapor pressure and the measuring temperature and some technical restrictions prevent its broader application, however. Static techniques usually work at constant temperature. The three different methods (1 through 3 above) were used to determine most of the vapor pressures of polymer solutions in the literature. All three methods have to solve the problems of establishing real thermodynamic equilibrium between liquid polymer solution and solvent vapor phase, long-time temperature constancy during the experiment, determination of the final polymer concentration, and determination of pressure and/or activity. Absolute vapor pressure measurement and differential vapor pressure methods were mostly used by early workers. Most recent measurements were done with the isopiestic sorption methods. Gas-liquid chromatography as IGC closes the gap at high polymer concentrations where vapor pressures cannot be measured with sufficient accuracy. HSGC can be considered as some combination of absolute vapor pressure measurement with GLC.

The following text (a short summary from the author's own review, 2000WOH) explains briefly the usual experimental equipment and the measuring procedures.

1. Absolute vapor pressure measurement

Absolute vapor pressure measurement may be considered the classical technique for our purpose, because one measures directly the vapor pressure above a solution of known polymer concentration. The literature gives a variety of absolute vapor pressure apparatuses developed and used by different authors. Vapor pressure measurement and solution equilibration were often made separately. A polymer sample is prepared by weighing, the sample flask is evacuated, degassed solvent is introduced into a sample flask that is sealed thereafter. Samples are equilibrated at elevated temperature in a thermostatted bath for some weeks. The flask with the equilibrated polymer solution is then connected with the pressure-measuring device at the measuring temperature. The vapor pressure is measured after reaching equilibrium and the final polymer concentration is obtained after correcting for the amount of evaporated solvent. Modern equipment applies electronic pressure sensors and digital technique to measure the vapor pressure. Data processing can then be made online by computers. A number of problems have to be solved during the experiment. The solution is usually of an amount of some cm^3 and may contain about 1g of polymer or even more. Degassing is absolutely necessary. All impurities in the pure solvent have to be eliminated. Equilibration of all prepared solutions is very time consuming (liquid oligomers need not so much time, of course). Increasing viscosity makes the preparation of concentrated solutions more and more difficult with further increasing amount of polymer. Solutions above 50-60 wt% can hardly be prepared (depending on the solvent/polymer pair under investigation).

The determination of the volume of solvent vaporized in the unoccupied space of the apparatus is difficult and can cause serious errors in the determination of the final solvent concentration. To circumvent the vapor phase correction, one can measure the concentration directly by means, for example, of a differential refractometer. The contact of solvent vapor with the Hg surface in older equipment may cause further errors. Complete thermostatting of the whole apparatus is necessary to avoid condensation of solvent vapor at colder spots. Since it is disadvantageous to thermostat Hg manometers at higher temperatures, null measurement instruments with pressure compensation were sometimes used. Modern electronic pressure sensors can be thermostatted within certain temperature ranges. If pressure measurement is made outside the thermostatted equilibrium cell, the connecting tubes must be heated slightly above the equilibrium temperature to avoid condensation.

The measurement of polymer solutions with lower polymer concentrations requires very precise pressure instruments, because the difference to the pure solvent vapor pressure becomes very small with decreasing amount of polymer. No one can really answer the question if real thermodynamic equilibrium is obtained or only a frozen non-equilibrium state is achieved. Non-equilibrium data can be detected from unusual shifts of the χ-function with some experience. Also some kind of hysteresis appearing in experimental data seems to point to non-equilibrium results. A common consistency test on the basis of the integrated Gibbs-Duhem equation does not work for vapor pressure data of binary polymer solutions because the vapor phase is pure solvent vapor. Thus, absolute vapor pressure measurements need very careful handling, plenty of time and an experienced experimentator. They are not the methods of choice for highly viscous polymer solutions.

2. Differential vapor pressure measurement

The differential method can be compared under some aspects with the absolute method, but it has some advantages. The measuring principle is to obtain the vapor pressure difference between the pure solvent and the polymer solution at the measuring temperature. Again, the polymer sample is filled, after weighing, into a sample flask, the apparatus is evacuated, a desired amount of degassed solvent is distilled into the sample flask to build the solution and the samples have to be equilibrated for a necessary duration of time. The complete apparatus is kept at constant measuring temperature and, after reaching equilibrium, the pressure difference is read from the manometer difference and the concentration is calculated after correcting the amount of vaporized solvent in the unoccupied space of the equipment. The pure solvent vapor pressure is usually precisely known from independent experiments.

Difference/differential manometers have some advantages from their construction: they are comparatively smaller and their resolution is much higher (modern pressure transducers can resolve differences of 0.1 Pa and less). However, there are the same disadvantages with sample/solution preparation (solutions of grams of polymer in some cm^3 volume, degassing, viscosity), long-time thermostatting of the complete apparatus because of long equilibrium times (increasing with polymer molar mass and concentration/viscosity of the solution), correction of unoccupied vapor space, impurities of the solvent, connection to the Hg surface in older equipment, and the problem of obtaining real thermodynamic equilibrium (or not) as explained above. Modern equipment uses electronic pressure sensors instead of Hg manometers and digital technique to measure the vapor pressure. Also, thermostatting is more precise in recent apparatuses. Problems caused by the determination of the unoccupied vapor space could be avoided by measuring the absolute vapor pressure as well.

Again, the concentration can be determined independently by using a differential refractometer and a normalized relation between concentration and refractive index. Degassing of the liquids remains a necessity. Time for establishing thermodynamic equilibrium could somewhat be shortened by intensive stirring. In comparison to absolute vapor pressure measurements, differential vapor pressure measurements with a high resolution for the pressure difference can be applied even for dilute polymer solutions where the solvent activity is very near to 1. They need more time than VPO measurements, however.

3. Isopiestic sorption/desorption methods

Isopiestic measurements allow a direct determination of solvent activity or vapor pressure in polymer solutions by using a reference system (a manometer may not have to be applied). There are two general principles for lowering the solvent activity in the reference system: concentration lowering or temperature lowering. Isopiestic measurements have to obey the condition that no polymer can vaporize (as might be the case for lower-molecular oligomers at higher temperatures).

Concentration lowering under isothermal conditions is the classical isopiestic technique, sometimes also called isothermal distillation. A number of solutions (two as the minimum) are in contact with each other via their common solvent vapor phase and solvent evaporates and condenses (this is the isothermal *distillation* process) between them as long as the chemical potential of the solvent is equal in all solutions. At least one solution serves as reference system, i.e., its solvent activity vs. solvent concentration dependence is precisely known. After an exact determination of the solvent concentration in all equilibrated solutions (usually by weighing), the solvent activity in all measured solutions is known from and equal to the activity of the reference solution. This method is almost exclusively used for aqueous polymer solutions, where salt solutions can be applied as reference systems. It is a standard method for inorganic salt systems.

Temperature lowering at specified isobaric or isochoric conditions is the most often used technique for the determination of solvent vapor pressures or activities in polymer solutions. The majority of all measurements are made using this kind of an isopiestic procedure where the pure solvent is used as the reference system. The equilibrium condition of equal chemical potential of the solvent in the polymer solution as well as in the reference system is realized by keeping the pure solvent at a lower temperature (T_1) than the measuring temperature (T_2) of the solution. In equilibrium, the vapor pressure of the pure solvent at the lower temperature is then equal to the partial pressure of the solvent in the polymer solution, i.e., $P_1^s(T_1) = P_1(T_2)$. Equilibrium is again established via the common vapor phase for both subsystems.

The vapor pressure of the pure solvent is either known from independent data or measured additionally in connection with the apparatus. The composition of the polymer solution can be altered by changing T_1 and a wide range of compositions can be studied (between 30-40 and 85-90 wt% polymer, depending on the solvent). Measurements above 85-90 wt% polymer are subject to increasing errors because of surface adsorption effects.

A broad variety of experimental equipment is based on this procedure. This isopiestic technique is the recommended one for most polymer solutions since it is advantageous in nearly all aspects of measurement. It covers the broadest concentration range. Only very small amounts of polymer are needed (about 30-50 mg with the classical quartz spring balance, about 100 μg with piezoelectric sorption detector or microbalance techniques, see below). It is much more rapid than all other methods explained above, because equilibrium time decreases drastically for such small amounts of polymer and polymer solution (about 12-24 hours for the quartz spring balance, about 3-4 hours for piezoelectric or microbalance techniques). The complete isotherm can be measured using a single loading of the apparatus. Equilibrium is easier to obtain since comparatively small amounts of solvent have to diffuse into the bulk sample solution. Equilibrium can be tested better by measuring sorption and desorption runs which must lead to equal results for thermodynamic absorption equilibrium.

Supercritical solvents can be investigated if the piezoelectric detector is used (otherwise buoyancy in dense fluids may cause serious problems). Much broader pressure and temperature ranges can be covered with relatively simple equipment, which may again be limited by the weighing system. Isopiestic sorption measurements can be automated and will also allow kinetic experiments. They have two disadvantages. First, isopiestic sorption measurements below about 30 wt% polymer are subject to increasing errors because very small temperature differences (vapor pressure changes) are connected with large changes in concentration. Second, problems may arise with precise thermostatting of both the solvent and the solution at different constant temperatures over a longer period of time.

The classical concept is the sorption method using a quartz spring balance that measures the extension of the quartz spring according to Hook's law (linear relationship, no hysteresis). In this method a weighed quantity of the (non-volatile) polymer is placed on the pan of the quartz spring balance within a measuring cell. The determination of spring extension vs. mass has to be made in advance as a calibration procedure. Reading of the spring extension is usually made by means of a cathetometer. The cell is sealed, evacuated and thermostatted to the measuring temperature (T_2) and the solvent is then introduced into the measuring cell as solvent vapor. The solvent vapor is absorbed by the polymer sample to form the polymer solution until thermodynamic equilibrium is reached. The solvent vapor is provided from a reservoir of either pure liquid solvent thermostatted at a lower temperature (T_1) or a reference liquid solution of known concentration/solvent partial pressure as in the case of the isothermal distillation procedure as described above. Such an apparatus was used widely in the author's work.

The following problems have to be solved during the experiment. The equilibrium cell has to be sealed carefully to avoid any air leakage over the complete duration of the measurements (to measure one isotherm takes about 14 days). Specially developed thin Teflon® sealing rings are preferred to grease. The polymer sample has to stand the temperature. Changes by thermal aging during the experiment must be avoided. The temperatures provided by the thermostats must not fluctuate more than ± 0.1 K. Condensation of solvent vapor at points that become colder than T_2 has to be avoided. As was stated by different experimentalists, additional measurement of the vapor pressure inside the isopiestic sorption apparatus seems to be necessary if there is some doubt about the real pressure or if no reliable pure solvent vapor pressure data exist for the investigated temperature range. This direct pressure measurement has the advantage that absolute pressures can be obtained and pressure fluctuations can be observed. More modern equipment applies electronic pressure sensors instead of Hg manometers to avoid the problems caused by the contact of solvent vapor with the mercury surface and to get a better resolution of the measuring pressure.

Isopiestic vapor sorption can be made using an electronic microbalance instead of the quartz spring balance. Electronic microbalances are commercially available from a number of producers. Their main advantages are their high resolution and their ability to allow kinetic measurements. Additionally, experiments using electronic microbalances can be automated easily and provide computing facilities.

The major disadvantage with some kinds of microbalances is that they cannot be used at high solvent vapor pressures and so are limited to a relatively small concentration range. However, since thin polymer films can be applied, this reduces both the time necessary to attain equilibrium (some hours) and the amount of polymer required, and equilibrium solvent absorption can be obtained also at polymer mass fractions approaching 1 (i.e., for small solvent concentrations). Depending on their construction, the balance head is situated inside or outside the measuring apparatus. Problems may arise when it is inside where the solvent vapor may come into contact with some electronic parts. Furthermore, all parts of the balance that are inside the apparatus have to be thermostatted to the measuring temperature to enable the correct equilibration of the polymer solution or even slightly above measuring temperature to avoid condensation of solvent vapor in parts of the balance. The allowed temperature range of the balance and its sensitivity to solvent corrosion determine the accessible measuring range of the complete apparatus. A magnetic suspension balance can be applied instead of an electronic microbalance. The magnetic suspension technique has the advantage that all sensitive parts of the balance are located outside the measuring cell because the balance and the polymer solution measuring cell are in separate chambers and connected by magnetic coupling only. This allows magnetic suspension balances to be used at temperatures up to about 500 K as well as at pressures up to about 200 MPa.

The most sensitive solvent vapor sorption method is the piezoelectric sorption detector. The amount of solvent vapor absorbed by a polymer is detected by a corresponding change in frequency of a piezoelectric quartz crystal coated with a thin film of the polymer because a frequency change is the response of a mass change at the surface of such a crystal. The frequency of the crystal decreases as mass increases when the crystal is placed in a gas or vapor medium. The frequency decrease is fairly linear. The polymer must be coated onto the crystal from a solution with some care to obtain a fairly uniform film. Measurements can be made at dynamic (vapor flow) or static conditions. With reasonable assumptions for the stability of the crystal's base frequency and the precision of the frequency counter employed, the piezoelectric method allows the detection of as few as 10 nanograms of solvent using a 10 MHz crystal. This greatly reduces both the time necessary to attain equilibrium (3-4 hours) and the amount of polymer required. Because very thin films are applied, equilibrium solvent absorption can be obtained also at polymer mass fractions approaching 1 (i.e., for small solvent concentrations). Sorption-desorption hysteresis has never been observed when using piezoelectric detectors. This demonstrates the effect of reducing the amount of polymer from about 50 mg for the quartz spring sorption technique by an order of 10^3 for the piezoelectric detector. However, measurements are limited to solvent concentrations well below the region where solution drops would be formed. On the other hand, measurements can be made also at higher temperatures and pressures. Limits are set by the stability of the electrical equipment and the construction of the measuring cell.

4. Gas-liquid chromatography (GLC)

GLC can be used to determine the activity coefficient of a solute in a (molten) polymer at essentially zero solute concentration. This type of activity coefficient is known as an infinite-dilution activity coefficient. Because the liquid polymer in the stationary phase acts as a *solvent* for the very small amount of an injected solute sample, this technique is often called *inverse* gas-liquid chromatography (IGC).

The equipment does not differ in principle very much from that used in analytical GLC. For operating at infinite dilution, the carrier gas flows directly to the column which is inserted into a thermostatted oil bath (to get a more precise temperature control than in a conventional GLC oven). The output of the column is measured with a flame ionization detector or alternately with a thermal conductivity detector. Helium is used today as the carrier gas (nitrogen was used in earlier work). From the difference between the retention time of the injected solvent sample and the retention time of a non-interacting gas (marker gas), thermodynamic equilibrium data can be obtained. Most experiments were done up to now with packed columns, but capillary columns were used too.

The experimental conditions must be chosen so that real thermodynamic data can be obtained, i.e., equilibrium bulk absorption conditions. Errors caused by unsuitable gas flow rates, unsuitable polymer loading percentages on the solid support material and support surface effects as well as any interactions between the injected sample and the solid support in packed columns, unsuitable sample size of the injected probes, carrier gas effects, and imprecise knowledge of the real amount of polymer in the column, can be sources of problems, whether data are nominally measured under real thermodynamic equilibrium conditions or not, and have to be eliminated. The sizeable pressure drop through the column must be measured and accounted for. Column preparation is the most difficult and time-consuming task within the IGC experiment. Two, three or even more columns must be prepared to test the reproducibility of the experimental results and to check any dependence on polymer loading and sometimes to filter out effects caused by the solid support. In addition, various tests regarding solvent sample size and carrier gas flow rate have to be done to find out correct experimental conditions. There is an additional condition for obtaining real thermodynamic equilibrium data that is caused by the nature of the polymer sample. Synthetic polymers are usually amorphous or semicrystalline products. Thermodynamic equilibrium data require the polymer to be in a molten state, however. This means that IGC measurements have to be performed for our purpose well above the glass transition temperature of the amorphous polymer or even above the melting temperature of the crystalline parts of a polymer sample. On the other hand, IGC can be applied to determine these temperatures. Only data at temperatures well above the glass transition temperature lead to real thermodynamic vapor-liquid equilibrium data. As a rule, the experimental temperature must exceed the glass transition temperature by about 50 K.

GLC can also be used to determine the partial pressure of a solute in a polymer solution at concentrations as great as 50 wt% solute. In this case of finite concentration IGC, a uniform background concentration of the solute is established in the carrier gas. The carrier gas is diverted to a saturator through a metering valve. In the saturator it passes through a diffuser in a well-stirred, temperature-controlled liquid bath. It leaves the separator with the solute equilibrium vapor pressure in the carrier gas. The solute concentration is varied by changing the saturator temperature. Precise control of the temperature bath is needed in order to obtain a constant plateau concentration. Upon leaving the saturator the gas flows to the injector block and then to the column. As in the infinite dilute case a small pulse of the solvent is then injected. This technique is known as elution on a plateau. Because finite concentration IGC is technically more complicated, few workers have applied it. Whereas the vapor sorption results are more accurate at higher concentrations, the reverse is true for finite concentration IGC since larger injection volumes have to be used, which strains the theory on which the calculations are based. Also, at large vapor concentrations the chromatographic peaks become more spread out, making the measurement of retention times less precise. Additionally, the concentration range is limited by the requirement that the saturator temperature must be below that of the column. Clearly, at higher measuring temperatures, higher solvent concentrations may be used. Finite concentration IGC can be extended to multicomponent systems. Data reduction is somewhat complicated, however.

VLE measurements for polymer solutions can be done by so-called headspace gas chromatography (HSGC), which is practically a combination of static vapor pressure measurement with gas chromatographic detection (97KOL). Again, polymer solutions have to be prepared in advance and have to be equilibrated at the measuring temperature for some weeks before measurement. HSGC experiments were carried out with an apparatus consisting of a headspace sampler and a normal gas chromatograph. The thermostatted headspace sampler samples a constant amount of gas phase and injects this mixture into the gas chromatograph. After separation of the components of the gaseous mixture in a capillary column, they are detected individually by a thermal conductivity detector. The signals are sent to an integrator which calculates the peak areas proportional to the amount of gas in the sample volume and consequently to the vapor pressure. Calibration can be done by measuring the signal of the pure solvent in dependence on temperature and comparing the data with the corresponding vapor pressure vs. temperature data. Measurements can be done between about 25 and 85 wt% polymer in the solution (again depending on temperature, solvent and polymer investigated). In order to guarantee thermodynamic equilibrium in the sampler, solutions have to be conditioned for at least 24 h at constant temperature in the headspace sampler before measurement. Additional degassing is not necessary and solvents have to be purified only to the extent that is necessary to prevent unfavorable interactions in the solution. The experimental error in the vapor pressures is typically of the order of 1-3%. One great advantage of HSGC is its capability to measure VLE data not only for binary polymer solutions but also for polymer solutions in mixed solvents since it provides a complete analysis of the vapor phase in equilibrium.

The *data reduction for infinite dilution IGC* starts with the usually obtained parameters of retention volume or net retention volume which have to be calculated from the measured retention times and the flow rate of the carrier gas at column conditions.

$$V_{net} = V_r - V_{dead} \tag{1}$$

where:

V_{net}	net retention volume
V_r	retention volume
V_{dead}	retention volume of the inert marker gas, dead retention, gas holdup

These net retention volumes are reduced to specific retention volumes, V_g^0, by division of equation (1) with the mass of the liquid (here the liquid is the molten copolymer). They are corrected for the pressure difference between column inlet and outlet pressure, and reduced to a temperature T_0 = 273.15 K.

$$V_g^0 = \left(\frac{V_{net}}{m_B}\right)\left(\frac{T_0}{T}\right)\frac{3(P_{in}/P_{out})^2-1}{2(P_{in}/P_{out})^3-1} \tag{2}$$

where:

V_g^0	specific retention volume corrected to 0°C = 273.15 K
m_B	mass of the copolymer in the liquid phase within the column
P_{in}	column inlet pressure
P_{out}	column outlet pressure
T	measuring temperature
T_0	reference temperature = 273.15 K

Theory of GLC provides the relation between V_g^0 and thermodynamic data for the low-molecular component (solvent A) at infinite dilution:

$$\left(\frac{P_A}{x_A^L}\right)^{\infty} = \frac{RT_0}{V_g^0 M_B} \quad \text{or} \quad \left(\frac{P_A}{w_A^L}\right)^{\infty} = \frac{RT_0}{V_g^0 M_A} \tag{3}$$

where:

M_A	molar mass of the solvent A
M_B	molar mass of the liquid (molten) polymer B
P_A	partial vapor pressure of the solvent A at temperature T
P_A^s	saturation vapor pressure of the pure liquid solvent A at temperature T
R	gas constant
x_A^L	mole fraction of solvent A in the liquid solution
w_A^L	mass fraction of solvent A in the liquid solution

The activity coefficients at infinite dilution read, if we neglect interactions to and between carrier gas molecules (which are normally helium):

$$\gamma_A^\infty = \left(\frac{RT_0}{V_g^0 M_B P_A^s}\right) \exp\left[\frac{(B_{AA} - V_A^L)(P - P_A^s)}{RT}\right] \tag{4}$$

$$\Omega_A^\infty = \left(\frac{RT_0}{V_g^0 M_A P_A^s}\right) \exp\left[\frac{(B_{AA} - V_A^L)(P - P_A^s)}{RT}\right] \tag{5}$$

where:

B_{AA}	second virial coefficient of the pure solvent A at temperature T
V_A^L	molar volume of the pure liquid solvent A at temperature T
γ_A	activity coefficient of the solvent A in the liquid phase with activity $a_A = x_A\gamma_A$
Ω_A	mass fraction-based activity coefficient of the solvent A in the liquid phase with activity $a_A = w_A\Omega_A$

The standard state pressure P has to be specified. It is common practice by many authors to define zero pressure as standard pressure since pressures are usually very low during GLC measurements. Then, equations (4 and 5) change to:

$$\gamma_A^\infty = \left(\frac{RT_0}{V_g^0 M_B P_A^s}\right) \exp\left[\frac{P^s{}_A(V_A^L - B_{AA})}{RT}\right] \tag{6}$$

$$\Omega_A^\infty = \left(\frac{RT_0}{V_g^0 M_A P_A^s}\right) \exp\left[\frac{P_A^s(V_A^L - B_{AA})}{RT}\right] \tag{7}$$

One should keep in mind that mole fraction-based activity coefficients γ_A become very small values for common polymer solutions and reach a value of zero for $M_B \rightarrow\infty$, which means a limited applicability at least to oligomer solutions. Therefore, the common literature provides only mass fraction-based activity coefficients for (high-molecular) polymer/(low-molecular) solvent pairs. The molar mass M_B of the polymeric liquid is an average value (M_n) according to the usual molar-mass distribution of polymers. Additionally, it is a second average if mixed stationary liquid phases are applied.

Furthermore, thermodynamic VLE data from GLC measurements are provided in the literature as values for $(P_A/w_A)^\infty$, see equation (3), i.e., classical mass fraction based Henry's constants (if assuming ideal gas phase behavior):

$$H_{A,B} = \left(\frac{P_A}{w_A^L}\right)^\infty = \frac{RT_0}{V_g^0 M_A} \tag{8}$$

Since $V_{net} = V_r - V_{dead}$, the marker gas is assumed to not be retained by the copolymer stationary phase and will elute at a retention time that is usually very small in comparison with those of the samples investigated. However, for small retention volumes, values for the mass fraction-based Henry's constants should be corrected for the solubility of the marker gas (76LIU). The apparent Henry's constant is obtained from equation (8) above.

$$H_{A,B} = H_{A,B}^{app}\left[1 + \frac{M_A H_{A,B}^{app}}{M_{ref} H_{A,ref}}\right]^{-1} \tag{9}$$

M_{ref} is the molar mass of the marker gas. The Henry's constant of the marker gas itself, determined by an independent experiment, need not be known very accurately, as it is usually much larger than the apparent Henry's constant of the sample.

5. Vapor-pressure osmometry (VPO)

Vapor-pressure osmometry is from its name comparable to membrane osmometry by allowing the vapor phase to act like a semipermeable membrane, but it is based on vapor pressure lowering or boiling temperature elevation. Since the direct measure of vapor pressure lowering of dilute polymer solutions is impractical because of the extreme sensitivity that is required, VPO is in widespread use for low-molecular and oligomer solutions (i.e., M_n less than 20,000 g/mol) by employing the thermoelectric method where two matched temperature-sensitive thermistors are placed in a chamber that is thermostatted to the measuring temperature and where the atmosphere is saturated with solvent vapor. If drops of pure solvent are placed on both thermistors, the thermistors will be at the same temperature (zero point calibration). If a solution drop is placed on one thermistor, a temperature difference ΔT which is caused by condensation of solvent vapor onto the solution drop occurs. From equilibrium thermodynamics it follows that this temperature increase has its theoretical limit when the vapor pressure of the solution is equal to that of the pure solvent, i.e., at infinite dilution. The obtained temperature difference is very small, about 10^{-5} K.

Because solvent transfer effects are measured, VPO is a dynamic method. This leads to a time-dependent measurement of ΔT. Depending on technical details of the equipment, sensitivity of the temperature detector, measuring temperature, solvent vapor pressure and polymer concentration in the solution drop, a steady state for ΔT can be obtained after some minutes. The value of ΔT^{st} is the basis for thermodynamic data reduction; see below. If measuring conditions do not allow a steady state, an extrapolation method to ΔT at zero measuring time can be employed for data reduction. Sometimes a value is used that is obtained after a predetermined time. However, this may lead to some problems with knowing the exact polymer concentration in the solution. The extrapolation method is somewhat more complicated and needs experience of the experimentator but gives an exact value of polymer concentration. Both methods are used within solvent activity measurements where polymer concentrations are higher and condensation is faster than in common polymer characterization experiments.

Experience has shown that careful selection of solvent and temperature is critical to the success of the VPO experiment. Nearly all common solvents, including water (usually, there are different thermistor sensors for organic solvents and for water), can be used with VPO. The measuring temperature should be chosen so that the vapor pressure of the solvent will be greater than 6,000 Pa, but not so high as to lead to problems with evaporation from the chamber. Solvent purity is critical, and volatile impurities and water must be avoided. Greater sensitivity can be achieved by using solvents with low enthalpies of vaporization. This means that not all desirable polymer/solvent pairs and not all temperature (pressure) ranges can be investigated by VPO. Additionally, VPO has some inherent sources of error. These can be attributed to the possible existence of surface films, to differences in diffusion coefficients in solutions, to appreciably different solution concentrations, to differences in heat conductivity, to problems with drop size and shape, to the occurrence of reactions in the solution, and to the presence of volatile solutes. Of course, most errors can be avoided to a good measure by careful laboratory practice and/or technical improvements, but they must be taken into account when measuring solvent activities.

The *data reduction of vapor-pressure osmometry (VPO)* uses the stationary temperature difference as the starting point for determining solvent activities. There is an analogy to the boiling point elevation in thermodynamic equilibrium. Therefore, in the steady-state period of the experiment, the following relation can be applied if one assumes that the steady state is sufficiently near the vapor-liquid equilibrium and linear non-equilibrium thermodynamics is valid:

$$\Delta T^{st} = -k_{VPO} \frac{RT^2}{\Delta_{LV} H_{0A}} \ln a_A \tag{10}$$

where:

a_A	activity of solvent A
k_{VPO}	VPO-specific constant (must be determined separately)
R	gas constant
T	measuring temperature (= temperature of the pure solvent drop)
ΔT^{st}	temperature difference between solution and solvent drops in the steady state
$\Delta_{LV}H_{0A}$	molar enthalpy of vaporization of the pure solvent A at temperature T

Liquid-liquid equilibrium (LLE) measurements

There are two different situations for the liquid-liquid equilibrium in copolymer/solvent systems: (i) the equilibrium between a dilute copolymer solution (sol) and a copolymer-rich solution (gel) and (ii) the equilibrium between the pure solvent and a swollen copolymer network (gel). Only case (i) is considered here. To understand the results of LLE experiments in copolymer/solvent systems, one has to take into account the strong influence of distribution functions on LLE, because fractionation occurs during demixing, both with respect to chemical distribution and to molar mass distribution. Fractionation during demixing leads to some effects by which the LLE phase behavior differs from that of an ordinary, strictly binary mixture, because a common copolymer solution is a multicomponent system. Cloud-point curves are measured instead of binodals; and per each individual feed concentration of the mixture, two parts of a coexistence curve occur below or above (UCST or LCST behavior) the cloud-point curve, i.e., to produce an infinite number of coexistence data.

Distribution functions of the feed copolymer belong only to cloud-point data. On the other hand, each pair of coexistence points is characterized by two new and different distribution functions in each coexisting phase. The critical concentration is the only feed concentration where both parts of the coexistence curve meet each other on the cloud-point curve at the critical point that belongs to the feed copolymer distribution function. The threshold (maximum or minimum corresponding to UCST or LCST behavior) temperature (or pressure) is not equal to the critical point, since the critical point is to be found at a shoulder of the cloud-point curve. Details were discussed by Koningsveld (68KON, 72KON). Thus, LLE data have to be specified in the tables as cloud-point or coexistence data, and coexistence data make sense only if the feed concentration is given.

Experimental methods can be divided into measurements of cloud-point curves, of real coexistence data, and of critical points.

Due to distinct changes in a number of physical properties at the phase transition border, a number of methods can be used to determine cloud-points. In many cases, the refractive index change is determined because refractive indices depend on concentration (with the rare exception of isorefractive phases) and the sample becomes cloudy when the highly dispersed droplets of the second phase appear at the beginning of phase separation. Simple experiments observe cloud-points visually. More sophisticated equipment applies laser techniques and light scattering, where changes in scattering pattern or intensity are recorded as a function of decreasing/increasing temperature or pressure. The point where first deviations from a basic line are detected is the cloud-point. Since demixing or phase homogenization requires some time (especially for highly viscous solutions), special care is to be applied to obtain good data. Around the critical point, large fluctuations occur (critical opalescence) and scattering data have to be measured at a 90° scattering angle. The determination of the critical point is to be made by independent methods; see below. Various other physical properties have been applied for detecting liquid-liquid phase separation, i.e., viscosity, ultrasonic absorption, thermal expansion, dielectric constant, differential thermal analysis (DTA) or differential scanning calorimetry (DSC), UV- or IR-spectroscopy, and size exclusion chromatography/gel permeation chromatography (SEC, GPC).

Real coexistence data were measured only in a small number of investigations. This is mainly due to very long equilibrium times (usually weeks) which are necessary for obtaining thermodynamically correct data. A common method is to cool homogeneous solutions in ampullas very slowly to the desired temperature in the LLE region, and equilibrium is reached after both phases are sharply separated and clear. After separating both phases, concentrations and distribution functions are measured. Acceptable results can be obtained for low copolymer concentrations (up to 20 wt%). Highly viscous copolymer solutions at higher concentrations can be investigated by a modified ultracentrifuge where the equilibrium is quickly established during cooling by action of gravitational forces. After some hours, concentrations, phase volume ratios and concentration differences can be determined.

Special methods are necessary to measure the critical point. Only for solutions of monodisperse copolymers, the critical point is the maximum (or minimum) of the binodal. Binodals of copolymer solutions can be rather broad and flat. Then, the exact position of the critical point can be obtained by the method of the rectilinear diameter:

$$\frac{(\varphi_B{}^{I} - \varphi_B{}^{II})}{2} - \varphi_B{}^{crit} \propto \left(1 - \frac{T}{T^{crit}}\right)^{1-\alpha} \tag{11}$$

where:

$\varphi_B{}^{I}$	volume fraction of the copolymer in coexisting phase I
$\varphi_B{}^{II}$	volume fraction of the copolymer in coexisting phase II
$\varphi_B{}^{crit}$	volume fraction of the copolymer at the critical point
T^{crit}	critical temperature
α	critical exponent

For solutions of polydisperse copolymers, such a procedure cannot be used because the critical concentration must be known in advance to measure its corresponding coexistence curve. Two different methods were developed to solve this problem: the phase-volume-ratio method (68KON) where one uses the fact that this ratio is exactly equal to one only at the critical point, and the coexistence concentration plot (69WOL) where an isoplethal diagram of values of $\varphi_B{}^I$ and $\varphi_B{}^{II}$ vs. φ_{0B} gives the critical point as the intersection of cloud-point and shadow curves. Details will not be discussed here. Treating copolymer solutions with divariate distribution functions by continuous thermodynamics is reviewed in 90RAE.

High-pressure phase equilibrium (HPPE) measurements

The experimental investigation of high-pressure fluid phase equilibria in copolymer solutions is confronted with the same problems discussed above insofar as the investigated phase equilibria correspond with a VLE-, LLE-, or VLLE-type behavior, which are the only cases considered here. The experimental equipment follows on the same techniques, however, extended to high pressure conditions, using high-pressure cells and autoclaves for turbidimetry, light scattering, viscometry, and others.

The solvents are in many cases supercritical fluids, i.e., gases/vapors above their critical temperature and pressure. Data were measured mainly for two kinds of solutions: solutions in supercritical CO_2 (and some other fluids) or solutions in supercritical monomers. There are recent reviews on phase behavior of polymers and copolymers in supercritical fluids (94MCH, 97KIR, and 99KIR) that summarize today's state of investigation.

Measurement of enthalpy changes in copolymer solutions

Experiments on enthalpy changes in binary copolymer solutions can be made within common microcalorimeters by applying one of the following three methods:

1. Measurement of the enthalpy change caused by solving a given amount of the solute copolymer in an (increasing) amount of solvent, i.e., the solution experiment
2. Measurement of the enthalpy change caused by mixing a given amount of a concentrated copolymer solution with an amount of pure solvent, i.e., the dilution experiment
3. Measurement of the enthalpy change caused by mixing a given amount of a liquid/molten copolymer with an amount of pure solvent, i.e., the mixing experiment

Care must be taken for polymer solutions with respect to the resolution of the instruments, which has to be higher than for common mixtures or solutions with larger enthalpic effects, and for all effects and necessary corrections when working with highly viscous media with long equilibration times. Thus, the usually employed calorimeters for such purposes are the Calvet-type calorimeters based on the heat flux principle, which need not be discussed further. Details can be found in 84HEM and 94MAR.

In particular, one has to distinguish between the following effects for polymer solutions. The (integral) enthalpy of mixing or the (integral) enthalpy of solution of a binary system is the amount of heat which must be supplied when n_A mole of pure solvent A and n_B mole of pure copolymer B are combined to form a homogeneous mixture/solution in order to keep the total system at constant temperature and pressure.

$$\Delta_M h = n_A H_A + n_B H_B - (n_A H_{0A} + n_B H_{0B}) \qquad (12a)$$
$$\Delta_{sol} h = n_A H_A + n_B H_B - (n_A H_{0A} + n_B H_{0B}) \qquad (12b)$$

where:

$\Delta_M h$	(integral) enthalpy of mixing
$\Delta_{sol} h$	(integral) enthalpy of solution
H_A	partial molar enthalpy of solvent A
H_B	partial molar enthalpy of copolymer B
H_{0A}	molar enthalpy of pure solvent A
H_{0B}	molar enthalpy of pure copolymer B
n_A	amount of substance of solvent A
n_B	amount of substance of copolymer B

The enthalpy effect might be positive (endothermic solution/mixture) or negative (exothermic solution/mixture) depending on the ratio n_A/n_B, i.e., the concentration of the total system. Unfortunately, in some of the older literature, the definition of the sign of the so-called *(integral) heat of solution* is reversed, compared to the enthalpy, occasionally causing some confusion. In principle, the enthalpy effect depends also on pressure, However, in the case of condensed systems this pressure dependence is relatively small. All values in this handbook usually refer to normal pressure. H_{0A} and H_{0B} are the molar enthalpies of pure solvent A and pure copolymer B and H_A and H_B the partial molar enthalpies of solvent and copolymer in the solution/mixture.

The value of the (integral) enthalpy of solution is dependent on the degree of crystallinity for semicrystalline copolymers and, usually to a lesser extent, on the thermal history of glassy copolymers. The (integral) enthalpy of mixing is independent of any crystalline or glassy aspects of the copolymer. Thus, the (integral) enthalpy of mixing can be obtained without difficulties only for liquid/molten copolymers mixed with a solvent. Otherwise, the melting enthalpy of the crystallites and/or the glass enthalpy have to be determined additionally by independent measurements. As such a procedure is rather difficult and might cause substantial errors, it is common to measure the (integral) intermediary enthalpy of dilution, i.e., the enthalpy effect obtained if solvent A is added to an existing homogeneous copolymer solution. The intermediary enthalpy of dilution is the difference between two values of the enthalpy of mixing/solution corresponding to the concentrations of the copolymer solution at the beginning and at the end of the dilution process. The term *integral* is often added to these enthalpy changes to describe changes where finite amounts of substances are mixed. Especially, the *integral enthalpy of solution/mixing for a copolymer B* is given in a number of literature sources by applying not the partial (molar or specific) quantities but the following two definitions:

• *per mole* copolymer B:

$$^{int}\Delta_{sol}H_B = \Delta_{sol}h/n_B = \Delta_{sol}H/x_B \quad (13a)$$
$$^{int}\Delta_M H_B = \Delta_M h/n_B = \Delta_M H/x_B \quad (13b)$$

• *per gram* copolymer B (where the intensive ΔHs are the specific ones):

$$^{int}\Delta_{sol}H_B = \Delta_{sol}h/m_B = \Delta_{sol}H/w_B \quad (13c)$$
$$^{int}\Delta_M H_B = \Delta_M h/m_B = \Delta_M H/w_B \quad (13d)$$

where:

$^{int}\Delta_{sol}H_B$	integral enthalpy of solution of copolymer B
$^{int}\Delta_M H_B$	integral enthalpy of mixing of copolymer B
m_B	mass of copolymer B
w_B	mass fraction of copolymer B
x_B	mole fraction of copolymer B

As stated above, the difference between ${}^{int}\Delta_{sol}H_B$ and ${}^{int}\Delta_M H_B$ is determined by any enthalpic effects caused from solid-liquid phase transition of the crystallites and/or from glass transition and is zero for liquid/molten copolymers.

The term *differential* is sometimes added to enthalpy changes where infinitesimal (i.e., very small) amounts were added to a very large amount of either solution or pure component. These enthalpy changes are usually called partial (molar or specific) enthalpies of solution/mixing. The mathematical definition of partial molar enthalpies of solution/mixing is given for the copolymer B by:

$$\Delta_{sol}H_B = (\partial \Delta_{sol} h / \partial n_B)_{P,T,n_j} = H_B - H_{0B} \tag{14a}$$

$$\Delta_M H_B = (\partial \Delta_M h / \partial n_B)_{P,T,n_j} = H_B - H_{0B} \tag{14b}$$

with a unit of J/mol. However, for copolymer solutions, $\Delta_{sol}H_B$ or $\Delta_M H_B$ is often expressed as the enthalpy change per unit mass of copolymer added which can be obtained from the following derivative:

$$\Delta_{sol}H_B = (\partial \Delta_{sol} h / \partial m_B)_{P,T,m_j} \tag{14c}$$

$$\Delta_M H_B = (\partial \Delta_M h / \partial m_B)_{P,T,m_j} \tag{14d}$$

where:

$\Delta_{sol}H_B$	partial molar (or specific) enthalpy of solution of the copolymer B
$\Delta_M H_B$	partial molar (or specific) enthalpy of mixing of the copolymer B

with a unit of J/g. Similar to these definitions one can find results related to one mole of monomers (or base units). The derivative is then made by applying the base mole fraction of the copolymer. The partial (molar or specific) enthalpy of solution of the copolymer B is equal to the so-called differential enthalpy of solution at finite concentrations which is, for finite concentrations, different from the ${}^{int}\Delta_{sol}H_B$ or ${}^{int}\Delta_M H_B$ data as defined above. For example, in the case of a binary mixture, one obtains the relation:

$$\Delta_M H_B = \Delta_M H + (1 - x_B)(\partial \Delta_M H / \partial x_B) \tag{15}$$

which results in different values to ${}^{int}\Delta_M H_B$. In the case of adding an infinitesimal amount of copolymer to the pure solvent, the partial (molar or specific) enthalpy of solution of the copolymer B is properly identified as the partial enthalpy of solution of the copolymer at infinite dilution, $\Delta_{sol}H_B^{\infty}$, or the partial enthalpy of mixing of the copolymer at infinite dilution, $\Delta_M H_B^{\infty}$. Its value at infinite dilution of the copolymer is equal to the so-called *first* integral enthalpy of solution (unfortunately, sometimes referred to more simply as the enthalpy of solution of the copolymer, but, as discussed above, identical values can only be obtained for infinite dilution). In practice, the partial (molar or specific) enthalpy of solution of the copolymer B is measured by mixing isothermally a large excess of pure solvent and a certain amount of the copolymer to form a homogeneous solution.

The state of the copolymer before dissolution can significantly affect the enthalpy of solution. An amorphous copolymer below its glass transition temperature T_g often dissolves with the release of heat. The enthalpy of solution of a glassy copolymer is usually dependent on temperature and, to some extent, on the thermal history of the glass-forming process. An amorphous copolymer above T_g can show endothermic or exothermic dissolution behavior depending on the nature of the solvent and the interaction energies involved as is the case for any enthalpy of mixing. The dissolving of a semicrystalline copolymer requires an additional amount of heat associated with the disordering of crystalline regions. Consequently, its enthalpy of solution is usually positive and depends on the degree of crystallinity of the given copolymer sample.

The mathematical definition for the partial molar enthalpies of solution/mixing is given for the solvent A by

$$\Delta_{sol}H_A = (\partial \Delta_{sol} h / \partial n_A)_{P,T,n_j} = H_A - H_{0A} \tag{16a}$$

$$\Delta_M H_A = (\partial \Delta_M h / \partial n_A)_{P,T,n_j} = H_A - H_{0A} \tag{16b}$$

where:

$\Delta_{sol}H_A$	partial molar enthalpy of solution of the solvent A
$\Delta_M H_A$	partial molar enthalpy of mixing of the solvent A (= differential enthalpy of dilution)
n_A	amount of substance of solvent A

again with a unit of J/mol. It is equal to the so-called differential enthalpy of dilution as a consequence of adding an infinitesimal amount of solvent to the solution/mixture. The integral enthalpy of dilution for the solvent is equivalent to the integral molar enthalpy of mixing for the solvent A as defined by:

$$^{int}\Delta_M H_A = \Delta_M h / n_A \tag{17}$$

and, in the case of adding a very small amount of solvent to the pure copolymer, the partial molar enthalpy of solution at infinite dilution of the solvent is obtained. Partial molar enthalpies of mixing (or dilution) of the solvent are included in this data collection only for cases where they were obtained from calorimetric experiments.

Generally, it is known that such partial molar enthalpies of mixing (or dilution) of the solvent can also be determined from the temperature dependence of the activity of the solvent:

$$\Delta_M H_A = R[\partial \ln a_A / \partial(1/T)]_P \tag{18}$$

where:

a_A	activity of solvent A
P	pressure
R	gas constant
T	temperature

Any thermodynamic equilibrium experiments that measure vapor-liquid equilibrium, osmotic equilibrium, swelling equilibrium, sedimentation equilibrium or static light scattering can serve as potential sources for such data. Even from measurements of the intrinsic viscosity, enthalpy data were derived applying some theoretical background. However, agreement between enthalpy changes measured by calorimetry and results determined from the temperature dependence of solvent activity data is often of limited quality. Therefore, such data are not included here, except for one instance. From engineering and also from scientific aspects, the partial molar enthalpy of mixing at infinite dilution of the solvent in the liquid/molten copolymer $\Delta_M H_A^\infty$ is of special importance. Therefore, data for $\Delta_M H_A^\infty$ determined by inverse gas-liquid chromatography (IGC) have been included here.

$$\Delta_M H_A^\infty = R[\partial \ln \Omega_A^\infty / \partial(1/T)]_P \tag{19}$$

where:

$\Delta_M H_A^\infty$	partial molar enthalpy of mixing at infinite dilution of the solvent A
P	pressure
R	gas constant
T	temperature
Ω_A^∞	mass fraction-based activity coefficient of the solvent A at infinite dilution

Additionally, the enthalpies of solution at infinite dilution $\Delta_{sol}H_{A(vap)}^\infty$ of gases or vapors in molten copolymers determined by IGC have been included since IGC is the best recommended method for such data.

$$\Delta_{sol}H_{A(vap)}^\infty = -R\,[\partial \ln V_g^0 / \partial(1/T)]_P \tag{20}$$

where:

$\Delta_{sol}H_{A(vap)}^\infty$	first integral enthalpy of solution of the vapor of solvent A (with $\Delta_{sol}H_{A(vap)}^\infty = \Delta_M H_A^\infty - \Delta_{LV}H_{0A}$)
V_g^0	specific retention volume corrected to 0°C

PVT measurement for the copolymer melt

There are two widely practiced methods for the *PVT* measurement of polymers and copolymers:

1. Piston-die technique
2. Confining fluid technique

which were described in detail by Zoller in papers and books (e.g., 86ZOL, 95ZOL). Thus, a short summary is sufficient here.

1. Piston-die technique

In the piston-die technique, the material is confined in a rigid die or cylinder, which it has to fill completely. A pressure is applied to the sample as a load on a piston, and the movement of the piston with pressure and temperature changes is used to calculate the specific volume of the sample. Experimental problems concerning solid samples need not be discussed here, since only data for the liquid/molten (equilibrium) state are taken into consideration for this handbook. A typical practical complication is leakage around the piston when low-viscosity melts or solutions are tested. Seals cause an amount of friction leading to uncertainties in the real pressure applied. There are commercial devices as well as laboratory-built machines which have been used in the literature.

2. Confining fluid technique

In the confining fluid technique, the material is surrounded at all times by a confining (inert) fluid, often mercury, and the combined volume changes of sample and fluid are measured by a suitable technique as a function of temperature and pressure. The volume change of the sample is determined by subtracting the volume change of the confining fluid. A problem with this technique lies in potential interactions between fluid and sample. Precise knowledge of the *PVT* properties of the confining fluid is additionally required. The above-mentioned problems for the piston-die technique can be avoided.

Commercial machines are available. Thus, most of the literature data from the last decade were measured with this technique.

For both techniques, the absolute specific volume of the sample must be known at a single condition of pressure and temperature. Normally, these conditions are chosen to be 298.15 K and normal pressure (101.325 kPa). There are a number of methods to determine specific volumes (or densities) under these conditions. For polymeric samples, hydrostatic weighing or density gradient columns were often used.

The tables in Chapter 6 provide specific volumes neither at or below the melting transition of semicrystalline materials nor at or below the glass transition of amorphous samples, since *PVT* data of solid polymer samples are non-equilibrium data and depend on sample history and experimental procedure (which will not be discussed here). Therefore, only equilibrium data for the liquid/molten state are tabulated. Their common accuracy (standard deviation) is about 0.001 cm^3/g in specific volume, 0.1 K in temperature and $0.005*P$ in pressure (95ZOL).

Determination of second virial coefficients A_2

There are a couple of methods for the experimental determination of the second virial coefficient: colligative properties (vapor pressure depression, freezing point depression, boiling point increase, membrane osmometry), scattering methods (classical light scattering, X-ray scattering, neutron scattering), sedimentation velocity and sedimentation equilibrium. Details of the special experiments can be found in many textbooks and will not be repeated here. See, for example, 72HUG, 74TOM, 75CAS, 75FUJ, 75GLO, 87ADA, 87BER, 87COO, 87KRA, 87WIG, 91CHU, 91MAY, 91MU2, 92HAR, and 99PET.

The *vapor pressure depression* of the solvent in a binary copolymer solution, i.e., the difference between the saturation vapor pressure of the pure solvent and the corresponding partial pressure in the solution, $\Delta P_A = P_A^s - P_A$, is expressed as:

$$\frac{\Delta P_A}{P_A} = V_A^L c_B \left[\frac{1}{M_n} + A_2 c_B + A_3 c_B^{\ 2} + \dots \right] \tag{21}$$

where:

$A_2, A_3, \dots$	second, third, ... osmotic virial coefficients at temperature T
c_B	(mass/volume) concentration at temperature T
M_n	number-average relative molar mass of the copolymer
ΔP_A	$P_A^s - P_A$, vapor pressure depression of the solvent A at temperature T
P_A	partial vapor pressure of the solvent A at temperature T
P_A^s	saturation vapor pressure of the pure liquid solvent A at temperature T
V_A^L	molar volume of the pure liquid solvent A at temperature T

The *freezing point depression*, $\Delta_{SL} T_A$, is:

$$\Delta_{SL} T_A = E_{SL} c_B \left[\frac{1}{M_n} + A_2 c_B + A_3 c_B^{\ 2} + \dots \right] \tag{22}$$

and the *boiling point increase*, $\Delta_{LV}T_A$, is:

$$\Delta_{LV}T_A = E_{LV}c_B\left[\frac{1}{M_n} + A_2c_B + A_3c_B^{\ 2} + \ldots\right] \tag{23}$$

where:

E_{LV}	ebullioscopic constant
E_{SL}	cryoscopic constant
$\Delta_{SL}T_A$	freezing point temperature difference between pure solvent and solution, i.e., ${}_{SL}T_A^{\ 0} - {}_{SL}T_A$
$\Delta_{LV}T_A$	boiling point temperature difference between solution and pure solvent, i.e., ${}_{LV}T_A - {}_{LV}T_A^{\ 0}$

The *osmotic pressure*, π, can be described as:

$$\frac{\pi}{c_B} = RT\left[\frac{1}{M_n} + A_2c_B + A_3c_B^{\ 2} + \ldots\right] \tag{24}$$

where:

R	gas constant
T	(measuring) temperature
π	osmotic pressure

In the *dilute concentration region*, the virial equation is usually truncated after the second virial coefficient which leads to a linear relationship. A linearized relation over a wider concentration range can be constructed if the Stockmayer-Casassa relation between A_2 and A_3 is applied:

$$A_3M_n = \left(\frac{A_2M_n}{2}\right)^2 \tag{25}$$

$$\left(\frac{\pi}{c_2}\right)^{0.5} = \left(\frac{RT}{M_n}\right)^{0.5}\left[1 + \frac{A_2M_n}{2}c_2\right] \tag{26}$$

Scattering methods enable the determination of A_2 via the common relation:

$$\frac{Kc_B}{R(q)} = \frac{1}{M_wP_z(q)} + 2A_2Q(q)c_B + \ldots \tag{27}$$

with

$$q = \frac{4\pi}{\lambda}\sin\frac{\theta}{2} \tag{28}$$

where:

K	a constant that summarizes the optical parameters of a scattering experiment
M_w	mass-average relative molar mass of the copolymer
$P_z(q)$	z-average of the scattering function
q	scattering vector
$Q(q)$	function for the q-dependence of A_2
$R(q)$	excess intensity of the scattered beam at the value q
λ	wavelength
θ	scattering angle

Depending on the chosen experiment (light, X-ray or neutron scattering), the constant K is to be calculated from different relations. For details see the corresponding textbooks (72HUG, 75CAS, 82GLA, 86HIG, 87BER, 87KRA, 87WIG, and 91CHU).

Thermodynamic data from the *ultracentrifuge* experiment can be obtained either from the sedimentation velocity (sedimentation coefficient) or from the sedimentation-diffusion equilibrium since the centrifugal forces are balanced by the activity gradient. The determination of sedimentation and diffusion coefficients yields the virial coefficients by:

$$\left(\frac{D}{s}\right)\left(1-\overline{\upsilon}_{B,spez}\rho_A\right) = RT\left(\frac{1}{M_B} + 2A_2c_B + 3A_3c_B^{\ 2} + \ldots\right) \tag{29}$$

where:

D	diffusion coefficient
M_B	molar mass of the polymer
s	sedimentation coefficient
$\overline{\upsilon}_{B,spez}$	partial specific volume of the polymer
ρ_A	density of the solvent

Sedimentation-diffusion equilibrium in an ultracentrifuge also gives a virial series:

$$\omega^2 h_D\left(1-\overline{\upsilon}_{B,spez}\rho_A\right)\left(\frac{\partial \ln c_B}{\partial h_D}\right) = RT\left(\frac{1}{M_B} + 2A_2c_B + 3A_3c_B^{\ 2} + \ldots\right) \tag{30}$$

where:

h_D	distance from the center of rotation
ω	angular velocity

Both equations are valid for monodisperse copolymers only, i.e., for one definite single component B. For all polydisperse copolymers, different averages were obtained for the sedimentation and the diffusion coefficients which depend on the applied measuring modes and the subsequent calculations. The averages of M_B correspond with averages of D and s and are mixed ones that have to be transformed into the desired common averages. For details, please see reviews 75FUJ, 91MU2, and 92HAR.

1.3. Guide to the data tables

Characterization of the copolymers

Copolymers vary by a number of characterization variables. Molar mass, chemical composition, and their distribution functions are the most important variables. However, tacticity, sequence distribution, branching, and end groups determine their thermodynamic behavior in solution too. Unfortunately, much less information is provided with respect to the copolymers that were applied in most of the thermodynamic investigations in the original literature. In many cases, the copolymer is characterized by its average chemical composition, some molar mass averages and some additional information (e.g., where it was synthesized) only. Sometimes even such an information is missed.

The molar mass averages are defined as follows:

number average M_n

$$M_n = \frac{\sum_i n_{B_i} M_{B_i}}{\sum_i n_{B_i}} = \frac{\sum_i w_{B_i}}{\sum_i w_{B_i} / M_{B_i}} \tag{31}$$

mass average M_w

$$M_w = \frac{\sum_i n_{B_i} M_{B_i}^2}{\sum_i n_{B_i} M_{B_i}} = \frac{\sum_i w_{B_i} M_{B_i}}{\sum_i w_{B_i}} \tag{32}$$

z-average M_z

$$M_z = \frac{\sum_i n_{B_i} M_{B_i}^3}{\sum_i n_{B_i} M_{B_i}^2} = \frac{\sum_i w_{B_i} M_{B_i}^2}{\sum_i w_{B_i} M_{B_i}} \tag{33}$$

viscosity average M_η

$$M_\eta = \left(\frac{\sum_i w_{B_i} M_{B_i}^a}{\sum_i w_{B_i}} \right)^{1/a} \tag{34}$$

where:

a	exponent in the viscosity - molar mass relationship
M_{Bi}	molar mass of the copolymer species B_i
n_{Bi}	amount of substance of copolymer species B_i
w_{Bi}	mass fraction of copolymer species B_i

The data tables of each chapter are provided below in order of the copolymers. The tables in all the following chapters always begin with a summary of the available characterization data for the corresponding samples of a given kind of copolymer. An acronym is defined for each copolymer sample corresponding to the comonomers. The average chemical composition of the copolymer is then added after a slash. For example, AN-B/51w describes an acrylonitrile/butadiene copolymer with an average chemical composition of 51 wt% acrylonitrile, and E-P/33x describes an ethylene/propylene copolymer with an average chemical composition of 33 mol% ethylene. Sometimes, a figure in parenthesis is added to distinguish between different samples characterized by equal average chemical composition. The acronyms are used in the subsequent presentation of the data tables to denote the copolymer samples. Appendix 8.1. provides a complete list of all copolymer acronyms used in this handbook together with the pages where data with corresponding copolymers are given.

Example (from Henry's constants data tables)

Copolymers from acrylonitrile and butadiene

Average chemical composition of the copolymers, acronyms and references:

Copolymer (B)	Acronym	Ref.
Acrylonitrile/butadiene copolymer		
18.0 wt% acrylonitrile	AN-B/18w	95TRI
34.0 wt% acrylonitrile	AN-B/34w(1)	93ISS
34.0 wt% acrylonitrile	AN-B/34w(1)	97SCH
34.0 wt% acrylonitrile	AN-B/34w(2)	95TRI
34.0 wt% acrylonitrile	AN-B/34w(3)	82LIP
41.0 wt% acrylonitrile	AN-B/41w	86IT2

Characterization:

Copolymer (B)	M_n/ g/mol	M_w/ g/mol	M_η/ g/mol	Further information
AN-B/18w			100000	Bayer India
AN-B/34w(1)	60000			Perbunan 3307, Bayer AG, Ger.
AN-B/34w(2)			111000	Bayer India
AN-B/34w(3)				Krynac, Polysar
AN-B/41w				no data

Further information indicates *no data* if nothing more than the type of the copolymer was stated in the corresponding literature source.

Measures for the copolymer concentration

The following concentration measures are used in the tables of this handbook (where B always denotes the copolymer, A denotes the solvent, and in ternary systems C denotes the third component):

mass/volume concentration:

$$c_A = m_A/V \qquad c_B = m_B/V \tag{35}$$

mass fraction:

$$w_A = m_A/\Sigma\, m_i \qquad w_B = m_B/\Sigma\, m_i \tag{36}$$

mole fraction:

$$x_A = n_A/\Sigma\, n_i \qquad x_B = n_B/\Sigma\, n_i \qquad \text{with} \qquad n_i = m_i/M_i \text{ and } M_B = M_n \tag{37}$$

volume fraction:

$$\varphi_A = (m_A/\rho_A)/\Sigma\, (m_i/\rho_i) \qquad \varphi_B = (m_B/\rho_B)/\Sigma\, (m_i/\rho_i) \tag{38}$$

segment fraction:

$$\psi_A = x_A r_A/\Sigma\, x_i r_i \qquad \psi_B = x_B r_B/\Sigma\, x_i r_i \qquad \text{usually with } r_A = 1 \tag{39}$$

base mole fraction:

$$z_A = x_A r_A/\Sigma\, x_i r_i \qquad z_B = x_B r_B/\Sigma\, x_i r_i \qquad \text{with } r_B = M_B/M_0 \text{ and } r_A = 1 \tag{40}$$

where:

c_A	(mass/volume) concentration of solvent A
c_B	(mass/volume) concentration of copolymer B
m_A	mass of solvent A
m_B	mass of copolymer B
M_A	molar mass of the solvent A
M_B	molar mass of the copolymer B
M_n	number-average relative molar mass
M_0	molar mass of a basic unit of the copolymer B
n_A	amount of substance of solvent A
n_B	amount of substance of copolymer B
r_A	segment number of the solvent A, usually $r_A = 1$
r_B	segment number of the copolymer B
V	volume of the liquid solution at temperature T
w_A	mass fraction of solvent A
w_B	mass fraction of copolymer B
x_A	mole fraction of solvent A
x_B	mole fraction of copolymer B
z_A	base mole fraction of solvent A
z_B	base mole fraction of copolymer B

φ_A	volume fraction of solvent A
φ_B	volume fraction of copolymer B
ρ_A	density of solvent A
ρ_B	density of copolymer B
ψ_A	segment fraction of solvent A
ψ_B	segment fraction of copolymer B

For high-molecular copolymers, a mole fraction is not an appropriate unit to characterize composition. However, for oligomeric products with rather low molar masses, mole fractions were sometimes used. In the common case of a distribution function for the molar mass, $M_B = M_n$ is to be chosen. Mass fraction and volume fraction can be considered as special cases of segment fractions depending on the way by which the segment size is actually determined: $r_i/r_A = M_i/M_A$ or $r_i/r_A = V_i/V_A = (M_i/\rho_i)/(M_A/\rho_A)$, respectively. Classical segment fractions are calculated by applying $r_i/r_A = V_i^{vdW}/V_A^{vdW}$ ratios where hard-core van der Waals volumes, V_i^{vdW}, are taken into account. Their special values depend on the chosen equation of state (or simply some group contribution schemes, e.g., 68BON, 90KRE) and have to be specified.

Volume fractions imply a temperature dependence and, as they are defined in equation (38), neglect excess volumes of mixing and, very often, the densities of the copolymer in the state of the solution are not known correctly. However, volume fractions can be calculated without the exact knowledge of the copolymer molar mass (or its averages). Base mole fractions are seldom applied for copolymer systems. The value for M_0, the molar mass of a basic unit of the copolymer, has to be determined according to the corresponding average chemical composition. Sometimes it is chosen arbitrarily, however, and has to be specified.

Experimental data tables

The data tables are sorted with respect to the copolymers. Within types of copolymers the individual samples are ordered by increasing average chemical composition as given in the head table with the copolymer acronyms. Subsequently, systems are ordered by their solvents. Solvents are listed alphabetically. When necessary, systems are ordered by increasing temperature. Each data set begins with two lines for the solution components, e.g.,

copolymer (B):	**AN-B/21w**		**95GUP**
solvent (A):	**acetonitrile**	**C_2H_3N**	**75-05-8**

where the copolymer sample is given in the first line together with the reference. The second line provides the solvent's chemical name, molecular formula and CAS-registry number. There are some exceptions from this type of presentation within the tables for Henry's constants, A_2 values, UCST/LCST data, and *PVT* data.

The originally measured data for each single system are then listed together with some comment lines if necessary. The data are usually given as published, but temperatures are always given in K. Pressures are sometimes recalculated into kPa or MPa. Enthalpy data are always recalculated into J or kJ, if necessary. Mass fraction-based Henry's constants are calculated from published specific retention volumes, if such data are not provided in the original source. They are always tabulated in MPa.

1.4. List of symbols

B	parameter of the Tait equation
B_{AA}	second virial coefficient of the pure solvent A at temperature T
c_A	(mass/volume) concentration of solvent A
c_B	(mass/volume) concentration of copolymer B
C	parameter of the Tait equation
E_{LV}	ebullioscopic constant
E_{SL}	cryoscopic constant
h_D	distance from the center of rotation
H^E	excess enthalpy = $\Delta_M H$ = enthalpy of mixing
H_A	partial molar enthalpy of solvent A
H_B	partial molar (or specific) enthalpy of copolymer B
H_{0A}	molar enthalpy of pure solvent A
H_{0B}	molar (or specific) enthalpy of pure copolymer B
$H_{A,B}$	classical mass fraction Henry's constant of solvent vapor A in molten copolymer B
$\Delta_{dil}H^{12}$	(integral) intermediary enthalpy of dilution (= $\Delta_M H^{(2)} - \Delta_M H^{(1)}$)
$\Delta_M H$	(integral) enthalpy of mixing
$\Delta_{sol}H$	(integral) enthalpy of solution
${}^{int}\Delta_M H_A$	integral enthalpy of mixing of solvent A (= integral enthalpy of dilution)
$\Delta_M H_A$	partial molar enthalpy of mixing of the solvent A (= differential enthalpy of dilution)
$\Delta_M H_A^\infty$	partial molar enthalpy of mixing at infinite dilution of the solvent A
${}^{int}\Delta_{sol}H_A$	integral enthalpy of solution of solvent A
$\Delta_{sol}H_A$	partial molar enthalpy of solution of the solvent A
$\Delta_{sol}H_A^\infty$	first integral enthalpy of solution of solvent A (= $\Delta_M H_A^\infty$ in the case of liquid/molten copolymers <u>and</u> a liquid solvent, i.e., it is different from the values for solutions of solvent vapors or gases in a liquid/molten copolymer $\Delta_{sol}H_{A(vap)}^\infty$)
$\Delta_{sol}H_{A(vap)}^\infty$	first integral enthalpy of solution of the vapor of solvent A (with $\Delta_{sol}H_{A(vap)}^\infty = \Delta_M H_A^\infty - \Delta_{LV}H_{0A}$)
$\Delta_{LV}H_{0A}$	molar enthalpy of vaporization of the pure solvent A at temperature T
${}^{int}\Delta_M H_B$	integral enthalpy of mixing of copolymer B
$\Delta_M H_B$	partial molar (or specific) enthalpy of mixing of copolymer B
$\Delta_M H_B^\infty$	partial molar (or specific) enthalpy of mixing at infinite dilution of copolymer B
${}^{int}\Delta_{sol}H_B$	integral enthalpy of solution of copolymer B
$\Delta_{sol}H_B$	partial molar (or specific) enthalpy of solution of copolymer B
$\Delta_{sol}H_B^\infty$	first integral enthalpy of solution of copolymer B ($\Delta_M H_B^\infty$ in the case of liquid/molten B)
k_{VPO}	VPO-specific constant (must be determined separately)
K	a constant that summarizes the optical parameters of a scattering experiment
m_A	mass of solvent A
m_B	mass of copolymer B
M	relative molar mass
M_A	molar mass of the solvent A
M_B	molar mass of the copolymer B
M_n	number-average relative molar mass
M_w	mass-average relative molar mass
M_η	viscosity-average relative molar mass
M_z	z-average relative molar mass
M_0	molar mass of a basic unit of the copolymer B
MI	melting index

n_A	amount of substance of solvent A
n_B	amount of substance of copolymer B
P	pressure
P_A	partial vapor pressure of the solvent A at temperature T
P_A^s	saturation vapor pressure of the pure liquid solvent A at temperature T
ΔP_A	$P_A^s - P_A$, vapor pressure depression of the solvent A at temperature T
P_{in}	column inlet pressure in IGC
P_{out}	column outlet pressure in IGC
$P_z(q)$	z-average of the scattering function
q	scattering vector
$Q(q)$	function for the q-dependence of A_2
R	gas constant
$R(q)$	excess intensity of the scattered beam at the value q
r_A	segment number of the solvent A, usually $r_A = 1$
r_B	segment number of the copolymer B
s	sedimentation coefficient
T	(measuring) temperature
T_g	glass transition temperature
T^{crit}	critical temperature
T_0	reference temperature (= 273.15 K)
ΔT^{st}	temperature difference between solution and solvent drops in VPO
$\Delta_{SL}T_A$	freezing point temperature difference between pure solvent and solution, i.e., ${}_{SL}T_A^0 - {}_{SL}T_A$
$\Delta_{LV}T_A$	boiling point temperature difference between solution and pure solvent, i.e., ${}_{LV}T_A - {}_{LV}T_A^0$
V	volume of the liquid solution at temperature T
V_A^L	molar volume of the pure liquid solvent A at temperature T
V_{net}	net retention volume in IGC
V_r	retention volume in IGC
V_{dead}	retention volume of the (inert) marker gas, dead retention, gas holdup in IGC
V_g^0	specific retention volume corrected to 0°C in IGC
V_{spez}	specific volume
w_A	mass fraction of solvent A
w_B	mass fraction of copolymer B
x_A	mole fraction of solvent A
x_B	mole fraction of copolymer B
z_A	base mole fraction of solvent A
z_B	base mole fraction of copolymer B
α	critical exponent
γ_A	activity coefficient of the solvent A in the liquid phase with activity $a_A = x_A\gamma_A$
λ	wavelength
φ_A	volume fraction of solvent A
φ_B	volume fraction of copolymer B
ρ_A	density of solvent A
ρ_B	density of copolymer B
ψ_A	segment fraction of solvent A
ψ_B	segment fraction of copolymer B
π	osmotic pressure
θ	scattering angle
ω	angular velocity
Ω_A	mass fraction-based activity coefficient of the solvent A in the liquid phase with activity $a_A = x_A\Omega_A$
Ω_A^∞	mass fraction-based activity coefficient of the solvent A at infinite dilution

1.5. References

68BON Bondi, A., *Physical Properties of Molecular Crystals, Liquids and Glasses*, J. Wiley & Sons, New York, 1968.

68KON Koningsveld, R. and Staverman, A.J., Liquid-liquid phase separation in multicomponent polymer solutions I and II, *J. Polym. Sci.*, Pt. A-2, 6, 305, 325, 1968.

69WOL Wolf, B.A., Zur Bestimmung der kritischen Konzentration von Polymerlösungen, *Makromol. Chem.*, 128, 284, 1969.

71YAM Yamakawa, H., *Modern Theory of Polymer Solutions*, Harper & Row, New York, 1971.

72HUG Huglin, M.B., Ed., *Light Scattering from Polymer Solutions*, Academic Press, New York, 1972.

72KON Koningsveld, R., Polymer Solutions and Fractionation, in *Polymer Science*, Jenkins, E.D., Ed., North-Holland, Amsterdam, 1972, 1047.

74TOM Tombs, M.P. and Peacock, A.R., *The Osmotic Pressure of Macromolecules*, Oxford University Press, London, 1974.

75BON Bonner, D.C., Vapor-liquid equilibria in concentrated polymer solutions, *Macromol. Sci. Rev. Macromol. Chem.*, C13, 263, 1975.

75CAS Casassa, E.F. and Berry, G.C., Light scattering from solutions of macromolecules, in *Polymer Molecular Weights*, Marcel Dekker, New York, 1975, Pt. 1, 161.

75FUJ Fujita, H., *Foundations of Ultracentrifugal Analysis*, J. Wiley & Sons, New York, 1975.

75GLO Glover, C.A., Absolute colligative property methods, in *Polymer Molecular Weights*, Marcel Dekker, New York, 1975, Pt. 1, 79.

76LIU Liu, D.D. and Prausnitz, J.M, Solubilities of gases and volatile liquids in polyethylene and in ethylene-vinyl acetate copolymers in the region 125-225 °C, *Ind. Eng. Chem. Fundam.*, 15, 330, 1976.

76NES Nesterov, A.E. and Lipatov, J.S., *Obrashchennaya Gasovaya Khromatografiya v Termodinamike Polimerov*, Naukova Dumka, Kiev, 1976.

82GLA Glatter, O. and Kratky, O., Eds., *Small-Angle X-Ray Scattering*, Academic Press, London, 1982.

84HEM Hemminger, W. and Höhne, G., *Calorimetry: Fundamentals and Practice*, Verlag Chemie, Weinheim, 1984.

86HIG Higgins, J.S. and Macconachie, A., Neutron scattering from macromolecules in solution, in *Polymer Solutions*, Forsman, W.C., Ed., Plenum Press, New York, 1986, 183.

86ZOL Zoller, P., Dilatometry, in *Encyclopedia of Polymer Science and Engineering*, Vol. 5, 2nd ed., Mark, H. et al., Eds., J. Wiley & Sons, New York, 1986, 69.

87ADA Adams, E.T., Osmometry, in *Encyclopedia of Polymer Science and Engineering*, Vol. 10, 2nd ed., Mark, H. et al., Eds., J. Wiley & Sons, New York, 1986, 636.

87BER Berry, G.C., Light scattering, in *Encyclopedia of Polymer Science and Engineering*, Vol. 8, 2nd ed., Mark, H. et al., Eds., J. Wiley & Sons, New York, 1986, 721.

87COO Cooper, A.R., Molecular weight determination, in *Encyclopedia of Polymer Science and Engineering*, Vol. 10, 2nd ed., Mark, H. et al., Eds., J. Wiley & Sons, New York, 1986, 1.

87KRA Kratochvil, P., *Classical Light Scattering from Polymer Solutions*, Elsevier, Amsterdam, 1987.

87WIG Wignall, G.D., Neutron scattering, in *Encyclopedia of Polymer Science and Engineering*, Vol. 10, 2nd ed., Mark, H. et al., Eds., J. Wiley & Sons, New York, 1986, 112.

88NES Nesterov, A.E., *Obrashchennaya Gasovaya Khromatografiya Polimerov*, Naukova Dumka, Kiev, 1988.

89LLO Lloyd, D.R., Ward, T.C., Schreiber, H.P., and Pizana, C.C., Eds., *Inverse Gas Chromatography*, ACS Symposium Series 391, American Chemical Society, Washington, 1989.

89VIL Vilcu, R. and Leca, M., *Polymer Thermodynamics by Gas Chromatography*, Elsevier, Amsterdam, 1989.

90BAR Barton, A.F.M., *CRC Handbook of Polymer-Liquid Interaction Parameters and Solubility Parameters*, CRC Press, Boca Raton, 1990.

90FUJ Fujita, H., *Polymer Solutions*, Elsevier, Amsterdam, 1990.

90KAM Kamide, K., *Thermodynamics of Polymer Solutions*, Elsevier, Amsterdam, 1990.

90KRE [Van] Krevelen, D.W., *Properties of Polymers*, 3rd ed., Elsevier, Amsterdam, 1990.

90RAE Raetzsch, M.D. and Wohlfarth, Ch., Continuous thermodynamics of copolymer systems, *Adv. Polym. Sci.*, 98, 49, 1990.

91CHU Chu, B., *Laser Light Scattering*, Academic Press, New York, 1991.

91MAY Mays, J.W. and Hadjichristidis, N., Measurement of molecular weight of polymers by osmometry, in *Modern Methods of Polymer Characterization*, Barth, H.G. and Mays, J.W., Eds., J. Wiley & Sons, New York, 1991, 201.

91MU1 Munk, P., Polymer characterization using inverse gas chromatography, in *Modern Methods of Polymer Characterization*, Barth, H.G. and Mays, J.W., Eds., J. Wiley & Sons, New York, 1991, 151.

91MU2 Munk, P., Polymer characterization using the ultracentrifuge, in *Modern Methods of Polymer Characterization*, Barth, H.G. and Mays, J.W., Eds., J. Wiley & Sons, New York, 1991, 271.

92HAR Harding, S.E., Rowe, A.J., and Horton, J.C., *Analytical Ultracentrifugation in Biochemistry and Polymer Science*, Royal Society of Chemistry, Cambridge, 1992.

92WEN Wen, H., Elbro, H.S., and Alessi, P., *Polymer Solution Data Collection.* I. Vapor-liquid equilibrium; II. Solvent activity coefficients at infinite dilution; III. Liquid-liquid equlibrium, Chemistry Data Series, Vol. 15, DECHEMA, Frankfurt am Main, 1992.

93DAN Danner, R.P. and High, M.S., *Handbook of Polymer Solution Thermodynamics*, American Institute of Chemical Engineers, New York, 1993.

94MAR Marsh, K.N., Ed., *Experimental Thermodynamics, Volume 4, Solution Calorimetry*, Blackwell Science, Oxford, 1994.

94MCH McHugh, M.A. and Krukonis, V.J., *Supercritical Fluid Extraction: Principles and Practice*, 2nd ed., Butterworth Publishing, Stoneham, 1994.

94WOH Wohlfarth, Ch., *Vapour-Liquid Equilibrium Data of Binary Polymer Solutions: Physical Science Data*, 44, Elsevier, Amsterdam, 1994.

95GUP Gupta, R.B., and Prausnitz, J.M., Vapor-liquid equilibria of copolymer + solvent and homopolymer + solvent binaries: new experimental data and their correlation, *J. Chem. Eng. Data*, 40, 784, 1995.

95ZOL Zoller, P. and Walsh, D.J., *Standard Pressure-Volume-Temperature Data for Polymers*, Technomic Publishing, Lancaster, 1995.

97KIR Kiran, E. and Zhuang, W., *Miscibility and Phase Separation of Polymers in Near- and Supercritical Fluids*, ACS Symposium Series 670, 2, 1997.

97KOL Kolb, B. and Ettre, L.S., *Static Headspace Gas Chromatography: Theory and Practice*, Wiley-VCH, Weinheim, 1997.

99BRA Brandrup, J., Immergut, E.H., and Grulke, E.A., Eds., *Polymer Handbook*, 4th ed., J. Wiley & Sons, New York, 1999.

99KIR Kirby, C.F. and McHugh, M.A., Phase behavior of polymers in supercritical fluid solvents, *Chem. Rev.*, 99, 565, 1999.

99PET Pethrick, R.A. and Dawkins, J.V., Eds., *Modern Techniques for Polymer Characterization*, J. Wiley & Sons, Chichester, 1999.

99PRA Prausnitz, J.M., Lichtenthaler, R.N., and de Azevedo, E.G., *Molecular Thermodynamics of Fluid Phase Equilibria*, 3rd ed., Prentice Hall, Upper Saddle River, NJ, 1999.

2000WOH Wohlfarth, Ch., Methods for the measurement of solvent activity of polymer solutions, in *Handbook of Solvents*, Wypych, G., Ed., ChemTec Publishing, Toronto, 2000, 146.

2. VAPOR-LIQUID EQUILIBRIUM (VLE) DATA OF BINARY COPOLYMER SOLUTIONS

2.1. Partial solvent vapor pressures or solvent activities for copolymer solutions

Copolymers from acrylonitrile and butadiene

Average chemical composition of the copolymers, acronyms and references:

Copolymer (B)	Acronym	Ref.
Acrylonitrile/butadiene copolymer		
21.0 wt% acrylonitrile	AN-B/21w	95GUP
33.0 wt% acrylonitrile	AN-B/33w	95GUP
51.0 wt% acrylonitrile	AN-B/51w	95GUP

Characterization:

Copolymer (B)	M_n/ g/mol	M_w/ g/mol	M_η/ g/mol	Further information
AN-B/21w				Sci. Polym. Products, catalog #523
AN-B/33w				Sci. Polym. Products, catalog #528
AN-B/51w				Sci. Polym. Products, catalog #530

Experimental VLE data:

copolymer (B): **AN-B/21w** **95GUP**
solvent (A): **acetonitrile** **C_2H_3N** **75-05-8**

T/K = 333.15

w_B	0.989	0.974	0.935	0.915	0.888	0.813
P_A/kPa	6.8	13.7	26.1	30.1	34.1	39.3

copolymer (B): **AN-B/21w** **95GUP**
solvent (A): **cyclohexane** **C_6H_{12}** **110-82-7**

T/K = 333.15

w_B	0.823	0.824	0.790	0.784	0.717
P_A/kPa	35.6	35.7	38.1	39.7	43.1

copolymer (B): **AN-B/21w** **95GUP**
solvent (A): **n-hexane** **C_6H_{14}** **110-54-3**

T/K = 333.15

w_B	0.975	0.956	0.918	0.896	0.866	0.830	0.764	0.749
P_A/kPa	12.9	25.6	38.8	44.5	50.8	57.5	63.7	64.9

copolymer (B): **AN-B/21w** **95GUP**
solvent (A): **n-octane** **C_8H_{18}** **111-65-9**

T/K = 333.15

w_B	0.964	0.948	0.936	0.921	0.889	0.842
P_A/kPa	2.90	4.19	5.08	5.91	6.91	8.41

copolymer (B): **AN-B/21w** **95GUP**
solvent (A): **n-pentane** **C_5H_{12}** **109-66-0**

T/K = 333.15

w_B	0.984	0.979	0.967	0.954	0.945	0.930	0.922	0.911
P_A/kPa	36.0	46.8	65.5	84.0	97.3	113.3	124.8	137.6

copolymer (B): **AN-B/33w** **95GUP**
solvent (A): **acetonitrile** **C_2H_3N** **75-05-8**

T/K = 333.15

w_B	0.977	0.957	0.935	0.909	0.852	0.794
P_A/kPa	6.7	13.3	18.3	23.2	32.5	37.9

copolymer (B): **AN-B/33w** **95GUP**
solvent (A): **n-hexane** **C_6H_{14}** **110-54-3**

T/K = 333.15

w_B	0.998	0.983	0.974	0.963
P_A/kPa	12.4	25.1	36.0	43.3

copolymer (B): **AN-B/33w** **95GUP**
solvent (A): **n-pentane** **C_5H_{12}** **109-66-0**

T/K = 333.15

w_B	0.986	0.975	0.973	0.968	0.964	0.956	0.950	0.945
P_A/kPa	26.9	53.5	64.7	77.5	89.7	101.7	115.7	130.1

copolymer (B): **AN-B/33w** **95GUP**
solvent (A): **trichloromethane** **$CHCl_3$** **67-66-3**

T/K = 333.15

w_B	0.920	0.830	0.828	0.738	0.628	0.576	0.515	0.437
P_A/kPa	13.6	26.8	26.9	39.6	52.8	59.9	66.3	73.3

copolymer (B): **AN-B/51w** **95GUP**
solvent (A): **acetonitrile** **C_2H_3N** **75-05-8**

T/K = 333.15

w_B	0.979	0.953	0.889	0.856	0.818	0.707
P_A/kPa	6.8	13.7	26.1	30.1	34.1	39.3

copolymer (B): **AN-B/51w** **95GUP**
solvent (A): **cyclohexane** **C_6H_{12}** **110-82-7**

T/K = 333.15

w_B	0.993	0.989	0.985	0.976	0.968	0.953	0.940	0.909
P_A/kPa	13.7	19.1	19.6	26.4	32.0	35.6	39.7	43.1

copolymer (B): **AN-B/51w** **95GUP**
solvent (A): **n-hexane** **C_6H_{14}** **110-54-3**

T/K = 333.15

w_B	0.982	0.980	0.971	0.967	0.959	0.956	0.938	0.938
P_A/kPa	12.9	25.6	38.8	44.5	50.8	57.5	63.7	64.9

copolymer (B): **AN-B/51w** **95GUP**
solvent (A): **n-octane** **C_8H_{18}** **111-65-9**

T/K = 333.15

w_B	0.993	0.990	0.987	0.985	0.979	0.974
P_A/kPa	2.85	4.19	5.08	5.91	6.91	8.41

copolymer (B):	**AN-B/51w**		**95GUP**
solvent (A):	**n-pentane**	C_5H_{12}	**109-66-0**

T/K = 333.15

w_B	0.991	0.987	0.981	0.978	0.978	0.974	0.973	0.966
P_A/kPa	36.0	46.8	65.5	84.0	97.3	113.3	124.8	137.6

Copolymers from acrylonitrile and styrene

Average chemical composition of the copolymers, acronyms and references:

Copolymer (B)	Acronym	Ref.
Acrylonitrile/styrene copolymer		
25.0 wt% acrylonitrile	S-AN/25w	98SCH
28.0 wt% acrylonitrile	S-AN/28w	96WO2
28.0 wt% acrylonitrile	S-AN/28w	97WO1
30.0 wt% acrylonitrile	S-AN/30w	95GUP

Characterization:

Copolymer (B)	M_n/ g/mol	M_w/ g/mol	M_η/ g/mol	Further information
S-AN/25w	90000	147000		Bayer AG Leverkusen, Ger.
S-AN/28w	46000	100000		synthesized in the laboratory
S-AN/30w		185000		Sci. Polym. Products, catalog #495

Experimental VLE data:

copolymer (B):	**S-AN/25w**		**98SCH**
solvent (A):	**toluene**	C_7H_8	**108-88-3**

T/K = 313.15

w_B	0.399	0.447	0.497	0.522	0.601	0.638	0.699	0.747	0.799
P_A/P_A^s	0.999	0.991	0.954	0.952	0.902	0.869	0.796	0.745	0.644
w_B	0.848	0.894	0.952						
P_A/P_A^s	0.544	0.476	0.331						

copolymer (B): **S-AN/25w** **98SCH**
solvent (A): **toluene** **C_7H_8** **108-88-3**

T/K = 323.15

w_B	0.352	0.448	0.499	0.524	0.603	0.643	0.702	0.750	0.802
P_A/P_A^s	0.996	0.986	0.934	0.918	0.877	0.848	0.779	0.717	0.628
w_B	0.852	0.897	0.955						
P_A/P_A^s	0.529	0.454	0.337						

copolymer (B): **S-AN/25w** **98SCH**
solvent (A): **toluene** **C_7H_8** **108-88-3**

T/K = 333.15

w_B	0.353	0.449	0.500	0.525	0.604	0.644	0.704	0.751	0.803
P_A/P_A^s	0.989	0.979	0.924	0.909	0.872	0.843	0.776	0.711	0.625
w_B	0.853	0.899	0.956						
P_A/P_A^s	0.523	0.446	0.332						

copolymer (B): **S-AN/28w** **96WO2, 97WO1**
solvent (A): **benzene** **C_6H_6** **71-43-2**

T/K = 343.15

w_B	0.8339	0.8053	0.7876	0.7501	0.7207	0.6967	0.6587	0.6384	0.6198
P_A/P_A^s	0.6433	0.7080	0.7415	0.8036	0.8454	0.8752	0.9097	0.9251	0.9369
w_B	0.5892	0.5564							
P_A/P_A^s	0.9598	0.9724							

copolymer (B): **S-AN/28w** **96WO2, 97WO1**
solvent (A): **propylbenzene** **C_9H_{12}** **103-65-1**

T/K = 398.15

w_B	0.8270	0.8132	0.8018	0.7781	0.7649	0.7346	0.7179	0.6834
P_A/P_A^s	0.7559	0.7834	0.8033	0.8423	0.8618	0.8991	0.9127	0.9372

copolymer (B): **S-AN/28w** **96WO2, 97WO1**
solvent (A): **toluene** **C_7H_8** **108-88-3**

T/K = 343.15

w_B	0.7962	0.7433	0.7192	0.6960	0.6607	0.6482	0.6199	0.5856
P_A/P_A^s	0.7257	0.8126	0.8458	0.8735	0.9090	0.9172	0.9377	0.5856

copolymer (B): **S-AN/28w** **96WO2, 97WO1**
solvent (A): **toluene** **C_7H_8** **108-88-3**

T/K = 373.15

w_B	0.8003	0.7632	0.7375	0.7157	0.7093	0.6831	0.6686	0.6388	0.6056
P_A/P_A^s	0.7071	0.7752	0.8147	0.8410	0.8523	0.8817	0.8951	0.9208	0.9427
w_B	0.5883	0.5654							
P_A/P_A^s	0.9567	0.9690							

copolymer (B): **S-AN/28w** **96WO2, 97WO1**
solvent (A): **o-xylene** **C_8H_{10}** **95-47-6**

T/K = 398.15

w_B	0.7606	0.7371	0.7107	0.6838	0.6593	0.6276	0.5960	0.5586
P_A/P_A^s	0.8048	0.8389	0.8722	0.9008	0.9228	0.9459	0.9642	0.9800

copolymer (B): **S-AN/28w** **96WO2, 97WO1**
solvent (A): **m-xylene** **C_8H_{10}** **108-38-3**

T/K = 398.15

w_B	0.7977	0.7611	0.7428	0.7167	0.6884	0.6460	0.6198	0.5894
P_A/P_A^s	0.7462	0.8162	0.8387	0.8690	0.9010	0.9405	0.9510	0.9701

copolymer (B): **S-AN/28w** **96WO2, 97WO1**
solvent (A): **p-xylene** **C_8H_{10}** **106-42-3**

T/K = 373.15

w_B	0.7632	0.7416	0.7204	0.6997	0.6566	0.6223	0.5917
P_A/P_A^s	0.8081	0.8378	0.8736	0.8883	0.9347	0.9547	0.9714

copolymer (B): **S-AN/28w** **96WO2, 97WO1**
solvent (A): **p-xylene** **C_8H_{10}** **106-42-3**

T/K = 398.15

w_B	0.8082	0.7896	0.7534	0.7263	0.7006	0.6731	0.6519	0.6292	0.6028
P_A/P_A^s	0.7188	0.7537	0.8142	0.8528	0.8828	0.9101	0.9286	0.9439	0.9609

copolymer (B): **S-AN/28w** **96WO2, 97WO1**
solvent (A): **p-xylene** **C_8H_{10}** **106-42-3**

T/K = 423.15

w_B	0.7738	0.7547	0.7259	0.6931	0.6674	0.6492	0.6211
P_A/P_A^s	0.7769	0.8077	0.8483	0.8846	0.9106	0.9253	0.9444

copolymer (B): **S-AN/30w** **95GUP**
solvent (A): **1,2-dichloroethane** **$C_2H_4Cl_2$** **107-06-2**

T/K = 343.15

w_B	0.861	0.804	0.757	0.719	0.704	0.661
P_A/kPa	30.8	36.5	40.7	43.9	46.0	49.1

copolymer (B): **S-AN/30w** **95GUP**
solvent (A): **1,2-dichloroethane** **$C_2H_4Cl_2$** **107-06-2**

T/K = 353.15

w_B	0.749	0.721	0.652	0.624	0.611
P_A/kPa	45.7	50.1	57.6	59.5	60.7

Copolymers from p-bromostyrene and p-methylstyrene

Average chemical composition of the copolymers, acronyms and references:

Copolymer (B)	Acronym	Ref.
p-Bromostyrene/p-methylstyrene copolymer		
16.7 mol% p-bromostyrene	pBS-pMS/17x	63COR
33.0 mol% p-bromostyrene	pBS-pMS/33x	63COR
50.0 mol% p-bromostyrene	pBS-pMS/50x	63COR
66.0 mol% p-bromostyrene	pBS-pMS/66x	63COR
83.3 mol% p-bromostyrene	pBS-pMS/83x	63COR

Characterization:

Copolymer (B)	M_n/ g/mol	M_w/ g/mol	M_η/ g/mol	Further information
pBS-pMS/17x	5000-10000			synthesized in the laboratory
pBS-pMS/33x	5000-10000			synthesized in the laboratory
pBS-pMS/50x	5000-10000			synthesized in the laboratory
pBS-pMS/66x	5000-10000			synthesized in the laboratory
pBS-pMS/83x	5000-10000			synthesized in the laboratory

Experimental VLE data:

copolymer (B): **pBS-pMS/17x** **63COR**
solvent (A): **toluene** **C_7H_8** **108-88-3**

T/K = 293.20

φ_B	0.1643	0.2633	0.3444	0.4641	0.5449	0.6416
a_A	0.9952	0.9865	0.9650	0.9259	0.8696	0.7859

copolymer (B): **pBS-pMS/33x** **63COR**
solvent (A): **toluene** **C_7H_8** **108-88-3**

T/K = 293.20

φ_B	0.1970	0.2792	0.3648	0.4799	0.5932	0.6473
a_A	0.9952	0.9860	0.9649	0.9255	0.8708	0.7858

copolymer (B): **pBS-pMS/50x** **63COR**
solvent (A): **toluene** **C_7H_8** **108-88-3**

T/K = 293.20

φ_B	0.1857	0.2672	0.3696	0.4870	0.5755	0.6629
a_A	0.9952	0.9865	0.9652	0.9250	0.8690	0.7855

copolymer (B): **pBS-pMS/66x** **63COR**
solvent (A): **toluene** **C_7H_8** **108-88-3**

T/K = 293.20

φ_B	0.1803	0.3066	0.3931	0.5074	0.5844	0.6742
a_A	0.9952	0.9865	0.9650	0.9250	0.8680	0.7852

copolymer (B): **pBS-pMS/83x** **63COR**
solvent (A): **toluene** **C_7H_8** **108-88-3**

T/K = 293.20

φ_B	0.2088	0.3489	0.4095	0.5417	0.6147	0.7028
a_A	0.9952	0.9868	0.9658	0.9256	0.8698	0.7859

Copolymers from butadiene and styrene

Average chemical composition of the copolymers, acronyms and references:

Copolymer (B)	Acronym	Ref.
Butadiene/styrene copolymer		
5.0 wt% styrene	S-BR/05w	95GUP
23.0 wt% styrene	S-BR/23w	95GUP
41.0 wt% styrene	S-BR/41w	96WO1
41.0 wt% styrene	S-BR/41w	97WO1
45.0 wt% styrene	S-BR/45w(1)	90IWA
45.0 wt% styrene	S-BR/45w(1)	91IWA
45.0 wt% styrene	S-BR/45w(2)	95GUP
Butadiene/styrene copolymer (block copolymer without specification)		
30.0 wt% styrene	S-?-B/30w	90IWA
30.0 wt% styrene	S-?-B/30w	91IWA
77.0 wt% styrene	S-?-B/77w	90IWA
77.0 wt% styrene	S-?-B/77w	91IWA
Polystyrene-b-polybutadiene-b-polystyrene triblock copolymer		
31.0 wt% styrene	S-B-S/31w	99FRE
Polystyrene-b-polybutadiene(hydrogenated)-b-polystyrene triblock copolymer		
13.0 wt% styrene	S-B(h)-S/13w	99FRE
31.0 wt% styrene	S-B(h)-S/31w	99FRE

Characterization:

Copolymer (B)	M_n/ g/mol	M_w/ g/mol	M_η/ g/mol	Further information
S-BR/05w				Sci. Polym. Products, catalog #199
S-BR/23w				Sci. Polym. Products, catalog #200
S-BR/41w	86300	181200		synthesized in the laboratory
S-BR/45w(1)	130430	600000		Sci. Polym. Products

continue

continue

Copolymer (B)	M_n/ g/mol	M_w/ g/mol	M_η/ g/mol	Further information
S-BR/45w(2)				Sci. Polym. Products, catalog #201
S-?-B/30w	108000	140000		Sci. Polym. Products
S-?-B/77w	117650	200000		Denki Kagaku Industry
S-B-S/31w	111000			
S-B(h)-S/13w	90000			synthesized in the laboratory
S-B(h)-S/31w	48000			synthesized in the laboratory

Experimental VLE data:

copolymer (B): **S-BR/05w** **95GUP**
solvent (A): **cyclohexane** **C_6H_{12}** **110-82-7**

T/K = 333.15

w_B	0.848	0.811	0.767	0.724	0.677
P_A/kPa	24.8	29.5	33.7	37.6	40.7

copolymer (B): **S-BR/05w** **95GUP**
solvent (A): **2-propanone** **C_3H_6O** **67-64-1**

T/K = 323.15

w_B	0.944	0.933	0.928	0.933	0.934	0.922	0.921	0.920	0.909
P_A/kPa	41.5	45.3	46.5	47.2	47.3	51.1	52.3	53.3	55.2
w_B	0.885	0.890	0.880						
P_A/kPa	59.6	59.7	61.3						

copolymer (B): **S-BR/05w** **95GUP**
solvent (A): **toluene** **C_7H_8** **108-88-3**

T/K = 308.15

w_B	0.957	0.901	0.816	0.681	0.584	0.487
P_A/kPa	0.93	2.08	3.04	4.36	4.83	5.45

copolymer (B): **S-BR/05w** **95GUP**
solvent (A): **trichloromethane** **$CHCl_3$** **67-66-3**

T/K = 323.15

w_B	0.960	0.894	0.796	0.728	0.659	0.543	0.511	0.463	0.420
P_A/kPa	8.5	16.3	25.5	31.6	38.3	45.5	47.6	50.0	53.7
w_B	0.362								
P_A/kPa	56.4								

copolymer (B): **S-BR/23w** **95GUP**
solvent (A): **cyclohexane** **C_6H_{12}** **110-82-7**

T/K = 296.65

w_B	0.803	0.724	0.672
P_A/kPa	7.64	9.03	10.03

copolymer (B): **S-BR/23w** **95GUP**
solvent (A): **cyclohexane** **C_6H_{12}** **110-82-7**

T/K = 333.15

w_B	0.864	0.832	0.792	0.756	0.713
P_A/kPa	24.8	29.5	33.7	37.6	40.7

copolymer (B): **S-BR/23w** **95GUP**
solvent (A): **n-hexane** **C_6H_{14}** **110-54-3**

T/K = 343.15

w_B	0.890	0.838	0.738
P_A/kPa	48.8	61.5	75.2

copolymer (B): **S-BR/23w** **95GUP**
solvent (A): **n-pentane** **C_5H_{12}** **109-66-0**

T/K = 333.15

w_B	0.954	0.945	0.930	0.922	0.911
P_A/kPa	84.0	97.3	113.3	124.8	137.6

copolymer (B): **S-BR/23w** **95GUP**
solvent (A): **2-propanone** **C_3H_6O** **67-64-1**

T/K = 333.15

w_B	0.979	0.963	0.938	0.908	0.872	0.853	0.817	0.769
P_A/kPa	26.4	40.7	54.8	68.5	81.3	84.8	91.5	97.9

copolymer (B): **S-BR/23w** **95GUP**
solvent (A): **2-propanone** **C_3H_6O** **67-64-1**

T/K = 323.15

w_B	0.951	0.939	0.929	0.925	0.915	0.912	0.909	0.900	0.875
P_A/kPa	36.3	41.5	45.3	46.5	51.1	52.3	53.3	55.2	59.6
w_B	0.880	0.870							
P_A/kPa	59.7	61.3							

copolymer (B): **S-BR/23w** **95GUP**
solvent (A): **toluene** **C_7H_8** **108-88-3**

T/K = 308.15

w_B	0.971	0.969	0.968	0.963	0.963	0.962
P_A/kPa	0.93	2.08	3.04	4.36	4.83	5.45

copolymer (B): **S-BR/23w** **95GUP**
solvent (A): **trichloromethane** **$CHCl_3$** **67-66-3**

T/K = 323.15

w_B	0.954	0.889	0.794	0.730	0.662	0.551	0.518	0.486	0.434
P_A/kPa	8.5	16.3	25.5	31.6	38.3	45.5	47.6	50.0	53.7
w_B	0.379								
P_A/kPa	56.4								

copolymer (B): **S-BR/41w** **96WO1, 97WO1**
solvent (A): **benzene** **C_6H_6** **71-43-2**

T/K = 343.15

w_B	0.7998	0.7553	0.7207	0.6864	0.6485	0.6039	0.5517	0.4916	0.4326
P_A/P_A^s	0.5091	0.5897	0.6418	0.6910	0.7388	0.7876	0.8357	0.8798	0.9143

copolymer (B): **S-BR/41w** **96WO1, 97WO1**
solvent (A): **cyclohexane** **C_6H_{12}** **110-82-7**

T/K = 343.15

w_B	0.7831	0.7564	0.7197	0.6712	0.6394	0.6008	0.5586	0.5234
P_A/P_A^s	0.6522	0.6966	0.7491	0.8083	0.8385	0.8693	0.8985	0.9176

copolymer (B): **S-BR/41w** **96WO1, 97WO1**
solvent (A): **ethylbenzene** **C_8H_{10}** **100-41-4**

T/K = 373.15

w_B	0.7917	0.7604	0.7285	0.6799	0.6412	0.5996	0.5616	0.5173	0.4845
P_A/P_A^s	0.5829	0.6358	0.6841	0.7471	0.7897	0.8271	0.8562	0.8854	0.9037

copolymer (B): **S-BR/41w** **96WO1, 97WO1**
solvent (A): **ethylbenzene** **C_8H_{10}** **100-41-4**

T/K = 398.15

w_B	0.6887	0.6528	0.6293	0.5818	0.5495	0.5106	0.4698
P_A/P_A^s	0.7292	0.7736	0.7973	0.8390	0.8629	0.8876	0.9108

copolymer (B): **S-BR/41w** **96WO1, 97WO1**
solvent (A): **mesitylene** **C_9H_{12}** **108-67-8**

T/K = 398.15

w_B	0.7694	0.7389	0.6908	0.6576	0.6232	0.5856	0.5407	0.5009
P_A/P_A^s	0.6520	0.6982	0.7617	0.7987	0.8311	0.8610	0.8911	0.9107

copolymer (B): **S-BR/41w** **96WO1, 97WO1**
solvent (A): **toluene** **C_7H_8** **108-88-3**

T/K = 343.15

w_B	0.7619	0.7198	0.6721	0.6276	0.5800	0.5317	0.4961	0.4533
P_A/P_A^s	0.5825	0.6476	0.7125	0.7634	0.8103	0.8497	0.8753	0.9011

copolymer (B): **S-BR/41w** **96WO1, 97WO1**
solvent (A): **toluene** **C_7H_8** **108-88-3**

T/K = 373.15

w_B	0.8168	0.7701	0.7387	0.6970	0.6319	0.5726	0.5063	0.4308
P_A/P_A^s	0.4844	0.5704	0.6214	0.6833	0.7612	0.8178	0.8691	0.9143

copolymer (B): **S-BR/41w** **96WO1, 97WO1**
solvent (A): **p-xylene** **C_8H_{10}** **106-42-3**

T/K = 398.15

w_B	0.8071	0.7593	0.7162	0.6784	0.6499	0.6075	0.5698	0.5201	0.4899
P_A/P_A^s	0.5536	0.6388	0.7027	0.7503	0.7803	0.8216	0.8520	0.8884	0.9050

copolymer (B): **S-BR/45w(2)** **95GUP**
solvent (A): **cyclohexane** **C_6H_{12}** **110-82-7**

T/K = 296.65

w_B	0.824	0.747	0.699
P_A/kPa	7.64	9.03	10.03

copolymer (B): **S-BR/45w(2)** **95GUP**
solvent (A): **cyclohexane** **C_6H_{12}** **110-82-7**

T/K = 333.15

w_B	0.870	0.837	0.800	0.760	0.718
P_A/kPa	24.8	29.5	33.7	37.6	40.7

copolymer (B): **S-BR/45w(1)** **90IWA**
solvent (A): **ethylbenzene** **C_8H_{10}** **100-41-4**

T/K = 373.15

w_B	0.9717	0.9589	0.9452	0.9448	0.9248	0.9112	0.9022	0.8830	0.8390
P_A/kPa	4.00	5.42	6.92	6.89	9.10	10.00	11.10	12.60	16.10
w_B	0.8380	0.0000							
P_A/kPa	16.10	34.25							

copolymer (B): **S-BR/45w(1)** **90IWA**
solvent (A): **ethylbenzene** **C_8H_{10}** **100-41-4**

T/K = 403.15

w_B	0.9791	0.9738	0.9723	0.9639	0.9587	0.9559	0.9539	0.9494	0.9407
P_A/kPa	7.14	8.79	9.45	11.40	12.60	13.80	14.30	15.70	18.10
w_B	0.9333	0.9332	0.9308	0.9241	0.9229	0.9110	0.9081	0.9036	0.8920
P_A/kPa	19.70	20.50	20.40	22.00	22.20	25.60	26.50	27.10	29.60
w_B	0.8850	0.8790	0.8450	0.0000					
P_A/kPa	31.30	32.00	38.70	85.55					

copolymer (B): **S-BR/45w(2)** **95GUP**
solvent (A): **n-hexane** **C_6H_{14}** **110-54-3**

T/K = 343.15

w_B	0.897	0.845	0.751
P_A/kPa	48.8	61.5	75.2

copolymer (B): **S-BR/45w(1)** **90IWA**
solvent (A): **n-nonane** **C_9H_{20}** **111-84-2**

T/K = 373.15

w_B	0.9832	0.9516	0.9428	0.9308	0.9093	0.9082	0.8820	0.8700	0.8460
P_A/kPa	2.87	6.45	7.36	8.51	10.20	10.20	12.00	12.70	14.00
w_B	0.8270	0.8260	0.0000						
P_A/kPa	14.90	14.90	20.72						

copolymer (B): **S-BR/45w(1)** **90IWA**
solvent (A): **n-nonane** **C_9H_{20}** **111-84-2**

T/K = 403.15

w_B	0.9819	0.9647	0.9589	0.9468	0.9286	0.9171	0.9098	0.8960	0.8820
P_A/kPa	6.86	12.10	13.80	17.40	21.40	24.10	25.40	28.00	30.50
w_B	0.8520	0.8290	0.0000						
P_A/kPa	35.00	37.40	55.85						

copolymer (B): **S-BR/45w(2)** **95GUP**
solvent (A): **n-pentane** **C_5H_{12}** **109-66-0**

T/K = 333.15

w_B	0.978	0.978	0.974	0.973	0.966
P_A/kPa	84.0	97.3	113.3	124.8	137.6

copolymer (B): **S-BR/45w(2)** **95GUP**
solvent (A): **2-propanone** **C_3H_6O** **67-64-1**

T/K = 323.15

w_B	0.954	0.939	0.926	0.922	0.926	0.929	0.912	0.908	0.905
P_A/kPa	36.3	41.5	45.3	46.5	47.2	47.3	51.1	52.3	53.3
w_B	0.895	0.863	0.870	0.859					
P_A/kPa	55.2	59.6	59.7	61.3					

copolymer (B): **S-BR/45w(2)** **95GUP**
solvent (A): **2-propanone** **C_3H_6O** **67-64-1**

T/K = 333.15

w_B	0.956	0.931	0.899	0.860	0.837	0.803	0.742
P_A/kPa	40.7	54.8	68.5	81.3	84.8	91.5	97.9

copolymer (B): **S-BR/45w(2)** **95GUP**
solvent (A): **trichloromethane** **$CHCl_3$** **67-66-3**

T/K = 323.15

w_B	0.957	0.893	0.795	0.728	0.657	0.537	0.500	0.466	0.408
P_A/kPa	8.5	16.3	25.5	31.6	38.3	45.5	47.6	50.0	53.7
w_B	0.350								
P_A/kPa	56.4								

copolymer (B): **S-?-B/30w** **90IWA**
solvent (A): **ethylbenzene** **C_8H_{10}** **100-41-4**

T/K = 373.15

w_B	0.9704	0.9528	0.9240	0.9098	0.9022	0.8760	0.8550	0.8360	0.8130
P_A/kPa	3.87	5.83	8.79	10.00	10.70	13.00	14.80	16.10	17.80
w_B	0.0000								
P_A/kPa	34.25								

copolymer (B): **S-?-B/30w** **90IWA**
solvent (A): **ethylbenzene** **C_8H_{10}** **100-41-4**

T/K = 403.15

w_B	0.9752	0.9692	0.9643	0.9582	0.9533	0.9509	0.9409	0.9368	0.9270
P_A/kPa	7.93	9.73	11.10	12.30	13.80	14.70	17.10	18.20	20.40
w_B	0.9196	0.9145	0.9055	0.9042	0.8950	0.8860	0.8700	0.8600	0.8510
P_A/kPa	22.20	23.50	25.60	25.70	27.90	29.60	33.00	34.80	37.10
w_B	0.8340	0.0000							
P_A/kPa	39.70	85.55							

copolymer (B): **S-?-B/30w** **90IWA**
solvent (A): **n-nonane** **C_9H_{20}** **111-84-2**

T/K = 373.15

w_B	0.9801	0.9673	0.9506	0.9489	0.9291	0.9229	0.9076	0.8970	0.8830
P_A/kPa	2.81	4.28	5.95	6.17	7.85	8.44	9.57	10.20	11.30
w_B	0.8690	0.8470	0.8360	0.0000					
P_A/kPa	12.00	13.20	13.60	20.72					

copolymer (B): **S-?-B/30w** **90IWA**
solvent (A): **n-nonane** **C_9H_{20}** **111-84-2**

T/K = 403.15

w_B	0.9867	0.9750	0.9584	0.9459	0.9294	0.9248	0.9224	0.9032	0.8950
P_A/kPa	4.62	8.12	12.70	15.90	20.20	20.80	21.40	25.30	26.80
w_B	0.8720	0.8700	0.8600	0.8400	0.0000				
P_A/kPa	30.60	30.70	32.40	34.90	55.85				

copolymer (B): **S-?-B/77w** **90IWA**
solvent (A): **ethylbenzene** **C_8H_{10}** **100-41-4**

T/K = 373.15

w_B	0.9780	0.9693	0.9614	0.9520	0.9319	0.9266	0.9140	0.8980	0.8810
P_A/kPa	3.85	4.85	5.81	7.17	9.10	9.69	11.10	12.60	14.70
w_B	0.8690	0.8460	0.8400	0.0000					
P_A/kPa	15.20	17.10	17.60	34.25					

copolymer (B): **S-?-B/77w** **90IWA**
solvent (A): **ethylbenzene** **C_8H_{10}** **100-41-4**

T/K = 403.15

w_B	0.9744	0.9685	0.9572	0.9461	0.9402	0.9358	0.9354	0.9232	0.9213
P_A/kPa	10.00	11.80	14.70	18.20	20.00	21.00	21.60	24.30	25.00
w_B	0.9026	0.9004	0.8940	0.8780	0.8560	0.0000			
P_A/kPa	29.60	30.40	32.10	35.00	40.80	85.55			

copolymer (B): **S-?-B/77w** **90IWA**
solvent (A): **n-nonane** **C_9H_{20}** **111-84-2**

T/K = 373.15

w_B	0.9847	0.9581	0.9514	0.9344	0.9275	0.9195	0.9037	0.8820	0.8490
P_A/kPa	3.41	7.32	7.88	10.20	10.90	11.30	12.80	14.40	16.40
w_B	0.8480	0.0000							
P_A/kPa	16.40	20.72							

copolymer (B): **S-?-B/77w** **90IWA**
solvent (A): **n-nonane** **C_9H_{20}** **111-84-2**

T/K = 403.15

w_B	0.9842	0.9765	0.9579	0.9578	0.9442	0.9429	0.9330	0.9120	0.9082
P_A/kPa	7.10	10.90	17.40	17.90	22.00	22.70	25.40	30.60	31.50

continue

continue

w_B	0.8780	0.8620	0.8300	0.0000
P_A/kPa	36.90	39.90	44.20	55.85

copolymer (B): **S-B-S/31w** **99FRE**
solvent (A): **cyclohexane** C_6H_{12} **110-82-7**

T/K = 323.85

w_B	0.9877	0.9866	0.9674	0.9627	0.9618	0.9614	0.9607	0.9539	0.9485
a_A	0.0533	0.0543	0.1324	0.1488	0.1536	0.1529	0.1568	0.1766	0.1988
w_B	0.9393	0.9373	0.9296	0.9131	0.9097	0.8989	0.8955	0.8857	0.8753
a_A	0.2288	0.2358	0.2612	0.3128	0.3314	0.3539	0.3563	0.3886	0.4153
w_B	0.8655	0.8625	0.8581	0.8508	0.8459	0.8414	0.8381	0.8352	0.8242
a_A	0.4398	0.4606	0.4569	0.4745	0.4854	0.4948	0.5035	0.5092	0.5327
w_B	0.8198	0.8198	0.8196						
a_A	0.5424	0.5424	0.5430						

copolymer (B): **S-B-S/31w** **99FRE**
solvent (A): **cyclohexane** C_6H_{12} **110-82-7**

T/K = 348.15

w_B	0.9820	0.9678	0.9530	0.9542	0.9246	0.9256	0.9260	0.8773	0.8810
a_A	0.0790	0.1282	0.1861	0.1832	0.2880	0.2805	0.2805	0.4209	0.4129
w_B	0.8815	0.7972							
a_A	0.4112	0.5942							

copolymer (B): **S-B-S/31w** **99FRE**
solvent (A): **cyclohexane** C_6H_{12} **110-82-7**

T/K = 373.15

w_B	0.9779	0.9481	0.9197	0.9167	0.8804
a_A	0.0915	0.1993	0.2907	0.2899	0.4042

copolymer (B): **S-B-S/31w** **99FRE**
solvent (A): **cyclohexane** C_6H_{12} **110-82-7**

T/K = 393.15

w_B	0.9845	0.9851	0.9850	0.9767	0.9765	0.9768	0.9628	0.9643	0.9489
a_A	0.0513	0.0513	0.0512	0.0778	0.0778	0.0780	0.1198	0.1192	0.1666
w_B	0.9492	0.9487	0.9486	0.9488	0.9255	0.9266	0.9264	0.8973	0.8972
a_A	0.1661	0.1657	0.1660	0.1664	0.2324	0.2327	0.2326	0.3173	0.3166
w_B	0.8920	0.8952	0.8954	0.8973					
a_A	0.3046	0.3186	0.3159	0.3166					

copolymer (B): **S-B(h)-S/13w** **99FRE**
solvent (A): **cyclohexane** **C_6H_{12}** **110-82-7**

T/K = 324.35

w_B	0.9963	0.9943	0.9929	0.9910	0.9888	0.9869	0.9868	0.9855	0.9839
a_A	0.0127	0.0204	0.0247	0.0312	0.0388	0.0474	0.0455	0.0484	0.0543
w_B	0.9807	0.9800	0.9710	0.9662	0.9555	0.9462	0.9460	0.9460	0.9452
a_A	0.0639	0.0666	0.0951	0.1102	0.1433	0.1732	0.1733	0.1712	0.1710
w_B	0.9404	0.9332	0.9129	0.9123	0.9118	0.9086	0.9078	0.9076	0.8396
a_A	0.1901	0.2098	0.2735	0.2841	0.2761	0.2833	0.2849	0.2874	0.5550
w_B	0.8381								
a_A	0.5602								

copolymer (B): **S-B(h)-S/31w** **99FRE**
solvent (A): **cyclohexane** **C_6H_{12}** **110-82-7**

T/K = 393.15

w_B	0.9935	0.9934	0.9906	0.9893	0.9891	0.9875	0.9867	0.9846	0.9810
a_A	0.0241	0.0247	0.0347	0.0400	0.0390	0.0441	0.0472	0.0556	0.0716
w_B	0.9771	0.9702	0.9670	0.9666	0.9635	0.9521	0.9499	0.9479	0.9472
a_A	0.0854	0.1088	0.1139	0.1156	0.1307	0.1748	0.1748	0.1782	0.1774
w_B	0.9409	0.9406	0.9294	0.9289	0.9139	0.9135	0.9084	0.9052	0.8832
a_A	0.1938	0.1966	0.2323	0.2339	0.2772	0.2785	0.2885	0.2967	0.3551
w_B	0.8830	0.8818	0.8808	0.8429	0.7884	0.7880	0.7410	0.7373	0.7368
a_A	0.3557	0.3570	0.3600	0.4446	0.5502	0.5491	0.6734	0.6751	0.6738

Copolymers from dimethylsiloxane and methylphenylsiloxane

Average chemical composition of the copolymers, acronyms and references:

Copolymer (B)	Acronym	Ref.
Dimethylsiloxane/methylphenylsiloxane copolymer 15 wt% methylphenylsiloxane	DMS-MPS/15w	98SCH

Characterization:

Copolymer (B)	M_n/ g/mol	M_w/ g/mol	M_η/ g/mol	Further information
DMS-MPS/15w	9100	41200		Dow Corning Corp., Midland, USA

Experimental VLE data:

copolymer (B): **DMS-MPS/15w** **98SCH**
solvent (A): **anisole** **C_7H_8O** **100-66-3**

T/K = 313.15

w_B	0.309	0.408	0.451	0.506	0.562	0.606	0.640	0.633	0.754
P_A/P_A^s	0.997	0.987	0.988	0.987	0.963	0.934	0.920	0.886	0.807
w_B	0.798	0.850	0.896	0.949					
P_A/P_A^s	0.734	0.631	0.505	0.301					

copolymer (B): **DMS-MPS/15w** **98SCH**
solvent (A): **anisole** **C_7H_8O** **100-66-3**

T/K = 323.15

w_B	0.406	0.449	0.503	0.558	0.604	0.638	0.631	0.753	0.798
P_A/P_A^s	0.992	0.997	0.987	0.968	0.950	0.918	0.926	0.813	0.733
w_B	0.850	0.896	0.949						
P_A/P_A^s	0.625	0.499	0.298						

copolymer (B): **DMS-MPS/15w** **98SCH**
solvent (A): **anisole** **C_7H_8O** **100-66-3**

T/K = 333.15

w_B	0.306	0.404	0.447	0.501	0.556	0.602	0.636	0.629	0.752
P_A/P_A^s	0.991	0.980	0.981	0.976	0.957	0.944	0.918	0.919	0.804
w_B	0.797	0.849	0.896	0.948					
P_A/P_A^s	0.724	0.610	0.486	0.289					

copolymer (B): **DMS-MPS/15w** **98SCH**
solvent (A): **anisole** **C_7H_8O** **100-66-3**

T/K = 343.15

w_B	0.306	0.404	0.447	0.501	0.556	0.602	0.636	0.629	0.752
P_A/P_A^s	0.988	0.977	0.971	0.966	0.942	0.925	0.896	0.903	0.793
w_B	0.797	0.849	0.896	0.948					
P_A/P_A^s	0.714	0.606	0.479	0.284					

copolymer (B): **DMS-MPS/15w** **98SCH**
solvent (A): **2-propanone** **C_3H_6O** **67-64-1**

T/K = 305.15

w_B	0.400	0.452	0.499	0.549	0.595	0.649	0.699	0.752	0.799
P_A/P_A^s	0.971	0.966	0.956	0.948	0.935	0.920	0.915	0.891	0.858
w_B	0.849	0.899	0.948						
P_A/P_A^s	0.785	0.683	0.489						

copolymer (B): **DMS-MPS/15w** **98SCH**
solvent (A): **2-propanone** **C_3H_6O** **67-64-1**

T/K = 309.15

w_B	0.401	0.453	0.500	0.551	0.596	0.651	0.701	0.753	0.801
P_A/P_A^s	0.980	0.973	0.976	0.961	0.947	0.932	0.880	0.846	0.823
w_B	0.850	0.900	0.949						
P_A/P_A^s	0.731	0.622	0.436						

copolymer (B): **DMS-MPS/15w** **98SCH**
solvent (A): **2-propanone** **C_3H_6O** **67-64-1**

T/K = 314.15

w_B	0.402	0.454	0.501	0.552	0.597	0.652	0.702	0.754	0.802
P_A/P_A^s	0.976	0.970	0.968	0.949	0.927	0.916	0.884	0.844	0.797
w_B	0.851	0.901	0.950						
P_A/P_A^s	0.718	0.599	0.411						

copolymer (B): **DMS-MPS/15w** **98SCH**
solvent (A): **2-propanone** **C_3H_6O** **67-64-1**

T/K = 318.15

w_B	0.404	0.456	0.502	0.553	0.599	0.653	0.703	0.756	0.804
P_A/P_A^s	0.975	0.975	0.963	0.947	0.923	0.886	0.862	0.823	0.776

continue

continue

w_B	0.853	0.902	0.951
P_A/P_A^s	0.693	0.575	0.374

copolymer (B): **DMS-MPS/15w** **98SCH**
solvent (A): **2-propanone** **C_3H_6O** **67-64-1**

T/K = 322.15

w_B	0.407	0.458	0.504	0.555	0.601	0.655	0.705	0.758	0.805
P_A/P_A^s	0.958	0.960	0.957	0.947	0.915	0.887	0.859	0.807	0.765
w_B	0.854	0.903	0.952						
P_A/P_A^s	0.687	0.565	0.372						

Copolymers from ethylene and vinyl acetate

Average chemical composition of the copolymers, acronyms and references:

Copolymer (B)	Acronym	Ref.
Ethylene/vinyl acetate copolymer		
9.0 wt% vinyl acetate	E-VA/09w	95GUP
15.6 wt% vinyl acetate	E-VA/16w(1)	73BEL
16.1 wt% vinyl acetate	E-VA/16w(2)	73BEL
16.6 wt% vinyl acetate	E-VA/17w(1)	73BEL
17.1 wt% vinyl acetate	E-VA/17w(2)	73BEL
25.0 wt% vinyl acetate	E-VA/25w	95GUP
39.4 wt% vinyl acetate	E-VA/39w	75BOB
39.5 wt% vinyl acetate	E-VA/40w	75BOB
41.4 wt% vinyl acetate	E-VA/41w	76KIE
41.8 wt% vinyl acetate	E-VA/42w	75BOB
42.6 wt% vinyl acetate	E-VA/43w	76KIE
50.0 wt% vinyl acetate	E-VA/50w	95GUP
70.0 wt% vinyl acetate	E-VA/70w	95GUP

Characterization:

Copolymer (B)	M_n/ g/mol	M_w/ g/mol	M_η/ g/mol	Further information
E-VA/09w				Sci. Polym. Products, catalog #506
E-VA/16w(1)	2774			fractionated in the laboratory
E-VA/16w(2)	4065			fractionated in the laboratory
E-VA/17w(1)	4231			fractionated in the laboratory
E-VA/17w(2)	4731			fractionated in the laboratory
E-VA/25w	20000	88000		Sci. Polym. Products, catalog #245
E-VA/39w	3650			fractionated in the laboratory
E-VA/40w	10100			fractionated in the laboratory
E-VA/41w	4620			fractionated in the laboratory
E-VA/42w	14550			fractionated in the laboratory
E-VA/43w	7180			fractionated in the laboratory
E-VA/50w	59500	250000		Sci. Polym. Products, catalog #785
E-VA/70w	50000	280000		Sci. Polym. Products, catalog #786

Experimental VLE data:

copolymer (B): **E-VA/16w(1)** **73BEL**
solvent (A): **benzene** **C_6H_6** **71-43-2**

T/K = 373.15

φ_B	0.0000	0.0982	0.1729	0.1895	0.2573	0.2839	0.3228	0.3641
P_A/Torr	1335.0	1327.9	1315.8	1313.0	1294.9	1287.9	1273.6	1254.1

copolymer (B): **E-VA/16w(2)** **73BEL**
solvent (A): **benzene** **C_6H_6** **71-43-2**

T/K = 373.15

φ_B	0.0000	0.0675	0.1307	0.1907	0.2503	0.3082	0.3310	0.3791	0.4668
P_A/Torr	1335.0	1331.6	1324.9	1315.1	1300.7	1284.4	1276.5	1255.0	1206.0

copolymer (B): **E-VA/17w(1)** **73BEL**
solvent (A): **benzene** **C_6H_6** **71-43-2**

T/K = 373.15

φ_B	0.0000	0.0775	0.1465	0.1884	0.2624	0.3308	0.3817
P_A/Torr	1335.0	1331.0	1323.3	1316.8	1300.3	1278.4	1257.8

copolymer (B): **E-VA/17w(2)** **73BEL**
solvent (A): **benzene** **C_6H_6** **71-43-2**

T/K = 373.15

φ_B	0.0000	0.0976	0.1647	0.2072	0.2609	0.3142	0.3670	0.3971
P_A/Torr	1335.0	1330.8	1323.6	1317.0	1307.9	1295.0	1286.7	1260.0

copolymer (B): **E-VA/39w** **75BOB**
solvent (A): **benzene** **C_6H_6** **71-43-2**

T/K = 303.15

w_B	0.000	0.244	0.292	0.333	0.498	0.516	0.536	0.590	0.622
P_A/Torr	118.5	118.3	118.2	116.4	110.1	106.5	105.0	101.6	95.5
w_B	0.739	0.760	0.835	0.872					
P_A/Torr	82.0	79.2	69.7	59.7					

copolymer (B): **E-VA/39w** **75BOB**
solvent (A): **benzene** **C_6H_6** **71-43-2**

T/K = 323.15

w_B	0.000	0.157	0.224	0.259	0.279	0.306	0.336	0.392	0.485
P_A/Torr	269.20	268.90	267.20	265.90	266.25	261.95	259.45	254.00	233.80
w_B	0.489	0.646	0.719	0.848	0.870				
P_A/Torr	215.40	210.10	189.80	167.70	155.15				

copolymer (B): **E-VA/40w** **75BOB**
solvent (A): **benzene** **C_6H_6** **71-43-2**

T/K = 303.15

w_B	0.000	0.247	0.299	0.318	0.349	0.396	0.415	0.478	0.497
P_A/Torr	118.50	118.20	117.40	117.70	116.90	115.95	114.10	110.75	109.55
w_B	0.518	0.622	0.672	0.712	0.779				
P_A/Torr	107.50	96.20	89.00	85.65	77.15				

copolymer (B): **E-VA/40w** **75BOB**
solvent (A): **benzene** **C_6H_6** **71-43-2**

T/K = 323.15

w_B	0.000	0.149	0.224	0.336	0.427	0.513	0.559	0.618	0.665
P_A/Torr	269.20	269.15	261.90	257.05	242.15	220.85	207.05	193.25	181.80
w_B	0.758	0.801	0.802	0.888	0.907				
P_A/Torr	150.90	121.35	124.35	106.00	70.20				

copolymer (B): **E-VA/42w** **75BOB**
solvent (A): **benzene** **C_6H_6** **71-43-2**

T/K = 323.15

w_B	0.000	0.307	0.325	0.348	0.435	0.495	0.527	0.557	0.621
P_A/Torr	269.20	265.80	258.25	256.10	243.20	234.60	227.00	217.90	203.35
w_B	0.692	0.730	0.749						
P_A/Torr	176.40	161.40	151.70						

copolymer (B): **E-VA/42w** **75BOB**
solvent (A): **benzene** **C_6H_6** **71-43-2**

T/K = 343.15

w_B	0.000	0.360	0.457	0.506	0.516	0.563	0.584	0.593	0.615
P_A/Torr	548.16	519.90	494.40	476.90	496.90	445.50	433.50	423.60	414.40
w_B	0.647	0.674	0.679						
P_A/Torr	388.10	372.10	366.60						

copolymer (B): **E-VA/09w** **95GUP**
solvent (A): **1-butanol** **$C_4H_{10}O$** **71-36-3**

T/K = 353.15

w_B	0.994	0.988	0.983	0.973	0.969	0.966	0.964
P_A/kPa	2.7	5.2	8.0	10.8	12.3	13.9	15.2

copolymer (B): **E-VA/41w** **76KIE**
solvent (A): **butyl acetate** **$C_6H_{12}O_2$** **123-86-4**

T/K = 323.15

w_B	0.1771	0.2141	0.2897	0.3163	0.3655	0.4005	0.4623	0.5462
P_A/Torr	41.80	40.30	38.00	36.80	34.80	33.80	31.30	28.00

copolymer (B): **E-VA/41w** **76KIE**
solvent (A): **butyl acetate** **$C_6H_{12}O_2$** **123-86-4**

T/K = 343.15

w_B	0.1802	0.2374	0.2580	0.3315	0.3655	0.4387	0.4892	0.5773
P_A/Torr	101.40	98.10	96.40	89.70	87.40	79.90	75.40	64.90

copolymer (B): **E-VA/41w** **76KIE**
solvent (A): **butyl acetate** **$C_6H_{12}O_2$** **123-86-4**

T/K = 363.15

w_B	0.1866	0.2591	0.3132	0.3767	0.4165	0.4571	0.5355	0.6488
P_A/Torr	219.80	209.60	201.10	190.00	181.50	175.20	158.30	134.80

copolymer (B): **E-VA/25w** **95GUP**
solvent (A): **cyclohexane** **C_6H_{12}** **110-82-7**

T/K = 353.15

w_B	0.860	0.772	0.762	0.723	0.700	0.631
P_A/kPa	39.7	54.9	55.5	60.7	64.1	70.4

copolymer (B): **E-VA/50w** **95GUP**
solvent (A): **cyclohexane** **C_6H_{12}** **110-82-7**

T/K = 353.15

w_B	0.896	0.841	0.830	0.795	0.775	0.713
P_A/kPa	39.7	54.9	55.5	60.7	64.1	70.4

copolymer (B): **E-VA/70w** **95GUP**
solvent (A): **cyclohexane** **C_6H_{12}** **110-82-7**

T/K = 353.15

w_B	0.974	0.919	0.877	0.867	0.844	0.809	0.758
P_A/kPa	14.4	39.7	54.9	55.5	60.7	64.1	70.4

copolymer (B): **E-VA/41w** **76KIE**
solvent (A): **ethyl acetate** **$C_4H_8O_2$** **141-78-6**

T/K = 303.15

w_B	0.2430	0.2805	0.3496	0.3565	0.3804	0.4400	0.5301
P_A/Torr	118.70	118.30	117.85	115.60	114.10	111.75	100.90

copolymer (B): **E-VA/41w** **76KIE**
solvent (A): **ethyl acetate** **$C_4H_8O_2$** **141-78-6**

T/K = 323.15

w_B	0.2554	0.2959	0.3241	0.3707	0.3817	0.4400	0.5704
P_A/Torr	281.20	279.10	274.90	272.80	263.10	257.20	230.20

copolymer (B): **E-VA/41w** **76KIE**
solvent (A): **ethyl acetate** **$C_4H_8O_2$** **141-78-6**

T/K = 343.15

w_B	0.2302	0.2885	0.3372	0.3651	0.4539	0.5754	0.7667
P_A/Torr	588.20	584.00	570.40	566.70	547.55	489.10	386.30

copolymer (B): **E-VA/43w** **76KIE**
solvent (A): **ethyl acetate** **$C_4H_8O_2$** **141-78-6**

T/K = 303.15

w_B	0.2143	0.2476	0.3038	0.3490	0.3654	0.4005	0.4487	0.4851
P_A/Torr	118.00	117.20	116.20	115.20	111.50	109.90	108.00	104.20

copolymer (B): **E-VA/43w** **76KIE**
solvent (A): **ethyl acetate** **$C_4H_8O_2$** **141-78-6**

T/K = 323.15

w_B	0.2208	0.2558	0.3170	0.3655	0.3849	0.4224	0.4764	0.5195
P_A/Torr	277.20	276.40	270.50	269.70	259.00	252.00	244.00	235.70

copolymer (B): **E-VA/43w** **76KIE**
solvent (A): **ethyl acetate** **$C_4H_8O_2$** **141-78-6**

T/K = 343.15

w_B	0.2475	0.2894	0.3003	0.3750	0.4399	0.5250	0.6110	0.6960
P_A/Torr	575.80	569.50	568.10	547.00	529.30	488.50	499.30	392.10

copolymer (B): **E-VA/41w** **76KIE**
solvent (A): **methyl acetate** **$C_3H_6O_2$** **79-20-9**

T/K = 303.15

w_B	0.2654	0.3257	0.3549	0.3651	0.4123	0.5102	0.5610	0.6130
P_A/Torr	265.75	261.75	260.05	258.75	255.25	240.75	232.15	215.75

copolymer (B): **E-VA/41w** **76KIE**
solvent (A): **methyl acetate** **$C_3H_6O_2$** **79-20-9**

T/K = 323.15

w_B	0.2561	0.3030	0.3762	0.3879	0.4387	0.4829	0.6341	0.7189
P_A/Torr	586.15	584.15	565.15	561.65	550.65	535.15	473.15	424.65

copolymer (B): **E-VA/43w** **76KIE**
solvent (A): **methyl acetate** **$C_3H_6O_2$** **79-20-9**

T/K = 303.15

w_B	0.1951	0.2632	0.3190	0.3725	0.3951	0.4839	0.5363	0.5744
P_A/Torr	265.75	263.95	262.75	258.25	257.05	244.75	234.15	220.75

copolymer (B): **E-VA/43w** **76KIE**
solvent (A): **methyl acetate** **$C_3H_6O_2$** **79-20-9**

T/K = 323.15

w_B	0.2154	0.2744	0.3706	0.4313	0.4775	0.5564	0.6653	0.7667
P_A/Torr	586.15	579.15	565.05	542.15	526.35	490.15	439.45	378.15

copolymer (B): **E-VA/41w** **76KIE**
solvent (A): **propyl acetate** **$C_5H_{10}O_2$** **109-60-4**

T/K = 303.15

w_B	0.1978	0.2330	0.2662	0.2815	0.3335	0.3865	0.4445	0.5075
P_A/Torr	42.70	42.10	41.50	40.70	39.50	38.20	36.20	34.00

copolymer (B): **E-VA/41w** **76KIE**
solvent (A): **propyl acetate** **$C_5H_{10}O_2$** **109-60-4**

T/K = 323.15

w_B	0.2048	0.2428	0.2671	0.2959	0.3249	0.3701	0.4180	0.4815
P_A/Torr	111.70	108.60	106.80	104.50	101.90	96.20	91.20	84.50

copolymer (B): **E-VA/41w** **76KIE**
solvent (A): **propyl acetate** **$C_5H_{10}O_2$** **109-60-4**

T/K = 343.15

w_B	0.2200	0.2644	0.2931	0.3387	0.3658	0.4052	0.4434	0.5749
P_A/Torr	250.60	243.00	237.50	227.90	220.90	211.00	202.70	168.00

copolymer (B): **E-VA/41w** **76KIE**
solvent (A): **propyl acetate** **$C_5H_{10}O_2$** **109-60-4**

T/K = 363.15

w_B	0.2523	0.3126	0.3569	0.4066	0.4427	0.5364	0.6073	0.8102
P_A/Torr	501.60	481.40	463.30	443.40	427.30	381.80	344.80	245.80

copolymer (B): **E-VA/16w(1)** **73BEL**
solvent (A): **toluene** **C_7H_8** **108-88-3**

T/K = 373.15

φ_B	0.0000	0.1459	0.2589	0.3228	0.3869	0.4668	0.4760	0.5356
P_A/Torr	558.60	542.50	533.10	518.50	499.90	472.60	470.80	444.60

copolymer (B): **E-VA/16w(2)** **73BEL**
solvent (A): **toluene** **C_7H_8** **108-88-3**

T/K = 373.15

φ_B	0.0000	0.0496	0.1272	0.2658	0.3599	0.3843	0.4432	0.5039
P_A/Torr	558.60	557.20	552.10	543.60	515.60	510.50	491.60	475.70

copolymer (B): **E-VA/17w(1)** **73BEL**
solvent (A): **toluene** **C_7H_8** **108-88-3**

T/K = 373.15

φ_B	0.0000	0.1209	0.2005	0.2113	0.3186	0.3711	0.4590	0.5357
P_A/Torr	558.60	553.00	545.60	544.40	527.90	513.60	493.60	473.80

copolymer (B): **E-VA/17w(2)** **73BEL**
solvent (A): **toluene** **C_7H_8** **108-88-3**

T/K = 373.15

φ_B	0.0000	0.1116	0.2037	0.2245	0.3482	0.3725	0.4318	0.4803	0.5302
P_A/Torr	558.60	555.60	547.20	544.40	528.20	522.90	509.00	497.10	485.10

copolymer (B): **E-VA/39w** **75BOB**
solvent (A): **toluene** **C_7H_8** **108-88-3**

T/K = 303.15

w_B	0.000	0.364	0.413	0.463	0.469	0.488	0.524	0.588	0.666
P_A/Torr	41.65	41.00	41.10	40.00	39.05	40.20	38.50	36.65	28.15
w_B	0.687								
P_A/Torr	30.05								

copolymer (B): **E-VA/39w** **75BOB**
solvent (A): **toluene** **C_7H_8** **108-88-3**

T/K = 323.15

w_B	0.000	0.389	0.437	0.501	0.520	0.571	0.642	0.661	0.668
P_A/Torr	97.25	96.30	93.25	90.35	89.30	87.35	81.50	78.00	74.20

continue

continue

w_B	0.718	0.730
P_A/Torr	62.25	63.75

copolymer (B): **E-VA/40w** **75BOB**
solvent (A): **toluene** **C_7H_8** **108-88-3**

T/K = 303.15

w_B	0.000	0.201	0.276	0.339	0.414	0.436	0.465	0.523	0.546
P_A/Torr	41.65	40.20	40.00	38.85	39.50	38.20	37.50	36.90	36.00
w_B	0.555	0.578	0.598	0.621					
P_A/Torr	37.40	35.50	33.90	32.30					

copolymer (B): **E-VA/40w** **75BOB**
solvent (A): **toluene** **C_7H_8** **108-88-3**

T/K = 323.15

w_B	0.000	0.237	0.311	0.328	0.438	0.495	0.515	0.568	0.604
P_A/Torr	97.25	96.15	95.00	95.95	94.60	94.00	91.00	89.35	80.60
w_B	0.630	0.673	0.694						
P_A/Torr	81.35	72.90	65.35						

copolymer (B): **E-VA/42w** **75BOB**
solvent (A): **toluene** **C_7H_8** **108-88-3**

T/K = 323.15

w_B	0.000	0.403	0.473	0.519	0.537	0.567	0.571	0.600	0.673
P_A/Torr	92.00	85.38	80.78	77.83	77.30	74.50	72.13	71.00	62.00
w_B	0.736	0.749	0.785						
P_A/Torr	52.60	51.50	44.90						

copolymer (B): **E-VA/42w** **75BOB**
solvent (A): **toluene** **C_7H_8** **108-88-3**

T/K = 343.15

w_B	0.000	0.375	0.448	0.504	0.543	0.619	0.649	0.652	0.672
P_A/Torr	204.05	191.80	182.60	176.70	169.80	154.30	149.80	147.30	139.20
w_B	0.694	0.728	0.765	0.795					
P_A/Torr	132.80	124.10	109.40	101.20					

copolymer (B): **E-VA/42w** **75BOB**
solvent (A): **toluene** **C_7H_8** **108-88-3**

T/K = 363.15

w_B	0.000	0.431	0.506	0.524	0.549	0.570	0.600	0.680	0.715
P_A/Torr	407.00	370.70	348.70	346.00	339.00	323.80	315.30	277.40	251.00
w_B	0.771								
P_A/Torr	220.00								

copolymer (B): **E-VA/09w** **95GUP**
solvent (A): **trichloromethane** **$CHCl_3$** **67-66-3**

T/K = 333.15

w_B	0.969	0.933	0.896	0.853	0.832	0.804	0.784	0.747
P_A/kPa	14.0	28.1	40.1	53.2	58.4	66.1	72.3	87.1

copolymer (B): **E-VA/25w** **95GUP**
solvent (A): **trichloromethane** **$CHCl_3$** **67-66-3**

T/K = 333.15

w_B	0.938	0.860	0.772	0.680	0.642	0.599	0.559	0.519
P_A/kPa	13.9	26.7	39.9	53.2	59.9	66.5	73.1	79.3

copolymer (B): **E-VA/50w** **95GUP**
solvent (A): **trichloromethane** **$CHCl_3$** **67-66-3**

T/K = 333.15

w_B	0.902	0.798	0.697	0.602	0.561	0.523	0.488	0.448
P_A/kPa	13.9	26.7	39.9	53.2	59.9	66.5	73.1	79.3

copolymer (B): **E-VA/70w** **95GUP**
solvent (A): **trichloromethane** **$CHCl_3$** **67-66-3**

T/K = 333.15

w_B	0.941	0.839	0.713	0.597	0.550	0.511	0.474	0.427
P_A/kPa	13.9	26.7	39.9	53.2	59.9	66.5	73.1	79.3

copolymer (B): **E-VA/39w** **75BOB**
solvent (A): **o-xylene** **C_8H_{10}** **95-47-6**

T/K = 323.15

w_B	0.000	0.157	0.298	0.363	0.376	0.381	0.457	0.503	0.628
P_A/Torr	25.76	25.30	25.25	25.10	25.30	25.30	24.20	23.20	17.80

continue

continue

w_B	0.630	0.643	0.730	0.817	0.852	0.867
P_A/Torr	17.50	18.35	14.80	11.90	10.40	8.45

copolymer (B): **E-VA/39w** **75BOB**
solvent (A): **o-xylene** **C_8H_{10}** **95-47-6**

T/K = 343.15

w_B	0.000	0.307	0.336	0.386	0.437	0.442	0.504	0.577	0.588
P_A/Torr	60.82	60.60	59.60	58.10	56.30	55.50	51.00	48.20	46.50
w_B	0.616	0.640	0.676	0.716	0.780				
P_A/Torr	45.25	43.30	42.00	39.80	35.55				

copolymer (B): **E-VA/42w** **75BOB**
solvent (A): **o-xylene** **C_8H_{10}** **95-47-6**

T/K = 363.15

w_B	0.000	0.380	0.479	0.487	0.581	0.676	0.724	0.790	0.800
P_A/Torr	139.00	129.10	120.90	120.65	108.85	93.10	85.10	71.15	63.10

copolymer (B): **E-VA/16w(1)** **73BEL**
solvent (A): **p-xylene** **C_8H_{10}** **106-42-3**

T/K = 373.15

φ_B	0.0000	0.1593	0.3009	0.3797	0.4441	0.4629	0.5091	0.5602
P_A/Torr	242.1	237.5	226.9	218.4	209.0	206.1	196.6	186.1

copolymer (B): **E-VA/16w(2)** **73BEL**
solvent (A): **p-xylene** **C_8H_{10}** **106-42-3**

T/K = 373.15

φ_B	0.0000	0.1767	0.2788	0.3282	0.3886	0.4585	0.5094	0.5728
P_A/Torr	242.1	237.0	229.4	225.0	217.7	206.9	199.0	181.3

copolymer (B): **E-VA/17w(1)** **73BEL**
solvent (A): **p-xylene** **C_8H_{10}** **106-42-3**

T/K = 373.15

φ_B	0.0000	0.1256	0.1878	0.2892	0.3569	0.4169	0.4525	0.5668
P_A/Torr	242.1	239.4	236.5	228.9	223.7	214.9	201.6	189.2

copolymer (B):	**E-VA/17w(2)**		**73BEL**
solvent (A):	**p-xylene**	**C_8H_{10}**	**106-42-3**

T/K = 373.15

φ_B	0.0000	0.1573	0.2319	0.3007	0.3691	0.4466	0.5104	0.5857
P_A/Torr	242.1	238.0	234.3	229.5	222.3	212.1	201.3	189.0

copolymer (B):	**E-VA/39w**		**75BOB**
solvent (A):	**p-xylene**	**C_8H_{10}**	**106-42-3**

T/K = 323.15

w_B	0.000	0.157	0.180	0.325	0.416	0.441	0.530	0.545	0.576
P_A/Torr	32.25	31.95	32.20	32.00	31.00	29.85	27.05	26.76	25.00
w_B	0.578	0.632							
P_A/Torr	25.80	23.20							

copolymer (B):	**E-VA/39w**		**75BOB**
solvent (A):	**p-xylene**	**C_8H_{10}**	**106-42-3**

T/K = 343.15

w_B	0.000	0.241	0.297	0.317	0.420	0.464	0.542	0.578	0.640
P_A/Torr	75.70	75.50	73.50	72.90	70.05	67.05	61.40	59.35	56.60
w_B	0.660	0.745	0.762						
P_A/Torr	51.90	46.65	43.60						

copolymer (B):	**E-VA/40w**		**75BOB**
solvent (A):	**p-xylene**	**C_8H_{10}**	**106-42-3**

T/K = 343.15

w_B	0.000	0.440	0.524	0.599	0.605	0.615	0.643	0.748	0.802
P_A/Torr	75.70	67.85	62.40	57.60	56.70	55.50	54.30	42.70	36.20

copolymer (B):	**E-VA/40w**		**75BOB**
solvent (A):	**p-xylene**	**C_8H_{10}**	**106-42-3**

T/K = 363.15

w_B	0.000	0.450	0.489	0.549	0.577	0.630	0.690	0.737	0.760
P_A/Torr	167.80	149.70	145.20	136.90	131.90	121.80	110.70	103.35	91.95
w_B	0.812								
P_A/Torr	77.20								

copolymer (B): **E-VA/42w** **75BOB**
solvent (A): **p-xylene** **C_8H_{10}** **106-42-3**

T/K = 323.15

w_B	0.000	0.451	0.512	0.597	0.634	0.649	0.732	0.737	0.800
P_A/Torr	32.25	28.60	26.85	24.45	23.05	22.75	19.30	18.30	14.95

copolymer (B): **E-VA/42w** **75BOB**
solvent (A): **p-xylene** **C_8H_{10}** **106-42-3**

T/K = 343.15

w_B	0.000	0.390	0.477	0.575	0.597	0.637	0.655	0.764	0.814
P_A/Torr	75.70	69.15	65.15	59.20	56.80	53.50	52.10	40.75	37.80

copolymer (B): **E-VA/42w** **75BOB**
solvent (A): **p-xylene** **C_8H_{10}** **106-42-3**

T/K = 363.15

w_B	0.000	0.450	0.509	0.558	0.600	0.641	0.681	0.717	0.732
P_A/Torr	167.80	149.70	142.95	135.90	127.50	120.80	114.80	104.00	109.70
w_B	0.773	0.827							
P_A/Torr	89.40	70.00							

Copolymers from ethylene oxide and *tert*-butyl methacrylate

Average chemical composition of the copolymers, acronyms and references:

Copolymer (B)	Acronym	Ref.
Poly(ethylene oxide)-b-poly(*tert*-butyl methacrylate) diblock copolymer		
49 mol% ethylene oxide	EO-b-tBMA/49x	94WOH

Characterization:

Copolymer (B)	M_n/ g/mol	Further information
EO-b-tBMA/49x	8200 (= 4000 + 4200)	synthesized in the laboratory

Experimental VLE data:

copolymer (B): **EO-b-tBMA/49x** **94WOH**
solvent (A): **toluene** **C_7H_8** **108-88-3**

T/K = 323.15

ψ_B	0.0000	0.4505	0.4669	0.4983	0.5444	0.5827	0.6132	0.6352	0.6469
P_A/Torr	92.12	86.46	85.81	83.88	81.36	78.58	76.25	74.12	72.88
ψ_B	0.6715	0.7086	0.7461	0.7622	0.7877				
P_A/Torr	69.51	65.20	60.40	57.28	53.11				

Comments: segment fraction is based on Bondi's volumina with r_B = 80

copolymer (B): **EO-b-tBMA/49x** **94WOH**
solvent (A): **toluene** **C_7H_8** **108-88-3**

T/K = 343.75

ψ_B	0.0000	0.5091	0.5128	0.5663	0.5976	0.6185	0.6439	0.6590	0.6879
P_A/Torr	208.40	184.38	182.95	175.99	170.90	167.55	162.12	158.66	151.55
ψ_B	0.6879	0.7232	0.7501	0.7501					
P_A/Torr	152.29	143.51	134.77	135.21					

Comments: segment fraction is based on Bondi's volumina with r_B = 80

Copolymers from ethylene oxide and methyl methacrylate

Average chemical composition of the copolymers, acronyms and references:

Copolymer (B)	Acronym	Ref.
Poly(ethylene oxide)-b-poly(methyl methacrylate) diblock copolymer 51.7 mol% ethylene oxide	EO-b-MMA/52x	86KRC
Poly(ethylene oxide)-b-poly(methyl methacrylate)-b-poly(ethylene oxide) triblock copolymer 61.0 mol% ethylene oxide	EO-MMA-EO/61x	86KRC

Characterization:

Copolymer (B)	M_n/ g/mol	Further information
EO-b-MMA/52x	8900 (= 4600 + 4300)	synthesized in the laboratory
EO-MMA-EO/61x	7860 (= 2400 + 3060 + 2400)	synthesized in the laboratory

Experimental VLE data:

copolymer (B): **EO-b-MMA/52x** **86KRC**
solvent (A): **toluene** C_7H_8 **108-88-3**

T/K = 323.41

ψ_B	0.536	0.562	0.648	0.747	0.766
P_A/kPa	10.626	10.226	9.613	7.933	7.773

Comments: segment fraction is based on Bondi's volumina with r_B = 83

copolymer (B): **EO-b-MMA/52x** **86KRC**
solvent (A): **toluene** C_7H_8 **108-88-3**

continue

continue

T/K = 343.27

ψ_B	0.522	0.562	0.567	0.615	0.641	0.644
P_A/kPa	23.878	22.638	22.691	21.878	21.252	20.132

Comments: segment fraction is based on Bondi's volumina with r_B = 83

copolymer (B): **EO-b-MMA/52x** **86KRC**
solvent (A): **toluene** **C_7H_8** **108-88-3**

T/K = 373.27

ψ_B	0.432	0.499	0.559	0.609	0.668	0.744	0.789	0.829	0.839
P_A/kPa	69.671	67.697	64.822	62.333	59.430	54.473	50.826	47.979	46.333
ψ_B	0.867								
P_A/kPa	44.703								

Comments: segment fraction is based on Bondi's volumina with r_B = 83

copolymer (B): **EO-MMA-EO/61x** **86KRC**
solvent (A): **toluene** **C_7H_8** **108-88-3**

T/K = 323.08

ψ_B	0.461	0.547	0.599	0.667	0.736	0.762
P_A/kPa	11.457	10.626	10.226	9.613	8.706	8.399

Comments: segment fraction is based on Bondi's volumina with r_B = 73.2

copolymer (B): **EO-MMA-EO/61x** **86KRC**
solvent (A): **toluene** **C_7H_8** **108-88-3**

T/K = 343.17

ψ_B	0.535	0.573	0.583	0.629	0.654	0.697
P_A/kPa	23.878	22.638	22.691	21.878	21.252	19.772

Comments: segment fraction is based on Bondi's volumina with r_B = 73.2

copolymer (B): **EO-MMA-EO/61x** **86KRC**
solvent (A): **toluene** **C_7H_8** **108-88-3**

T/K = 373.26

ψ_B	0.554	0.612	0.628	0.665	0.685	0.729	0.785	0.876
P_A/kPa	69.010	66.786	65.349	63.299	61.341	57.274	56.992	54.174

Comments: segment fraction is based on Bondi's volumina with r_B = 73.2

Copolymers from ethylene oxide and propylene oxide

Average chemical composition of the copolymers, acronyms and references:

Copolymer (B)	Acronym	Ref.
Poly(ethylene oxide)-b-poly(propylene oxide) diblock copolymer		
40.0 mol% propylene oxide	EO-b-PO/40x	94WOH
58.0 wt% propylene oxide	EO-b-PO/58w	89MOE
Poly(ethylene oxide)-b-poly(propylene oxide)-b-poly(ethylene oxide) triblock copolymer		
50.0 mol% propylene oxide	EO-PO-EO/50x	94WOH
67.0 wt% propylene oxide	EO-PO-EO/67w	89MOE
87.0 wt% propylene oxide	EO-PO-EO/87w	89MOE
93.0 wt% propylene oxide	EO-PO-EO/93w	89MOE
Poly(propylene oxide)-b-poly(ethylene oxide)-b-poly(propylene oxide) triblock copolymer		
50.0 mol% propylene oxide	PO-EO-PO/50x	94WOH

Characterization:

Copolymer (B)	M_n/ g/mol	M_w/ g/mol	M_η/ g/mol	Further information
EO-b-PO/40x	4240	4700		synthesized in the laboratory
EO-b-PO/58w	1700			
EO-PO-EO/50x	10800			synthesized in the laboratory
EO-PO-EO/67w	1900			
EO-PO-EO/87w	950			
EO-PO-EO/93w	1800			
PO-EO-PO/50x	10800			synthesized in the laboratory

Experimental VLE data:

copolymer (B): **EO-b-PO/40x** **94WOH**
solvent (A): **ethylbenzene** **C_8H_{10}** **100-41-4**

T/K = 343.75

w_B	0.1075	0.1277	0.1496	0.1859	0.2135	0.2315	0.2626	0.3182	0.5333
P_A/P_A^s	0.9961	0.9959	0.9922	0.9892	0.9851	0.9839	0.9795	0.9637	0.9057
w_B	0.6201	0.6639	0.6987	0.7048	0.7339	0.7655	0.7805	0.8040	0.8290
P_A/P_A^s	0.8528	0.8135	0.7823	0.7750	0.7431	0.7072	0.6805	0.6532	0.6036

copolymer (B): **EO-b-PO/58w** **89MOE**
solvent (A): **tetrachloromethane** **CCl_4** **56-23-5**

T/K = 303.15

w_B	0.1004	0.2011	0.3019	0.4026	0.5038	0.6038	0.7032	0.8031	0.8971
a_A	0.9830	0.9466	0.8884	0.8088	0.7088	0.5925	0.4620	0.3163	0.1704

copolymer (B): **EO-PO-EO/50x** **94WOH**
solvent (A): **toluene** **C_7H_8** **108-88-3**

T/K = 323.35

ψ_B	0.0000	0.4191	0.4480	0.4565	0.4877	0.5584	0.5834	0.6325	0.6906
P_A/Torr	92.90	84.97	83.29	82.84	80.76	75.51	73.03	68.18	60.69
ψ_B	0.7184	0.7604	0.7856	0.8077					
P_A/Torr	57.08	50.74	46.62	42.56					

Comments: segment fraction is based on Bondi's volumina with r_B = 103

copolymer (B): **EO-PO-EO/50x** **94WOH**
solvent (A): **toluene** **C_7H_8** **108-88-3**

T/K = 343.75

ψ_B	0.0000	0.4471	0.4635	0.4811	0.5474	0.5629	0.5818	0.6436	0.6717
P_A/Torr	208.40	178.31	176.28	173.77	163.06	160.68	157.33	144.30	138.30
ψ_B	0.6986	0.7356	0.7556						
P_A/Torr	130.92	121.29	114.87						

Comments: segment fraction is based on Bondi's volumina with r_B = 103

copolymer (B): **EO-PO-EO/67w** **89MOE**
solvent (A): **tetrachloromethane** **CCl_4** **56-23-5**

T/K = 303.15

w_B	0.0995	0.1993	0.2994	0.3992	0.5004	0.6004	0.7021	0.8014	0.9032
a_A	0.9919	0.9620	0.9114	0.8383	0.7416	0.6264	0.4910	0.3423	0.1755

copolymer (B): **EO-PO-EO/87w** **89MOE**
solvent (A): **tetrachloromethane** **CCl_4** **56-23-5**

T/K = 303.15

w_B	0.0991	0.1987	0.2988	0.3986	0.4991	0.5998	0.7004	0.8007	0.9025
a_A	0.9810	0.9453	0.8912	0.8162	0.7204	0.6054	0.4746	0.3281	0.1677

copolymer (B): **EO-PO-EO/93w** **89MOE**
solvent (A): **tetrachloromethane** **CCl_4** **56-23-5**

T/K = 303.15

w_B	0.0991	0.1988	0.2986	0.3994	0.4998	0.6002	0.7009	0.8017	0.8981
a_A	0.9861	0.9518	0.8964	0.8173	0.7179	0.6010	0.4674	0.3219	0.1712

copolymer (B): **PO-EO-PO/50x** **94WOH**
solvent (A): **toluene** **C_7H_8** **108-88-3**

T/K = 323.05

ψ_B	0.0000	0.3885	0.4300	0.4341	0.4682	0.5138	0.5604	0.5761	0.6269
P_A/Torr	91.72	84.97	83.29	82.84	80.76	77.99	74.37	73.03	68.18
ψ_B	0.7369	0.7580	0.7838						
P_A/Torr	54.35	50.74	46.62						

Comments: segment fraction is based on Bondi's volumina with r_B = 103

copolymer (B): **PO-EO-PO/50x** **94WOH**
solvent (A): **toluene** **C_7H_8** **108-88-3**

T/K = 342.65

ψ_B	0.0000	0.3794	0.4077	0.4516	0.5346	0.5629	0.6045	0.6148	0.6320
P_A/Torr	200.10	181.02	178.31	173.77	163.06	158.86	152.05	150.12	147.40
ψ_B	0.6646	0.7063	0.7262	0.7695	0.7898				
P_A/Torr	140.64	130.92	125.98	114.87	108.41				

Comments: segment fraction is based on Bondi's volumina with r_B = 103

Copolymers from ethylene oxide and styrene

Average chemical composition of the copolymers, acronyms and references:

Copolymer (B)	Acronym	Ref.
Polystyrene-b-poly(ethylene oxide) diblockcopolymer blocks are not further specified	S-b-EO/?	94WOH
Poly(ethylene oxide)-b-polystyrene-b-poly(ethylene oxide) triblock copolymer blocks are not further specified	EO-S-EO/?	94WOH

Characterization:

Copolymer (B)	M_n/ g/mol	M_w/ g/mol	M_η/ g/mol	Further information
S-b-EO/?	45000			synthesized in the laboratory
EO-S-EO/?	45000			synthesized in the laboratory

Experimental VLE data:

copolymer (B):	**S-b-EO/?**		**94WOH**
solvent (A):	**toluene**	C_7H_8	**108-88-3**

T/K = 322.95

w_B	0.0000	0.5773	0.6233	0.6728	0.6772	0.6981	0.7255	0.7387	0.7551
P_A/Torr	91.33	82.40	79.77	76.10	75.51	73.38	70.90	69.71	67.48
w_B	0.7842	0.7903	0.8222						
P_A/Torr	65.05	62.82	58.71						

copolymer (B): **S-b-EO/?** **94WOH**
solvent (A): **toluene** **C_7H_8** **108-88-3**

T/K = 342.65

w_B	0.0000	0.5628	0.5805	0.6064	0.6212	0.6532	0.6839	0.7077	0.7280
P_A/Torr	200.10	182.80	179.74	178.06	175.64	170.71	165.13	160.98	156.29
w_B	0.7385	0.7439	0.7467	0.7841	0.8155				
P_A/Torr	154.61	152.14	153.18	144.15	135.01				

copolymer (B): **EO-S-EO/?** **94WOH**
solvent (A): **toluene** **C_7H_8** **108-88-3**

T/K = 323.35

w_B	0.0000	0.5825	0.6191	0.6634	0.6865	0.7081	0.7210	0.7379	0.7592
P_A/Torr	92.90	82.40	79.77	75.51	73.38	70.90	69.71	67.48	65.05
w_B	0.7666	0.7704	0.8017	0.8269					
P_A/Torr	64.11	62.82	58.71	54.00					

copolymer (B): **EO-S-EO/?** **94WOH**
solvent (A): **toluene** **C_7H_8** **108-88-3**

T/K = 343.75

w_B	0.0000	0.5576	0.5759	0.5978	0.6141	0.6165	0.6390	0.6717	0.6863
P_A/Torr	208.40	182.80	179.74	178.06	175.64	174.51	170.71	165.13	160.98
w_B	0.7144	0.7144	0.7209	0.7343	0.7591	0.7856	0.8057		
P_A/Torr	154.61	156.29	153.18	152.14	144.15	135.01	127.11		

Copolymers from 3-hydroxybutanoic acid and 3-hydroxypentanoic acid

Average chemical composition of the copolymers, acronyms and references:

Copolymer (B)	Acronym	Ref.
3-Hydroxybutanoic acid/3-hydroxypentanoic acid copolymer		
14 wt% 3-hydroxypentanoic acid	HBA-HPA/14w	99WOH
22 wt% 3-hydroxypentanoic acid	HBA-HPA/22w	99WOH

Characterization:

Copolymer (B)	M_n/ g/mol	M_w/ g/mol	M_η/ g/mol	Further information
HBA-HPA/14w	153000	454000		Aldrich Chem.Co., catalog #34739-6
HBA-HPA/22w	188000	529000		Aldrich Chem.Co., catalog #34741-8

Experimental VLE data:

copolymer (B): **HBA-HPA/14w** **99WOH**
solvent (A): **chlorobenzene** **C_6H_5Cl** **108-90-7**

T/K = 373.15

w_B	0.7854	0.7495	0.7028	0.6566	0.6205	0.5746	0.5311	0.0000
P_A/Pa	23780.0	26315.0	28811.7	30649.7	32403.5	33602.1	34540.3	38293.0

copolymer (B): **HBA-HPA/14w** **99WOH**
solvent (A): **chlorobenzene** **C_6H_5Cl** **108-90-7**

T/K = 388.15

w_B	0.7312	0.6970	0.6627	0.6245	0.5989	0.5581	0.5128	0.4885	0.0000
P_A/Pa	43492.0	46497.2	49109.0	51272.2	52716.4	54775.2	56514.4	57104.4	61455.4

copolymer (B): **HBA-HPA/14w** **99WOH**
solvent (A): **chlorobenzene** **C_6H_5Cl** **108-90-7**

T/K = 403.15

w_B	0.8103	0.7537	0.7093	0.6687	0.6329	0.5908	0.5617	0.5113	0.0000
P_A/Pa	53051.4	62960.1	68848.4	74082.4	77856.2	81961.9	83886.7	86475.3	94819.4

copolymer (B): **HBA-HPA/w** **99WOH**
solvent (A): **chlorobenzene** **C_6H_5Cl** **108-90-7**

T/K = 373.15

w_B	0.7663	0.7015	0.6779	0.6482	0.6289	0.5963	0.5340	0.4813	0.4220
P_A/Pa	23538.7	27283.8	28260.2	29592.8	30354.9	31698.9	33418.3	34578.6	35777.2

copolymer (B):	**HBA-HPA/w**		**99WOH**
solvent (A):	**chlorobenzene**	**C_6H_5Cl**	**108-90-7**

T/K = 388.15

w_B	0.0000	0.4112	0.5127	0.5600	0.6122	0.6354	0.6708	0.7046	0.7352
P_A/Pa	61455.4	57983.2	54984.1	52833.2	50178.3	48605.1	45882.6	43246.2	40658.9
w_B	0.7740	0.7970							
P_A/Pa	37039.2	34556.4							

copolymer (B):	**HBA-HPA/w**		**99WOH**
solvent (A):	**chlorobenzene**	**C_6H_5Cl**	**108-90-7**

T/K = 403.15

w_B	0.0000	0.5091	0.5586	0.5997	0.6354	0.6731	0.7182	0.7501	0.7877
P_A/Pa	94819.4	85015.1	81582.6	78690.6	74376.3	71199.9	64941.8	60722.3	54435.8

Copolymers from styrene and butyl methacrylate

Average chemical composition of the copolymers, acronyms and references:

Copolymer (B)	Acronym	Ref.
Styrene/butyl methacrylate copolymer 50.0 wt% styrene	S-BMA/50w	95GUP

Characterization:

Copolymer (B)	M_n/ g/mol	M_w/ g/mol	M_η/ g/mol	Further information
S-BMA/50w	105300	200000		Sci. Polym. Products, catalog #595

Experimental VLE data:

copolymer (B):	**S-BMA/50w**		**95GUP**
solvent (A):	**2-propanone**	**C_3H_6O**	**67-64-1**

T/K = 333.15

w_B	0.978	0.956	0.947	0.929	0.910	0.867	0.856	0.836	0.799
P_A/kPa	23.5	43.3	51.6	60.3	67.6	81.7	85.3	90.4	95.9
w_B	0.776	0.746							
P_A/kPa	98.7	100.4							

copolymer (B):	**S-BMA/50w**		**95GUP**
solvent (A):	**trichloromethane**	**$CHCl_3$**	**67-66-3**

T/K = 343.15

w_B	0.922	0.852	0.778	0.708	0.631	0.545	0.488
P_A/kPa	13.3	26.4	40.1	53.6	65.5	78.8	86.0

Copolymers from styrene and docosyl maleate

Average chemical composition of the copolymers, acronyms and references:

Copolymer (B)	Acronym	Ref.
Styrene/docosyl maleate copolymer 50.0 mol% styrene	S-DCM/50x	97MIO

Characterization:

Copolymer (B)	M_n/ g/mol	M_w/ g/mol	M_η/ g/mol	Further information
S-DCM/50x	244500	293400		synthesized in the laboratory

Experimental VLE data:

copolymer (B): **S-DCM/50x** **97MIO**
solvent (A): **cyclohexane** **C_6H_{12}** **110-82-7**

T/K = 333.15

w_B	0.976	0.967	0.943	0.929	0.906	0.881	0.846	0.809	0.765
P_A/Torr	50.5	76.5	106.0	131.5	157.5	188.5	218.0	251.5	279.0
w_B	0.710	0.630	0.000						
P_A/Torr	300.5	362.5	390.0						

copolymer (B): **S-DCM/50x** **97MIO**
solvent (A): **methanol** **CH_4O** **67-56-1**

T/K = 333.15

w_B	0.982	0.975	0.965	0.954	0.940	0.920	0.895	0.879	0.000
P_A/Torr	128.0	190.0	247.5	296.0	354.5	400.5	457.5	489.5	635.0

copolymer (B): **S-DCM/50x** **97MIO**
solvent (A): **2-propanone** **C_3H_6O** **67-64-1**

T/K = 323.15

w_B	0.974	0.965	0.957	0.945	0.927	0.904	0.874	0.846	0.799
P_A/Torr	106.0	156.0	196.0	243.0	295.5	346.5	406.0	448.0	492.0
w_B	0.769	0.000							
P_A/Torr	513.0	612.0							

copolymer (B): **S-DCM/50x** **97MIO**
solvent (A): **2-propanone** **C_3H_6O** **67-64-1**

T/K = 343.15

w_B	0.976	0.964	0.950	0.937	0.915	0.890	0.858	0.827	0.806
P_A/Torr	230.0	317.5	408.5	494.5	587.0	679.5	786.0	857.5	896.0
w_B	0.000								
P_A/Torr	1196.0								

Copolymers from styrene and dodecyl maleate

Average chemical composition of the copolymers, acronyms and references:

Copolymer (B)	Acronym	Ref.
Styrene/dodecyl maleate copolymer 50.0 mol% styrene	S-DDM/50x	97MIO

Characterization:

Copolymer (B)	M_n/ g/mol	M_w/ g/mol	M_η/ g/mol	Further information
S-DDM/50x	177500	213000		synthesized in the laboratory

Experimental VLE data:

copolymer (B): **S-DDM/50x** **97MIO**
solvent (A): **cyclohexane** **C_6H_{12}** **110-82-7**

T/K = 333.15

w_B	0.990	0.978	0.963	0.941	0.935	0.914	0.887	0.856	0.821
P_A/Torr	50.5	76.5	106.0	131.5	157.5	188.5	218.0	251.5	279.0
w_B	0.769	0.692	0.000						
P_A/Torr	300.5	362.5	390.0						

copolymer (B): **S-DDM/50x** **97MIO**
solvent (A): **methanol** **CH_4O** **67-56-1**

T/K = 333.15

w_B	0.976	0.966	0.950	0.931	0.907	0.835	0.808	0.000
P_A/Torr	128.0	190.0	247.5	296.0	354.5	457.5	489.5	635.0

copolymer (B):	**S-DDM/50x**		**97MIO**
solvent (A):	**2-propanone**	**C_3H_6O**	**67-64-1**

T/K = 323.15

w_B	0.965	0.950	0.941	0.921	0.893	0.861	0.826	0.786	0.712
P_A/Torr	106.0	156.0	196.0	243.0	295.5	346.5	406.0	448.0	492.0
w_B	0.665	0.000							
P_A/Torr	513.0	612.0							

copolymer (B):	**S-DDM/50x**		**97MIO**
solvent (A):	**2-propanone**	**C_3H_6O**	**67-64-1**

T/K = 343.15

w_B	0.978	0.961	0.947	0.927	0.898	0.863	0.817	0.778	0.746
P_A/Torr	230.0	317.5	408.5	494.5	587.0	679.5	786.0	857.5	896.0
w_B	0.000								
P_A/Torr	1196.0								

Copolymers from styrene and isoprene

Average chemical composition of the copolymers, acronyms and references:

Copolymer (B)	Acronym	Ref.
Polystyrene-b-polyisoprene-b-polystyrene triblock copolymer		
15.0 wt% styrene	S-I-S/15w	99FRE
24.2 wt% styrene	S-I-S/24w	99FRE
30.0 wt% styrene	S-I-S/30w	99FRE

Characterization:

Copolymer (B)	M_n/ g/mol	M_w/ g/mol	M_η/ g/mol	Further information
S-I-S/15w	159000			synthesized in the laboratory
S-I-S/24w	113000			synthesized in the laboratory
S-I-S/30w	105000			synthesized in the laboratory

Experimental VLE data:

copolymer (B): **S-I-S/15w** **99FRE**
solvent (A): **cyclohexane** **C_6H_{12}** **110-82-7**

T/K = 323.45

w_B	0.9941	0.9922	0.9921	0.9883	0.9833	0.9826	0.9791	0.9709	0.9649
a_A	0.0242	0.0298	0.0291	0.0449	0.0646	0.0661	0.0798	0.1097	0.1306
w_B	0.9531	0.9209	0.8960	0.8959	0.8680	0.8679	0.8447	0.8376	0.8364
a_A	0.1712	0.2792	0.3526	0.3529	0.4409	0.4412	0.5140	0.5505	0.5546
w_B	0.8359	0.8195	0.8183						
a_A	0.5547	0.6841	0.6941						

copolymer (B): **S-I-S/24w** **99FRE**
solvent (A): **cyclohexane** **C_6H_{12}** **110-82-7**

T/K = 323.55

w_B	0.9916	0.9839	0.9806	0.9724	0.9581	0.9579	0.9445	0.9421	0.8907
a_A	0.0325	0.0602	0.0700	0.1007	0.1496	0.1503	0.1948	0.2032	0.3552
w_B	0.8894	0.8570	0.8158						
a_A	0.3606	0.4376	0.5434						

copolymer (B): **S-I-S/30w** **99FRE**
solvent (A): **cyclohexane** **C_6H_{12}** **110-82-7**

T/K = 323.95

w_B	0.9922	0.9921	0.9892	0.9891	0.9814	0.9785	0.9768	0.9723	0.9670
a_A	0.0327	0.0321	0.0440	0.0439	0.0737	0.0854	0.0912	0.1086	0.1294
w_B	0.9574	0.9492	0.9305	0.9231	0.9226	0.8606	0.7565		
a_A	0.1644	0.1925	0.2565	0.2791	0.2817	0.4405	0.6721		

Copolymers from styrene and maleic anhydride

Average chemical composition of the copolymers, acronyms and references:

Copolymer (B)	Acronym	Ref.
Styrene/maleic anhydride copolymer 50.0 mol% styrene	S-MAH/50x	97MIO

Characterization:

Copolymer (B)	M_n/ g/mol	M_w/ g/mol	M_η/ g/mol	Further information
S-MAH/50x	96700	116000		

Experimental VLE data:

copolymer (B): **S-MAH/50x** **97MIO**
solvent (A): **methanol** **CH_4O** **67-56-1**

T/K = 333.15

w_B	0.932	0.906	0.878	0.830	0.794	0.761	0.737	0.700	0.657
P_A/Torr	127.5	175.5	229.5	286.5	339.0	393.0	434.5	484.5	505.0
w_B	0.000								
P_A/Torr	635.0								

copolymer (B): **S-MAH/50x** **97MIO**
solvent (A): **2-propanone** **C_3H_6O** **67-64-1**

T/K = 323.15

w_B	0.902	0.856	0.819	0.783	0.771	0.714	0.621	0.000
P_A/Torr	215.5	278.0	336.5	389.0	400.0	454.0	489.0	612.0

Copolymers from styrene and methyl methacrylate

Average chemical composition of the copolymers, acronyms and references:

Copolymer (B)	Acronym	Ref.
Styrene/methyl methacrylate copolymer		
44.0 wt% styrene	S-MMA/44w	97WO2
50.0 wt% styrene	S-MMA/50w	97TAN
Polystyrene-b-poly(methyl methacrylate) diblock copolymer		
47.8 wt% polystyrene	S-b-MMA/48w	98WOH

Characterization:

Copolymer (B)	M_n/ g/mol	M_w/ g/mol	Further information
S-MMA/44w	50600	85000	synthesized in the laboratory
S-MMA/50w	91000	100000	Polysciences Inc.
S-b-MMA/48w	18000	20500 (= 9800-b-10700)	Polysciences Inc.

Experimental VLE data:

copolymer (B): S-MMA/44w 97WO2
solvent (A): benzene C_6H_6 71-43-2

T/K = 343.15

w_B	0.0000	0.4664	0.5157	0.5797	0.6218	0.6566	0.7079	0.7615	0.8101
P_A/Pa	73128.3	67029.4	65054.9	61786.1	59190.0	56652.5	52147.8	46421.8	40015.8

copolymer (B): S-MMA/44w 97WO2
solvent (A): ethylbenzene C_8H_{10} 100-41-4

continue

continue

T/K = 373.15

w_B	0.0000	0.5735	0.5905	0.6201	0.6489	0.6717	0.7056	0.7339	0.7653
P_A/Pa	34300.5	31378.1	31172.3	30668.1	29937.5	29296.1	28253.3	27196.9	25704.8
w_B	0.8097								
P_A/Pa	23190.6								

copolymer (B): **S-MMA/44w** **97WO2**
solvent (A): **ethylbenzene** **C_8H_{10}** **100-41-4**

T/K = 398.15

w_B	0.0000	0.5496	0.6039	0.6328	0.6600	0.7014	0.7221	0.7575	0.7930
P_A/Pa	74321.2	68866.0	67022.9	65618.2	64243.2	61493.4	59895.5	56603.0	52693.7
w_B	0.8143								
P_A/Pa	49661.4								

copolymer (B): **S-MMA/44w** **97WO2**
solvent (A): **mesitylene** **C_9H_{12}** **108-67-8**

T/K = 398.15

w_B	0.0000	0.6365	0.6783	0.7010	0.7321	0.7598	0.7834	0.8066	0.8293
P_A/Pa	32215.7	29635.2	28936.1	28614.0	27708.7	26822.8	25982.0	24703.0	23085.8

copolymer (B): **S-MMA/44w** **97WO2**
solvent (A): **toluene** **C_7H_8** **108-88-3**

T/K = 343.15

w_B	0.0000	0.4775	0.5203	0.5674	0.6186	0.6607	0.7091	0.7520	0.7997
P_A/Pa	27250.3	24991.3	24459.9	23707.8	22857.6	21868.4	20484.1	19042.5	17017.8

copolymer (B): **S-MMA/44w** **97WO2**
solvent (A): **toluene** **C_7H_8** **108-88-3**

T/K = 373.15

w_B	0.0000	0.5029	0.5586	0.6111	0.6508	0.6930	0.7366	0.7823
P_A/Pa	74464.2	67256.1	65118.9	62594.6	59765.0	57464.0	53673.8	48736.8

copolymer (B): **S-MMA/44w** **97WO2**
solvent (A): **p-xylene** **C_8H_{10}** **106-42-3**

T/K = 398.15

w_B	0.0000	0.6109	0.6393	0.6708	0.6954	0.7112	0.7467	0.7799	0.8083
P_A/Pa	69803.4	64896.2	63967.8	63179.1	61985.4	60987.2	58209.1	54893.4	51598.7
w_B	0.8363								
P_A/Pa	47368.6								

copolymer (B): **S-MMA/50w** **97TAN**
solvent (A): **methyl acetate** **$C_3H_6O_2$** **79-20-9**

T/K = 323.15

w_B	0.919	0.879	0.826	0.754	0.679	0.000
P_A/kPa	43.10	52.81	61.39	68.83	74.73	78.69

copolymer (B): **S-MMA/50w** **97TAN**
solvent (A): **2-propanone** **C_3H_6O** **67-64-1**

T/K = 323.15

w_B	0.926	0.874	0.769	0.555	0.000
P_A/kPa	51.98	64.30	76.06	81.66	81.67

copolymer (B): **S-MMA/50w** **97TAN**
solvent (A): **trichloromethane** **$CHCl_3$** **67-66-3**

T/K = 323.15

w_B	0.884	0.749	0.626	0.492	0.376	0.000
P_A/kPa	14.36	27.96	40.43	51.86	60.79	70.17

copolymer (B): **S-b-MMA/48w** **98WOH**
solvent (A): **benzene** **C_6H_6** **71-43-2**

T/K = 343.15

w_B	0.0000	0.4524	0.5067	0.5491	0.5889	0.6291	0.6674	0.7127	0.7567
P_A/Pa	73128.3	67614.5	65530.0	63512.0	61303.5	58751.0	55877.5	51840.5	46977.5

copolymer (B): **S-b-MMA/48w** **98WOH**
solvent (A): **ethylbenzene** **C_8H_{10}** **100-41-4**

continue

continue

T/K = 373.15

w_B	0.0000	0.5801	0.6199	0.6512	0.6990	0.7293	0.7617	0.8014
P_A/Pa	34300.5	31591.0	30891.0	30243.0	28843.5	27807.5	26384.0	24151.0

copolymer (B): **S-b-MMA/48w** **98WOH**
solvent (A): **ethylbenzene** **C_8H_{10}** **100-41-4**

T/K = 398.15

w_B	0.0000	0.5794	0.6262	0.6508	0.6833	0.7158	0.7485	0.7771	0.8144
P_A/Pa	74321.2	68279.0	66465.5	65410.0	63500.0	61137.0	58201.0	54886.0	50211.5

copolymer (B): **S-b-MMA/48w** **98WOH**
solvent (A): **mesitylene** **C_9H_{12}** **108-67-8**

T/K = 398.15

w_B	0.0000	0.6714	0.6997	0.7223	0.7591	0.7982	0.8245
P_A/Pa	32215.7	29603.0	28965.0	28430.5	26984.0	25295.5	23646.5

copolymer (B): **S-b-MMA/48w** **98WOH**
solvent (A): **toluene** **C_7H_8** **108-88-3**

T/K = 343.15

w_B	0.0000	0.5017	0.5340	0.5611	0.5960	0.6326	0.6723	0.7138	0.7469
P_A/Pa	27250.3	24751.5	24362.0	23947.5	23372.5	22699.5	21511.5	20585.0	19418.5
w_B	0.7782								
P_A/Pa	18277.0								

copolymer (B): **S-b-MMA/48w** **98WOH**
solvent (A): **toluene** **C_7H_8** **108-88-3**

T/K = 373.15

w_B	0.0000	0.4832	0.5364	0.5796	0.6298	0.6663	0.6997	0.7462	0.7897
P_A/Pa	74464.2	68224.0	66668.0	64694.5	61939.0	59750.0	57404.5	53197.0	48424.0

copolymer (B): **S-b-MMA/48w** **98WOH**
solvent (A): **p-xylene** **C_8H_{10}** **106-42-3**

T/K = 398.15

w_B	0.0000	0.5998	0.6476	0.6895	0.7192	0.7465	0.7839	0.8124
P_A/Pa	69803.4	64910.0	63870.0	62292.5	60268.0	58328.0	54314.0	51424.0

Copolymers from styrene and pentyl maleate

Average chemical composition of the copolymers, acronyms and references:

Copolymer (B)	Acronym	Ref.
Styrene/pentyl maleate copolymer 50.0 mol% styrene	S-PM/50x	97MIO

Characterization:

Copolymer (B)	M_n/ g/mol	M_w/ g/mol	M_η/ g/mol	Further information
S-PM/50x	130000	156000		synthesized in the laboratory

Experimental VLE data:

copolymer (B): **S-PM/50x** **97MIO**
solvent (A): **cyclohexane** **C_6H_{12}** **110-82-7**

T/K = 333.15

w_B	0.996	0.994	0.982	0.981	0.975	0.953	0.946	0.934	0.892
P_A/Torr	76.5	106.0	131.5	157.5	188.5	218.0	251.5	279.0	300.5
w_B	0.874	0.000							
P_A/Torr	362.5	390.0							

copolymer (B): **S-PM/50x** **97MIO**
solvent (A): **methanol** **CH_4O** **67-56-1**

T/K = 333.15

w_B	0.970	0.956	0.940	0.915	0.882	0.847	0.783	0.734	0.000
P_A/Torr	128.0	190.0	247.5	296.0	354.5	400.5	457.5	489.5	635.0

copolymer (B): **S-PM/50x** **97MIO**
solvent (A): **2-propanone** **C_3H_6O** **67-64-1**

T/K = 323.15

w_B	0.967	0.942	0.932	0.914	0.888	0.848	0.806	0.759	0.675
P_A/Torr	106.0	156.0	196.0	243.0	295.5	346.5	406.0	448.0	492.0
w_B	0.631	0.000							
P_A/Torr	513.0	612.0							

copolymer (B): **S-PM/50x** **97MIO**
solvent (A): **2-propanone** **C_3H_6O** **67-64-1**

T/K = 343.15

w_B	0.985	0.968	0.950	0.924	0.891	0.849	0.794	0.745	0.706
P_A/Torr	230.0	317.5	408.5	494.5	587.0	679.5	786.0	857.5	896.0
w_B	0.000								
P_A/Torr	1196.0								

Copolymers from vinyl acetate and vinyl chloride

Average chemical composition of the copolymers, acronyms and references:

Copolymer (B)	Acronym	Ref.
Vinyl acetate/vinyl chloride copolymer		
10.0 wt% vinyl acetate	VA-VC/10w	95GUP
12.0 wt% vinyl acetate	VA-VC/12w(1)	98KIM
12.0 wt% vinyl acetate	VA-VC/12w(2)	95GUP

Characterization:

Copolymer (B)	M_n/ g/mol	M_w/ g/mol	M_η/ g/mol	Further information
VA-VC/10w		115000		Sci. Polym. Products, catalog #068
VA-VC/12w(1)	7800	22000		Sci. Polym. Products
VA-VC/12w(2)	14300	30000		Sci. Polym. Products, catalog #063

Experimental VLE data:

copolymer (B):	**VA-VC/12w(1)**		**98KIM**
solvent (A):	**benzene**	C_6H_6	**71-43-2**

T/K = 398.15

w_B	0.9747	0.9689	0.9610	0.9503	0.9421	0.9309	0.9190
a_A	0.1084	0.1384	0.1648	0.1991	0.2099	0.2593	0.2911

copolymer (B):	**VA-VC/12w(1)**		**98KIM**
solvent (A):	**benzene**	C_6H_6	**71-43-2**

T/K = 418.15

w_B	0.9690	0.9630	0.9561	0.9482	0.9400	0.9318	0.9289
a_A	0.1120	0.1371	0.1614	0.1846	0.2166	0.2774	0.2869

copolymer (B):	**VA-VC/12w(2)**		**95GUP**
solvent (A):	**1-butanol**	$C_4H_{10}O$	**71-36-3**

T/K = 353.15

w_B	0.993	0.984	0.978	0.967	0.960	0.956	0.954
P_A/kPa	2.7	5.2	8.0	10.8	12.3	13.9	15.2

copolymer (B):	**VA-VC/12w(1)**		**98KIM**
solvent (A):	**chlorobenzene**	C_6H_5Cl	**108-90-7**

T/K = 398.15

w_B	0.9880	0.9841	0.9794	0.9738	0.9674	0.9600	0.9516
a_A	0.0828	0.0927	0.1159	0.1517	0.1990	0.2268	0.2642

copolymer (B):	**VA-VC/12w(1)**		**98KIM**
solvent (A):	**ethylbenzene**	C_8H_{10}	**100-41-4**

T/K = 398.15

w_B	0.9703	0.9611	0.9510	0.9384	0.9225	0.9010	0.8716
a_A	0.2431	0.3024	0.3670	0.4353	0.5054	0.5755	0.6955

copolymer (B):	**VA-VC/12w(1)**		**98KIM**
solvent (A):	**ethylbenzene**	C_8H_{10}	**100-41-4**

T/K = 418.15

w_B	0.9794	0.9748	0.9691	0.9622	0.9547	0.9460	0.9364
a_A	0.1793	0.2100	0.2508	0.2973	0.3446	0.3957	0.4476

copolymer (B): **VA-VC/12w(1)** **98KIM**
solvent (A): **ethylbenzene** **C_8H_{10}** **100-41-4**

T/K = 428.15

w_B	0.9814	0.9773	0.9727	0.9676	0.9613	0.9538	0.9446
a_A	0.1406	0.1709	0.2030	0.2360	0.2690	0.3260	0.3890

copolymer (B): **VA-VC/12w(2)** **95GUP**
solvent (A): **methanol** **CH_4O** **67-56-1**

T/K = 353.15

w_B	0.971	0.970	0.968	0.963	0.963	0.962
P_A/kPa	13.6	28.4	39.9	52.5	65.7	79.1

copolymer (B): **VA-VC/12w(1)** **98KIM**
solvent (A): **n-octane** **C_8H_{18}** **111-65-9**

T/K = 398.15

w_B	0.9846	0.9828	0.9809	0.9789	0.9766	0.9742	0.9718
a_A	0.1859	0.2023	0.2188	0.2355	0.2521	0.2686	0.2849

copolymer (B): **VA-VC/12w(1)** **98KIM**
solvent (A): **n-octane** **C_8H_{18}** **111-65-9**

T/K = 418.15

w_B	0.9823	0.9804	0.9786	0.9766	0.9744	0.9722	0.9701
a_A	0.2006	0.2268	0.2496	0.2606	0.2738	0.3038	0.3296

copolymer (B): **VA-VC/12w(1)** **98KIM**
solvent (A): **p-xylene** **C_8H_{10}** **106-42-3**

T/K = 398.15

w_B	0.9884	0.9843	0.9791	0.9729	0.9649	0.9549	0.9432
a_A	0.0927	0.1472	0.1970	0.2368	0.2890	0.3436	0.4189

copolymer (B): **VA-VC/12w(1)** **98KIM**
solvent (A): **p-xylene** **C_8H_{10}** **106-42-3**

T/K = 418.15

w_B	0.9864	0.9815	0.9752	0.9675	0.9583	0.9476	0.9355
a_A	0.1262	0.1677	0.2148	0.2765	0.3420	0.4001	0.4566

Copolymers from vinyl acetate and vinyl alcohol

Average chemical composition of the copolymers, acronyms and references:

Copolymer (B)	Acronym	Ref.
Vinyl acetate/vinyl alcohol copolymer		
8.0 mol% vinyl acetate	VA-VAL/08x	2000NAG
17.5 mol% vinyl acetate	VA-VAL/18x	2000NAG

Characterization:

Copolymer (B)	M_n/ g/mol	M_w/ g/mol	M_η/ g/mol	Further information
VA-VAL/08x	70000			ρ_B (298 K) = 1.241 g/cm^3
VA-VAL/18x	85000			ρ_B (298 K) = 1.234 g/cm^3

Experimental VLE data:

copolymer (B): **VA-VAL/08x** **2000NAG**
solvent (A): **water** **H_2O** **7732-18-5**

T/K = 298.15

φ_B	0.0039	0.0075	0.0083	0.0113	0.0125	0.0151	0.0164	0.0206
a_A	0.999968	0.999963	0.999959	0.999945	0.999917	0.999896	0.999867	0.999839
φ_B	0.0302	0.0330	0.0360	0.0409	0.0505	0.0575	0.0635	0.0727
a_A	0.999998	0.999996	0.999995	0.999992	0.999991	0.999990	0.999987	0.999980
φ_B	0.0747	0.0883	0.0910	0.1065				
a_A	0.999819	0.999746	0.999738	0.999627				

Comments: $\varphi_B = (w_B/1.241)\,(0.99705 + 0.245\,w_B)$

copolymer (B):	**VA-VAL/18x**		**2000NAG**
solvent (A):	**water**	**H_2O**	**7732-18-5**

T/K = 298.15

φ_B	0.0039	0.0076	0.0114	0.0150	0.0222	0.0305	0.0363	0.0494
a_A	0.999999	0.999997	0.999994	0.999991	0.999983	0.999971	0.999961	0.999928
φ_B	0.0626	0.0756	0.0878					
a_A	0.999883	0.999832	0.999767					

Comments: $\varphi_B = (w_B/1.234)\,(0.99705 + 0.237\,w_B)$

2.2. Classical mass-fraction Henry's constants of solvent vapors in molten copolymers

Copolymers from acrylonitrile and butadiene

Average chemical composition of the copolymers, acronyms and references:

Copolymer (B)	Acronym	Ref.
Acrylonitrile/butadiene copolymer		
18.0 wt% acrylonitrile	AN-B/18w	95TRI
34.0 wt% acrylonitrile	AN-B/34w(1)	93ISS, 97SCH
34.0 wt% acrylonitrile	AN-B/34w(2)	95TRI
34.0 wt% acrylonitrile	AN-B/34w(3)	82LIP
41.0 wt% acrylonitrile	AN-B/41w	86IT2

Characterization:

Copolymer (B)	M_n/ g/mol	M_w/ g/mol	M_η/ g/mol	Further information
AN-B/18w			100000	Bayer India
AN-B/34w(1)	60000			Perbunan 3307, Bayer AG, Ger.
AN-B/34w(2)			111000	Bayer India
AN-B/34w(3)				Krynac, Polysar
AN-B/41w				no data

Experimental data:

copolymer (B): AN-B/18w 95TRI

Comments: Henry's constants were not corrected for solubility of the marker gas

Solvent (A)	T/ K	$H_{A,B}$/ MPa	Solvent (A)	T/ K	$H_{A,B}$/ MPa
benzene	353.15	0.2525	n-nonane	353.15	0.06916
n-decane	353.15	0.02913	n-octane	353.15	0.1708
1,4-dioxane	353.15	0.1291	2-propanone	353.15	1.124
ethyl acetate	353.15	0.4129	tetrahydrofuran	353.15	0.3937
n-heptane	353.15	0.4353	toluene	353.15	0.09431
n-hexane	353.15	1.110	trichloromethane	353.15	0.1960

copolymer (B): AN-B/34w(1) 93ISS

Comments: Henry's constants were not corrected for solubility of the marker gas

Solvent (A)	T/ K	$H_{A,B}$/ MPa	Solvent (A)	T/ K	$H_{A,B}$/ MPa
benzene	333.15	0.2622	n-nonane	373.15	0.8153
benzene	353.15	0.4657	n-nonane	393.15	1.414
benzene	373.15	0.8845	n-octane	333.15	0.5001
benzene	393.15	1.388	n-octane	353.15	0.8916
n-heptane	333.15	1.308	n-octane	373.15	1.638
n-heptane	353.15	2.705	n-octane	393.15	2.603
n-heptane	373.15	3.603	toluene	333.15	0.1051
n-heptane	393.15	5.475	toluene	353.15	0.2108
n-nonane	333.15	0.2007	toluene	373.15	0.3969
n-nonane	353.15	0.4117	toluene	393.15	0.6886

copolymer (B): AN-B/34w(1) **97SCH**

Comments: Henry's constants were not corrected for solubility of the marker gas

Solvent (A)	T/ K	$H_{A,B}$/ MPa	Solvent (A)	T/ K	$H_{A,B}$/ MPa
benzene	330.15	0.1954	n-hexane	330.15	0.4091
cyclohexane	330.15	0.1920	toluene	330.15	0.06506
n-heptane	330.15	0.1420			

copolymer (B): AN-B/34w(2) **95TRI**

Comments: Henry's constants were not corrected for solubility of the marker gas

Solvent (A)	T/ K	$H_{A,B}$/ MPa	Solvent (A)	T/ K	$H_{A,B}$/ MPa
benzene	353.15	0.3038	n-nonane	353.15	0.1434
n-decane	353.15	0.06299	n-octane	353.15	0.3348
1,4-dioxane	353.15	0.09262	2-propanone	353.15	0.7357
ethyl acetate	353.15	0.4173	tetrahydrofuran	353.15	0.3442
n-heptane	353.15	0.7986	toluene	353.15	0.1289
n-hexane	353.15	2.370	trichloromethane	353.15	0.2131

copolymer (B): AN-B/34w(3) **82LIP**

Comments: Henry's constants were not corrected for solubility of the marker gas

Solvent (A)	T/ K	$H_{A,B}$/ MPa	Solvent (A)	T/ K	$H_{A,B}$/ MPa
benzene	348.15	0.2346	1-octene	348.15	0.2997
butylcyclohexane	348.15	0.02746	2-pentanone	348.15	0.1336
1-chlorobutane	348.15	0.3311	tetrachloromethane	348.15	0.2014
n-decane	348.15	0.0385	toluene	348.15	0.0921
n-octane	348.15	0.2395	trichloromethane	348.15	0.1714

copolymer (B): AN-B/41w 86IT2

Comments: Henry's constants were corrected for solubility of the marker gas N_2

Solvent (A)	T/ K	$H_{A,B}$/ MPa	Solvent (A)	T/ K	$H_{A,B}$/ MPa
benzene	398.15	1.24	ethylbenzene	398.15	0.333
benzene	423.15	2.15	ethylbenzene	423.15	0.657
benzene	448.15	3.47	ethylbenzene	448.15	1.19
benzene	473.15	5.17	ethylbenzene	473.15	1.94
1,3-butadiene	398.15	17.1	n-hexane	398.15	6.70
1,3-butadiene	423.15	24.1	n-hexane	423.15	10.5
1,3-butadiene	448.15	33.6	n-hexane	448.15	15.2
1,3-butadiene	473.15	44.3	n-hexane	473.15	21.2
n-butane	398.15	30.3	mesitylene	398.15	0.155
n-butane	423.15	40.2	mesitylene	423.15	0.337
n-butane	448.15	51.6	mesitylene	448.15	0.655
n-butane	473.15	60.9	mesitylene	473.15	1.14
cyclohexane	398.15	3.08	2-methylpropane	398.15	38.1
cyclohexane	423.15	4.84	2-methylpropane	423.15	47.1
cyclohexane	448.15	7.18	2-methylpropane	448.15	57.1
cyclohexane	473.15	10.1	2-methylpropane	473.15	64.1
n-decane	398.15	0.498	n-octane	398.15	1.78
n-decane	423.15	0.995	n-octane	423.15	3.13
n-decane	448.15	1.83	n-octane	448.15	5.08
n-decane	473.15	2.97	n-octane	473.15	7.47
2,2-dimethylbutane	398.15	9.84	toluene	398.15	0.599
2,2-dimethylbutane	423.15	13.9	toluene	423.15	1.12
2,2-dimethylbutane	448.15	19.1	toluene	448.15	1.92
2,2-dimethylbutane	473.15	25.1	toluene	473.15	3.01

Copolymers from acrylonitrile and 2-(3-methyl-3-phenylcyclobutyl)-2-hydroxyethyl methacrylate

Average chemical composition of the copolymers, acronyms and references:

Copolymer (B)	Acronym	Ref.
Acrylonitrile/2-(3-methyl-3-phenylcyclobutyl)-2-hydroxyethyl methacrylate copolymer 45.0 mol% acrylonitrile	AN-MPCHEMA/45x	2000KAY

Characterization:

Copolymer (B)	M_n/ g/mol	M_w/ g/mol	M_η/ g/mol	Further information
AN-MPCHEMA/45x				ρ_B = 1.075 g/cm³, synthesized in the laboratory

Experimental data:

copolymer (B): AN-MPCHEMA/45x 2000KAY

Comments: Henry's constants were not corrected for solubility of the marker gas

Solvent (A)	T/ K	$H_{A,B}$/ MPa	Solvent (A)	T/ K	$H_{A,B}$/ MPa
benzene	423.15	3.603	methyl acetate	423.15	4.813
benzene	433.15	4.288	methyl acetate	433.15	5.514
benzene	443.15	4.536	methyl acetate	443.15	5.873
benzene	453.15	4.798	methyl acetate	453.15	6.231
benzene	463.15	5.137	methyl acetate	463.15	6.269
2-butanone	423.15	4.074	n-nonane	423.15	1.970
2-butanone	433.15	4.532	n-nonane	433.15	2.236
2-butanone	443.15	4.845	n-nonane	443.15	2.446
2-butanone	453.15	5.459	n-nonane	453.15	2.687
2-butanone	463.15	5.876	n-nonane	463.15	3.128
n-decane	423.15	1.319	n-octane	423.15	2.661
n-decane	433.15	1.559	n-octane	433.15	2.945
n-decane	443.15	1.772	n-octane	443.15	3.151
n-decane	453.15	1.988	n-octane	453.15	3.488
n-decane	463.15	2.323	n-octane	463.15	3.602
n-dodecane	423.15	0.567	2-propanone	423.15	5.784
n-dodecane	433.15	0.719	2-propanone	433.15	6.410
n-dodecane	443.15	0.911	2-propanone	443.15	6.824
n-dodecane	453.15	1.046	2-propanone	453.15	7.228
n-dodecane	463.15	1.333	2-propanone	463.15	8.302
ethanol	423.15	7.938	toluene	423.15	2.565
ethanol	433.15	8.649	toluene	433.15	2.931
ethanol	443.15	9.354	toluene	443.15	3.124
ethanol	453.15	9.666	toluene	453.15	3.349
ethanol	463.15	10.38	toluene	463.15	3.975

continue

continue

Solvent (A)	$T/$ K	$H_{A,B}/$ MPa	Solvent (A)	$T/$ K	$H_{A,B}/$ MPa
ethyl acetate	423.15	3.646	n-undecane	423.15	0.8903
ethyl acetate	433.15	4.198	n-undecane	433.15	1.093
ethyl acetate	443.15	4.475	n-undecane	443.15	1.326
ethyl acetate	453.15	4.845	n-undecane	453.15	1.549
ethyl acetate	463.15	5.064	n-undecane	463.15	1.834
methanol	423.15	12.18	o-xylene	423.15	1.493
methanol	433.15	13.22	o-xylene	433.15	1.802
methanol	443.15	13.58	o-xylene	443.15	2.128
methanol	453.15	14.52	o-xylene	453.15	2.340
methanol	463.15	15.44	o-xylene	463.15	3.473

Copolymers from acrylonitrile and styrene

Average chemical composition of the copolymers, acronyms and references:

Copolymer (B)	Acronym	Ref.
Acrylonitrile/styrene copolymer		
28.0 wt% acrylonitrile	S-AN/28w	97WO1
62.0 mol% acrylonitrile	S-AN/62x	84MAT

Characterization:

Copolymer (B)	$M_n/$ g/mol	$M_w/$ g/mol	$M_\eta/$ g/mol	Further information
S-AN/28w	46000	100000		synthesized in the laboratory
S-AN/62x		86000		synthesized in the laboratory

Experimental data:

copolymer (B): **S-AN/28w** **97WO1**

Comments: Henry's constants were not corrected for solubility of the marker gas

Solvent (A)	T/ K	$H_{A,B}$/ MPa	Solvent (A)	T/ K	$H_{A,B}$/ MPa
benzene	423.15	3.50	o-xylene	423.15	0.80
benzene	448.15	5.42	o-xylene	448.15	1.40
benzene	473.15	8.10	o-xylene	473.15	2.40
propylbenzene	423.15	0.55	m-xylene	423.15	0.85
propylbenzene	448.15	1.055	m-xylene	448.15	1.53
propylbenzene	473.15	1.80	m-xylene	473.15	2.50
toluene	423.15	1.65	p-xylene	423.15	0.90
toluene	448.15	2.75	p-xylene	448.15	1.55
toluene	473.15	4.30	p-xylene	473.15	2.60

copolymer (B): **S-AN/62x** **84MAT**

Comments: Henry's constants were not corrected for solubility of the marker gas

Solvent (A)	T/ K	$H_{A,B}$/ MPa	Solvent (A)	T/ K	$H_{A,B}$/ MPa
benzene	393.15	2.47	toluene	393.15	1.14
benzene	423.15	4.96	toluene	423.15	2.57
benzene	453.15	7.63	toluene	453.15	4.99
ethyl acetate	393.15	2.62	trichloromethane	393.15	1.94
ethyl acetate	423.15	7.92	trichloromethane	423.15	4.30
ethyl acetate	453.15	12.1	trichloromethane	453.15	7.04

Copolymers from acrylonitrile and vinylidene chloride

Average chemical composition of the copolymers, acronyms and references:

Copolymer (B)	Acronym	Ref.
Acrylonitrile/vinylidene chloride copolymer 20 wt% acrylonitrile	AN-VdC/20w	86DEM

Characterization:

Copolymer (B)	M_n/ g/mol	M_w/ g/mol	M_η/ g/mol	Further information
AN-VdC/20w				Polysciences Inc.

Experimental data:

copolymer (B): AN-VdC/20w 86DEM

Comments: Henry's constants were not corrected for solubility of the marker gas

Solvent (A)	T/ K	$H_{A,B}$/ MPa	Solvent (A)	T/ K	$H_{A,B}$/ MPa
1,1-dichloroethene	291.0	1.498	1,1-dichloroethene	314.0	2.984
1,1-dichloroethene	298.0	1.847	1,1-dichloroethene	323.0	3.991
1,1-dichloroethene	306.0	2.437			

Copolymers from butadiene and styrene

Average chemical composition of the copolymers, acronyms and references:

Copolymer (B)	Acronym	Ref.
Butadiene/styrene copolymer		
20.0 mol% styrene	S-BR/20x	84MAT
25.0 mol% styrene	S-BR/25x	87OEN
15.0 wt% styrene	S-BR/15w(1)	95PET
15.0 wt% styrene	S-BR/15w(2)	93ISS, 97SCH
23.0 wt% styrene	S-BR/23x	93ISS
23.0 wt% styrene	S-BR/23x	97SCH
23.5 wt% styrene	S-BR/24w	95PET
25.0 wt% styrene	S-BR/25w	95PET
30.0 wt% styrene	S-BR/30w	86IT1
40.0 wt% styrene	S-BR/40w(1)	95PET
40.0 wt% styrene	S-BR/40w(2)	93ISS, 97SCH
41.0 wt% styrene	S-BR/41w	97WO1
45.0 wt% styrene	S-BR/45w	86IT1
77.0 wt% styrene	S-BR/77w	86IT1
Polystyrene-b-polybutadiene-b-polystyrene triblock copolymer		
17.0 wt% styrene	S-B-S/17w	92ROM
31.0 wt% styrene	S-B-S/31w	92ROM

Characterization:

Copolymer (B)	M_n/ g/mol	M_w/ g/mol	M_η/ g/mol	Further information
S-BR/20x		60000		
S-BR/25x	170000			Petkim Petrokimya A.S. Yarimca, Turkey
S-BR/15w(1)		170000		18 wt% 1,2-vinyl, 58 wt% trans
S-BR/15w(2)	240000			Buna EM BT98, Buna Werke Huels/Marl
S-BR/23w	210000	400000		Buna EM 1500, Buna Werke Huels/Marl
S-BR/24w		170000		15 wt% 1,2-vinyl, 53.5 wt% trans
S-BR/25w		200000		73 wt% 1,2-vinyl
S-BR/30w	107700	140000		
S-BR/40w(1)		155000		12 wt% 1,2-vinyl, 42 wt% trans
S-BR/40w(2)	190000			Buna EM 1516, Buna Werke Huels/Marl

continue

continue

Copolymer (B)	M_n/ g/mol	M_w/ g/mol	M_η/ g/mol	Further information
S-BR/41w	86300	181200		synthesized in the laboratory
S-BR/45w	130000	600000		
S-BR/77w	117600	200000		
S-B-S/17w	114800	188300		Kraton D-1301X
S-B-S/31w	110800	139600		Kraton D-1101

Experimental data:

copolymer (B): S-BR/20x **84MAT**

Comments: Henry's constants were not corrected for solubility of the marker gas

Solvent (A)	T/ K	$H_{A,B}$/ MPa	Solvent (A)	T/ K	$H_{A,B}$/ MPa
benzene	393.15	1.77	toluene	393.15	0.77
benzene	423.15	3.71	toluene	423.15	1.72
benzene	453.15	6.68	toluene	453.15	3.33
cyclohexane	393.15	2.19	trichloromethane	393.15	1.54
cyclohexane	423.15	4.16	trichloromethane	423.15	3.25
cyclohexane	453.15	7.37	trichloromethane	453.15	5.01

copolymer (B): S-BR/25x **87OEN**

Comments: Henry's constants were not corrected for solubility of the marker gas

Solvent (A)	T/ K	$H_{A,B}$/ MPa	Solvent (A)	T/ K	$H_{A,B}$/ MPa
benzene	343.3	0.3031	cyclohexane	358.3	0.7113
benzene	353.3	0.3906	cyclohexane	363.2	0.8285
benzene	358.3	0.4549	n-hexane	343.3	0.9960
benzene	363.2	0.5285	n-hexane	353.1	1.3191
cyclohexane	343.3	0.4648	n-hexane	358.3	1.6120
cyclohexane	353.1	0.6189	n-hexane	363.2	1.8290

copolymer (B): S-BR/15w(1) 95PET

Comments: Henry's constants were not corrected for solubility of the marker gas

Solvent (A)	T/ K	$H_{A,B}$/ MPa	Solvent (A)	T/ K	$H_{A,B}$/ MPa
benzene	323.15	0.1145	n-heptane	393.15	0.9323
benzene	353.15	0.3358	n-hexane	323.15	0.3059
benzene	373.15	0.6294	n-hexane	353.15	0.7950
benzene	393.15	0.9492	n-hexane	373.15	1.342
cyclohexane	323.15	0.1382	n-hexane	393.15	2.160
cyclohexane	353.15	0.3790	n-nonane	323.15	0.01156
cyclohexane	373.15	0.6534	n-nonane	353.15	0.04699
cyclohexane	393.15	1.061	n-nonane	373.15	0.1027
n-decane	353.15	0.01920	n-nonane	393.15	0.2020
n-decane	373.15	0.04503	n-octane	323.15	0.03392
n-decane	393.15	0.09523	n-octane	353.15	0.1177
n-heptane	323.15	0.1020	n-octane	373.15	0.2424
n-heptane	353.15	0.3119	n-octane	393.15	0.4342
n-heptane	373.15	0.5884			

copolymer (B): S-BR/15w(2) 93ISS

Comments: Henry's constants were not corrected for solubility of the marker gas

Solvent (A)	T/ K	$H_{A,B}$/ MPa	Solvent (A)	T/ K	$H_{A,B}$/ MPa
benzene	333.15	0.1707	n-nonane	333.15	0.01888
benzene	353.15	0.3303	n-nonane	353.15	0.04760
benzene	373.15	0.5873	n-nonane	373.15	0.1051
benzene	393.15	0.9477	n-nonane	393.15	0.2045
n-heptane	333.15	0.1523	n-octane	333.15	0.05246
n-heptane	353.15	0.3056	n-octane	353.15	0.1185
n-heptane	373.15	0.5617	n-octane	373.15	0.2412
n-heptane	393.15	0.9551	n-octane	393.15	0.4346
n-hexane	333.15	0.4348	toluene	333.15	0.05511
n-hexane	353.15	0.7965	toluene	353.15	0.1199
n-hexane	373.15	1.363	toluene	373.15	0.2310
n-hexane	393.15	2.182	toluene	393.15	0.3993

copolymer (B): **S-BR/15w(2)** **97SCH**

Comments: Henry's constants were not corrected for solubility of the marker gas

Solvent (A)	T/ K	$H_{A,B}$/ MPa	Solvent (A)	T/ K	$H_{A,B}$/ MPa
benzene	330.15	0.1707	n-hexane	330.15	0.4348
cyclohexane	330.15	0.1988	toluene	330.15	0.05510
n-heptane	330.15	0.1523			

copolymer (B): **S-BR/23w** **93ISS**

Comments: Henry's constants were not corrected for solubility of the marker gas

Solvent (A)	T/ K	$H_{A,B}$/ MPa	Solvent (A)	T/ K	$H_{A,B}$/ MPa
benzene	333.15	0.1749	n-nonane	333.15	0.02070
benzene	353.15	0.3401	n-nonane	353.15	0.05108
benzene	373.15	0.6074	n-nonane	373.15	0.1102
benzene	393.15	1.006	n-nonane	393.15	0.2146
n-heptane	333.15	0.1597	n-octane	333.15	0.05686
n-heptane	353.15	0.3277	n-octane	353.15	0.1268
n-heptane	373.15	0.5969	n-octane	373.15	0.2555
n-heptane	393.15	1.036	n-octane	393.15	0.4582
n-hexane	333.15	0.4596	toluene	333.15	0.05687
n-hexane	353.15	0.8464	toluene	353.15	0.1225
n-hexane	373.15	1.455	toluene	373.15	0.2362
n-hexane	393.15	2.359	toluene	393.15	0.4241

copolymer (B): **S-BR/23w** **97SCH**

Comments: Henry's constants were not corrected for solubility of the marker gas

Solvent (A)	T/ K	$H_{A,B}$/ MPa	Solvent (A)	T/ K	$H_{A,B}$/ MPa
benzene	330.15	0.1748	n-hexane	330.15	0.4595
cyclohexane	330.15	0.2033	toluene	330.15	0.05687
n-heptane	330.15	0.1601			

copolymer (B): **S-BR/24w** **95PET**

Comments: Henry's constants were not corrected for solubility of the marker gas

Solvent (A)	T/ K	$H_{A,B}$/ MPa	Solvent (A)	T/ K	$H_{A,B}$/ MPa
benzene	323.15	0.1154	n-heptane	393.15	1.093
benzene	353.15	0.3266	n-hexane	323.15	0.3247
benzene	373.15	0.5766	n-hexane	353.15	0.8192
benzene	393.15	0.9507	n-hexane	373.15	1.399
cyclohexane	323.15	0.1428	n-hexane	393.15	2.218
cyclohexane	353.15	0.3860	n-nonane	323.15	0.01248
cyclohexane	373.15	0.6726	n-nonane	353.15	0.04973
cyclohexane	393.15	1.080	n-nonane	373.15	0.1070
n-decane	353.15	0.02042	n-nonane	393.15	0.2111
n-decane	373.15	0.04752	n-octane	323.15	0.03580
n-decane	393.15	0.09996	n-octane	353.15	0.1245
n-heptane	323.15	0.1062	n-octane	373.15	0.2504
n-heptane	353.15	0.3140	n-octane	393.15	0.4554
n-heptane	373.15	0.5861			

copolymer (B): **S-BR/25w** **95PET**

Comments: Henry's constants were not corrected for solubility of the marker gas

Solvent (A)	T/ K	$H_{A,B}$/ MPa	Solvent (A)	T/ K	$H_{A,B}$/ MPa
benzene	323.15	0.1248	n-heptane	393.15	1.046
benzene	353.15	0.3537	n-hexane	323.15	0.3333
benzene	373.15	0.6222	n-hexane	353.15	0.8652
benzene	393.15	1.048	n-hexane	373.15	1.443
cyclohexane	323.15	0.1490	n-hexane	393.15	2.338
cyclohexane	353.15	0.4046	n-nonane	323.15	0.01247
cyclohexane	373.15	0.6998	n-nonane	353.15	0.05058
cyclohexane	393.15	1.152	n-nonane	373.15	0.1086
n-decane	353.15	0.02067	n-nonane	393.15	0.2130
n-decane	373.15	0.04801	n-octane	323.15	0.03630
n-decane	393.15	0.1009	n-octane	353.15	0.1270
n-heptane	323.15	0.1085	n-octane	373.15	0.2524
n-heptane	353.15	0.3266	n-octane	393.15	0.4627
n-heptane	373.15	0.6031			

copolymer (B): **S-BR/30w** **86IT1**

Comments: Henry's constants were not corrected for solubility of the marker gas

Solvent (A)	T/ K	$H_{A,B}$/ MPa	Solvent (A)	T/ K	$H_{A,B}$/ MPa
benzene	423.15	2.07	2,2-dimethylbutane	498.15	16.7
benzene	448.15	3.26	ethylbenzene	423.15	0.525
benzene	473.15	4.87	ethylbenzene	448.15	0.940
benzene	473.15	4.98	ethylbenzene	473.15	1.57
benzene	498.15	7.38	ethylbenzene	498.15	2.57
benzene	498.15	7.14	n-hexane	423.15	4.44
1,3-butadiene	423.15	17.7	n-hexane	448.15	6.64
1,3-butadiene	448.15	24.3	n-hexane	473.15	9.75
1,3-butadiene	473.15	29.6	n-hexane	498.15	13.5
1,3-butadiene	498.15	41.2	mesitylene	423.15	0.241
n-butane	423.15	19.1	mesitylene	448.15	0.473
n-butane	448.15	26.1	mesitylene	473.15	0.842
n-butane	473.15	33.0	mesitylene	498.15	1.49
n-butane	498.15	43.4	2-methylpropane	423.15	24.5
cyclohexane	423.15	2.32	2-methylpropane	448.15	31.9
cyclohexane	448.15	3.69	2-methylpropane	473.15	39.5
cyclohexane	473.15	5.44	2-methylpropane	498.15	54.7
cyclohexane	498.15	7.84	n-octane	423.15	1.14
n-decane	423.15	0.311	n-octane	448.15	2.00
n-decane	448.15	0.619	n-octane	473.15	3.23
n-decane	473.15	1.12	n-octane	498.15	5.22
n-decane	498.15	1.98	toluene	423.15	0.981
2,2-dimethylbutane	423.15	5.91	toluene	448.15	1.66
2,2-dimethylbutane	448.15	8.81	toluene	473.15	2.64
2,2-dimethylbutane	473.15	12.0	toluene	498.15	4.16

copolymer (B): **S-BR/40w(1)** **95PET**

Comments: Henry's constants were not corrected for solubility of the marker gas

Solvent (A)	T/ K	$H_{A,B}$/ MPa	Solvent (A)	T/ K	$H_{A,B}$/ MPa
benzene	323.15	0.1248	n-heptane	393.15	1.046
benzene	353.15	0.3537	n-hexane	323.15	0.3333
benzene	373.15	0.6222	n-hexane	353.15	0.8652
benzene	393.15	1.048	n-hexane	373.15	1.443
cyclohexane	323.15	0.1490	n-hexane	393.15	2.338
cyclohexane	353.15	0.4046	n-nonane	323.15	0.01247

continue

continue

Solvent (A)	$T/$ K	$H_{A,B}/$ MPa	Solvent (A)	$T/$ K	$H_{A,B}/$ MPa
cyclohexane	373.15	0.6998	n-nonane	353.15	0.05058
cyclohexane	393.15	1.152	n-nonane	373.15	0.1086
n-decane	353.15	0.02067	n-nonane	393.15	0.2130
n-decane	373.15	0.04801	n-octane	323.15	0.03630
n-decane	393.15	0.1009	n-octane	353.15	0.1270
n-heptane	323.15	0.1085	n-octane	373.15	0.2524
n-heptane	353.15	0.3266	n-octane	393.15	0.4627
n-heptane	373.15	0.6031			

copolymer (B): **S-BR/40w(2)** **93ISS**

Comments: Henry's constants were not corrected for solubility of the marker gas

Solvent (A)	$T/$ K	$H_{A,B}/$ MPa	Solvent (A)	$T/$ K	$H_{A,B}/$ MPa
benzene	333.15	0.1838	n-nonane	333.15	0.02543
benzene	353.15	0.3518	n-nonane	353.15	0.06000
benzene	373.15	0.6209	n-nonane	373.15	0.1282
benzene	393.15	1.052	n-nonane	393.15	0.2562
n-heptane	333.15	0.1947	n-octane	333.15	0.06921
n-heptane	353.15	0.3826	n-octane	353.15	0.1485
n-heptane	373.15	0.6910	n-octane	373.15	0.2923
n-heptane	393.15	1.192	n-octane	393.15	0.5413
n-hexane	333.15	0.5441	toluene	333.15	0.06121
n-hexane	353.15	0.9938	toluene	353.15	0.1269
n-hexane	373.15	1.660	toluene	373.15	0.2433
n-hexane	393.15	2.698	toluene	393.15	0.4344

copolymer (B): **S-BR/40w(2)** **97SCH**

Comments: Henry's constants were not corrected for solubility of the marker gas

Solvent (A)	$T/$ K	$H_{A,B}/$ MPa	Solvent (A)	$T/$ K	$H_{A,B}/$ MPa
benzene	330.15	0.1838	n-hexane	330.15	0.5440
cyclohexane	330.15	0.2323	toluene	330.15	0.06121
n-heptane	330.15	0.1947			

copolymer (B): S-BR/41w 97WO1

Comments: Henry's constants were not corrected for solubility of the marker gas

Solvent (A)	T/ K	$H_{A,B}$/ MPa	Solvent (A)	T/ K	$H_{A,B}$/ MPa
benzene	373.15	0.670	mesitylene	398.15	0.150
benzene	398.15	1.24	mesitylene	423.15	0.311
benzene	423.15	2.05	mesitylene	448.15	0.569
benzene	448.15	3.23	mesitylene	473.15	0.970
benzene	473.15	4.85	styrene	373.15	0.115
1,3-butadiene	373.15	11.0	styrene	398.15	0.240
1,3-butadiene	398.15	17.7	styrene	423.15	0.463
1,3-butadiene	423.15	24.0	styrene	448.15	0.825
cyclohexane	373.15	0.853	styrene	473.15	1.35
cyclohexane	398.15	1.35	toluene	373.15	0.270
cyclohexane	423.15	2.40	toluene	398.15	0.535
cyclohexane	448.15	3.83	toluene	423.15	0.987
cyclohexane	473.15	5.64	toluene	448.15	1.56
ethylbenzene	373.15	0.138	toluene	473.15	2.50
ethylbenzene	398.15	0.303	p-xylene	373.15	0.140
ethylbenzene	423.15	0.560	p-xylene	398.15	0.298
ethylbenzene	448.15	1.02	p-xylene	423.15	0.580
ethylbenzene	473.15	1.65	p-xylene	448.15	0.998
mesitylene	373.15	0.065	p-xylene	473.15	1.60

copolymer (B): S-BR/45w 86IT1

Comments: Henry's constants were not corrected for solubility of the marker gas

Solvent (A)	T/ K	$H_{A,B}$/ MPa	Solvent (A)	T/ K	$H_{A,B}$/ MPa
benzene	423.15	2.03	ethylbenzene	423.15	0.512
benzene	448.15	3.41	ethylbenzene	448.15	0.980
benzene	473.15	5.07	ethylbenzene	473.15	1.63
benzene	498.15	7.55	ethylbenzene	498.15	2.64
1,3-butadiene	423.15	19.6	n-hexane	423.15	4.68
1,3-butadiene	448.15	27.3	n-hexane	448.15	7.13
1,3-butadiene	473.15	38.0	n-hexane	473.15	10.6
1,3-butadiene	498.15	50.0	n-hexane	498.15	15.3
n-butane	423.15	21.0	mesitylene	423.15	0.237
n-butane	448.15	29.3	mesitylene	448.15	0.496
n-butane	473.15	39.8	mesitylene	473.15	0.878
n-butane	498.15	53.9	mesitylene	498.15	1.51

continue

continue

Solvent (A)	T/ K	$H_{A,B}$/ MPa	Solvent (A)	T/ K	$H_{A,B}$/ MPa
cyclohexane	423.15	2.38	2-methylpropane	423.15	26.6
cyclohexane	448.15	3.88	2-methylpropane	448.15	36.0
cyclohexane	473.15	5.76	2-methylpropane	473.15	45.2
cyclohexane	498.15	8.25	2-methylpropane	498.15	57.2
n-decane	423.15	0.319	n-octane	423.15	1.18
n-decane	448.15	0.675	n-octane	448.15	2.18
n-decane	473.15	1.20	n-octane	473.15	3.43
n-decane	498.15	2.07	n-octane	498.15	5.45
2,2-dimethylbutane	423.15	6.68	toluene	423.15	0.964
2,2-dimethylbutane	448.15	9.73	toluene	448.15	1.74
2,2-dimethylbutane	473.15	14.2	toluene	473.15	2.74
2,2-dimethylbutane	498.15	19.6	toluene	498.15	4.26

copolymer (B): **S-BR/77w** **86IT1**

Comments: Henry's constants were not corrected for solubility of the marker gas

Solvent (A)	T/ K	$H_{A,B}$/ MPa	Solvent (A)	T/ K	$H_{A,B}$/ MPa
benzene	423.15	2.38	ethylbenzene	423.15	0.619
benzene	448.15	3.80	ethylbenzene	448.15	1.12
benzene	473.15	5.77	ethylbenzene	473.15	1.88
benzene	498.15	8.22	ethylbenzene	498.15	2.96
1,3-butadiene	423.15	22.8	n-hexane	423.15	5.84
1,3-butadiene	448.15	31.8	n-hexane	448.15	8.85
1,3-butadiene	473.15	41.0	n-hexane	473.15	13.2
1,3-butadiene	498.15	53.5	n-hexane	498.15	18.0
n-butane	423.15	27.0	mesitylene	423.15	0.296
n-butane	448.15	36.0	mesitylene	448.15	0.578
n-butane	473.15	46.8	mesitylene	473.15	1.05
n-butane	498.15	59.9	mesitylene	498.15	1.76
cyclohexane	423.15	3.10	2-methylpropane	423.15	29.8
cyclohexane	448.15	4.78	2-methylpropane	448.15	40.4
cyclohexane	473.15	7.08	2-methylpropane	473.15	55.1
cyclohexane	498.15	9.74	2-methylpropane	498.15	69.9
n-decane	423.15	0.469	n-octane	423.15	1.67
n-decane	448.15	0.899	n-octane	448.15	2.84
n-decane	473.15	1.63	n-octane	473.15	4.58
n-decane	498.15	2.68	n-octane	498.15	6.81
2,2-dimethylbutane	423.15	8.28	toluene	423.15	1.14
2,2-dimethylbutane	448.15	12.4	toluene	448.15	1.96
2,2-dimethylbutane	473.15	16.6	toluene	473.15	3.15
2,2-dimethylbutane	498.15	22.1	toluene	498.15	4.69

copolymer (B): **S-B-S/17w** **92ROM**

Comments: Henry's constants were not corrected for solubility of the marker gas

Solvent (A)	T/ K	$H_{A,B}$/ MPa	Solvent (A)	T/ K	$H_{A,B}$/ MPa
benzene	308.15	0.07361	n-heptane	348.15	0.2593
benzene	328.15	0.1499	n-hexane	308.15	0.2012
benzene	348.15	0.2823	n-hexane	328.15	0.3881
2-butanone	308.15	0.1731	n-hexane	348.15	0.6846
2-butanone	328.15	0.3592	toluene	308.15	0.02207
2-butanone	348.15	0.6673	toluene	328.15	0.04990
cyclohexane	308.15	0.09241	toluene	348.15	0.1014
cyclohexane	328.15	0.1799	trichloromethane	308.15	0.07124
cyclohexane	348.15	0.3255	trichloromethane	328.15	0.1452
ethylbenzene	308.15	0.00796	trichloromethane	348.15	0.2694
ethylbenzene	328.15	0.01963	p-xylene	308.15	0.00784
ethylbenzene	348.15	0.04270	p-xylene	328.15	0.01940
n-heptane	308.15	0.06244	p-xylene	348.15	0.04245
n-heptane	328.15	0.1349			

copolymer (B): **S-B-S/31w** **92ROM**

Comments: Henry's constants were not corrected for solubility of the marker gas

Solvent (A)	T/ K	$H_{A,B}$/ MPa	Solvent (A)	T/ K	$H_{A,B}$/ MPa
benzene	308.15	0.07733	n-heptane	348.15	0.3059
benzene	328.15	0.1589	n-hexane	308.15	0.2292
benzene	348.15	0.3038	n-hexane	328.15	0.4552
2-butanone	308.15	0.1740	n-hexane	348.15	0.7962
2-butanone	328.15	0.3754	toluene	308.15	0.02354
2-butanone	348.15	0.7142	toluene	328.15	0.05347
cyclohexane	308.15	0.1079	toluene	348.15	0.1081
cyclohexane	328.15	0.2108	trichloromethane	308.15	0.06818
cyclohexane	348.15	0.3732	trichloromethane	328.15	0.1522
ethylbenzene	308.15	0.00851	trichloromethane	348.15	0.2822
ethylbenzene	328.15	0.02157	p-xylene	308.15	0.00844
ethylbenzene	348.15	0.04671	p-xylene	328.15	0.02127
n-heptane	308.15	0.07359	p-xylene	348.15	0.04591
n-heptane	328.15	0.1574			

Copolymers from dimethyl siloxane and bisphenol-A carbonate

Average chemical composition of the copolymers, acronyms and references:

Copolymer (B)	Acronym	Ref.
Dimethylsiloxane/bisphenol-A carbonate random copolymer		
49.0 wt% dimethylsiloxane	DMS-BAC/49w	81WAR
Dimethylsiloxane/bisphenol-A carbonate alternating block copolymer		
32.2 wt% dimethylsiloxane	DMS-BAC/32w	81WAR
41.0 wt% dimethylsiloxane	DMS-BAC/41w	81WAR
42.0 wt% dimethylsiloxane	DMS-BAC/42w	81WAR
57.2 wt% dimethylsiloxane	DMS-BAC/57w	81WAR
69.7 wt% dimethylsiloxane	DMS-BAC/70w	81WAR

Characterization:

Copolymer (B)	M_n/ g/mol	
DMS-BAC/32w	50000-100000	(M_n - PDMS-blocks = 10000, M_n - BAC-blocks = 19000)
DMS-BAC/41w	50000-100000	(M_n - PDMS-blocks = 5000, M_n - BAC-blocks = 6300)
DMS-BAC/42w	50000-100000	(M_n - PDMS-blocks = 10000, M_n - BAC-blocks = 11500)
DMS-BAC/49w	50000-100000	(statist. react. with PDMS-2000-oligomers)
DMS-BAC/57w	50000-100000	(M_n - PDMS-blocks = 5000, M_n - BAC-blocks = 3400)
DMS-BAC/70w	50000-100000	(M_n - PDMS-blocks = 10000, M_n - BAC-blocks = 3400)

Experimental data:

copolymer (B): DMS-BAC/32w **81WAR**

Comments: Henry's constants were not corrected for solubility of the marker gas

Solvent (A)	T/ K	$H_{A,B}$/ MPa	Solvent (A)	T/ K	$H_{A,B}$/ MPa
chlorobenzene	433.15	0.7975	o-dichlorobenzene	433.15	0.2116
chlorobenzene	453.15	1.223	o-dichlorobenzene	453.15	0.3395
n-decane	433.15	0.7824	toluene	433.15	1.813
n-decane	453.15	1.228			

copolymer (B): DMS-BAC/41w **81WAR**

Comments: Henry's constants were not corrected for solubility of the marker gas

Solvent (A)	T/ K	$H_{A,B}$/ MPa	Solvent (A)	T/ K	$H_{A,B}$/ MPa
chlorobenzene	433.15	0.7975	o-dichlorobenzene	453.15	0.3456
chlorobenzene	453.15	1.187	toluene	433.15	1.724
o-dichlorobenzene	433.15	0.2088			

copolymer (B): DMS-BAC/42w **81WAR**

Comments: Henry's constants were not corrected for solubility of the marker gas

Solvent (A)	T/ K	$H_{A,B}$/ MPa	Solvent (A)	T/ K	$H_{A,B}$/ MPa
chlorobenzene	433.15	0.8269	n-decane	468.15	1.438
chlorobenzene	453.15	1.230	o-dichlorobenzene	433.15	0.2164
chlorobenzene	468.15	1.654	o-dichlorobenzene	453.15	0.3568
n-decane	433.15	0.6542	o-dichlorobenzene	468.15	0.5082
n-decane	453.15	1.308	toluene	433.15	1.786

copolymer (B): DMS-BAC/57w **81WAR**

Comments: Henry's constants were not corrected for solubility of the marker gas

Solvent (A)	T/ K	$H_{A,B}$/ MPa	Solvent (A)	T/ K	$H_{A,B}$/ MPa
chlorobenzene	433.15	0.7760	o-dichlorobenzene	453.15	0.3652
chlorobenzene	453.15	1.173	o-dichlorobenzene	468.15	0.4710
chlorobenzene	468.15	1.529	toluene	433.15	1.611
o-dichlorobenzene	433.15	0.2232			

copolymer (B): **DMS-BAC/49w** **81WAR**

Comments: Henry's constants were not corrected for solubility of the marker gas

Solvent (A)	T/ K	$H_{A,B}$/ MPa	Solvent (A)	T/ K	$H_{A,B}$/ MPa
chlorobenzene	433.15	0.8006	n-decane	433.15	0.5934
chlorobenzene	453.15	1.245	n-decane	453.15	0.9558

copolymer (B): **DMS-BAC/70w** **81WAR**

Comments: Henry's constants were not corrected for solubility of the marker gas

Solvent (A)	T/ K	$H_{A,B}$/ MPa	Solvent (A)	T/ K	$H_{A,B}$/ MPa
chlorobenzene	433.15	0.7701	o-dichlorobenzene	433.15	0.2149
chlorobenzene	453.15	1.160	o-dichlorobenzene	453.15	0.3593
n-decane	433.15	0.4256	toluene	433.15	1.541
n-decane	453.15	0.6596			

Copolymers from ethylene and carbon monoxide

Average chemical composition of the copolymers, acronyms and references:

Copolymer (B)	Acronym	Ref.
Ethylene/carbon monoxide copolymer 10.5 wt% carbon monoxide	E-CO/11w	78DIP

Characterization:

Copolymer (B)	M_n/ g/mol	M_w/ g/mol	M_η/ g/mol	Further information
E-CO/11w				no data

Experimental data:

copolymer (B): E-CO/11w **78DIP**

Comments: Henry's constants were not corrected for solubility of the marker gas

Solvent (A)	T/ K	$H_{A,B}$/ MPa	Solvent (A)	T/ K	$H_{A,B}$/ MPa
n-octane	381.23	0.3391	n-octane	397.23	0.5281
n-octane	383.21	0.3589	n-octane	399.24	0.5518
n-octane	385.20	0.3811	n-octane	401.24	0.5914
n-octane	387.20	0.4028	n-octane	403.23	0.6135
n-octane	389.21	0.4236	n-octane	405.23	0.6538
n-octane	391.22	0.4508	n-octane	407.23	0.6717
n-octane	393.23	0.4767	n-octane	409.23	0.7204
n-octane	395.24	0.4961	n-octane	411.23	0.7422

Copolymers from ethylene and propylene

Average chemical composition of the copolymers, acronyms and references:

Copolymer (B)	Acronym	Ref.
Ethylene/propylene copolymer		
unspecified contents of propylene	E-P/?	86PRI
unspecified contents of propylene	E-P/?	87PRI
Ethylene/propylene/diene terpolymer		
50 mol% propylene	E-P-D/50x	93ISS
50 mol% propylene	E-P-D/50x	97SCH

Characterization:

Copolymer (B)	M_n/ g/mol	M_w/ g/mol	M_η/ g/mol	Further information
E-P/?	24500			low molecular weight EPR from Aldrich
E-P-D/50x	90000	200000		Buna AP341, Buna-Werke Huels/Marl

Experimental data:

copolymer (B): E-P/? 86PRI, 87PRI

Comments: Henry's constants were not corrected for solubility of the marker gas

Solvent (A)	T/ K	$H_{A,B}$/ MPa	Solvent (A)	T/ K	$H_{A,B}$/ MPa
benzene	298.15	0.06423	n-octane	298.15	0.01007
cyclohexane	298.15	0.05695	n-pentane	298.15	0.4342
dichloromethane	298.15	0.1833	tetrachloromethane	298.15	0.03278
n-heptane	298.15	0.03579	toluene	298.15	0.01806
n-hexane	298.15	0.1212	trichloromethane	298.15	0.06564

copolymer (B): E-P-D/50x 86PRI, 87PRI

Comments: Henry's constants were not corrected for solubility of the marker gas

Solvent (A)	T/ K	$H_{A,B}$/ MPa	Solvent (A)	T/ K	$H_{A,B}$/ MPa
benzene	333.15	0.2204	n-nonane	333.15	0.01394
benzene	353.15	0.4018	n-nonane	353.15	0.03662
benzene	373.15	0.6913	n-nonane	373.15	0.08031
benzene	393.15	1.111	n-nonane	393.15	0.1592
n-heptane	333.15	0.1179	n-octane	333.15	0.04125
n-heptane	353.15	0.2409	n-octane	353.15	0.09344
n-heptane	373.15	0.4517	n-octane	373.15	0.1895
n-heptane	393.15	0.7611	n-octane	393.15	0.3479
n-hexane	333.15	0.3467	toluene	333.15	0.06840
n-hexane	353.15	0.6417	toluene	353.15	0.1425
n-hexane	373.15	1.082	toluene	373.15	0.2675
n-hexane	393.15	1.723	toluene	393.15	0.4577

Copolymers from ethylene and vinyl acetate

Average chemical composition of the copolymers, acronyms and references:

Copolymer (B)	Acronym	Ref.
Ethylene/vinyl acetate copolymer		
3.95 wt% vinyl acetate	E-VA/04w	76LIU
9.2 wt% vinyl acetate	E-VA/09w	76LIU
15.4 wt% vinyl acetate	E-VA/15w	99GUO
29.0 wt% vinyl acetate	E-VA/29w	78DIN
30.3 wt% vinyl acetate	E-VA/30w	76LIU
40.0 wt% vinyl acetate	E-VA/40w	82LI1
40.0 wt% vinyl acetate	E-VA/40w	82LI2
45.0 wt% vinyl acetate	E-VA/45w	83WAL
48.5 wt% vinyl acetate	E-VA/49w	99GUO

Characterization:

Copolymer (B)	M_n/ g/mol	M_w/ g/mol	M_η/ g/mol	M_z/ g/mol	Further information
E-VA/04w	36600	147900		392700	Union Carbide, DXM-228, $MI = 0.91$
E-VA/09w	36600	415500		2793000	Union Carbide, DXM-231, $MI = 2.9$
E-VA/15w			11000		
E-VA/29w	43200				Union Carbide, DX-31034, $MI = 15.0$
E-VA/30w	42300	325800		2238000	Union Carbide, DXM-196, $MI = 14.0$
E-VA/40w					Aldrich Chem. Co.
E-VA/45w	37700	194500			Levapren 45N, Bayer AG, Ger.
E-VA/49w	81500	220000			

Experimental data:

copolymer (B): E-VA/04w **76LIU**

Comments: Henry's constants were corrected for solubility of the marker gas

Solvent (A)	T/ K	$H_{A,B}$/ MPa	Solvent (A)	T/ K	$H_{A,B}$/ MPa
2-butanone	398.15	3.192	methyl chloride	473.15	37.1
2-butanone	423.15	4.864	methyl chloride	498.15	44.6
2-butanone	448.15	6.991	2-propanol	398.15	7.60
2-butanone	473.15	9.828	2-propanol	423.15	10.44
2-butanone	498.15	12.56	2-propanol	448.15	13.68
carbon dioxide	398.15	126.6	2-propanol	473.15	17.73
carbon dioxide	423.15	139.8	2-propanol	498.15	22.60
carbon dioxide	448.15	150.0	2-propanone	398.15	8.511
carbon dioxide	473.15	160.0	2-propanone	423.15	11.96
carbon dioxide	498.15	169.0	2-propanone	448.15	15.50
ethane	398.15	102.0	2-propanone	473.15	20.26
ethane	423.15	121.6	2-propanone	498.15	25.84
ethane	448.15	137.8	sulfur dioxide	398.15	22.49
ethane	473.15	156.0	sulfur dioxide	423.15	28.0
ethane	498.15	172.0	sulfur dioxide	448.15	33.4
ethene	398.15	142.0	sulfur dioxide	473.15	40.9
ethene	423.15	154.0	sulfur dioxide	498.15	48.4
ethene	448.15	172.0	vinyl acetate	398.15	3.242
ethene	473.15	189.5	vinyl acetate	423.15	4.914
ethene	498.15	202.0	vinyl acetate	448.15	7.042
methyl chloride	398.15	20.26	vinyl acetate	473.15	9.524
methyl chloride	423.15	25.9	vinyl acetate	498.15	12.46
methyl chloride	448.15	31.6			

copolymer (B): E-VA/09w **76LIU**

Comments: Henry's constants were corrected for solubility of the marker gas

Solvent (A)	T/ K	$H_{A,B}$/ MPa	Solvent (A)	T/ K	$H_{A,B}$/ MPa
2-butanone	398.15	2.786	methyl chloride	473.15	36.48
2-butanone	423.15	4.408	methyl chloride	498.15	42.96
2-butanone	448.15	6.637	2-propanol	398.15	5.978

continue

continue

Solvent (A)	T/ K	$H_{A,B}$/ MPa	Solvent (A)	T/ K	$H_{A,B}$/ MPa
2-butanone	473.15	8.714	2-propanol	423.15	8.815
2-butanone	498.15	11.35	2-propanol	448.15	12.56
carbon dioxide	398.15	114.5	2-propanol	473.15	17.22
carbon dioxide	423.15	130.7	2-propanol	498.15	22.29
carbon dioxide	448.15	145.9	2-propanone	398.15	7.397
carbon dioxide	473.15	158.0	2-propanone	423.15	10.54
carbon dioxide	498.15	172.0	2-propanone	448.15	14.18
ethane	398.15	108.4	2-propanone	473.15	18.44
ethane	423.15	122.6	2-propanone	498.15	23.91
ethane	448.15	138.8	sulfur dioxide	398.15	18.44
ethane	473.15	150.0	sulfur dioxide	423.15	24.42
ethane	498.15	172.0	sulfur dioxide	448.15	30.70
ethene	398.15	134.8	sulfur dioxide	473.15	36.07
ethene	423.15	154.0	sulfur dioxide	498.15	42.66
ethene	448.15	174.0	vinyl acetate	398.15	2.837
ethene	473.15	195.5	vinyl acetate	423.15	4.458
ethene	498.15	210.0	vinyl acetate	448.15	6.485
methyl chloride	398.15	19.25	vinyl acetate	473.15	8.714
methyl chloride	423.15	24.32	vinyl acetate	498.15	12.16
methyl chloride	448.15	30.40			

copolymer (B): **E-VA/15w** **99GUO**

Comments: Henry's constants were not corrected for solubility of the marker gas

Solvent (A)	T/ K	$H_{A,B}$/ MPa	Solvent (A)	T/ K	$H_{A,B}$/ MPa
benzene	393.15	0.8667	n-octane	393.15	0.3434
cyclohexane	393.15	0.9070	tetrachloromethane	393.15	0.4701
1,4-dioxane	393.15	0.7078	tetrahydrofuran	393.15	1.384
ethyl acetate	393.15	1.759	toluene	393.15	0.3710

copolymer (B): **E-VA/29w** **78DIN**

Comments: Henry's constants were not corrected for solubility of the marker gas

Solvent (A)	T/ K	$H_{A,B}$/ MPa	Solvent (A)	T/ K	$H_{A,B}$/ MPa
acetaldehyde	423.61	18.48	n-heptane	433.68	2.327
acetaldehyde	433.68	12.61	n-hexane	423.61	3.9045
acetic acid	423.61	2.157	n-hexane	433.68	4.600
acetic acid	433.68	2.543	methanol	423.61	24.03
acetonitrile	423.61	9.298	methanol	433.68	26.45
acetonitrile	433.68	11.18	methylcyclohexane	423.61	1.271
acrylonitrile	423.61	6.285	methylcyclohexane	433.68	1.554
acrylonitrile	433.68	7.698	nitroethane	423.61	1.943
benzene	423.61	1.859	nitroethane	433.68	2.424
benzene	433.68	2.286	nitromethane	423.61	3.695
1-bromobutane	423.61	0.788	nitromethane	433.68	4.543
1-bromobutane	433.68	1.008	1-nitropropane	423.61	1.376
2-bromobutane	423.61	1.006	1-nitropropane	433.68	1.712
2-bromobutane	433.68	1.255	2-nitropropane	423.61	1.387
1-butanol	423.61	1.960	2-nitropropane	433.68	1.756
1-butanol	433.68	2.493	n-octane	423.61	0.9287
2-butanol	423.61	2.995	n-octane	433.68	1.198
2-butanol	433.68	3.584	1-octene	423.61	0.9643
2-butanone	423.61	3.688	1-octene	433.68	1.230
2-butanone	433.68	4.412	n-pentane	423.61	8.197
chlorobenzene	423.61	0.3937	n-pentane	433.68	9.177
chlorobenzene	433.68	0.5156	3-pentanone	423.61	1.765
cyclohexane	423.61	2.052	3-pentanone	433.68	2.238
cyclohexane	433.68	2.489	2-propanol	423.61	7.411
1,2-dichloroethane	423.61	1.372	2-propanol	433.68	8.325
1,2-dichloroethane	433.68	1.808	2-propanone	423.61	8.670
dichloromethane	423.61	3.633	2-propanone	433.68	10.29
dichloromethane	433.68	4.210	propionitrile	423.61	4.581
diethyl ether	423.61	7.076	propionitrile	433.68	5.579
diethyl ether	433.68	8.881	propyl acetate	423.61	1.666
1,4-dioxane	423.61	1.399	propyl acetate	433.68	2.094
1,4-dioxane	433.68	1.732	tetrachloromethane	423.61	1.045
dipropyl ether	423.61	2.010	tetrachloromethane	433.68	1.274
dipropyl ether	433.68	2.526	tetrahydrofuran	423.61	2.963
ethanol	423.61	11.38	tetrahydrofuran	433.68	3.629
ethanol	433.68	13.14	trichloromethane	423.61	1.434
ethyl acetate	423.61	3.326	trichloromethane	433.68	1.748
ethyl acetate	433.68	4.192	2,2,2-trifluoroethanol	423.61	3.598
formic acid	423.61	2.911	2,2,2-trifluoroethanol	433.68	4.514
formic acid	433.68	3.562	water	423.61	37.00
furan	423.61	6.404	water	433.68	53.01
furan	433.68	7.414	m-xylene	423.61	0.4208
n-heptane	423.61	1.892	m-xylene	433.68	0.5640

copolymer (B): E-VA/30w 76LIU

Comments: Henry's constants were corrected for solubility of the marker gas

Solvent (A)	T/ K	$H_{A,B}$/ MPa	Solvent (A)	T/ K	$H_{A,B}$/ MPa
2-butanone	398.15	2.148	methyl chloride	473.15	35.06
2-butanone	423.15	3.445	methyl chloride	498.15	43.87
2-butanone	448.15	5.218	2-propanol	398.15	4.053
2-butanone	473.15	7.40	2-propanol	423.15	6.434
2-butanone	498.15	10.23	2-propanol	448.15	9.727
carbon dioxide	398.15	101.3	2-propanol	473.15	13.78
carbon dioxide	423.15	120.5	2-propanol	498.15	18.64
carbon dioxide	448.15	138.8	2-propanone	398.15	5.168
carbon dioxide	473.15	157.0	2-propanone	423.15	8.005
carbon dioxide	498.15	176.0	2-propanone	448.15	11.45
ethane	398.15	109.0	2-propanone	473.15	16.01
ethane	423.15	130.0	2-propanone	498.15	19.66
ethane	448.15	150.0	sulfur dioxide	398.15	11.35
ethane	473.15	177.0	sulfur dioxide	423.15	16.72
ethane	498.15	203.0	sulfur dioxide	448.15	22.80
ethene	398.15	133.7	sulfur dioxide	473.15	33.44
ethene	423.15	152.0	sulfur dioxide	498.15	39.31
ethene	448.15	178.0	vinyl acetate	398.15	2.280
ethene	473.15	203.0	vinyl acetate	423.15	3.597
ethene	498.15	226.0	vinyl acetate	448.15	5.370
methyl chloride	398.15	15.30	vinyl acetate	473.15	7.40
methyl chloride	423.15	21.28	vinyl acetate	498.15	10.44
methyl chloride	448.15	27.36			

copolymer (B): E-VA/40w 82LI1

Comments: Henry's constants were not corrected for solubility of the marker gas

Solvent (A)	T/ K	$H_{A,B}$/ MPa	Solvent (A)	T/ K	$H_{A,B}$/ MPa
benzene	348.15	0.2830	n-octane	348.15	0.1263
1-butanol	348.15	0.1335	1-octene	348.15	0.1255
chlorobenzene	348.15	0.03648	n-pentane	348.15	2.533
1-chlorobutane	348.15	0.3290	2-pentanone	348.15	0.1987
n-decane	348.15	0.02067	toluene	348.15	0.1042
n-hexane	348.15	0.8660	trichloromethane	348.15	0.1806

copolymer (B): E-VA/40w **82LI1**

Comments: Henry's constants were corrected for solubility of the marker gas

Solvent (A)	T/ K	$H_{A,B}$/ MPa	Solvent (A)	T/ K	$H_{A,B}$/ MPa
benzene	338.15	0.2006	n-octane	338.15	0.0859
benzene	348.15	0.2827	n-octane	348.15	0.1267
benzene	358.15	0.3830	n-octane	358.15	0.1793
1-butanol	338.15	0.0851	1-octene	338.15	0.0843
1-butanol	348.15	0.1337	1-octene	348.15	0.1256
1-butanol	358.15	0.2037	1-octene	358.15	0.1773
chlorobenzene	338.15	0.02371	n-pentane	338.15	1.996
chlorobenzene	348.15	0.0365	n-pentane	348.15	0.2503
chlorobenzene	358.15	0.0532	n-pentane	358.15	2.888
1-chlorobutane	338.15	0.2320	2-pentanone	338.15	0.1388
1-chlorobutane	348.15	0.3293	2-pentanone	348.15	0.1986
1-chlorobutane	358.15	0.4367	2-pentanone	358.15	0.2807
n-decane	338.15	0.01277	toluene	338.15	0.0697
n-decane	348.15	0.02067	toluene	348.15	0.1044
n-decane	358.15	0.0317	toluene	358.15	0.1459
n-hexane	338.15	0.6424	trichloromethane	338.15	0.1236
n-hexane	348.15	0.8663	trichloromethane	348.15	0.1803
n-hexane	358.15	1.297	trichloromethane	358.15	0.2472

copolymer (B): E-VA/45w **83WAL**

Comments: Henry's constants were not corrected for solubility of the marker gas

Solvent (A)	T/ K	$H_{A,B}$/ MPa	Solvent (A)	T/ K	$H_{A,B}$/ MPa
2-butanone	343.15	0.7612	methanol	373.15	18.36
2-butanone	373.15	2.647	n-pentane	343.15	3.9346
dichloromethane	343.15	0.8012	n-pentane	373.15	10.49
dichloromethane	373.15	2.785	2-propanone	343.15	2.444
diethyl ether	343.15	2.980	2-propanone	373.15	4.888
diethyl ether	373.15	8.418	tetrahydrofuran	343.15	0.7283
ethyl acetate	343.15	0.6898	tetrahydrofuran	373.15	2.421
ethyl acetate	373.15	2.322	trichloromethane	343.15	0.2615
methanol	343.15	7.081	trichloromethane	373.15	0.8889

copolymer (B): E-VA/49w 99GUO

Comments: Henry's constants were not corrected for solubility of the marker gas

Solvent (A)	T/ K	$H_{A,B}$/ MPa	Solvent (A)	T/ K	$H_{A,B}$/ MPa
benzene	393.15	0.8176	n-octane	393.15	0.5209
cyclohexane	393.15	1.231	tetrachloromethane	393.15	0.5045
1,4-dioxane	393.15	0.5046	tetrahydrofuran	393.15	1.160
ethyl acetate	393.15	1.253	toluene	393.15	0.3619

Copolymers from ethyl methacrylate and methacrylic acid

Average chemical composition of the copolymers, acronyms and references:

Copolymer (B)	Acronym	Ref.
Ethyl methacrylate/methacrylic acid copolymer 36.6 mol% methacrylic acid	EMA-MA/37x	92CHE

Characterization:

Copolymer (B)	M_n/ g/mol	M_w/ g/mol	M_η/ g/mol	Further information
EMA-MA/37x				synthesized in the laboratory

Experimental data:

copolymer (B): EMA-MA/37x **92CHE**

Comments: Henry's constants were not corrected for solubility of the marker gas

Solvent (A)	T/ K	$H_{A,B}$/ MPa	Solvent (A)	T/ K	$H_{A,B}$/ MPa
benzene	413.15	4.838	n-decane	413.15	1.497
benzene	423.15	6.045	n-decane	423.15	1.689
benzene	433.15	7.966	n-decane	433.15	2.245
benzene	443.15	9.757	n-decane	443.15	2.705
benzene	453.15	11.68	n-decane	453.15	3.653

Copolymers from isobutyl methacrylate and acrylic acid

Average chemical composition of the copolymers, acronyms and references:

Copolymer (B)	Acronym	Ref.
Isobutyl methacrylate/acrylic acid copolymer		
10.5 mol% acrylic acid	IBMA-AA/11x	96DJA
22.0 mol% acrylic acid	IBMA-AA/22x	96DJA

Characterization:

Copolymer (B)	M_n/ g/mol	M_w/ g/mol	M_η/ g/mol	Further information
IBMA-AA/11x				synthesized in the laboratory
IBMA-AA/22x				synthesized in the laboratory

Experimental data:

copolymer (B): IBMA-AA/11x 96DJA

Comments: Henry's constants were not corrected for solubility of the marker gas

Solvent (A)	T/ K	$H_{A,B}$/ MPa	Solvent (A)	T/ K	$H_{A,B}$/ MPa
benzene	423.15	3.917	n-decane	423.15	0.6869
benzene	433.15	4.705	n-decane	433.15	0.9445
benzene	443.15	5.752	n-decane	443.15	1.210
benzene	453.15	7.537	n-decane	453.15	1.558

copolymer (B): IBMA-AA/22x 96DJA

Comments: Henry's constants were not corrected for solubility of the marker gas

Solvent (A)	T/ K	$H_{A,B}$/ MPa	Solvent (A)	T/ K	$H_{A,B}$/ MPa
benzene	423.15	4.290	n-decane	423.15	0.8616
benzene	433.15	5.239	n-decane	433.15	1.105
benzene	443.15	6.937	n-decane	443.15	1.455
benzene	453.15	8.613	n-decane	453.15	2.006

Copolymers from styrene and acrylic acid

Average chemical composition of the copolymers, acronyms and references:

Copolymer (B)	Acronym	Ref.
Styrene/acrylic acid copolymer 20.0 mol% acrylic acid	S-AA/20x	91CHE

Characterization:

Copolymer (B)	M_n/ g/mol	M_w/ g/mol	M_η/ g/mol	Further information
S-AA/20x				synthesized in the laboratory

Experimental data:

copolymer (B): **S-AA/20x** **91CHE**

Comments: Henry's constants were not corrected for solubility of the marker gas

Solvent (A)	T/ K	$H_{A,B}$/ MPa	Solvent (A)	T/ K	$H_{A,B}$/ MPa
benzene	423.15	4.095	n-decane	423.15	1.125
benzene	433.15	4.743	n-decane	433.15	1.355
benzene	443.15	5.850	n-decane	443.15	1.900
benzene	453.15	7.287	n-decane	453.15	2.258

Copolymers from styrene and butyl methacrylate

Average chemical composition of the copolymers, acronyms and references:

Copolymer (B)	Acronym	Ref.
Styrene/butyl methacrylate copolymer 58.0 wt% styrene	S-BMA/58w	81DIP

Characterization:

Copolymer (B)	M_n/ g/mol	M_w/ g/mol	M_η/ g/mol	Further information
S-BMA/58w	32500	72500		Xerox Corp., Can.

Experimental data:

copolymer (B): S-BMA/58w 81DIP

Comments: Henry's constants were not corrected for solubility of the marker gas

Solvent (A)	T/ K	$H_{A,B}$/ MPa	Solvent (A)	T/ K	$H_{A,B}$/ MPa
benzene	413.15	1.900	n-octane	413.15	1.442
1-chlorobutane	413.15	2.282	tetrachloromethane	413.15	1.202
cyclohexane	413.15	2.802	trichloromethane	413.15	1.411
n-decane	413.15	0.3873	2,2,4-trimethylpentane	413.15	2.727
dichloromethane	413.15	3.481	3,4,5-trimethylheptane	413.15	0.5778
tert-butyl-benzene	413.15	0.1987			

Copolymers from styrene and isobutyl methacrylate

Average chemical composition of the copolymers, acronyms and references:

Copolymer (B)	Acronym	Ref.
Styrene/isobutyl methacrylate copolymer 80.0 wt% styrene	S-IBMA/80w	81DIP

Characterization:

Copolymer (B)	M_n/ g/mol	M_w/ g/mol	M_η/ g/mol	Further information
S-IBMA/80w		70000		Xerox Corp., Can.

Experimental data:

copolymer (B): S-IBMA/80w **81DIP**

Comments: Henry's constants were not corrected for solubility of the marker gas

Solvent (A)	T/ K	$H_{A,B}$/ MPa	Solvent (A)	T/ K	$H_{A,B}$/ MPa
benzene	413.15	2.030	n-octane	413.15	1.669
cyclohexane	413.15	3.116	tetrachloromethane	413.15	1.318
n-decane	413.15	0.4494	trichloromethane	413.15	1.677
1-chlorobutane	413.15	2.553	2,2,4-trimethylpentane	413.15	3.458
dichloromethane	413.15	3.869	3,4,5-trimethylheptane	413.15	0.6947
tert-butyl-benzene	413.15	0.2257			

Copolymers from styrene and N,N-dimethyl aminoethyl methacrylate

Average chemical composition of the copolymers, acronyms and references:

Copolymer (B)	Acronym	Ref.
Styrene/N,N-dimethyl aminoethyl methacrylate copolymer		
6.0 mol% methacrylate	S-DMAM/06x	96DJA
12.0 mol% methacrylate	S-DMAM/12x	96DJA

Characterization:

Copolymer (B)	M_n/ g/mol	M_w/ g/mol	M_η/ g/mol	Further information
S-DMAM/06x				synthesized in the laboratory
S-DMAM/12x				synthesized in the laboratory

Experimental data:

copolymer (B): S-DMAM/06x 96DJA

Comments: Henry's constants were not corrected for solubility of the marker gas

Solvent (A)	T/ K	$H_{A,B}$/ MPa	Solvent (A)	T/ K	$H_{A,B}$/ MPa
benzene	423.15	8.189	n-decane	423.15	2.150
benzene	433.15	9.040	n-decane	433.15	2.693
benzene	443.15	10.25	n-decane	443.15	3.320
benzene	453.15	12.37	n-decane	453.15	4.138

copolymer (B): S-DMAM/12x 96DJA

Comments: Henry's constants were not corrected for solubility of the marker gas

Solvent (A)	T/ K	$H_{A,B}$/ MPa	Solvent (A)	T/ K	$H_{A,B}$/ MPa
benzene	423.15	4.022	n-decane	423.15	1.089
benzene	433.15	5.239	n-decane	433.15	1.376
benzene	443.15	6.551	n-decane	443.15	1.811
benzene	453.15	8.316	n-decane	453.15	2.283

Copolymers from styrene and 2-(3-methyl-3-phenylcyclobutyl)-2-hydroxyethyl methacrylate

Average chemical composition of the copolymers, acronyms and references:

Copolymer (B)	Acronym	Ref.
Styrene/2-(3-methyl-3-phenylcyclobutyl)-2-hydroxyethyl methacrylate copolymer 40.0 mol% styrene	S-MPCHEMA/40x	2000KAY

Characterization:

Copolymer (B)	M_n/ g/mol	M_w/ g/mol	M_η/ g/mol	Further information
S-MPCHEMA/40x				$\rho_B = 1.045$ g/cm^3, synthesized in the laboratory

Experimental data:

copolymer (B): S-MPCHEMA/40x 2000KAY

Comments: Henry's constants were not corrected for solubility of the marker gas

Solvent (A)	T/ K	$H_{A,B}$/ MPa	Solvent (A)	T/ K	$H_{A,B}$/ MPa
benzene	423.15	4.600	methyl acetate	423.15	8.734
benzene	433.15	4.106	methyl acetate	433.15	8.734
benzene	443.15	4.536	methyl acetate	443.15	9.491
benzene	453.15	5.210	methyl acetate	453.15	9.889

continue

continue

Solvent (A)	T/ K	$H_{A,B}$/ MPa	Solvent (A)	T/ K	$H_{A,B}$/ MPa
2-butanone	423.15	6.389	n-nonane	423.15	1.375
2-butanone	433.15	6.773	n-nonane	433.15	1.489
2-butanone	443.15	7.758	n-nonane	443.15	1.712
2-butanone	453.15	7.994	n-nonane	453.15	1.906
n-decane	423.15	0.8922	n-octane	423.15	2.436
n-decane	433.15	1.015	n-octane	433.15	2.808
n-decane	443.15	1.188	n-octane	443.15	3.078
n-decane	453.15	1.342	n-octane	453.15	3.319
n-dodecane	423.15	0.4777	2-propanone	423.15	8.670
n-dodecane	433.15	0.5196	2-propanone	433.15	9.655
n-dodecane	443.15	0.6539	2-propanone	443.15	10.80
n-dodecane	453.15	0.7974	2-propanone	453.15	12.61
ethanol	423.15	10.27	toluene	423.15	2.676
ethanol	433.15	12.17	toluene	433.15	2.896
ethanol	443.15	13.32	toluene	443.15	3.506
ethanol	453.15	13.40	toluene	453.15	3.349
ethyl acetate	423.15	5.872	n-undecane	423.15	0.5249
ethyl acetate	433.15	5.715	n-undecane	433.15	0.5877
ethyl acetate	443.15	6.559	n-undecane	443.15	0.7918
ethyl acetate	453.15	6.509	n-undecane	453.15	1.044
methanol	423.15	12.89	o-xylene	423.15	0.9177
methanol	433.15	15.61	o-xylene	433.15	1.131
methanol	443.15	19.31	o-xylene	443.15	1.364
methanol	453.15	20.91	o-xylene	453.15	1.618

Copolymers from styrene and nonyl methacrylate

Average chemical composition of the copolymers, acronyms and references:

Copolymer (B)	Acronym	Ref.
Styrene/nonyl methacrylate copolymer 80 mol% styrene	S-NMA/80x	95BOG

Characterization:

Copolymer (B)	M_n/ g/mol	M_w/ g/mol	M_η/ g/mol	Further information
S-NMA/80x	29000			synthesized in the laboratory

Experimental data:

copolymer (B): S-NMA/80x **95BOG**

Comments: Henry's constants were not corrected for solubility of the marker gas

Solvent (A)	T/ K	$H_{A,B}$/ MPa	Solvent (A)	T/ K	$H_{A,B}$/ MPa
1-butanol	526.6	20.43	n-heptane	526.6	7.083
dibutyl ether	526.6	10.90	n-octane	526.6	4.418
1,4-dioxane	526.6	13.57			

Copolymers from tetrafluoroethylene and 2,2-bis(trifluoromethyl)-4,5-difluoro-1,3-dioxole

Average chemical composition of the copolymers, acronyms and references:

Copolymer (B)	Acronym	Ref.
Tetrafluoroethylene/2,2-bis(trifluoromethyl)-4,5-difluoro-1,3-dioxole copolymer 13 mol% tetrafluoroethylene	TFE-PFD/13x	99BON

Characterization:

Copolymer (B)	M_n/ g/mol	M_w/ g/mol	M_η/ g/mol	Further information
TFE-PFD/13x				AF2400, DuPont, Wilmington

Experimental data:

copolymer (B): TFE-PFD/13x 99BON

Comments: Henry's constants were not corrected for solubility of the marker gas

Solvent (A)	T/ K	$H_{A,B}$/ MPa	Solvent (A)	T/ K	$H_{A,B}$/ MPa
benzene	373.15	0.5591	n-heptane	373.15	0.3192
benzene	423.15	1.211	n-heptane	423.15	0.9066
n-hexane	373.15	0.7530	n-pentane	423.15	5.246
n-hexane	423.15	2.635			

Copolymers from vinyl acetate and vinyl alcohol

Average chemical composition of the copolymers, acronyms and references:

Copolymer (B)	Acronym	Ref.
Vinyl acetate/vinyl alcohol copolymer		
43.4 mol% vinyl acetate	VA-VAL/43x	86CAS
60.9 mol% vinyl acetate	VA-VAL/61x	86CAS
74.4 mol% vinyl acetate	VA-VAL/74x	86CAS
94.8 mol% vinyl acetate	VA-VAL/95x	86CAS

Characterization:

Copolymer (B)	M_n/ g/mol	M_w/ g/mol	M_η/ g/mol	Further information
VA-VAL/43x				synthesized in the laboratory
VA-VAL/61x				synthesized in the laboratory
VA-VAL/74x				synthesized in the laboratory
VA-VAL/95x				synthesized in the laboratory

Experimental data:

copolymer (B): VA-VAL/43x 86CAS

Comments: Henry's constants were not corrected for solubility of the marker gas

Solvent (A)	T/ K	$H_{A,B}$/ MPa	Solvent (A)	T/ K	$H_{A,B}$/ MPa
benzene	393.15	4.263	1-hexanol	415.15	0.6594
benzene	400.15	4.953	1-hexanol	423.15	0.8636
benzene	408.15	5.646	*tert*-butanol	393.15	1.463
benzene	415.15	6.548	*tert*-butanol	400.15	1.804
benzene	423.15	7.552	*tert*-butanol	408.15	2.255
1-butanol	393.15	1.019	*tert*-butanol	415.15	2.750
1-butanol	400.15	1.290	*tert*-butanol	423.15	3.374
1-butanol	408.15	1.651	n-nonane	393.15	5.731
1-butanol	415.15	2.039	n-nonane	400.15	6.534
1-butanol	423.15	2.560	n-nonane	408.15	7.767
2-butanol	393.15	1.926	n-nonane	415.15	8.898
2-butanol	400.15	2.309	n-nonane	423.15	10.67
2-butanol	408.15	2.972	1-octanol	393.15	0.1033
2-butanol	415.15	3.575	1-octanol	400.15	0.1335
2-butanol	423.15	4.365	1-octanol	408.15	0.1793
cyclohexanol	393.15	0.1780	1-octanol	415.15	0.2359
cyclohexanol	400.15	0.2347	1-octanol	423.15	0.3135
cyclohexanol	408.15	0.3115	1-pentanol	393.15	0.5446
cyclohexanol	415.15	0.3926	1-pentanol	400.15	0.6965
cyclohexanol	423.15	0.5059	1-pentanol	408.15	0.9049
n-decane	393.15	3.136	1-pentanol	415.15	1.134
n-decane	400.15	3.782	1-pentanol	423.15	1.444
n-decane	408.15	4.613	2-pentanol	393.15	1.069
n-decane	415.15	5.393	2-pentanol	400.15	1.312
n-decane	423.15	6.489	2-pentanol	408.15	1.715

continue

continue

Solvent (A)	T/ K	$H_{A,B}$/ MPa	Solvent (A)	T/ K	$H_{A,B}$/ MPa
1-decanol	393.15	0.03609	2-pentanol	415.15	2.108
1-decanol	400.15	0.04897	2-pentanol	423.15	2.640
1-decanol	408.15	0.06846	1-propanol	393.15	2.073
1-decanol	415.15	0.08973	1-propanol	400.15	2.520
1-decanol	423.15	0.1262	1-propanol	408.15	3.179
n-dodecane	393.15	0.9768	1-propanol	415.15	3.825
n-dodecane	400.15	1.239	1-propanol	423.15	4.672
n-dodecane	408.15	1.597	2-propanol	393.15	3.630
n-dodecane	415.15	1.975	2-propanol	400.15	4.339
n-dodecane	423.15	2.502	2-propanol	408.15	5.446
ethylbenzene	393.15	1.370	2-propanol	415.15	6.472
ethylbenzene	400.15	1.646	2-propanol	423.15	7.793
ethylbenzene	408.15	2.003	n-tetradecane	393.15	0.3877
ethylbenzene	415.15	2.377	n-tetradecane	400.15	0.5024
ethylbenzene	423.15	2.849	n-tetradecane	408.15	0.6538
1-heptanol	393.15	0.1767	n-tetradecane	415.15	0.8248
1-heptanol	400.15	0.2337	n-tetradecane	423.15	1.078
1-heptanol	408.15	0.3127	toluene	393.15	2.293
1-heptanol	415.15	0.4008	toluene	400.15	2.691
1-heptanol	423.15	0.5250	toluene	408.15	3.231
n-hexadecane	393.15	0.1361	toluene	415.15	3.781
n-hexadecane	400.15	0.1840	toluene	423.15	4.410
n-hexadecane	408.15	0.2588	n-undecane	393.15	1.755
n-hexadecane	415.15	0.3406	n-undecane	400.15	2.195
n-hexadecane	423.15	0.4580	n-undecane	408.15	2.784
1-hexanol	393.15	0.2946	n-undecane	415.15	3.340
1-hexanol	400.15	0.3850	n-undecane	423.15	4.093
1-hexanol	408.15	0.5211			

copolymer (B): VA-VAL/61x **86CAS**

Comments: Henry's constants were not corrected for solubility of the marker gas

Solvent (A)	T/ K	$H_{A,B}$/ MPa	Solvent (A)	T/ K	$H_{A,B}$/ MPa
benzene	393.15	3.153	1-hexanol	415.15	0.5736
benzene	400.15	3.516	1-hexanol	423.15	0.7324
benzene	408.15	4.165	*tert*-butanol	393.15	1.314

continue

continue

Solvent (A)	$T/$ K	$H_{A,B}/$ MPa	Solvent (A)	$T/$ K	$H_{A,B}/$ MPa
benzene	415.15	4.774	*tert*-butanol	400.15	1.607
benzene	423.15	5.549	*tert*-butanol	408.15	2.009
1-butanol	393.15	0.9730	*tert*-butanol	415.15	2.432
1-butanol	400.15	1.203	*tert*-butanol	423.15	2.978
1-butanol	408.15	1.517	n-nonane	393.15	3.768
1-butanol	415.15	1.839	n-nonane	400.15	4.319
1-butanol	423.15	2.258	n-nonane	408.15	5.045
2-butanol	393.15	1.720	n-nonane	415.15	5.864
2-butanol	400.15	2.103	n-nonane	423.15	6.864
2-butanol	408.15	2.603	1-octanol	393.15	0.0891
2-butanol	415.15	3.123	1-octanol	400.15	0.1182
2-butanol	423.15	3.806	1-octanol	408.15	0.1609
cyclohexanol	393.15	0.1748	1-octanol	415.15	0.2078
cyclohexanol	400.15	0.2230	1-octanol	423.15	0.2739
cyclohexanol	408.15	0.2921	1-pentanol	393.15	0.5055
cyclohexanol	415.15	0.3662	1-pentanol	400.15	0.6332
cyclohexanol	423.15	0.4697	1-pentanol	408.15	0.8150
n-decane	393.15	2.111	1-pentanol	415.15	1.002
n-decane	400.15	2.486	1-pentanol	423.15	1.257
n-decane	408.15	3.000	2-pentanol	393.15	0.9507
n-decane	415.15	3.508	2-pentanol	400.15	1.177
n-decane	423.15	4.178	2-pentanol	408.15	1.478
1-decanol	393.15	0.03048	2-pentanol	415.15	1.798
1-decanol	400.15	0.04152	2-pentanol	423.15	2.213
1-decanol	408.15	0.05856	1-propanol	393.15	1.946
1-decanol	415.15	0.07819	1-propanol	400.15	2.381
1-decanol	423.15	0.1067	1-propanol	408.15	2.943
n-dodecane	393.15	0.7032	1-propanol	415.15	3.549
n-dodecane	400.15	0.8680	1-propanol	423.15	4.374
n-dodecane	408.15	1.100	2-propanol	393.15	3.433
n-dodecane	415.15	1.331	2-propanol	400.15	4.008
n-dodecane	423.15	1.669	2-propanol	408.15	4.915
ethylbenzene	393.15	0.9689	2-propanol	415.15	5.814
ethylbenzene	400.15	1.160	2-propanol	423.15	6.897
ethylbenzene	408.15	1.409	n-tetradecane	393.15	0.2482
ethylbenzene	415.15	1.667	n-tetradecane	400.15	0.3131
ethylbenzene	423.15	1.997	n-tetradecane	408.15	0.4135
1-heptanol	393.15	0.1546	n-tetradecane	415.15	0.5171
1-heptanol	400.15	0.1999	n-tetradecane	423.15	0.6618
1-heptanol	408.15	0.2663	toluene	393.15	1.632
1-heptanol	415.15	0.3389	toluene	400.15	1.926
1-heptanol	423.15	0.4385	toluene	408.15	2.310
n-hexadecane	393.15	0.08317	toluene	415.15	2.679

continue

continue

Solvent (A)	T/ K	$H_{A,B}$/ MPa	Solvent (A)	T/ K	$H_{A,B}$/ MPa
n-hexadecane	400.15	0.1096	toluene	423.15	3.164
n-hexadecane	408.15	0.1535	n-undecane	393.15	1.231
n-hexadecane	415.15	0.1999	n-undecane	400.15	1.483
n-hexadecane	423.15	0.2669	n-undecane	408.15	1.821
1-hexanol	393.15	0.2767	n-undecane	415.15	2.178
1-hexanol	400.15	0.3401	n-undecane	423.15	2.661
1-hexanol	408.15	0.4588			

copolymer (B): VA-VAL/74x 86CAS

Comments: Henry's constants were not corrected for solubility of the marker gas

Solvent (A)	T/ K	$H_{A,B}$/ MPa	Solvent (A)	T/ K	$H_{A,B}$/ MPa
benzene	393.15	2.159	1-hexanol	415.15	0.4725
benzene	400.15	2.578	1-hexanol	423.15	0.6012
benzene	408.15	2.982	*tert*-butanol	393.15	1.117
benzene	415.15	3.449	*tert*-butanol	400.15	1.355
benzene	423.15	4.142	*tert*-butanol	408.15	1.683
1-butanol	393.15	0.8476	*tert*-butanol	415.15	2.013
1-butanol	400.15	1.032	*tert*-butanol	423.15	2.457
1-butanol	408.15	1.297	1-octanol	393.15	0.0705
1-butanol	415.15	1.568	1-octanol	400.15	0.0927
1-butanol	423.15	1.926	1-octanol	408.15	0.1248
2-butanol	393.15	1.468	1-octanol	415.15	0.1538
2-butanol	400.15	1.771	1-octanol	423.15	0.2019
2-butanol	408.15	2.185	n-nonane	393.15	2.374
2-butanol	415.15	2.599	n-nonane	400.15	2.784
2-butanol	423.15	3.146	n-nonane	408.15	3.291
cyclohexanol	393.15	0.1526	n-nonane	415.15	3.752
cyclohexanol	400.15	0.1935	n-nonane	423.15	4.427
cyclohexanol	408.15	0.2518	1-pentanol	393.15	0.4277
cyclohexanol	415.15	0.3142	1-pentanol	400.15	0.5330
cyclohexanol	423.15	0.3994	1-pentanol	408.15	0.6834
n-decane	393.15	1.294	1-pentanol	415.15	0.8400
n-decane	400.15	1.579	1-pentanol	423.15	1.052
n-decane	408.15	1.896	2-pentanol	393.15	0.7840
n-decane	415.15	2.229	2-pentanol	400.15	0.9653

continue

continue

Solvent (A)	T/ K	$H_{A,B}$/ MPa	Solvent (A)	T/ K	$H_{A,B}$/ MPa
n-decane	423.15	2.683	2-pentanol	408.15	1.212
1-decanol	393.15	0.02294	2-pentanol	415.15	1.468
1-decanol	400.15	0.03104	2-pentanol	423.15	1.804
1-decanol	408.15	0.04320	1-propanol	393.15	1.752
1-decanol	415.15	0.05603	1-propanol	400.15	2.122
1-decanol	423.15	0.07616	1-propanol	408.15	2.614
n-dodecane	393.15	0.4346	1-propanol	415.15	3.113
n-dodecane	400.15	0.5340	1-propanol	423.15	3.749
n-dodecane	408.15	0.6754	2-propanol	393.15	2.983
n-dodecane	415.15	0.8251	2-propanol	400.15	3.630
n-dodecane	423.15	1.020	2-propanol	408.15	4.199
ethylbenzene	393.15	0.6586	2-propanol	415.15	4.934
ethylbenzene	400.15	0.7947	2-propanol	423.15	6.226
ethylbenzene	408.15	0.9759	n-tetradecane	393.15	0.1461
ethylbenzene	415.15	1.161	n-tetradecane	400.15	0.1873
ethylbenzene	423.15	1.401	n-tetradecane	408.15	0.2452
1-heptanol	393.15	0.1254	n-tetradecane	415.15	0.3049
1-heptanol	400.15	0.1621	n-tetradecane	423.15	0.3889
1-heptanol	408.15	0.2154	toluene	393.15	1.129
1-heptanol	415.15	0.2730	toluene	400.15	1.341
1-heptanol	423.15	0.3546	toluene	408.15	1.624
n-hexadecane	393.15	0.04975	toluene	415.15	1.912
n-hexadecane	400.15	0.06664	toluene	423.15	2.253
n-hexadecane	408.15	0.09052	n-undecane	393.15	0.7474
n-hexadecane	415.15	0.1171	n-undecane	400.15	0.9156
n-hexadecane	423.15	0.1546	n-undecane	408.15	1.126
1-hexanol	393.15	0.2295	n-undecane	415.15	1.344
1-hexanol	400.15	0.2907	n-undecane	423.15	1.638
1-hexanol	408.15	0.3788			

copolymer (B): **VA-VAL/95x** **86CAS**

Comments: Henry's constants were not corrected for solubility of the marker gas

Solvent (A)	T/ K	$H_{A,B}$/ MPa	Solvent (A)	T/ K	$H_{A,B}$/ MPa
benzene	393.15	1.565	1-hexanol	415.15	0.4028
benzene	400.15	1.959	1-hexanol	423.15	0.5095

continue

continue

Solvent (A)	T/ K	$H_{A,B}$/ MPa	Solvent (A)	T/ K	$H_{A,B}$/ MPa
benzene	408.15	2.322	*tert*-butanol	393.15	1.036
benzene	415.15	2.790	*tert*-butanol	400.15	1.254
benzene	423.15	3.271	*tert*-butanol	408.15	1.543
1-butanol	393.15	0.7631	*tert*-butanol	415.15	1.839
1-butanol	400.15	0.9229	*tert*-butanol	423.15	2.232
1-butanol	408.15	1.147	n-nonane	393.15	1.511
1-butanol	415.15	1.383	n-nonane	400.15	1.800
1-butanol	423.15	1.700	n-nonane	408.15	2.208
2-butanol	393.15	1.384	n-nonane	415.15	2.623
2-butanol	400.15	1.646	n-nonane	423.15	3.208
2-butanol	408.15	2.021	1-octanol	393.15	0.0560
2-butanol	415.15	2.384	1-octanol	400.15	0.0730
2-butanol	423.15	2.872	1-octanol	408.15	0.0979
cyclohexanol	393.15	0.1404	1-octanol	415.15	0.1255
cyclohexanol	400.15	0.1760	1-octanol	423.15	0.1697
cyclohexanol	408.15	0.2291	1-pentanol	393.15	0.3785
cyclohexanol	415.15	0.2659	1-pentanol	400.15	0.4703
cyclohexanol	423.15	0.3417	1-pentanol	408.15	0.5954
n-decane	393.15	0.8262	1-pentanol	415.15	0.7288
n-decane	400.15	0.9939	1-pentanol	423.15	0.9084
n-decane	408.15	1.234	2-pentanol	393.15	1.164
n-decane	415.15	1.486	2-pentanol	400.15	1.384
n-decane	423.15	1.837	2-pentanol	408.15	1.699
1-decanol	393.15	0.01760	2-pentanol	415.15	2.005
1-decanol	400.15	0.02378	2-pentanol	423.15	2.415
1-decanol	408.15	0.03340	1-propanol	393.15	1.696
1-decanol	415.15	0.04401	1-propanol	400.15	2.051
1-decanol	423.15	0.05880	1-propanol	408.15	2.490
n-dodecane	393.15	0.2769	1-propanol	415.15	3.033
n-dodecane	400.15	0.3432	1-propanol	423.15	3.687
n-dodecane	408.15	0.4293	2-propanol	393.15	2.937
n-dodecane	415.15	0.5289	2-propanol	400.15	3.458
n-dodecane	423.15	0.6539	2-propanol	408.15	4.130
ethylbenzene	393.15	0.4671	2-propanol	415.15	4.979
ethylbenzene	400.15	0.5645	2-propanol	423.15	5.990
ethylbenzene	408.15	0.6968	n-tetradecane	393.15	0.09415
ethylbenzene	415.15	0.8324	n-tetradecane	400.15	0.1201
ethylbenzene	423.15	1.009	n-tetradecane	408.15	0.1580
1-heptanol	393.15	0.1085	n-tetradecane	415.15	0.1967
1-heptanol	400.15	0.1387	n-tetradecane	423.15	0.2513
1-heptanol	408.15	0.1833	toluene	393.15	0.8586
1-heptanol	415.15	0.2283	toluene	400.15	1.025
1-heptanol	423.15	0.2941	toluene	408.15	1.241

continue

continue

Solvent (A)	T/ K	$H_{A,B}$/ MPa	Solvent (A)	T/ K	$H_{A,B}$/ MPa
n-hexadecane	393.15	0.02648	toluene	415.15	1.459
n-hexadecane	400.15	0.03700	toluene	423.15	1.741
n-hexadecane	408.15	0.04973	n-undecane	393.15	0.5100
n-hexadecane	415.15	0.06603	n-undecane	400.15	0.6146
n-hexadecane	423.15	0.08915	n-undecane	408.15	0.7627
1-hexanol	393.15	0.2017	n-undecane	415.15	0.9133
1-hexanol	400.15	0.2529	n-undecane	423.15	1.110
1-hexanol	408.15	0.3273			

Copolymers from vinyl acetate and vinyl chloride

Average chemical composition of the copolymers, acronyms and references:

Copolymer (B)	Acronym	Ref.
Vinyl acetate/vinyl chloride copolymer		
3 wt% vinyl acetate	VA-VC/03w	79HEI
3 wt% vinyl acetate	VA-VC/03w	80MER
10 wt% vinyl acetate	VA-VC/10w(1)	79HEI
10 wt% vinyl acetate	VA-VC/10w(2)	88SAT
15 wt% vinyl acetate	VA-VC/15w	78OIS
17 wt% vinyl acetate	VA-VC/17w	79HEI
17 wt% vinyl acetate	VA-VC/17w	80MER
Vinyl acetate/vinyl chloride/glycidyl methacrylate terpolymer		
10.5 wt% VA, 80 wt% VC, 9.5 wt% GMA	VA-VC-GMA/10w	78OIS
Vinyl acetate/vinyl chloride/hydroxypropyl acrylate terpolymer		
5 wt% VA, 80 wt% VC, 15 wt% HPA	VA-VC-HPA/05w	78OIS
34 wt% VA, 58 wt% VC, 8 wt% HPA	VA-VC-HPA/34w	78OIS

Characterization:

Copolymer (B)	M_n/ g/mol	M_w/ g/mol	M_η/ g/mol	Further information
VA-VC/03w				Polysciences, Inc.
VA-VC/10w(1)				Polysciences, Inc.
VA-VC/10w(2)				no data
VA-VC/15w	26000			Union Carbide Corp.
VA-VC/17w				Polysciences, Inc.
VA-VC-GMA/10w	8400			Union Carbide Corp.
VA-VC-HPA/05w	10000			Union Carbide Corp.
VA-VC-HPA/34w	12000			Union Carbide Corp.

Experimental data:

copolymer (B): **VA-VC/03w** **79HEI**

Comments: Henry's constants were corrected for solubility of the marker gas

Solvent (A)	T/ K	$H_{A,B}$/ MPa	Solvent (A)	T/ K	$H_{A,B}$/ MPa
benzene	398.15	2.24	nitroethane	398.15	0.97
benzene	413.15	3.02	nitroethane	413.15	1.39
cyclohexane	398.15	5.09	n-heptane	398.15	6.02
cyclohexane	413.15	6.48	n-heptane	413.15	7.50
1,2-dichloroethane	398.15	1.53	2-propanol	398.15	6.60
1,2-dichloroethane	413.15	2.13	2-propanol	413.15	8.57
1,4-dioxane	398.15	0.85	toluene	398.15	1.03
1,4-dioxane	413.15	1.26	toluene	413.15	1.46

copolymer (B): **VA-VC/03w** **80MER**

Comments: Henry's constants were not corrected for solubility of the marker gas

Solvent (A)	T/ K	$H_{A,B}$/ MPa	Solvent (A)	T/ K	$H_{A,B}$/ MPa
acetaldehyde	398.15	15.16	1,4-dioxane	413.15	1.245
acetaldehyde	413.15	18.41	n-heptane	398.15	5.964
acetic acid	398.15	3.232	n-heptane	413.15	7.555
acetic acid	413.15	3.940	methanol	398.15	13.63
acetonitrile	398.15	4.009	methanol	413.15	17.72
acetonitrile	413.15	5.371	nitroethane	398.15	0.9919
benzene	398.15	2.219	nitroethane	413.15	1.363
benzene	413.15	2.997	2-propanol	398.15	6.516
2-butanone	398.15	2.812	2-propanol	413.15	8.590
2-butanone	413.15	3.387	2-propanone	398.15	5.356
cyclohexane	398.15	5.140	2-propanone	413.15	6.517
cyclohexane	413.15	6.425	tetrahydrofuran	398.15	3.088
1,2-dichloroethane	398.15	1.550	tetrahydrofuran	413.15	3.351
1,2-dichloroethane	413.15	2.105	toluene	398.15	1.058
1,4-dioxane	398.15	0.8536	toluene	413.15	1.425

copolymer (B): **VA-VC/10w(1)** **80MER**

Comments: Henry's constants were not corrected for solubility of the marker gas

Solvent (A)	T/ K	$H_{A,B}$/ MPa	Solvent (A)	T/ K	$H_{A,B}$/ MPa
acetaldehyde	398.15	12.57	1,4-dioxane	413.15	1.199
acetaldehyde	413.15	16.63	n-heptane	398.15	5.528
acetic acid	398.15	2.055	n-heptane	413.15	7.311
acetic acid	413.15	2.909	methanol	398.15	12.66
acetonitrile	398.15	3.952	methanol	413.15	15.08
acetonitrile	413.15	5.170	nitroethane	398.15	0.8474
benzene	398.15	2.033	nitroethane	413.15	1.293
benzene	413.15	2.769	2-propanol	398.15	5.814
2-butanone	398.15	2.368	2-propanol	413.15	7.874
2-butanone	413.15	3.088	2-propanone	398.15	4.769
cyclohexane	398.15	4.653	2-propanone	413.15	6.110
cyclohexane	413.15	6.133	tetrahydrofuran	398.15	2.250
1,2-dichloroethane	398.15	1.350	tetrahydrofuran	413.15	3.150
1,2-dichloroethane	413.15	1.896	toluene	398.15	0.9481
1,4-dioxane	398.15	0.7695	toluene	413.15	1.369

copolymer (B): **VA-VC/10w(1)** **79HEI**

Comments: Henry's constants were corrected for solubility of the marker gas

Solvent (A)	T/ K	$H_{A,B}$/ MPa	Solvent (A)	T/ K	$H_{A,B}$/ MPa
benzene	398.15	2.05	nitroethane	398.15	0.85
benzene	413.15	2.84	nitroethane	413.15	1.29
cyclohexane	398.15	4.69	n-heptane	398.15	5.58
cyclohexane	413.15	6.18	n-heptane	413.15	7.37
1,2-dichloroethane	398.15	1.40	2-propanol	398.15	5.89
1,2-dichloroethane	413.15	1.91	2-propanol	413.15	7.95
1,4-dioxane	398.15	0.78	toluene	398.15	0.98
1,4-dioxane	413.15	1.21	toluene	413.15	1.38

copolymer (B): **VA-VC/10w(2)** **88SAT**

Comments: Henry's constants were corrected for solubility of the marker gas

Solvent (A)	T/ K	$H_{A,B}$/ MPa	Solvent (A)	T/ K	$H_{A,B}$/ MPa
acrylonitrile	388.15	2.64	n-nonane	388.15	1.32
acrylonitrile	398.15	3.29	n-nonane	398.15	1.60
acrylonitrile	408.15	4.10	n-nonane	408.15	1.95
acrylonitrile	423.15	5.38	n-nonane	423.15	2.60
benzene	388.15	1.67	2-propanone	388.15	3.82
benzene	398.15	2.02	2-propanone	398.15	4.67
benzene	408.15	2.74	2-propanone	408.15	5.80
benzene	423.15	3.42	2-propanone	423.15	7.52
2-butanone	388.15	1.90	styrene	388.15	0.224
2-butanone	398.15	2.19	styrene	398.15	0.301
2-butanone	408.15	2.76	styrene	408.15	0.405
2-butanone	423.15	3.74	styrene	423.15	0.602
chlorobenzene	388.15	0.293	tetrachloromethane	388.15	1.81
chlorobenzene	398.15	0.382	tetrachloromethane	398.15	1.88
chlorobenzene	408.15	0.501	tetrachloromethane	408.15	2.18
chlorobenzene	423.15	0.719	tetrachloromethane	423.15	2.72
n-decane	388.15	0.679	toluene	388.15	0.854
n-decane	398.15	0.849	toluene	398.15	0.962
n-decane	408.15	1.03	toluene	408.15	1.20
n-decane	423.15	1.49	toluene	423.15	1.60
ethylbenzene	388.15	0.413	trichloromethane	388.15	1.69

continue

continue

Solvent (A)	T/ K	$H_{A,B}$/ MPa	Solvent (A)	T/ K	$H_{A,B}$/ MPa
ethylbenzene	398.15	0.540	trichloromethane	398.15	1.97
ethylbenzene	408.15	0.692	trichloromethane	408.15	2.36
ethylbenzene	423.15	0.999	trichloromethane	423.15	3.11
fluorobenzene	388.15	1.36	vinyl acetate	388.15	2.68
fluorobenzene	398.15	1.67	vinyl acetate	398.15	3.24
fluorobenzene	408.15	2.08	p-xylene	388.15	0.350

copolymer (B): VA-VC/15w **78OIS**

Comments: Henry's constants were corrected for solubility of the marker gas

Solvent (A)	T/ K	$H_{A,B}$/ MPa	Solvent (A)	T/ K	$H_{A,B}$/ MPa
2-propanone	383.15	4.276	propyl acetate	423.15	3.668
2-propanone	403.15	6.647	vinyl chloride	383.15	21.68
2-propanone	423.15	9.859	vinyl chloride	403.15	30.50
propyl acetate	383.15	1.216	vinyl chloride	423.15	41.14
propyl acetate	403.15	2.178			

copolymer (B): VA-VC/17w **80MER**

Comments: Henry's constants were not corrected for solubility of the marker gas

Solvent (A)	T/ K	$H_{A,B}$/ MPa	Solvent (A)	T/ K	$H_{A,B}$/ MPa
acetaldehyde	398.15	12.28	1,4-dioxane	413.15	1.161
acetaldehyde	413.15	15.16	n-heptane	398.15	5.396
acetic acid	398.15	1.513	n-heptane	413.15	7.311
acetic acid	413.15	2.292	methanol	398.15	11.07
acetonitrile	398.15	3.436	methanol	413.15	13.90
acetonitrile	413.15	4.688	nitroethane	398.15	0.8003
benzene	398.15	1.978	nitroethane	413.15	1.255
benzene	413.15	2.769	2-propanol	398.15	5.558

continue

continue

Solvent (A)	T/ K	$H_{A,B}$/ MPa	Solvent (A)	T/ K	$H_{A,B}$/ MPa
2-butanone	398.15	1.909	2-propanol	413.15	7.559
2-butanone	413.15	2.787	2-propanone	398.15	4.031
cyclohexane	398.15	4.653	2-propanone	413.15	5.507
cyclohexane	413.15	5.970	tetrahydrofuran	398.15	2.059
1,2-dichloroethane	398.15	1.311	tetrahydrofuran	413.15	2.917
1,2-dichloroethane	413.15	1.779	toluene	398.15	0.9518
1,4-dioxane	398.15	0.7582	toluene	413.15	1.354

copolymer (B): VA-VC/17w **79HEI**

Comments: Henry's constants were corrected for solubility of the marker gas

Solvent (A)	T/ K	$H_{A,B}$/ MPa	Solvent (A)	T/ K	$H_{A,B}$/ MPa
benzene	398.15	2.00	n-heptane	398.15	5.45
benzene	413.15	2.79	n-heptane	413.15	7.37
cyclohexane	398.15	4.70	nitroethane	398.15	0.79
cyclohexane	413.15	6.03	nitroethane	413.15	1.27
1,2-dichloroethane	398.15	1.29	2-propanol	398.15	5.64
1,2-dichloroethane	413.15	1.83	2-propanol	413.15	7.65
1,4-dioxane	398.15	0.76	toluene	398.15	0.93
1,4-dioxane	413.15	1.18	toluene	413.15	1.38

copolymer (B): VA-VC-GMA/10w **78OIS**

Comments: Henry's constants were corrected for solubility of the marker gas

Solvent (A)	T/ K	$H_{A,B}$/ MPa	Solvent (A)	T/ K	$H_{A,B}$/ MPa
2-propanone	383.15	4.327	vinyl acetate	383.15	3.131
2-propanone	403.15	6.809	vinyl acetate	403.15	5.107
2-propanone	423.15	10.23	vinyl acetate	423.15	7.984
propyl acetate	383.15	1.246	vinyl chloride	383.15	22.39
propyl acetate	403.15	2.189	vinyl chloride	403.15	30.90
propyl acetate	423.15	3.678	vinyl chloride	423.15	41.04

copolymer (B): **VA-VC-HPA/05w** **78OIS**

Comments: Henry's constants were corrected for solubility of the marker gas

Solvent (A)	T/ K	$H_{A,B}$/ MPa	Solvent (A)	T/ K	$H_{A,B}$/ MPa
2-propanone	383.15	3.992	propyl acetate	423.15	3.698
2-propanone	403.15	6.606	vinyl chloride	383.15	23.10
2-propanone	423.15	10.44	vinyl chloride	403.15	31.61
propyl acetate	383.15	1.185	vinyl chloride	423.15	41.64
propyl acetate	403.15	2.138			

copolymer (B): **VA-VC-HPA/34w** **78OIS**

Comments: Henry's constants were corrected for solubility of the marker gas

Solvent (A)	T/ K	$H_{A,B}$/ MPa	Solvent (A)	T/ K	$H_{A,B}$/ MPa
2-propanone	383.15	3.719	vinyl chloride	383.15	18.44
2-propanone	403.15	6.090	vinyl chloride	403.15	26.04
2-propanone	423.15	9.433	vinyl chloride	423.15	35.77

Copolymers from vinyl chloride and vinylidene chloride

Average chemical composition of the copolymers, acronyms and references:

Copolymer (B)	Acronym	Ref.
Vinyl chloride/vinylidene chloride copolymer		
20 wt% vinyl chloride	VC-VdC/20w(1)	86DEM
20 wt% vinyl chloride	VC-VdC/20w(2)	91DEM

Characterization:

Copolymer (B)	M_n/ g/mol	M_w/ g/mol	M_η/ g/mol	Further information
VC-VdC/20w(1)	90000			Polysciences Inc.
VC-VdC/20w(2)				Borden Chem. Corp., USA

Experimental data:

copolymer (B): VC-VdC/20w(1) 86DEM

Comments: Henry's constants were not corrected for solubility of the marker gas

Solvent (A)	T/ K	$H_{A,B}$/ MPa	Solvent (A)	T/ K	$H_{A,B}$/ MPa
1,1-dichloroethene	291.15	1.628	1,1-dichloroethene	314.15	3.576
1,1-dichloroethene	298.15	1.982	1,1-dichloroethene	323.15	4.244
1,1-dichloroethene	306.15	2.766			

copolymer (B): VC-VdC/20w(2) 91DEM

Comments: Henry's constants were not corrected for solubility of the marker gas

Solvent (A)	T/ K	$H_{A,B}$/ MPa	Solvent (A)	T/ K	$H_{A,B}$/ MPa
benzene	363.15	2.823	n-octane	363.15	0.06041
cyclohexane	363.15	9.637	tetrachloromethane	363.15	3.601
n-decane	363.15	0.5149	tetrahydrofuran	363.15	2.739
n-heptane	363.15	6.296	toluene	363.15	1.049
n-nonane	363.15	1.157	trichloromethane	363.15	3.523

2.3. References

63COR Corneliussen, R., Rice, S.A., and Yamakawa, H., On the thermodynamic properties of solutions of polar polymers. A comparison of experiment and theory, *J. Chem. Phys.*, 38, 1768, 1963.

72PEI Peinze, K., Dampfdruckmessungen in binären Lösungen aus Aromaten und Copolymeren, *Diploma paper*, TH Leuna-Merseburg, 1972.

73BEL Belorussow, J., Freie Exzessenthalpie binärer Lösungen aus Aromaten und Ethylen-Vinylacetat-Copolymeren, *Diploma paper*, TH Leuna-Merseburg, 1973.

75BOB Boblenz, M. and Glindemann, D., Abhängigkeit der freien Exzessenthalpie vom Lösungsmittel in binären Mischungen mit Ethylen-Vinylacetat-Copolymerisat, *Diploma paper*, TH Leuna-Merseburg, 1975.

76KIE Kiessling, D., Untersuchungen des Flüssig-Dampf-Gleichgewichts von Polyethylenvinyl-acetat-Essigester-Systemen, *Diploma paper*, TH Leuna-Merseburg, 1976.

76LIU Liu, D.D. and Prausnitz, J.M., Solubilities of gases and volatile liquids in polyethylene and in ethylene-vinyl acetate copolymers in the region 125-225 °C, *Ind. Eng. Chem. Fundam.*, 15, 330, 1976.

78DIN Dincer, S. and Bonner, D.C., Thermodynamic analysis of ethylene and vinylacetate copolymer with various solvents by gas chromatography, *Macromolecules*, 11, 107, 1978.

78DIP DiPaola-Baranyi, G., Braun, J.-M., and Guillet, J.E., Partial molar heats of mixing of small molecules with polymers by gas chromatography, *Macromolecules*, 11, 224, 1978.

78OIS Oishi, T. and Prausnitz, J.M., Solubilities of some volatile polar fluids in copolymers containing vinyl chloride, vinyl acetate, glycidyl methacrylate, and hydroxypropyl acrylate, *Ind. Eng. Chem. Fundam.*, 17, 109, 1978.

79HEI Heintz, A., Lichtenthaler, R.N., and Prausnitz, J.M., Solubilities of volatile solvents in polyvinylacetate, polyvinylchloride, and their random copolymers, *Ber. Bunsenges. Phys. Chem.* 83, 926, 1979.

80MER Merk, W., Lichtenthaler, R.N., and Prausnitz, J.M., Solubilities of fifteen solvents in copolymers of poly(vinyl acetate) and poly(vinyl chloride) from gas-liquid chromatography. Estimation of polymer solubility parameters, *J. Phys. Chem.*, 84, 1694, 1980.

81DIP DiPaola-Baranyi, G., Thermodynamic miscibility of various solutes with styrene-butyl methacrylate polymers and copolymers, *Macromolecules*, 14, 683, 1981.

81WAR Ward, T.C., Sheeny, D.P., Riffle, J.S., and McGrath, J.E., Inverse gas chromatography studies of poly(dimethylsiloxane)-polycarbonate copolymers and blends, *Macromolecules*, 14, 1791, 1981.

82LI1 Lipson, J.E.G. and Guillet, J.E., Solubilities and solubility parameters of polymers from inverse gas chromatography, in *Macromolecular Solutions, Solvent Prop. Relat. Polym. Pap. Symp.*, 1982, 14.

82LI2 Lipson, J.E.G. and Guillet, J.E., Measurement of solubility and solubility parameters for small organic solutes in polymer films by gas chromatography, *J. Coat. Technol.*, 54, 89, 1982.

83WAL Walsh, D.J., Higgins, J.S., Rostami, S., and Weerape-Ruma, K., Compatibility of ethylene-vinylacetate copolymers with chlorinated polyethylenes. 2. Investigation of the thermodynamic parameters, *Macromolecules*, 16, 391, 1983.

84MAT Matsumara, K., Measurement of infinite dilution activity coefficient of solvent in polymer by gas-liquid chromatography, *Kagaku Kogaku Ronbunshu*, 10, 516, 1984.

86CAS Castells, R.S. and Massa, G.D., Study of polymer-solvent interactions by gas chromatography. Copolymers of vinyl acetate and vinyl alcohol with hydrocarbons and alcohols, *J. Appl. Polym. Sci.*, 32, 5917, 1986.

86DEM Demertzis, P.G. and Kontominas, M.G., Interaction between vinylidene chloride and copolymers of vinylidene chloride by inverse gas chromatography, *Lebensm.-Wiss. Technol.*, 19, 249, 1986.

86IT1 Itsuno, S., Iwai, Y., and Arai, Y., Measurement and correlation of solubilities of hydrocarbon gases in styrene-butadiene copolymers, *Kagaku Kogaku Ronbunshu*, 12, 349, 1986.

86IT2 Itsuno, S., Ohyama, K., Iwai, Y., and Arai, Y., Measurement and correlation of solubilities for hydrocarbon gases in acrylonitrile-butadiene copolymer, *Kagaku Kogaku Ronbunshu*, 12, 734, 1986.

86KRC Krcek, M., *Diploma paper*, Verdampfungsgleichgewicht in Systemen aus Polymermischungen oder Blockcopolymeren und einem Lösungsmittel, TH Leuna-Merseburg, 1986.

86PRI Price, G.J., Guillet, J.E., and Purnell, J.H., Measurement of solubility parameters by gas-liquid chromatography, *J. Chromatogr.*, 369, 273, 1986.

87OEN Oener, M. and Dincer, S., Thermodynamical properties of polymer-probe pairs by gas chromatography, *Polymer*, 28, 279, 1987.

87PRI Price, G.J. and Guillet, J.E., The use of gas chromatography to study solubility in polymeric systems, *J. Solution Chem.*, 16, 605, 1987.

88SAT Sato, Y., Inomata, H., and Arai, K., Solubilities of fifteen organic substances in poly(vinyl chloride), poly(vinyl acetate), and vinyl chloride-vinyl acetate copolymer, *Kobunshi Ronbunshu*, 45, 287, 1988.

89MOE Moeller, F., Energetik der Wechselwirkungen von Oligomeren des Ethylen- und Propylenoxids und deren Cooligomeren in Mischung mit CCl_4, *PhD-Thesis*, TU München, 1989.

90IWA Iwai, Y., Miyamoto, S., Nakano, K., and Arai, Y., Measurement and prediction of solubilities of ethylbenzene vapor in styrene-butadiene rubbers, *J. Chem. Eng. Japan*, 23, 508, 1990.

91CHE Cherrak, D.-E. and Djadoun, S., A study of mixtures of poly(ethylene oxide) with polystyrene or poly(styrene-co-acrylic acid) by inverse gas chromatography and viscosimetry, *Polym. Bull.*, 27, 289, 1991.

91DEM Demertzis, P.G., Riganakos, A., and Akrida-Demertzi, K., An inverse gas chromatographic study of compatibility of food grade PVdC copolymer and low volatility plasticizers, *Polym. Internat.*, 25, 229, 1991.

91IWA Iwai, Y., Ishidao, T., Miyamoto, S., Ikeda, H., and Arai, Y., Solubilities of nonane vapor in styrene-butadiene copolymers at 100 and 130 °C, *Fluid Phase Equil.*, 68, 197, 1991.

92CHE Cherrak, D.-E. and Djadoun, S., Inverse gas chromatography of poly(ethylene oxide) with poly(ethyl methacrylate) or poly(ethyl methacrylate-co-methacrylic acid), *J. Appl. Polym. Sci., Appl. Polym. Symp.*, 51, 209, 1992.

92ROM Romdhane, I.H., Plana, A., Hwang, S., and Danner, R.P., Thermodynamic interactions of solvents with styrene-butadiene-styrene triblock copolymers, *J. Appl. Polym. Sci.*, 45, 2049, 1992.

93ISS Issel, H.-M., Thermodynamische und rheologische Steuerung der Materialeigenschaften von Elastomeren durch trans-Poly(octenylen), *PhD-Thesis*, Univ. Hannover, 1993.

94WOH Wohlfarth, Ch., *Vapour-Liquid Equilibrium Data of Binary Polymer Solutions: Physical Science Data*, 44, Elsevier, Amsterdam, 1994.

95BOG Bogillo, V.I. and Voelkel, A., Solution properties of amorphous co- and terpolymers of styrene as examined by inverse gas chromatography, *J. Chromatogr. A*, 715, 127, 1995.

95GUP Gupta, R.B. and Prausnitz, J.M., Vapor-liquid equilibria of copolymer + solvent and homopolymer + solvent binaries: new experimental data and their correlation, *J. Chem. Eng. Data*, 40, 784, 1995.

95PET Peterseim, V., Quantitative Betrachtung der Verträglichkeit von Kautschuken und Untersuchungen zur Rußdistribution und Rußdispersion, *PhD-Thesis*, Univ. Hannover, 1995.

95TRI Tripathi, V.S., Lal, D., and Sen, A.K., Studies on thermodynamic compatibility of nitrile rubber and polybutadiene by inverse gas chromatography, *J. Appl. Polym. Sci.*, 58, 1681, 1995.

96DJA Djadoun, S., Karasz, F.E., and Hamou, A.S.H., Blends of poly(isobutyl methacrylate) with poly(styrene-co-acrylic acid) and of poly(isobutyl methacrylate-co-acrylic acid) with poly(styrene-co-N,N-dimethyl-aminoethyl methacrylate), *Thermochim. Acta*, 282/283, 399, 1996.

96WO1 Wohlfarth, Ch., Isopiestic vapor sorption measurements for solutions of butadiene/styrene random copolymer in cyclohexane, benzene, toluene, ethylbenzene, 1,4-dimethylbenzene, or 1,3,5-trimethylbenzene at (343.15 - 398.15) K, *ELDATA: Int. Electron. J. Phys.-Chem. Data*, 2, 13, 1996.

96WO2 Wohlfarth, Ch., Isopiestic vapor sorption measurements for solutions of acrylonitrile/ styrene random copolymer in benzene, toluene, 1,2-dimethylbenzene, 1,3-dimethylbenzene, 1,4-dimethylbenzene, or propylbenzene at (343.15 - 423.15) K, *ELDATA: Int. Electron. J. Phys.-Chem. Data*, 2, 163, 1996.

97MIO Mio, C., Jayachandran, K.N., and Prausnitz, J.M., Vapor-liquid equilibria for binary solutions of some comb polymers based on poly(styrene-co-maleic anhydride) in acetone, methanol and cyclohexane, *Fluid Phase Equil.*, 141, 165, 1997.

97SCH Schuster, R.H., Issel, H.M., and Peterseim, V., Charakterisierung der Kautschuk - Lösungsmittel Wechselwirkung mittels inverser Gaschromatographie, *Kautschuk Gummi Kunststoffe*, 50, 890, 1997.

97TAN Tanbonliong, J.O. and Prausnitz, J.M., Vapor-liquid equilibria for some binary and ternary polymer solutions, *Polymer*, 38, 5775, 1997.

97WO1 Wohlfarth, Ch., Vapour-liquid equilibria of concentrated poly(styrene-co-butadiene) and poly(styrene-co-acrylonitrile) solutions, *Macromol. Chem. Phys.*, 198, 2689, 1997.

97WO2 Wohlfarth, Ch., Isopiestic vapor sorption measurements for solutions of poly(methyl methacrylate-co-styrene) in benzene, toluene, ethylbenzene, 1,4-dimethylbenzene, or 1,3,5-trimethylbenzene at (343.15 - 398.15) K, *ELDATA: Int. Electron. J. Phys.-Chem. Data*, 3, 47, 1997.

98KIM Kim, N.H., Kim, S.J., Won, Y.S., and Choi, J.S., Prediction of activities of solvents in concentrated copolymer solutions, *Korean J. Chem. Eng.*, 15, 141, 1998.

98SCH Schneider, A., Homopolymer- und Copolymerlösungen im Vergleich: Wechselwirkungsparameter und Grenzflächenspannung, *PhD-Thesis*, Univ. Mainz, 1998.

98WOH Wohlfarth, Ch., Isopiestic vapor sorption measurements for solutions of a polystyrene-block-poly(methyl methacrylate) diblock copolymer in benzene, toluene, ethylbenzene, 1,4-dimethylbenzene, or 1,3,5-trimethylbenzene at (343.15 - 398.15) K, *ELDATA: Int. Electron. J. Phys.-Chem. Data*, 4, 83, 1998.

99BON Bondar, V.I., Freeman, B.D., and Yampolskii, Yu.P., Sorption of gases and vapors in an amorphous glassy perfluorodioxol copolymer, *Macromolecules*, 32, 6163, 1999.

99FRE French, R.N. and Koplos, G.J., Activity coefficients of solvents in elastomers measured with a quartz crystal microbalance, *Fluid Phase Equil.*, 158-160, 879, 1999.

99GUO Guo, Y., Gu, B., Lu, Z., and Du, Q., Study on segmental interaction parameter between ethylene and vinyl acetate monomer units of EVA by inverse gas chromatography, *J. Appl. Polym. Sci.*, 71, 693, 1999.

99WOH Wohlfarth, Ch., Isopiestic vapor sorption measurements for solutions of poly(3-hydroxybutanoic acid-co-3-hydroxypentanoic acid) in chlorobenzene at 373.15, 388.15 and 403.15 K, *ELDATA: Int. Electron. J. Phys.-Chem. Data*, 5, 5, 1999.

2000KAY Kaya, I. and Demirelli, K., Determination of thermodynamic properties of poly[2-(3-methyl-3-phenyl-cyclobutyl)-2-hydroxyethylmethacrylate] and its copolymers at infinite dilution using inverse gas chromatography, *Polymer*, 41, 2855, 2000.

2000NAG Nagy, M., A comparative and systematic thermodynamic study of aqueous solutions and hydrogels of homo- and copolymers of poly(vinyl alcohol), *Phys. Chem. Chem. Phys.*, 2, 2613, 2000.

3. LIQUID-LIQUID EQUILIBRIUM (LLE) DATA OF QUASIBINARY OR QUASITERNARY COPOLYMER SOLUTIONS

3.1. Cloud-point and/or coexistence curves of quasibinary solutions

Copolymers from acrylonitrile and styrene

Average chemical composition of the copolymers, acronyms and references:

Copolymer (B)	Acronym	Ref.
Acrylonitrile/styrene copolymer		
25.0 wt% acrylonitrile	S-AN/25w	98SCH
25.0 wt% acrylonitrile	S-AN/25w	2000SCH

Characterization:

Copolymer (B)	M_n/ g/mol	M_w/ g/mol	M_η/ g/mol	Further information
S-AN/25w	90000	147000		Bayer AG Leverkusen, Germany

Experimental LLE data:

copolymer (B):	**S-AN/25w**		**98SCH, 2000SCH**
solvent (A):	**toluene**	**C_7H_8**	**108-88-3**

Type of data: cloud-points (UCST-behavior)

w_B	0.006	0.010	0.020	0.042	0.061	0.080	0.081	0.101	0.141
T/K	324.35	326.85	326.15	321.35	316.85	313.15	313.15	309.35	300.25

copolymer (B): **S-AN/25w** **98SCH, 2000SCH**
solvent (A): **toluene** **C_7H_8** **108-88-3**

Type of data: critical point (UCST)

w_B	0.080
T/K	313.15

copolymer (B): **S-AN/25w** **98SCH, 2000SCH**
solvent (A): **toluene** **C_7H_8** **108-88-3**

Type of data: coexistence data (tie lines)

The total feed concentration of the copolymer in the homogeneous system is: $w_B = 0.080$

This is the critical concentration, i.e., sol phase and gel phase are identical phases at the critical point, see the first line in the table below.

Demixing temperature	w_B sol phase	w_B gel phase	Fractionation during demixing sol phase M_n/ g/mol	sol phase M_w/ g/mol	gel phase M_n/ g/mol	gel phase M_w/ g/mol
T/ K						
313.15	0.080	0.080	90000	147000	90000	147000
311.65	0.055	0.125	62000	124000	80000	170000
309.35	0.043	0.151	48500	105000	74500	163000
307.05	0.029	0.156	56500	187000	84000	167000
305.15	0.046	0.147	54000	110000	91000	174000
302.95	0.035	0.168	47000	89000	90000	171000
300.95	0.042	0.181	48000	90000	90000	175000
299.15	0.033	0.178	80000	170000	79000	169000

Comments: apparent M_w and M_n values were determined via polystyrene standard

Copolymers from N-cyclopropylacrylamide and vinylferrocene

Average chemical composition of the copolymers, acronyms and references:

Copolymer (B)	Acronym	Ref.
N-Cyclopropylacrylamide/vinylferrocene copolymer		
0.0 mol% vinylferrocene	NCPAM-VFe/00x	98KUR
1.0 mol% vinylferrocene	NCPAM-VFe/01x	98KUR
3.0 mol% vinylferrocene	NCPAM-VFe/03x	98KUR

Characterization:

Copolymer (B)	M_n/ g/mol	M_w/ g/mol	M_η/ g/mol	Further information
NCPAM-VFe/00x				synthesized in the laboratory
NCPAM-VFe/01x				synthesized in the laboratory
NCPAM-VFe/03x				synthesized in the laboratory

Experimental LLE data:

copolymer (B): NCPAM-VFe/00x **98KUR**
solvent (A): water H_2O **7732-18-5**

Type of data: cloud-point (LCST-behavior)

c_B/(g/l) 1.0 T/K 320.25

copolymer (B): NCPAM-VFe/01x **98KUR**
solvent (A): water H_2O **7732-18-5**

Type of data: cloud-point (LCST-behavior)

c_B/(g/l) 1.0 T/K 314.85

copolymer (B): NCPAM-VFe/03x **98KUR**
solvent (A): water H_2O **7732-18-5**

Type of data: cloud-point (LCST-behavior)

c_B/(g/l) 1.0 T/K 296.95

Copolymers from N,N-diethylacrylamide and vinylferrocene

Average chemical composition of the copolymers, acronyms and references:

Copolymer (B)	Acronym	Ref.
N,N-Diethylacrylamide/vinylferrocene copolymer		
0.0 mol% vinylferrocene	DEA-VFe/00x	97KUR
1.0 mol% vinylferrocene	DEA-VFe/01x	97KUR
3.0 mol% vinylferrocene	DEA-VFe/03x	97KUR

Characterization:

Copolymer (B)	M_n/ g/mol	M_w/ g/mol	M_η/ g/mol	Further information
DEA-VFe/00x				synthesized in the laboratory
DEA-VFe/01x				synthesized in the laboratory
DEA-VFe/03x				synthesized in the laboratory

Experimental LLE data:

copolymer (B):	**DEA-VFe/00x**		**97KUR**
solvent (A):	**water**	**H_2O**	**7732-18-5**

Type of data: cloud-point (LCST-behavior)

c_B/(g/l) 1.0 T/K 304.65

copolymer (B):	**DEA-VFe/01x**		**97KUR**
solvent (A):	**water**	**H_2O**	**7732-18-5**

Type of data: cloud-point (LCST-behavior)

c_B/(g/l) 1.0 T/K 300.35

copolymer (B):	**DEA-VFe/03x**		**97KUR**
solvent (A):	**water**	**H_2O**	**7732-18-5**

Type of data: cloud-point (LCST-behavior)

c_B/(g/l) 1.0 T/K 293.55

Copolymers from dimethylsiloxane and methylphenylsiloxane

Average chemical composition of the copolymers, acronyms and references:

Copolymer (B)	Acronym	Ref.
Dimethylsiloxane/methylphenylsiloxane copolymer		
15 wt% methylphenylsiloxane	DMS-MPS/15w	98SCH
15 wt% methylphenylsiloxane	DMS-MPS/15w	2000SCH

Characterization:

Copolymer (B)	M_n/ g/mol	M_w/ g/mol	M_η/ g/mol	Further information
DMS-MPS/15w	9100	41200		Dow Corning Corp., Midland, USA

Experimental LLE data:

copolymer (B): DMS-MPS/15w 98SCH, 2000SCH
solvent (A): anisole C_7H_8O 100-66-3

Type of data: cloud-points (UCST-behavior)

w_B	0.010	0.030	0.050	0.100	0.150	0.200	0.226	0.239	0.240
T/K	309.85	309.75	308.65	304.05	299.35	291.55	293.15	291.75	291.45

w_B	0.250	0.300	0.351
T/K	291.15	287.85	284.35

copolymer (B): DMS-MPS/15w 98SCH, 2000SCH
solvent (A): anisole C_7H_8O 100-66-3

Type of data: critical point (UCST)

w_B	0.240
T/K	291.45

copolymer (B):	**DMS-MPS/15w**		**98SCH, 2000SCH**
solvent (A):	**anisole**	**C_7H_8O**	**100-66-3**

Type of data: coexistence data (tie lines)

The total feed concentration of the copolymer in the homogeneous system is: w_B = 0.250

This is a concentration above the critical concentration where the feed phase is a gel phase.

Demixing temperature	w_B sol phase	gel phase	Fractionation during demixing sol phase		gel phase	
T/K			M_n/g/mol	M_w/g/mol	M_n/g/mol	M_w/g/mol
290.65	0.195	0.298	8300	25200	9400	42600
290.29	0.204	0.282	9000	27200	10900	41200
289.38	0.165	0.327	7800	16600	5200	40100
288.27	0.165	0.328	4700	21800	5500	35000
287.00	0.189	0.345	8300	19600	12100	43000
285.96	0.137	0.357	4400	18600	5700	39000
284.83	0.115	0.391	4100	13700	5500	39500
283.36	0.128	0.399	7300	15600	11800	40600

Comments: apparent M_w and M_n values were determined via polystyrene standard

copolymer (B):	**DMS-MPS/15w**		**98SCH, 2000SCH**
solvent (A):	**2-propanone**	**C_3H_6O**	**67-64-1**

Type of data: cloud-points (UCST-behavior)

w_B	0.010	0.044	0.064	0.080	0.098	0.152	0.198	0.248	0.298
T/K	296.75	298.15	297.25	296.35	295.15	291.55	288.45	285.25	282.65
w_B	0.310								
T/K	282.45								

copolymer (B):	**DMS-MPS/15w**		**98SCH, 2000SCH**
solvent (A):	**2-propanone**	**C_3H_6O**	**67-64-1**

Type of data: critical point (UCST)

w_B	0.310
T/K	282.45

copolymer (B): **DMS-MPS/15w** **98SCH, 2000SCH**
solvent (A): **2-propanone** **C_3H_6O** **67-64-1**

Type of data: coexistence data (tie lines)

The total feed concentration of the copolymer in the homogeneous system is: $w_B = 0.320$

This is a concentration above the critical concentration where the feed phase is a gel phase.

Demixing temperature	w_B sol phase	w_B gel phase	Fractionation during demixing sol phase		gel phase	
T/K			M_n/ g/mol	M_w/ g/mol	M_n/ g/mol	M_w/ g/mol
281.00	0.229	0.413	8700	22400	13600	52400
280.05	0.186	0.453	8100	18100	13600	51100
279.35	0.191	0.409	9000	22600	13200	53600
278.05	0.183	0.473				
277.25	0.179	0.460	7800	18600	14200	52600
276.05	0.167	0.465	7700	16000	13100	50900

Comments: apparent M_w and M_n values were determined via polystyrene standard

Copolymers from N-ethylacrylamide and vinylferrocene

Average chemical composition of the copolymers, acronyms and references:

Copolymer (B)	Acronym	Ref.
N-Ethylacrylamide/vinylferrocene copolymer		
0.0 mol% vinylferrocene	NEAM-VFe/00x	97KUR
1.0 mol% vinylferrocene	NEAM-VFe/01x	97KUR
3.0 mol% vinylferrocene	NEAM-VFe/03x	97KUR

Characterization:

Copolymer (B)	M_n/ g/mol	M_w/ g/mol	M_η/ g/mol	Further information
NEAM-VFe/00x				synthesized in the laboratory
NEAM-VFe/01x				synthesized in the laboratory
NEAM-VFe/03x				synthesized in the laboratory

Experimental LLE data:

copolymer (B): NEAM-VFe/00x **97KUR**
solvent (A): water H_2O **7732-18-5**

Type of data: cloud-point (LCST-behavior)

c_B/(g/l)	1.0
T/K	317.85

copolymer (B): NEAM-VFe/01x **97KUR**
solvent (A): water H_2O **7732-18-5**

Type of data: cloud-point (LCST-behavior)

c_B/(g/l)	1.0
T/K	335.65

copolymer (B): NEAM-VFe/03x **97KUR**
solvent (A): water H_2O **7732-18-5**

Type of data: cloud-point (LCST-behavior)

c_B/(g/l)	1.0
T/K	317.85

Copolymers from ethylene and vinyl acetate

Average chemical composition of the copolymers, acronyms and references:

Copolymer (B)	Acronym	Ref.
Ethylene/vinyl acetate copolymer 42.6 mol% vinyl acetate	E-VA/43x	86RAE

Characterization:

Copolymer (B)	M_n/ g/mol	M_w/ g/mol	M_z/ g/mol	Further information
E-VA/43x	14800	41500	79200	ρ_B (293.15 K) = 1.241 g/cm^3

Experimental LLE data:

copolymer (B): E-VA/43x 86RAE
solvent (A): methyl acetate $C_3H_6O_2$ 79-20-9

Type of data: cloud-points (UCST-behavior)

φ_B	0.0051	0.0117	0.0152	0.0300	0.0453	0.0579	0.0740	0.0817	0.0843
T/K	311.1	314.9	315.9	314.9	313.9	312.4	310.3	309.3	309.0
φ_B	0.0875	0.0934	0.0992	0.1019	0.1031	0.1052	0.1225	0.1278	0.1303
T/K	308.7	308.0	307.5	307.2	307.0	306.8	305.7	305.3	305.7
φ_B	0.1500	0.1650	0.1947	0.2570					
T/K	304.4	303.7	302.2	296.3					

copolymer (B): E-VA/43x 86RAE
solvent (A): methyl acetate $C_3H_6O_2$ 79-20-9

Type of data: critical point (UCST)

φ_B	0.1031
T/K	307.0

Copolymers from ethylene oxide and propylene oxide

Average chemical composition of the copolymers, acronyms and references:

Copolymer (B)	Acronym	Ref.
Poly(ethylene oxide)-b-poly(propylene oxide)-b-poly(ethylene oxide) triblock copolymer		
20.0 wt% propylene oxide	EO-PO-EO/20w	95KIM

Characterization:

Copolymer (B)	M_n/ g/mol	M_w/ g/mol	M_η/ g/mol	Further information
EO-PO-EO/20w	2500			

Experimental LLE data:

copolymer (B): **EO-PO-EO/20w** **95KIM**
solvent (A): **water** **H_2O** **7732-18-5**

Type of data: cloud-points (LCST-behavior)

w_B	0.005	0.010	0.025	0.050	0.075	0.100
T/K	308.15	305.15	301.15	299.15	297.15	296.15

Copolymers from N-isopropylacrylamide and itaconic acid

Average chemical composition of the copolymers, acronyms and references:

Copolymer (B)	Acronym	Ref.
N-Isopropylacrylamide/itaconic acid copolymer		
9.8 mol% itaconic acid	NIPAM-IA/10x	99ERB
23.0 mol% itaconic acid	NIPAM-IA/23x	99ERB

Characterization:

Copolymer (B)	M_n/ g/mol	M_w/ g/mol	M_η/ g/mol	Further information
NIPAM-IA/10x				synthesized in the laboratory
NIPAM-IA/23x				synthesized in the laboratory

Experimental LLE data:

copolymer (B): NIPAM-IA/10x 99ERB
solvent (A): water H_2O 7732-18-5

Type of data: cloud-points (LCST-behavior)

w_B	0.005	0.005	0.005	0.005	0.005	0.005
pH	2.01	2.30	3.09	3.32	4.08	5.03
T/K	307.55	308.15	309.25	310.15	312.65	332.55

copolymer (B): NIPAM-IA/23x 99ERB
solvent (A): water H_2O 7732-18-5

Type of data: cloud-points (LCST-behavior)

w_B	0.005	0.005	0.005
pH	1.00	2.02	3.14
T/K	310.35	310.65	315.55

Copolymers from N-isopropylacrylamide and vinylferrocene

Average chemical composition of the copolymers, acronyms and references:

Copolymer (B)	Acronym	Ref.
N-Isopropylacrylamide/vinylferrocene copolymer		
0.0 mol% vinylferrocene	NIPAM-VFe/00x	98KUR
1.0 mol% vinylferrocene	NIPAM-VFe/01x	98KUR
3.0 mol% vinylferrocene	NIPAM-VFe/03x	98KUR

Characterization:

Copolymer (B)	M_n/ g/mol	M_w/ g/mol	M_η/ g/mol	Further information
NIPAM-VFe/00x				synthesized in the laboratory
NIPAM-VFe/01x				synthesized in the laboratory
NIPAM-VFe/03x				synthesized in the laboratory

Experimental LLE data:

copolymer (B): **NIPAM-VFe/00x** **98KUR**
solvent (A): **water** **H_2O** **7732-18-5**

Type of data: cloud-point (LCST-behavior)

c_B/(g/l) 1.0 T/K 306.45

copolymer (B): **NIPAM-VFe/01x** **98KUR**
solvent (A): **water** **H_2O** **7732-18-5**

Type of data: cloud-point (LCST-behavior)

c_B/(g/l) 1.0 T/K 304.45

copolymer (B): **NIPAM-VFe/03x** **98KUR**
solvent (A): **water** **H_2O** **7732-18-5**

Type of data: cloud-point (LCST-behavior)

c_B/(g/l) 1.0 T/K 299.15

Copolymers from styrene and α-methylstyrene

Average chemical composition of the copolymers, acronyms and references:

Copolymer (B)	Acronym	Ref.
Styrene/α-methylstyrene copolymer 20.0 mol% styrene	S-αMS/20x	95PFO

Characterization:

Copolymer (B)	M_n/ g/mol	M_w/ g/mol	M_η/ g/mol	Further information
S-αMS/20x	100000	114000		Sci. Polym. Products

Experimental LLE data:

copolymer (B): **S-αMS/20x** **95PFO**
solvent (A): **butyl acetate** **$C_6H_{12}O_2$** **123-86-4**

Type of data: cloud-points (UCST-behavior)

w_B	0.0050	0.0100	0.0202	0.0298	0.0403	0.0501	0.0705	0.0893	0.1187
T/K	274.85	283.65	288.35	288.75	288.65	288.85	288.55	286.75	281.95
w_B	0.1405								
T/K	279.95								

copolymer (B): **S-αMS/20x** **95PFO**
solvent (A): **butyl acetate** **$C_6H_{12}O_2$** **123-86-4**

Type of data: cloud-points (LCST-behavior)

w_B	0.0050	0.0100	0.0202	0.0298	0.0403	0.0501	0.0705	0.0893	0.1187
T/K	468.45	463.55	458.15	456.85	453.05	454.45	453.95	454.45	455.05
w_B	0.1405								
T/K	456.55								

copolymer (B): **S-αMS/20x** **95PFO**
solvent (A): **cyclohexane** **C_6H_{12}** **110-82-7**

Type of data: cloud-points (UCST-behavior)

w_B	0.0050	0.0101	0.0196	0.0297	0.0398	0.0505	0.0532	0.0703	0.1008
T/K	279.35	282.15	284.55	284.55	285.05	285.65	285.85	285.55	285.35
w_B	0.1521								
T/K	283.55								

copolymer (B): **S-αMS/20x** **95PFO**
solvent (A): **cyclohexane** **C_6H_{12}** **110-82-7**

Type of data: cloud-points (LCST-behavior)

w_B	0.0050	0.0101	0.0196	0.0297	0.0398	0.0505	0.0532	0.0703	0.1008
T/K	494.65	490.55	486.45	484.85	486.55	485.15	487.05	486.35	485.95
w_B	0.1521								
T/K	489.25								

copolymer (B): **S-αMS/20x** **95PFO**
solvent (A): **cyclopentane** **C_5H_{10}** **287-92-3**

Type of data: cloud-points (UCST-behavior)

w_B	0.0053	0.0104	0.0241	0.0364	0.0482	0.0526	0.0913	0.1437	0.1876
T/K	281.65	284.65	288.95	290.15	290.95	290.85	290.65	289.65	288.45

copolymer (B): **S-αMS/20x** **95PFO**
solvent (A): **cyclopentane** **C_5H_{10}** **287-92-3**

Type of data: cloud-points (LCST-behavior)

w_B	0.0053	0.0104	0.0241	0.0364	0.0482	0.0526	0.0913	0.1437	0.1876
T/K	431.35	427.55	422.65	422.45	421.55	421.05	421.25	422.25	423.35

copolymer (B): **S-αMS/20x** **95PFO**
solvent (A): ***trans*-decahydronaphthalene** **$C_{10}H_{18}$** **493-02-7**

Type of data: cloud-points (UCST-behavior)

w_B	0.0050	0.0099	0.0197	0.0301	0.0402	0.0473	0.0687	0.0845	0.1177
T/K	254.45	259.55	262.05	263.05	263.75	263.75	264.05	264.15	263.95
w_B	0.1385	0.1893							
T/K	263.55	263.65							

copolymer (B): **S-αMS/20x** **95PFO**
solvent (A): **hexyl acetate** $C_8H_{16}O_2$ **142-92-7**

Type of data: cloud-points (UCST-behavior)

w_B	0.0050	0.0100	0.0202	0.0301	0.0393	0.0496	0.0703	0.0899	0.1192
T/K	277.05	284.25	287.35	288.25	287.35	288.55	287.15	286.75	282.95
w_B	0.1486								
T/K	280.95								

copolymer (B): **S-αMS/20x** **95PFO**
solvent (A): **hexyl acetate** $C_8H_{16}O_2$ **142-92-7**

Type of data: cloud-points (LCST-behavior)

w_B	0.0050	0.0100	0.0202	0.0301	0.0393	0.0496	0.0703	0.0899	0.1192
T/K	526.55	524.15	515.85	515.05	514.55	514.45	514.65	514.25	514.45
w_B	0.1486								
T/K	514.05								

copolymer (B): **S-αMS/20x** **95PFO**
solvent (A): **pentyl acetate** $C_7H_{14}O_2$ **628-63-7**

Type of data: cloud-points (UCST-behavior)

w_B	0.0010	0.0050	0.0107	0.0201	0.0301	0.0401	0.0505	0.0689	0.0840
T/K	274.55	289.75	296.25	298.25	304.25	302.25	300.45	300.65	299.95
w_B	0.1093	0.1527							
T/K	297.15	294.05							

copolymer (B): **S-αMS/20x** **95PFO**
solvent (A): **pentyl acetate** $C_7H_{14}O_2$ **628-63-7**

Type of data: cloud-points (LCST-behavior)

w_B	0.0010	0.0050	0.0107	0.0201	0.0301	0.0401	0.0505	0.0689	0.0840
T/K	502.75	493.55	489.55	482.95	481.55	482.45	480.65	481.35	481.55
w_B	0.1093	0.1527							
T/K	483.05	485.95							

3.2. Table of systems where binary LLE data were published only in graphical form as phase diagrams or related figures

Copolymer (B)	Solvent (A)	Ref.
Acrylamide/N,N-dimethylaminoethyl methacrylate copolymer	water	97CHO
Acrylic acid/N-isopropylacrylamide copolymer	water	2000BOK
Acrylic acid/methyl acrylate copolymer	water	93TAG
Acrylic acid/2-methyl-5-vinylpyridine copolymer	water	85VED
Acrylic acid/nonyl acrylate copolymer	bisphenol-A-diglycidyl ether	99MIK
N-Acryloylpyrrolidine/vinylferrocene copolymer	water	94KUR
tert-Butylstyrene/dimethylsiloxane triblock copolymer	2-butanone	86KUE
	1-nitropropane	86KUE
N,N-Dimethylacrylamide/2-methoxyethyl acrylate copolymer	water	96ELE
	water	97ELE
N,N-Dimethylacrylamide/N-phenylacrylamide copolymer	water	96MIY
Dimethylsiloxane/methylphenylsiloxane copolymer	anisole	2000SCH
	2-propanone	2000SCH
Dimethylsiloxane/1,1,3,3-tetramethyldisiloxanyl-ethylene diblock copolymer	ethoxybenzene	2000AZU

Copolymer (B)	**Solvent (A)**	**Ref.**
Ethylene/propylene copolymer	n-hexane	86IRA
	2-methylbutane	86IRA
	methylcyclopentane	86IRA
	3-methylpentane	86IRA
	n-pentane	86IRA
Ethylene/vinyl acetate copolymer	diphenyl ether	90VAN
Ethylene/vinyl alcohol copolymer	2-propanol	90CHE
	water	71SHI
Ethylene glycol/N-isopropylacrylamide copolymer	water	97TOP
Ethylene oxide/1,2-butylene oxide copolymer	water	2000SAH
Ethylene oxide/dimethylsiloxane triblock copolymer	water	92YAN
Ethylene oxide/propylene oxide copolymer	acetic acid	93JOH
	butanoic acid	93JOH
	propionic acid	93JOH
	water	81MED
	water	87ANA
	water	91LOU
	water	93JOH
	water	94ZHA
2-Ethyl-2-oxazoline/ε-caprolactone diblock copolymer	water	2000KIM
	water	2000KIM
N-Isopropylacrylamide/butyl methacrylate copolymer	water	93FEI
Maleic anhydride/diethylene glycol copolymer	styrene	92LEC
Poly(ethylene oxide)-b-poly(propylene oxide) diblock copolymer	formamide	90SAM
	water	90SAM
	water	90SIM

Copolymer (B)	**Solvent (A)**	**Ref.**
Polystyrene-b-poly(butyl methacrylate) diblock copolymer	2-propanol	94SIQ
Styrene/acrylonitrile copolymer	2-butanone	68TER
	cyclohexane	68TER
	ethyl acetate	82MAN
	toluene	72TE1
	toluene	72TE2
	toluene	2000SCH
Styrene/butyl methacrylate copolymer	2-butanone	90KYO
Styrene/cellulose diacetate graft copolymer	2-propanone	82GOL
	N,N-dimethylformamide	82GOL
	tetrahydrofuran	82GOL
N-Vinylacetamide/N-vinylisobutyramide copolymer	water	97SUW
Vinyl acetate /vinyl alcohol copolymer	water	97PAE
Vinyl acetate/vinyl caprolactam copolymer	water	93PAS
Vinyl acetate/1-vinyl-2-pyrrolidinone copolymer	water	74TAN
Vinyl alcohol/vinyl butyrate copolymer	water	74TAN
	water	84SHI
Vinyl amine/vinyl caprolactam copolymer	water	91TAG
	water	94TAG
Vinyl caprolactam/N-vinyl-N-methylacetamide copolymer	water	93PAS
N-Vinylisobutyramide/N-vinylvaleramide copolymer	water	97SUW

3.3. Cloud-point and/or coexistence curves of quasiternary solutions containing at least one copolymer

Copolymers from dimethylsiloxane and styrene

Average chemical composition of the copolymers, acronyms and references:

Copolymer (B)	Acronym	Ref.
Poly(dimethylsiloxane)-b-polystyrene diblock copolymer		
50.0 wt% styrene	DMS-b-S/50w	97SCH

Characterization:

Copolymer (B)	M_n/ g/mol	M_w/ g/mol	M_η/ g/mol	Further information
DMS-b-S/50w		55000		synthesized in the laboratory

Experimental LLE data:

copolymer (B):	**DMS-b-S/50w**		**97SCH**
solvent (A):	**cyclohexane**	**C_6H_{12}**	**110-82-7**
polymer (C):	**polystyrene**	**M_w/g.mol^{-1} = 96000 (M_w/M_n = 1.04)**	

Type of data: cloud-points
Comments: the weight ratio of polystyrene : DMS-b-S/50w was kept constant at 7.5 : 1

$w_B + w_C$	0.011	0.040	0.070	0.100	0.126	0.150	0.179
T/K	290.32	293.57	295.07	296.99	301.85	310.82	340.68

Comments: experimental data by personal communication with A. Schneider

Copolymers from ethylene oxide and propylene oxide

Average chemical composition of the copolymers, acronyms and references:

Copolymer (B)	Acronym	Ref.
Ethylene oxide/propylene oxide copolymer		
33.3 mol% ethylene oxide	EO-PO/33x	98LI1
50.0 mol% ethylene oxide	EO-PO/50x(1)	99LIW
50.0 mol% ethylene oxide	EO-PO/50x(2)	99LIW
50.0 mol% ethylene oxide	EO-PO/50x(3)	98LI1
50.0 mol% ethylene oxide	EO-PO/50x(4)	99LIW
50.0 mol% ethylene oxide	EO-PO/50x(5)	98LI2

Characterization:

Copolymer (B)	M_n/ g/mol	M_w/ g/mol	M_η/ g/mol	Further information
EO-PO/33x	3000	3330		Zhejiang Univ. Chem. Factory, PR China
EO-PO/50x(1)	780	865		Zhejiang Univ. Chem. Factory, PR China
EO-PO/50x(2)	2340	2480		Zhejiang Univ. Chem. Factory, PR China
EO-PO/50x(3)	3500	3900		Zhejiang Univ. Chem. Factory, PR China
EO-PO/50x(4)	3640	4040		Zhejiang Univ. Chem. Factory, PR China
EO-PO/50x(5)	4200	4650		Zhejiang Univ. Chem. Factory, PR China

Experimental LLE data:

copolymer (B):	**EO-PO/33x**		**98LI1**
solvent (A):	**water**	**H_2O**	**7732-18-5**
polymer (C):	**hydroxypropyl starch**	**M_n = 10000**	**(Reppe Glykos AB, Sweden)**

Type of data: coexistence data (tie lines)

T/K = 298.15

continue

continue

Total system			Bottom phase			Top phase		
w_A	w_B	w_C	w_A	w_B	w_C	w_A	w_B	w_C
0.8500	0.1000	0.0500	0.7729	0.0250	0.2021	0.8571	0.1080	0.0349
0.8399	0.0999	0.0602	0.7618	0.0222	0.2160	0.8533	0.1139	0.0328
0.8300	0.1000	0.0700	0.7494	0.0204	0.2302	0.8498	0.1200	0.0302
0.8208	0.0995	0.0797	0.7405	0.0190	0.2405	0.8460	0.1260	0.0280
0.8099	0.0999	0.0902	0.7276	0.0172	0.2552	0.8420	0.1321	0.0259
0.8002	0.0999	0.0999	0.7169	0.0159	0.2672	0.8370	0.1380	0.0250
0.7901	0.1000	0.1099	0.7077	0.0152	0.2771	0.8320	0.1441	0.0239
0.7801	0.1000	0.1199	0.6950	0.0140	0.2909	0.8280	0.1489	0.0231

Plait-point composition: $w_A = 0.827 + w_B = 0.063 + w_C = 0.110$

copolymer (B): **EO-PO/33x** **98LI1**
solvent (A): **water** **H_2O** **7732-18-5**
polymer (C): **hydroxypropyl starch** **$M_n = 20000$** **(Reppe Glykos AB, Sweden)**

Type of data: coexistence data (tie lines)

T/K = 298.15

Total system			Bottom phase			Top phase		
w_A	w_B	w_C	w_A	w_B	w_C	w_A	w_B	w_C
0.8503	0.0999	0.0498	0.7872	0.0220	0.1908	0.8597	0.1118	0.0285
0.8401	0.1000	0.0599	0.7699	0.0171	0.2130	0.8561	0.1200	0.0239
0.8299	0.1001	0.0700	0.7567	0.0122	0.2311	0.8519	0.1271	0.0210
0.8201	0.1000	0.0799	0.7480	0.0100	0.2420	0.8469	0.1340	0.0190
0.8099	0.1000	0.0901	0.7336	0.0062	0.2602	0.8431	0.1390	0.0179
0.7998	0.1000	0.1002	0.7235	0.0050	0.2715	0.8361	0.1481	0.0158
0.7902	0.1000	0.1098	0.7145	0.0050	0.2805	0.8323	0.1529	0.0148
0.7796	0.1002	0.1202	0.7068	0.0050	0.2882	0.8260	0.1600	0.0140

Plait-point composition: $w_A = 0.832 + w_B = 0.058 + w_C = 0.110$

copolymer (B):	**EO-PO/50x(3)**		**98LI1**
solvent (A):	**water**	**H_2O**	**7732-18-5**
polymer (C):	**hydroxypropyl starch**	**M_n = 10000**	**(Reppe Glykos AB, Sweden)**

Type of data: coexistence data (tie lines)

T/K = 298.15

Total system			Bottom phase			Top phase		
w_A	w_B	w_C	w_A	w_B	w_C	w_A	w_B	w_C
0.8498	0.1000	0.0502	0.7895	0.0303	0.1802	0.8549	0.1070	0.0381
0.8399	0.0999	0.0602	0.7767	0.0282	0.1951	0.8517	0.1141	0.0342
0.8296	0.1000	0.0704	0.7679	0.0251	0.2070	0.8476	0.1202	0.0322
0.8199	0.1001	0.0800	0.7516	0.0222	0.2262	0.8419	0.1281	0.0300
0.8099	0.1002	0.0899	0.7447	0.0212	0.2341	0.8378	0.1342	0.0280
0.8000	0.1000	0.1000	0.7326	0.0201	0.2473	0.8335	0.1403	0.0262
0.7901	0.1000	0.1099	0.7207	0.0201	0.2592	0.8299	0.1451	0.0250
0.7801	0.1000	0.1199	0.7090	0.0200	0.2710	0.8248	0.1511	0.0241

Plait-point composition: w_A = 0.835 + w_B = 0.067 + w_C = 0.098

copolymer (B):	**EO-PO/50x(3)**		**98LI1**
solvent (A):	**water**	**H_2O**	**7732-18-5**
polymer (C):	**hydroxypropyl starch**	**M_n = 20000**	**(Reppe Glykos AB, Sweden)**

Type of data: coexistence data (tie lines)

T/K = 298.15

Total system			Bottom phase			Top phase		
w_A	w_B	w_C	w_A	w_B	w_C	w_A	w_B	w_C
0.8403	0.0999	0.0598	0.7753	0.0241	0.2006	0.8627	0.1022	0.0351
0.8499	0.1001	0.0500	0.7620	0.0198	0.2182	0.8605	0.1090	0.0305
0.8403	0.0999	0.0704	0.7448	0.0142	0.2410	0.8570	0.1160	0.0270
0.8302	0.0999	0.0598	0.7290	0.0110	0.2600	0.8534	0.1230	0.0236
0.8201	0.1001	0.0699	0.7203	0.0092	0.2705	0.8495	0.1275	0.0230
0.8099	0.1001	0.0798	0.7089	0.0063	0.2848	0.8448	0.1322	0.0230
0.7996	0.1001	0.0899	0.6973	0.0052	0.2975	0.8410	0.1370	0.0220
0.7900	0.1001	0.1099	0.6892	0.0050	0.3058	0.8357	0.1423	0.0220
0.7802	0.0999	0.1199	0.6750	0.0050	0.3200	0.8320	0.1462	0.0218

continue

continue

Plait-point composition: $w_A = 0.829 + w_B = 0.063 + w_C = 0.108$

copolymer (B): **EO-PO/50x(1)** **99LIW**
solvent (A): **water** **H_2O** **7732-18-5**
component (C): **ammonium sulfate** **$(NH_4)_2SO_4$** **7783-20-2**

Type of data: coexistence data (tie lines)

T/K = 298.15

Total system			Bottom phase			Top phase		
w_A	w_B	w_C	w_A	w_B	w_C	w_A	w_B	w_C
0.6996	0.2005	0.0999	0.8007	0.0290	0.1703	0.5869	0.3839	0.0292
0.6894	0.2001	0.1105	0.7939	0.0193	0.1868	0.5553	0.4221	0.0226
0.6775	0.2017	0.1208	0.7847	0.0176	0.1977	0.5254	0.4576	0.0170
0.6587	0.2006	0.1407	0.7669	0.0103	0.2228	0.4774	0.5086	0.0140
0.6406	0.2003	0.1591	0.7479	0.0143	0.2378	0.4355	0.5570	0.0075

Plait-point composition: $w_A = 0.778 + w_B = 0.145 + w_C = 0.082$

copolymer (B): **EO-PO/50x(2)** **99LIW**
solvent (A): **water** **H_2O** **7732-18-5**
component (C): **ammonium sulfate** **$(NH_4)_2SO_4$** **7783-20-2**

Type of data: coexistence data (tie lines)

T/K = 298.15

Total system			Bottom phase			Top phase		
w_A	w_B	w_C	w_A	w_B	w_C	w_A	w_B	w_C
0.7670	0.1512	0.0818	0.8624	0.0221	0.1155	0.6510	0.3224	0.0266
0.7594	0.1505	0.0901	0.8566	0.0151	0.1283	0.6160	0.3630	0.0210
0.7502	0.1498	0.1000	0.8440	0.0119	0.1441	0.5832	0.4008	0.0160
0.7308	0.1495	0.1197	0.8264	0.0086	0.1650	0.5152	0.4749	0.0099
0.7095	0.1503	0.1402	0.8045	0.0049	0.1906	0.4590	0.5362	0.0048

Plait-point composition: $w_A = 0.800 + w_B = 0.138 + w_C = 0.062$

copolymer (B): **EO-PO/50x(4)** **99LIW**
solvent (A): **water** **H_2O** **7732-18-5**
component (C): **ammonium sulfate** **$(NH_4)_2SO_4$** **7783-20-2**

Type of data: coexistence data (tie lines)

T/K = 298.15

Total system			Bottom phase			Top phase		
w_A	w_B	w_C	w_A	w_B	w_C	w_A	w_B	w_C
0.8090	0.1205	0.0705	0.8750	0.0247	0.1003	0.6631	0.3147	0.0222
0.8003	0.1196	0.0801	0.8667	0.0115	0.1218	0.6179	0.3642	0.0179
0.7743	0.1206	0.1051	0.8453	0.0077	0.1470	0.5386	0.4513	0.0101
0.7502	0.1197	0.1301	0.8251	0.0039	0.1710	0.4751	0.5190	0.0059
0.7319	0.1200	0.1481	0.8067	0.0041	0.1892	0.4213	0.5752	0.0036

Plait-point composition: $w_A = 0.815 + w_B = 0.124 + w_C = 0.051$

copolymer (B): **EO-PO/50x(5)** **98LI2**
solvent (A): **water** **H_2O** **7732-18-5**
component (C): **dipotassium phosphate** **K_2HPO_4** **7758-11-4**

Type of data: coexistence data (tie lines)

T/K = 275.15

Total system			Bottom phase			Top phase		
w_A	w_B	w_C	w_A	w_B	w_C	w_A	w_B	w_C
0.7147	0.1801	0.1052	0.8168	0.0010	0.1822	0.5958	0.3916	0.0141
0.7449	0.1652	0.0899	0.8408	0.0052	0.1540	0.6497	0.3298	0.0205
0.7653	0.1548	0.0799	0.8568	0.0096	0.1336	0.6744	0.3014	0.0242
0.7847	0.1453	0.0700	0.8631	0.0231	0.1138	0.7275	0.2395	0.0330

Plait-point composition: $w_A = 0.821 + w_B = 0.113 + w_C = 0.066$

copolymer (B): **EO-PO/50x(5)** **98LI2**
solvent (A): **water** **H_2O** **7732-18-5**
component (C): **dipotassium phosphate** **K_2HPO_4** **7758-11-4**

Type of data: coexistence data (tie lines)

T/K = 298.15

Total system			Bottom phase			Top phase		
w_A	w_B	w_C	w_A	w_B	w_C	w_A	w_B	w_C
0.7148	0.1800	0.1052	0.8350	0.0021	0.1629	0.5137	0.4793	0.0070
0.7252	0.1749	0.0999	0.8381	0.0031	0.1588	0.5420	0.4501	0.0079
0.7450	0.1651	0.0899	0.8534	0.0021	0.1445	0.5573	0.4341	0.0086
0.7652	0.1549	0.0799	0.8719	0.0028	0.1253	0.5855	0.4037	0.0108
0.7848	0.1452	0.0700	0.8835	0.0053	0.1112	0.6552	0.3300	0.0148
0.8050	0.1350	0.0600	0.8986	0.0103	0.0911	0.6773	0.3056	0.0171

Plait-point composition: $w_A = 0.827 + w_B = 0.131 + w_C = 0.042$

copolymer (B): **EO-PO/50x(5)** **98LI2**
solvent (A): **water** **H_2O** **7732-18-5**
component (C): **potassium dihydrogen phosphate** **KH_2PO_4** **7778-77-0**

Type of data: coexistence data (tie lines)

T/K = 298.15

Total system			Bottom phase			Top phase		
w_A	w_B	w_C	w_A	w_B	w_C	w_A	w_B	w_C
0.7100	0.1950	0.0950	0.8382	0.0100	0.1518	0.5003	0.4850	0.0147
0.7250	0.1850	0.0900	0.8424	0.0097	0.1479	0.5818	0.3960	0.0222
0.7399	0.1751	0.0850	0.8477	0.0109	0.1414	0.6273	0.3435	0.0292
0.7700	0.1550	0.0750	0.8630	0.0216	0.1154	0.6721	0.2933	0.0346
0.7851	0.1450	0.0699	0.8610	0.0394	0.0996	0.6980	0.2653	0.0367

Plait-point composition: $w_A = 0.802 + w_B = 0.138 + w_C = 0.060$

copolymer (B):	**EO-PO/50x(5)**		**98LI2**
solvent (A):	**water**	$\mathbf{H_2O}$	**7732-18-5**
component (C):	**sodium sulfate**	$\mathbf{Na_2SO_4}$	**7757-82-6**

Type of data: coexistence data (tie lines)

T/K = 298.15

Total system			Bottom phase			Top phase		
w_A	w_B	w_C	w_A	w_B	w_C	w_A	w_B	w_C
0.7750	0.1550	0.0700	0.8827	0.0041	0.1132	0.5800	0.4183	0.0017
0.7900	0.1450	0.0650	0.8973	0.0062	0.0965	0.6163	0.3770	0.0067
0.8049	0.1350	0.0601	0.9058	0.0094	0.0848	0.6309	0.3634	0.0057
0.8200	0.1252	0.0550	0.9082	0.0128	0.0790	0.6724	0.3165	0.0111
0.8350	0.1150	0.0500	0.9089	0.0223	0.0688	0.7160	0.2714	0.0126

Plait-point composition: $w_A = 0.838 + w_B = 0.132 + w_C = 0.030$

Copolymers from ethylene terephthalate and p-hydroxybenzoic acid

Average chemical composition of the copolymers, acronyms and references:

Copolymer (B)	Acronym	Ref.
Ethylene terephthalate/p-hydroxybenzoic acid copolymer		
27.0 mol% p-hydroxybenzoic acid	ET-pHBA/27x	88SCH
35.0 mol% p-hydroxybenzoic acid	ET-pHBA/35x	88SCH

Characterization:

Copolymer (B)	M_n/ g/mol	M_w/ g/mol	M_η/ g/mol	Further information
ET-pHBA/27x	6000			synthesized in the laboratory
ET-pHBA/35x	6000			synthesized in the laboratory

Experimental LLE data:

copolymer (B): ET-pHBA/27x 88SCH
solvent (A): trichloromethane $CHCl_3$ 67-66-3
polymer (C): polycarbonate-bisphenol-A M_n = 19300, M_w = 32800 g/mol

Type of data: cloud-points

T/K = 295.15

φ_A	0.8588	0.8554	0.8777	0.8931	0.9095	0.9096	0.9208	0.9218	0.9234
φ_B	0.1412	0.1367	0.1021	0.0782	0.0537	0.0444	0.0371	0.0308	0.0282
φ_C	0.0000	0.0079	0.0202	0.0287	0.0368	0.0459	0.0421	0.0474	0.0484
φ_A	0.9172	0.9161	0.9037	0.8628					
φ_B	0.0265	0.0227	0.0186	0.0129					
φ_C	0.0563	0.0612	0.0777	0.1243					

copolymer (B): ET-pHBA/35x 88SCH
solvent (A): trichloromethane $CHCl_3$ 67-66-3
polymer (C): polycarbonate-bisphenol-A M_n = 19300, M_w = 32800 g/mol

Type of data: cloud-points

T/K = 295.15

φ_A	0.9021	0.9005	0.9021	0.9052	0.9386	0.9493	0.9583	0.9673	0.9692
φ_B	0.0979	0.0973	0.0932	0.0844	0.0442	0.0333	0.0225	0.0126	0.0083
φ_C	0.0000	0.0025	0.0047	0.0104	0.0172	0.0174	0.0192	0.0201	0.0225
φ_A	0.9473	0.9292	0.9234						
φ_B	0.0091	0.0094	0.0623						
φ_C	0.0435	0.0614	0.0143						

Comments: experimental data by personal communication with M. Hess

3.4. Table of systems where ternary LLE data were published only in graphical form as phase diagrams or related figures

Copolymer (B)	Second and third component	Ref.
Arylene sulfonoxide/butadiene polyblock copolymer		
	polybutadiene and trichloromethane	88ROG
	polybutadiene and 1,1,2,2-tetrachloroethane	88ROG
Ethyl acrylate/4-vinylpyridine copolymer		
	polystyrene and tetrahydrofuran	92WAN
Ethylene/vinyl alcohol copolymer		
	dimethylsulfoxide and water	97YOU
	2-propanol and water	98YOU
Ethylene oxide/propylene oxide copolymer		
	dextran and water	86ALB
	hydroxypropyl starch and water	94MOD
	poly(ethylene oxide) and water	94ZH1
	poly(propylene oxide) and water	94ZH1
Ethylene terephthalate/p-hydroxybenzoic acid copolymer		
	polycarbonate-bisphenol-A and trichloromethane	88SCH
Poly(ethylene oxide)-b-poly(propylene oxide) -b-poly(ethylene oxide) triblock copolymer		
	1-butanol and water	97HOL
	dextran and water	86ALB
	dextran and water	95SVE
Poly(ethylene oxide)-b-poly(tetrahydrofuran) -b-poly(ethylene oxide) triblock copolymer		
	1-butanol and water	97HOL

Copolymer (B)	Second and third component	Ref.
Methyl methacrylate/4-vinylpyridine copolymer		
	styrene/4-vinylpyridine copolymer and 1,4-dioxane	77DJA
	styrene/4-vinylpyridine copolymer and 2-butanone	77DJA
	styrene/4-vinylpyridine copolymer and trichloromethane	77DJA
Polystyrene-b-polybutadiene-b-polystyrene triblock copolymer		
	polybutadiene and 4-methyl-2-pentanone	97HWA
Polystyrene-b-poly(dimethylsiloxane) diblock copolymer		
	polystyrene and cyclohexane	97SCH
Polystyrene-b-polyisoprene diblock copolymer (hydrogenated PIP)		
	N,N-dimethylformamide and methylcyclohexane	93PO1
Styrene/acrylamide copolymer		
	vinyl acetate/vinyl alcohol copolymer and tetrahydrofuran	82LE1
Styrene/acrylic acid copolymer		
	vinyl acetate/vinyl alcohol copolymer and tetrahydrofuran	82LE1
Styrene/acrylonitrile copolymer		
	poly(methyl methacrylate) and dimethyl phthalate	77BER
Styrene/butadiene copolymer		
	polybutadiene and toluene	89INO
	polystyrene and tetrahydrofuran	72WHI
	polystyrene and styrene	75WHI
Styrene/methacrylic acid copolymer		
	poly(ethylene oxide) and benzene	92LE2
	poly(ethylene oxide) and tetrahydrofuran	92LE2
	poly(ethylene oxide) and trichloromethane	92LE2

Copolymer (B)	Second and third component	Ref.
Styrene/2-methoxyethyl methacrylate copolymer		
	1,1,2,2-tetrachloroethene and dimethyl sulfoxide	93PO2
Styrene/methyl methacrylate copolymer		
	nitroethane and decahydronaphthalene	94KAW
Styrene/4-vinylpyridine copolymer		
	methyl methacrylate/4-vinylpyridine copolymer and 2-butanone	77DJA
	methyl methacrylate/4-vinylpyridine copolymer and 1,4-dioxane	77DJA
	methyl methacrylate/4-vinylpyridine copolymer and trichloromethane	77DJA
	vinyl acetate/methacrylic acid copolymer and tetrahydrofuran	82LE1
	vinyl acetate/vinyl alcohol copolymer and tetrahydrofuran	82LE1
Styrene/1-vinyl-2-pyrrolidinone copolymer		
	vinyl acetate/vinyl alcohol copolymer and tetrahydrofuran	82LE1
Vinyl acetate/methacrylic acid copolymer		
	styrene/4-vinylpyridine copolymer and tetrahydrofuran	82LE1
Vinyl acetate/vinyl alcohol copolymer		
	poly(ethylene glycol) and water	84HEF
	poly(vinyl alcohol) and water	97PAE
	styrene/acrylamide copolymer and tetrahydrofuran	82LE1
	styrene/acrylic acid copolymer and tetrahydrofuran	82LE1
	styrene/4-vinylpyridine copolymer and tetrahydrofuran	82LE1
	styrene/1-vinyl-2-pyrrolidinone copolymer and tetrahydrofuran	82LE1

3.5. Lower critical (LCST) and/or upper critical (UCST) solution temperatures of copolymer solutions

Average chemical composition of the copolymers, acronyms and references:

Copolymer (B)	Acronym	Ref.
N,N-Dimethylacrylamide/2-butoxyethyl acrylate copolymer		
50 wt% 2-butoxyethyl acrylate	DMA-BOEA/50w	92MUE
N,N-Dimethylacrylamide/butyl acrylate copolymer		
15 wt% butyl acrylate	DMA-BA/15w	92MUE
20 wt% butyl acrylate	DMA-BA/20w	92MUE
30 wt% butyl acrylate	DMA-BA/30w	92MUE
35 wt% butyl acrylate	DMA-BA/35w	92MUE
N,N-Dimethylacrylamide/2-ethoxyethyl acrylate copolymer		
50 wt% 2-ethoxyethyl acrylate	DMA-EOEA/50w	92MUE
75 wt% 2-ethoxyethyl acrylate	DMA-EOEA/75w	92MUE
N,N-Dimethylacrylamide/ethyl acrylate copolymer		
25 wt% ethyl acrylate	DMA-EA/25w	92MUE
30 wt% ethyl acrylate	DMA-EA/30w	92MUE
50 wt% ethyl acrylate	DMA-EA/50w	92MUE
55 wt% ethyl acrylate	DMA-EA/55w	92MUE
N,N-Dimethylacrylamide/2-methoxyethyl acrylate copolymer		
38 mol% 2-methoxyethyl acrylate	DMA-MOEA/38x	96ELE
45 mol% 2-methoxyethyl acrylate	DMA-MOEA/45x	96ELE
50 wt% 2-methoxyethyl acrylate	DMA-MOEA/50w	92MUE
55 mol% 2-methoxyethyl acrylate	DMA-MOEA/55x	96ELE
68 mol% 2-methoxyethyl acrylate	DMA-MOEA/68x	96ELE
70 wt% 2-methoxyethyl acrylate	DMA-MOEA/70w	92MUE
75 wt% 2-methoxyethyl acrylate	DMA-MOEA/75w	92MUE
80 wt% 2-methoxyethyl acrylate	DMA-MOEA/80w	92MUE
82 mol% 2-methoxyethyl acrylate	DMA-MOEA/82x	96ELE
92 mol% 2-methoxyethyl acrylate	DMA-MOEA/92x	96ELE
N,N-Dimethylacrylamide/propyl acrylate copolymer		
20 wt% propyl acrylate	DMA-PA/20w	92MUE
30 wt% propyl acrylate	DMA-PA/30w	92MUE

Copolymer (B)	Acronym	Ref.
N,N-Dimethylacrylamide/propyl acrylate copolymer		
40 wt% propyl acrylate	DMA-PA/40w	92MUE
50 wt% propyl acrylate	DMA-PA/50w	92MUE
N,N-Dimethylacrylamide/methyl acrylate copolymer		
30 wt% methyl acrylate	DMA-MA/30w	92MUE
40 wt% methyl acrylate	DMA-MA/40w	92MUE
50 wt% methyl acrylate	DMA-MA/50w	92MUE
55 wt% methyl acrylate	DMA-MA/55w	92MUE
60 wt% methyl acrylate	DMA-MA/60w	92MUE
70 wt% methyl acrylate	DMA-MA/70w	92MUE
Dimethylsiloxane/methylphenylsiloxane copolymer		
15 wt% methylphenylsiloxane	DMS-MPS/15w	98SCH
15 wt% methylphenylsiloxane	DMS-MPS/15w	2000SCH
Ethylene/propylene copolymer		
33 mol% ethylene	E-P/33x	81CHA
43 mol% ethylene	E-P/43x	86IRA
53 mol% ethylene	E-P/53x	81CHA
63 mol% ethylene	E-P/63x	81CHA
75 mol% ethylene	E-P/75x	81CHA
81 mol% ethylene	E-P/81x	81CHA
Ethylene/vinyl acetate copolymer		
2.3 wt% vinyl acetate	E-VA/02w	90VAN
4.0 wt% vinyl acetate	E-VA/04w	90VAN
7.1 wt% vinyl acetate	E-VA/07w	90VAN
9.5 wt% vinyl acetate	E-VA/09w	90VAN
9.7 wt% vinyl acetate	E-VA/10w	90VAN
12.1 wt% vinyl acetate	E-VA/12w	90VAN
42.6 mol% vinyl acetate	E-VA/43x	86RAE
Ethylene/vinyl alcohol copolymer		
87.2 mol% vinyl alcohol	E-VAL/87x	71SHI
88.9 mol% vinyl alcohol	E-VAL/89x	71SHI
91.0 mol% vinyl alcohol	E-VAL/91x	71SHI
94.1 mol% vinyl alcohol	E-VAL/94x	71SHI
Ethylene oxide/propylene oxide copolymer		
72.4 mol% ethylene oxide	EO-PO/72x	91LOU
79.5 mol% ethylene oxide	EO-PO/80x	91LOU
86.6 mol% ethylene oxide	EO-PO/87x	91LOU

Copolymer (B)	**Acronym**	**Ref.**
N-Isopropylacrylamide/acrylamide copolymer		
15 mol% acrylamide	NIPAM-AM/15x	94MUM
30 mol% acrylamide	NIPAM-AM/30x	94MUM
45 mol% acrylamide	NIPAM-AM/45x	94MUM
Maleic anhydride/diethylene glycol copolymer		
1:1, 50 mol% maleic anhydride	MAH-DEG/50x	92LEC
Styrene/acrylonitrile copolymer		
21.1 wt% acrylonitrile	S-AN/21w	72TE2
23.2 wt% acrylonitrile	S-AN/23w	72TE2
25.0 wt% acrylonitrile	S-AN/25w	98SCH
25.0 wt% acrylonitrile	S-AN/25w	2000SCH
51.0 mol% acrylonitrile	S-AN/51x	82MAN
Styrene/methyl methacrylate copolymer		
52.0 mol% styrene	S-MMA/52x	70KOT
Styrene/α-methylstyrene copolymer		
20 mol% styrene	S-αMS/20x	95PFO
Styrene/cellulose diacetate graft copolymer		
77.4 wt% grafted polystyrene	gS-CDA/77w	82GOL
Tetrafluoroethylene/trifluoronitrosomethane copolymer		
1:1, alternating	TFE-TFNM/50x	61MOR
Vinyl alcohol/vinyl butyrate copolymer		
7.5 mol% butyralized PVA	VAL-VBU/08x	84SHI
9.9 mol% butyralized PVA	VAL-VBU/10x	84SHI
12.7 mol% butyralized PVA	VAL-VBU/13x	84SHI
Vinyl amine/vinyl caprolactam copolymer		
3 mol% vinyl amine	VAMN-VCPL/03x	91TAG
3.8 mol% vinyl amine	VAMN-VCPL/04x	94TAG

Copolymer-solvent systems and corresponding UCST and/or LCST values

Copolymer (B)	M_n/ g/mol	M_w/ g/mol	M_η/ g/mol	Solvent	UCST/ K	LCST/ K
DMA-BA/15w				water		346.2
DMA-BA/20w				water		323.2
DMA-BA/30w				water		294.2
DMA-BA/35w				water		281.2
DMA-BOEA/50w				water		< 273.2
DMA-EOEA/50w				water		319.2
DMA-EOEA/75w				water		285.2
DMA-EA/30w				water		347.2
DMA-EA/40w				water		334.2
DMA-EA/50w				water		287.2
DMA-EA/55w				water		< 273.2
DMA-MOEA/38x				water		353.2
DMA-MOEA/45x				water		333.2
DMA-MOEA/50w				water		343.2
DMA-MOEA/55x				water		315.2
DMA-MOEA/68x				water		305.2
DMA-MOEA/70w				water		313.2
DMA-MOEA/75w				water		309.2
DMA-MOEA/80w				water		303.2
DMA-MOEA/82x				water		288.2
DMA-MOEA/92x				water		283.2
DMA-MA/30w				water		371.2
DMA-MA/40w				water		338.2
DMA-MA/50w				water		314.2
DMA-MA/55w				water		294.2
DMA-MA/60w				water		279.2
DMA-MA/70w				water		< 273.2
DMA-PA/20w				water		353.2
DMA-PA/30w				water		337.2
DMA-PA/40w				water		294.2
DMA-PA/50w				water		281.2

Copolymer (B)	M_n/ g/mol	M_w/ g/mol	M_η/ g/mol	Solvent	UCST/ K	LCST/ K
DMS-MPS/15w	9100	41200		2-propanone	282.45	
DMS-MPS/15w	9100	41200		anisole	291.45	
E-P/33x			145000	cyclohexane		534.
E-P/33x			145000	cyclopentane		490.
E-P/33x			145000	2,2-dimethylbutane		428.
E-P/33x			145000	2,3-dimethylbutane		452.
E-P/33x			145000	3,4-dimethylhexane		541.
E-P/33x			145000	2,2-dimethylpentane		472.
E-P/33x			145000	2,3-dimethylpentane		500.
E-P/33x			145000	2,4-dimethylpentane		464.
E-P/33x			145000	3-ethylpentane		511.
E-P/33x			145000	n-heptane		502.
E-P/33x			145000	n-hexane		455.
E-P/33x			145000	2-methylbutane		396.
E-P/33x			145000	methylcyclohexane		558.
E-P/33x			145000	methylcyclopentane		512.
E-P/33x			145000	2-methylhexane		486.
E-P/33x			145000	n-nonane		558.
E-P/33x			145000	n-octane		528.
E-P/33x			145000	n-pentane		409.
E-P/33x			145000	2,2,4,4-tetramethylpentane		539.
E-P/33x			145000	2,2,3-trimethylbutane		500.
E-P/33x			145000	2,2,4-trimethylpentane		503.
E-P/43x	70000	140000		n-hexane		436.
E-P/43x	70000	140000		2-methylpentane		474.
E-P/43x	70000	140000		n-pentane		441.
E-P/53x			154000	2,2-dimethylbutane		407.
E-P/53x			154000	2,3-dimethylbutane		437.
E-P/53x			154000	2,2-dimethylpentane		453.
E-P/53x			154000	2,3-dimethylpentane		488.
E-P/53x			154000	2,4-dimethylpentane		445.
E-P/53x			154000	3-ethylpentane		500.
E-P/53x			154000	n-heptane		493.
E-P/53x			154000	n-hexane		443.
E-P/53x			154000	n-pentane		395.
E-P/53x			154000	2,2,3-trimethylbutane		488.
E-P/53x			154000	2,3,4-trimethylhexane		565.
E-P/53x			154000	2,2,4-trimethylpentane		484.
E-P/63x			236000	cyclohexane		526.
E-P/63x			236000	cyclopentane		481.

Copolymer (B)	M_n/ g/mol	M_w/ g/mol	M_η/ g/mol	Solvent	UCST/ K	LCST/ K
E-P/63x			236000	2,3-dimethylbutane		429.
E-P/63x			236000	3,4-dimethylhexane		530.
E-P/63x			236000	2,2-dimethylpentane		444.
E-P/63x			236000	2,3-dimethylpentane		482.
E-P/63x			236000	2,4-dimethylpentane		434.
E-P/63x			236000	3-ethylpentane		492.
E-P/63x			236000	n-heptane		485.
E-P/63x			236000	n-hexane		436.
E-P/63x			236000	2-methylbutane		348.
E-P/63x			236000	methylcyclopentane		498.
E-P/63x			236000	n-nonane		547.
E-P/63x			236000	n-octane		512.
E-P/63x			236000	n-pentane		387.
E-P/63x			236000	2,2,4,4-tetramethylpentane		528.
E-P/63x			236000	2,2,3-trimethylbutane		479.
E-P/63x			236000	2,2,4-trimethylpentane		479.
E-P/75x			109000	2,2-dimethylpentane		431.
E-P/75x			109000	2,4-dimethylpentane		425.
E-P/75x			109000	n-heptane		475.
E-P/75x			109000	n-hexane		427.
E-P/75x			109000	n-nonane		542.
E-P/75x			109000	n-octane		509.
E-P/75x			109000	n-pentane		378.
E-P/75x			109000	2,2,4,4-tetramethylpentane		523.
E-P/75x			109000	2,2,4-trimethylpentane		469.
E-P/81x			195000	cyclohexane		522.
E-P/81x			195000	cyclopentane		474.
E-P/81x			195000	2,2-dimethylbutane		381.
E-P/81x			195000	2,3-dimethylbutane		413.
E-P/81x			195000	2,4-dimethylhexane		478.
E-P/81x			195000	2,5-dimethylhexane		466.
E-P/81x			195000	3,4-dimethylhexane		522.
E-P/81x			195000	2,2-dimethylpentane		425.
E-P/81x			195000	2,3-dimethylpentane		471.
E-P/81x			195000	2,4-dimethylpentane		420.
E-P/81x			195000	3-ethylpentane		478.
E-P/81x			195000	n-heptane		468.
E-P/81x			195000	n-hexane		425.
E-P/81x			195000	2-methylbutane		327.
E-P/81x			195000	methylcyclohexane		541.
E-P/81x			195000	methylcyclopentane		493.
E-P/81x			195000	2-methylhexane		453.

Copolymer (B)	M_n/ g/mol	M_w/ g/mol	M_η/ g/mol	**Solvent**	**UCST/ K**	**LCST/ K**
E-P/81x			195000	3-methylhexane		459.
E-P/81x			195000	n-nonane		540.
E-P/81x			195000	n-octane		506.
E-P/81x			195000	n-pentane		370.
E-P/81x			195000	2,2,4,4-tetramethylpentane		519.
E-P/81x			195000	2,2,3-trimethylbutane		461.
E-P/81x			195000	2,2,4-trimethylpentane		495.
E-VA/02w	52000	465000		diphenyl ether	404.2	
E-VA/04w	47000	280000		diphenyl ether	392.5	
E-VA/07w	34000	460000		diphenyl ether	378.2	
E-VA/09w	53000	350000		diphenyl ether	367.3	
E-VA/10w	55000	490000		diphenyl ether	370.8	
E-VA/12w	66000	300000		diphenyl ether	360.4	
E-VA/43x	14800	41500		methyl acetate	307.0	
E-VAL/87x	∞	∞		water	463.55	285.65
E-VAL/89x	∞	∞		water	449.15	290.75
E-VAL/91x	∞	∞		water	428.45	302.95
E-VAL/94x	∞	∞		water	389.25	324.45
EO-PO/72x		36000		water		333.
EO-PO/80x		30800		water		345.
EO-PO/80x		32500		water		345.
EO-PO/87x		30100		water		355.5
NIPAM-AM/15x		3100000		water		315.15
NIPAM-AM/30x		4500000		water		326.15
NIPAM-AM/45x		3900000		water		347.15
MAH-DEG/50x	2200	4200		styrene	352.15	
MAH-DEG/50x	2380	4590		styrene	339.15	
S-AN/21w	∞	∞		toluene	325.4	
S-AN/23w	∞	∞		toluene	355.1	
S-AN/25w	90000	147000		toluene	313.15	
S-AN/51x		347000		ethyl acetate		344.15
S-AN/51x		457000		ethyl acetate		337.35
S-AN/51x		794000		ethyl acetate		328.65
S-AN/51x		912000		ethyl acetate		326.65

Copolymer (B)	M_n/ g/mol	M_w/ g/mol	M_η/ g/mol	Solvent	UCST/ K	LCST/ K
S-AN/51x		1365000		ethyl acetate		323.95
S-AN/51x		2240000		ethyl acetate		320.15
S-MMA/52x	∞	∞		cyclohexanol	334.65	
S-αMS/20x	100000	114000		butyl acetate	288.85	453.05
S-αMS/20x	100000	114000		cyclohexane	285.85	484.85
S-αMS/20x	100000	114000		cyclopentane	290.95	421.05
S-αMS/20x	100000	114000		*trans*-decahydro-naphthalene	264.15	
S-αMS/20x	100000	114000		hexyl acetate	288.55	514.15
S-αMS/20x	100000	114000		pentyl acetate	303.15	480.65
gS-CDA/77w		480000		N,N-dimethyl-formamide	257.	402.
gS-CDA/77w		750000		N,N-dimethyl-formamide	262.	399.
gS-CDA/77w		1100000		N,N-dimethyl-formamide	266.	389.
gS-CDA/77w		1200000		N,N-dimethyl-formamide	271.	366.
gS-CDA/77w		480000		tetrahydrofuran		367.
gS-CDA/77w		750000		tetrahydrofuran		363.
gS-CDA/77w		1100000		tetrahydrofuran		361.
gS-CDA/77w		1200000		tetrahydrofuran		355.
TFE-TFNM/50x	∞	∞		1,1,2-trichloro-trifluoroethane	301.6	
VAL-VBU/08x	∞	∞		water	408.0	298.25
VAL-VBU/10x	∞	∞		water		296.45
VAL-VBU/13x	∞	∞		water		287.55
VCPL-VAMN/03x	∞	∞		water		308.8
VCPL-VAMN/04x			160000	water		308.8

3.6. References

61MOR Morneau, G.A., Roth, P.I., and Shultz, A.R., Trifluoronitrosomethane/tetrafluoroethylene elastomers dilute solution properties and molecular weight, *J. Polym. Sci.*, 55, 609, 1961.

68TER Teramachi, S. and Nagasawa, M., The fractionation of copolymers by chemical composition, *J. Macromol. Sci.-Chem. A*, 2, 1169, 1968.

70KOT Kotaka, T., Tanaka, T., Ohnuma, H., Murakami, Y., and Inagaki, H., Dilute solution properties of styrene-methyl methacrylate copolymers with different architecture, *Polym. J.*, 1, 245, 1970.

71SHI Shibatani, K. and Oyanagi, Y., Solution properties of vinyl alcohol-ethylene copolymers in water, *Kobunshi Kagaku*, 28, 361, 1971.

72TE1 Teramachi, S., Tomioka, H., and Mamoru, S., Phase-separation phenomena of copolymer solutions and fractionation of copolymers by chemical composition, *J. Macromol. Sci.-Chem. A*, 6, 97, 1972.

72TE2 Teramachi, S. and Fujikawa, T., Phase-separation phenomena of random copolymer solution: SAN random copolymer-toluene systems, *J. Macromol. Sci.-Chem. A*, 6, 1393, 1972.

72WHI White, J.L., Salladay, D.G., Quisenberry, D.O., and Mac Lean, D.L., Gel permeation and thin-layer chromatographic characterization and solution properties of butadiene and styrene homopolymers and copolymers, *J. Appl. Polym. Sci.*, 16, 2811, 1972.

74TAN Taniguchi, Y., Suzuki, K., and Enomoto, T., The effect of pressure on the cloud point of aqueous polymer solutions, *J. Coll. Interface Sci.*, 46, 511, 1974.

75WHI White, J.L. and Patel, R.D., Phase separation conditions in polystyrene-styrene-(butadiene-styrene) copolymer solutions, *J. Appl. Polym. Sci.*, 19, 1775, 1975.

77BER Bernstein, R.E., Cruz, C.A., Paul, D.R., and Barlow, J.W., LCST behavior in polymer blends, *Macromolecules*, 10, 681, 1977.

77DJA Djadoun, S., Goldberg, R.N., and Morawetz, H., Ternary systems containing an acidic copolymer, a basic copolymer, and a solvent 1, *Macromolecules*, 10, 1015, 1977.

81CHA Charlet, G. and Delmas, G., Thermodynamic properties of polyolefine solutions at high temperature, *Polymer*, 22, 1181, 1981.

81MED Medved, Z.N., Penzel, U., Denosiva, T.A., and Lebedev, V.S., Study of phase equilibrium in mixtures of water and ethylene oxide-propylene oxide cooligomers (Russ.), *Vysokomol. Soedin., Ser. B*, 23, 276, 1981.

82GOL Goloborod'ko, V.I., Valatin, S.M., and Tashmukhamedov, I.P., Vliyanie sostava i molekulyarnoi massy privitykh sopolimerov na fasovoe ravnovesie ikh rastvorov, *Uzb. Khim. Zh.*, (3), 33, 1982.

82LE1 Lecourtier, J., Lafuma, F., and Quivoron, C., Compatibilization of copolymers in solution through interchain hydrogen bonding, *Eur. Polym. J.*, 18, 241, 1982.

82LE2 Lecourtier, J., Lafuma, F., and Quivoron, C., Study of polymer compatibilization in solution through polymer/polymer interactions: Ternary systems poly(ethylene oxide)/styrene-methacrylic acid copolymers/solvent, *Makromol. Chem.*, 183, 2021, 1982.

82MAN Mangalam, P. V. and Kalpagam, V., Styrene-acrylonitrile random copolymer in ethyl acetate, *J. Polym. Sci., Polym. Phys. Ed.*, 20, 773, 1982.

84HEF Hefford, R.J., Polymer mixing in aqueous solution, *Polymer*, 25, 979, 1984.

84SHI Shiomi, T., Imai, K., Watanabe, C., and Miya, M., Thermodynamic and conformational properties of partially butyralized poly(vinyl alcohol) in aqueous solution, *J. Polym. Sci., Polym. Phys. Ed.*, 22, 1305, 1984.

85VED Vedikhina, L.I., Kurmaeva, A.I., and Barabanov, V.P., Phase separation in solutions of polyampholytes of acrylic acid-2-methyl-5-vinylpyridine copolymers (Russ.), *Vysokomol. Soedin., Ser. A*, 27, 2131, 1985.

86ALB Albertson, P.-A., *Partition of Cell Particles and Macromolecules*, 3rd ed., J. Wiley & Sons, New York, 1986.

86IRA Irani, C.A. and Cozewith, C., Lower critical solution temperature behavior of ethylene-propylene copolymers in multicomponent solvents, *J. Appl. Polym. Sci.*, 31, 1879, 1986.

86KUE Kuecuekyavruz, Z. and Kuecuekyavruz, S., Theta-behaviour of poly(p-*tert*-butylstyrene)-b-poly(dimethylsiloxane)-b-poly(p-*tert*-butylstyrene), *Makromol. Chem.*, 187, 2469, 1986.

86RAE Raetzsch, M.T., Kehlen, H., Browarzik, D., and Schirutschke, M., Cloud-point curve for the system copoly(ethylene-vinyl acetate) + methyl acetate. Measurement and prediction by continuous thermodynamics, *J. Macromol. Sci.-Chem. A*, 23, 1349, 1986.

87ANA Ananthapadmanabhan, K.P. and Goddard, E.D., The relationship between clouding and aqueous biphase formation in polymer solutions, *Coll. Surfaces*, 25, 393, 1987.

88ROG Rogovina, L.Z., Nikiforova, G.G., Martirosov, V.A., Shilov, V.V., Gomza, Yu.P., and Slonimskii, G.L., Parametry termodinamicheskoi nesovmestimosti v sisteme poliarilensul'f-oksid-polibutadien i ikh svyaz' so svoistvami poliblochnykh sopolimerov na ikh osnove, *Vysokomol. Soedin., Ser. A*, 30, 598, 1988.

88SCH Schubert, F., Friedrich, K., Hess, M., and Kosfeld, R., Investigations on phase diagrams of a coil-polymer (PC) and a semiflexible thermotropic mainchain polymer (PET-co-PHB) in solution, *Mol. Cryst. Liq. Cryst.*, 155, 477, 1988.

89INO Inoue, T. and Ougizawa, T., Characterization of phase behavior in polymer blends by light scattering, *J. Macromol. Sci.-Chem. A*, 26, 147, 1989.

90CHE Chen, L.-W. and Young, T.H., EVAL membranes for blood dialysis, *Makromol. Chem., Macromol. Symp.*, 33, 183, 1990.

90KYO Kyoumen, M., Baba, Y., Kagemoto, A., and Beatty, C.L., Determination of the consolute temperature of poly[styrene-co-(butyl methacrylate)] solutions by simultaneous measurement of differential thermal analysis and laser transmittance, *Macromolecules*, 23, 1085, 1990.

90SAM Samii, A.A., Lindman, B., and Karstrom, G., Phase behavior of some nonionic polymers in non-aqueous solvents, *Prog. Coll. Polym. Sci.*, 82, 280, 1990.

90SIM Simek, L., Petrik, S., Hadobas, F., and Bohdanecky, M., Solubility in water of triblock (PEP) copolymers of ethylene and propylene oxides, *Eur. Polym. J.*, 26, 375, 1990.

90VAN Van der Haegen, R. and Van Opstal, L., Thermodynamic investigation of the phase separation behaviour of ethylene-vinyl acetate copolymers in diphenyl ether, *Makromol. Chem.*, 191, 1871, 1990.

91LOU Louai, A., Sarazin, D., Pollet, G., Francois, J., and Moreaux, F., Properties of ethylene oxide-propylene oxide statistical copolymers in aqueous solution, *Polymer*, 32, 703, 1991.

91TAG Tager, A.A., Safronov, A.P., Berezyuk, E.A., and Galaev, I.Yu., Hydrophobic interactions and lower critical solution temperatures of aqueous polymer solutions (Russ.), *Vysokomol. Soedin., Ser. B*, 33, 572, 1991.

92LEC Lecointe, J.P., Pascault, J.P., Suspene, L., and Yang, Y.S., Cloud-point curves and interaction parameters of unsaturated polyester-styrene solutions, *Polymer*, 33, 3226, 1992.

92MUE Mueller, K.F, Thermotropic aqueous gels and solutions of N,N-dimethylacrylamide-acrylate copolymers, *Polymer*, 33, 3470, 1992.

92WAN Wang, J., Khokhlov, A., Peiffer, D.G., and Chu, B., Phase equilibria in the ternary system zinc sulfonated polystyrene/poly(ethyl acrylate-4-vinylpyridine)/tetrahydrofuran, *Macromolecules*, 25, 2566, 1992.

92YAN Yang, J., Wegner, G., and Koningsveld, R., Phase behavior of ethylene oxide-dimethyl-siloxane PEO-PDMS-PEO triblock copolymers with water, *Colloid Polym. Sci.*, 270, 1080, 1992.

93FEI Feil, H., Bae, Y.H., Feifen, J., and Kim, S.W., Effect of comonomer hydrophobicity and ionization on the lower critical solution temperature of N-isopropylacrylamide copolymers, *Macromolecules*, 26, 2496, 1993.

93JOH Johansson, H.O., Karlstroem, G., and Tjerneld, F., Experimental and theoretical study of phase separation in aqueous solutions of clouding polymers and carboxylic acids, *Macromolecules*, 26, 4478, 1993.

93PAS Pashkin, I.I., Kirsh, Yu.E., Zubov, V.P., Anisimova, T.V., Kuz'kina, I.F., Voloshina, Ya.P., and Krylov, A.V., Synthesis of water-soluble N-vinylcaprolactam-based copolymers and physicochemical properties of their aqueous solutions (Russ.), *Vysokomol. Soedin., Ser. A*, 35, 481, 1993.

93PO1 Podesva, J., Stejskal, J., and Kratochvil, P., Fractionation of a diblock copolymer in a demixing-solvent system, *J. Appl. Polym. Sci.*, 49, 1265, 1993.

93PO2 Podesva, J., Stejskal, J., Prochazka, O., Spacek, P., and Enders, S., Fractionation of a statistical copolymer in a demixing-solvent system: Theory and experiment, *J. Appl. Polym. Sci.*, 48, 1127, 1993.

93TAG Tager, A.A., Klyuzhin, E.S., Adamova, L.V., and Safronov, A.P., Thermodynamics of dissolution of acrylic acid-methyl acrylate copolymers in water (Russ.), *Vysokomol. Soedin., Ser. B*, 35, 1357, 1993.

94KAW Kawai, T., Terajima, T., Ogaswara, T., Baba, H., and Teramachi, S., Phase separation phenomena in quasi-ternary systems of demixing solvents and poly(styrene-*stat*-methyl methacrylate) and composition fractionation, *Kogakuin Daigaku Kenkyu Hokoku*, 76, 23, 1994.

94KUR Kuramoto, N., Shishida, Y., and Nagai, K., Preparation of thermally responsive electroactive poly(N-acryloylpyrrolidine-co-vinylferrocene), *Macromol. Rapid Commun.*, 15, 441, 1994.

94MOD Modlin, R.F., Alred, P.A., and Tjerneld, F., Utilization of temperature-induced phase separation for the purification of ecdysone and 20-hydroxyecdysone from spinach, *J. Chromatogr. A*, 668, 229, 1994.

94MUM Mumick, P.S. and McCormick, C.L., Water soluble copolymers. 54: N-isopropylacrylamide-co-acrylamide copolymers in drag reduction: synthesis, characterization, and dilute solution behavior, *Polym. Eng. Sci.*, 34, 1419, 1994.

94SIQ Siqueira, D.F., Nunes, S.P., and Wolf, B.A., Solution properties of a diblock copolymer in a selective solvent of marginal quality 1, *Macromolecules*, 27, 234, 1994.

94TAG Tager, A.A., Safronov, A.P., Berezyuk, E.A., and Galaev, I.Yu., Lower critical solution temperature and hydrophobic hydration in aqueous polymer solutions, *Coll. Polym. Sci.*, 272, 1234, 1994.

94ZH1 Zhang, K., Carlsson, M., Linse, P., and Lindman, B., Phase behavior of copolymer-homopolymer mixtures in aqueous solution, *J. Phys. Chem.*, 98, 2452, 1994.

94ZH2 Zhang, K., Karlstroem, G., and Lindman, B., Ternary aqueous mixtures of a nonionic polymer with a surfactant or a second polymer, *J. Phys. Chem.*, 98, 4411, 1994.

95KIM Kim, Y.-H., Kwon, C., Bae, Y.H., and Kim, S.W., Saccharide effect on the lower critical solution temperature of thermosensitive polymers, *Macromolecules*, 28, 939, 1995.

95PFO Pfohl, O., Hino, T., and Prausnitz, J.M., Solubilities of styrene-based polymers and copolymers in common solvents, *Polymer*, 36, 2065, 1995.

95SVE Svensson, M., Linse, P., and Tjerneld, F., Phase behavior in aqueous two-phase systems containing micelle-forming block copolymers, *Macromolecules*, 28, 3597, 1995.

96ELE El-Ejmi, A.A.S. and Huglin, M.B., Characterization of N,N-dimethylacrylamide/2-methoxyethylacrylate copolymers and phase behaviour of their thermotropic aqueous solutions, *Polym. Internat.*, 39, 113, 1996.

96MIY Miyazaki, H. and Kataoka, K., Preparation of polyacrylamide derivatives showing thermo-reversible coacervate formation and their potential application to two-phase separation processes, *Polymer*, 37, 681, 1996.

97CHO Cho, S.H., Jhon, M.S., Yuk, S.H., and Lee, H.B., Temperature-induced phase transition of poly(N,N-dimethylaminoethyl methacrylate-co-acrylamide), *J. Polym. Sci., Part B: Polym. Phys.*, 35, 595, 1997.

97ELE El-Ejmi, A.A.S. and Huglin, M.B., Behaviour of poly(N,N-dimethylacrylamide-co-2-methoxyethylacrylate) in non-aqueous solution and LCST behaviour in water, *Eur. Polym. J.*, 33, 1281, 1997.

97HOL Holmqvist, P., Alexandridis, P., and Lindman, B., Phase behavior and structure of ternary amphiphilic block copolymer-alkanol-water systems: Comparison of poly(ethylene oxide)/

poly(propylene oxide) to poly(ethylene xide)/poly(tetrahydro-furan) copolymers, *Langmuir*, 13, 2471, 1997.

97HWA Hwang, J.H., Lee, K.H., and Lee, D.C., Study of phase behavior in SBS triblock copolymer/ polybutadiene/methyl isobutyl ketone ternary system, *Pollimo*, 21, 745, 1997.

97KUR Kuramoto, N., Shishido, Y., and Nagai, K., Thermosensitive and redox-active polymers: preparation and properties of poly(N-ethylacrylamide-co-vinylferrocene) and poly(N,N-diethylacrylamide-co-vinylferrocene), *J. Polym. Sci., Part A: Polym. Chem.*, 35, 1967, 1997.

97PAE Pae, B.J., Moon, T.J., Lee, C.H., Ko, M.B., Park, M., Lim S., Kim, J., and Choe, C.R., Phase behavior in PVA/water solution, *Korea Polym. J.*, 5, 126, 1997.

97SCH Schneider, A. and Wolf, B.A., Interfacial tension of demixed polymer solutions: Augmentation by polymer additives, *Macromol. Rapid Commun.*, 19, 561, 1997.

97SUW Suwa, K., Morishita, K., Kishida, A., and Akashi, M., Synthesis and functionalities of poly(N-vinylalkylamide), *J. Polym. Sci., Part A: Polym. Chem.*, 35, 3087, 1997.

97TOP Topp, M.D.C., Dijkstra, P.J., Talsma, H., and Feijen, J., Thermosensitive micelle-forming block copolymers of poly(ethylene glycol) and poly(N-isopropylacrylamide), *Macromolecules*, 30, 8518, 1997.

97YOU Young, T.-H., Lai, J.-Y., You, W.-M., and Cheng, L.-P., Equilibrium phase behavior of the membrane forming water-DMSO-EVAL copolymer system, *J. Membr. Sci.*, 128, 55, 1997.

98KUR Kuramoto, N. and Shishido, Y., Property of thermo-sensitive and redox-active poly(N-cyclopropylacrylamide-co-vinylferrocene) and poly(N-isopropylacrylamide-co-vinylferrocene), *Polymer*, 39, 669, 1998.

98LI1 Li, M., Zhu, Z.-Q., Wu, Y.-T., and Lin, D.-Q., Measurement of phase diagrams for new aqueous two-phase systems and prediction by a generalized multicomponent osmotic virial equation, *Chem. Eng. Sci.*, 53, 2577, 1998.

98LI2 Li, M., Wang, Y.-X., and Zhu, Z.-Q., Liquid-liquid equilibria for random copolymer of ethylene oxide + propylene oxide + salt + water, *J. Chem. Eng. Data*, 43, 93, 1998.

98SCH Schneider, A., Homopolymer- und Copolymerlösungen im Vergleich: Wechselwirkungsparameter und Grenzflächenspannung, *PhD-Thesis*, Univ. Mainz, 1998.

98YOU Young, T.-H., Cheng, L.-P., Hsieh, C.-C., and Chen, L.-W., Phase behavior of EVAL polymers in water-2-propanol cosolvent, *Macromolecules*, 31, 1229, 1998.

99ERB Erbil, C., Akpinar, F.D., and Uyanik, N., Investigation of the thermal aggregations in aqueous poly(N-isopropylacrylamide-co-itaconic acid) solutions, *Macromol. Chem. Phys.*, 200, 2448, 1999.

99LIW Li, W., Zhu, Z.-Q., and Lin, D.-Q., Liquid-liquid equilibria of aqueous two-phase systems containing ethylene oxide-propylene oxide random copolymer and ammonium sulfate, *J. Chem. Eng. Data*, 44, 921, 1999.

99MIK Mikhailov, Yu.M., Ganina, L.V., Makhonina, L.I., Smirnov, V.S., and Shapayeva, N.V., Mutual diffusion and phase equilibrium in copolymers of nonyl acrylate and acrylic acid - diglycidyl ether of bisphenol-A systems, *J. Appl. Polym. Sci.*, 74, 2353, 1999.

2000AZU Azuma, T., Tyagi, O.S., and Nose, T., Static and dynamic properties of block-copolymer solution in poor solvent I, *Polym. J.*, 32, 151, 2000.

2000BOK Bokias, G., Staikos, G., and Iliopoulos, I., Solution properties and phase behaviour of copolymers of acrylic acid with N-isopropylacrylamide, *Polymer*, 41, 7399, 2000.

2000KIM Kim, C., Lee, S.C., Kang, S.W., Kwon, I.C., and Jeong, S.Y., Phase-transition characteristics of amphiphilic poly(2-ethyl-2-oxazoline)/poly(ε-caprolactone) block copolymers in aqueous solutions *J. Polym. Sci., Part B: Polym. Phys.*, 38, 2400, 2000.

2000SAH Sahakaro, K., Chaibundit, C., Kaligradaki, Z., Mai, S.-M., Heatley, F., Booth, C., Padget, J.C., and Shirley, I.M., Clouding of aqueous solutions of difunctional tapered statistical copolymers of ethylene oxide and 1,2-butylene oxide, *Eur. Polym. J.*, 36, 1835, 2000.

2000SCH Schneider, A. and Wolf, B.A., Specific features of the interfacial tension in the case of phase separated solutions of random copolymers, *Polymer*, 41, 4089, 2000.

4. HIGH-PRESSURE PHASE EQUILIBRIUM (HPPE) DATA OF COPOLYMER SOLUTIONS IN SUPERCRITICAL FLUIDS

4.1. Experimental data of quasibinary copolymer solutions

Copolymers from butylene oxide and ethylene oxide

Average chemical composition of the copolymers, acronyms and references:

Copolymer (B)	Acronym	Ref.
poly(butylene oxide)-b-poly(ethylene oxide) diblock copolymer 44.4 mol% ethylene oxide	BO-EO/44x	98ONE

Characterization:

Copolymer (B)	M_n/ g/mol	Further information
BO-EO/44x	1620 (= 960 + 660)	SAM 185, hydroxyl terminated

Experimental HPPE data:

copolymer (B):	**BO-EO/44x**		**98ONE**
solvent (A):	**carbon dioxide**	**CO_2**	**124-38-9**

Type of data: cloud-points

w_B	0.001
T/K	308.15
P/MPa	27.58

Copolymers from ethylene and acrylic acid

Average chemical composition of the copolymers, acronyms and references:

Copolymer (B)	Acronym	Ref.
Ethylene/acrylic acid copolymer		
3.0 wt% acrylic acid	E-AA/03w	92WIN
6.0 wt% acrylic acid	E-AA/06w	92WIN
7.3 wt% acrylic acid	E-AA/07w	92WIN

Characterization:

Copolymer (B)	M_n/ g/mol	M_w/ g/mol	M_η/ g/mol	Further information
E-AA/03w	37500	183000		synthesized in the laboratory
E-AA/06w	30000	150000		synthesized in the laboratory
E-AA/07w	25000	126000		synthesized in the laboratory

Experimental HPPE data:

copolymer (B): **E-AA/03w** **92WIN**
solvent (A): **ethene** **C_2H_4** **74-85-1**

Type of data: cloud-points

w_B	0.019	0.019	0.019	0.019	0.019	0.082	0.082	0.082	0.082
T/K	413.15	433.15	453.15	473.15	493.15	413.15	433.15	453.15	473.15
P/MPa	196.8	178.4	163.1	152.2	141.7	186.7	166.6	153.1	143.3
w_B	0.082	0.128	0.128	0.128	0.128	0.128	0.186	0.186	0.186
T/K	493.15	413.15	433.15	453.15	473.15	493.15	413.15	433.15	453.15
P/MPa	135.4	181.2	162.8	150.1	140.3	132.2	172.1	154.3	142.5

continue

continue

w_B	0.186	0.186	0.231	0.231	0.231	0.231	0.231	0.295	0.295
T/K	473.15	493.15	413.15	433.15	453.15	473.15	493.15	413.15	433.15
P/MPa	134.0	128.5	165.6	148.9	137.7	129.8	124.2	153.1	139.8
w_B	0.295	0.295	0.295	0.318	0.318	0.318	0.318	0.318	0.338
T/K	453.15	473.15	493.15	413.15	433.15	453.15	473.15	493.15	413.15
P/MPa	130.2	123.1	118.0	150.4	136.6	127.9	121.3	115.8	146.0
w_B	0.338	0.338	0.338	0.338	0.428	0.428	0.428	0.428	0.428
T/K	433.15	453.15	473.15	493.15	413.15	433.15	453.15	473.15	493.15
P/MPa	134.1	125.3	118.9	113.4	127.5	119.4	113.4	108.7	105.1
w_B	0.510	0.510	0.510	0.510	0.510				
T/K	413.15	433.15	453.15	473.15	493.15				
P/MPa	109.8	105.5	102.0	99.0	96.0				

copolymer (B): E-AA/03w 92WIN
solvent (A): ethene C_2H_4 74-85-1

Type of data: coexistence data

T/K = 403.15

Total feed concentration of the copolymer in the homogeneous system: w_B = 0.155

Demixing pressure	w_A bottom phase	w_A top phase	Fractionation during demixing bottom phase			top phase		
P/ MPa			M_n/ g/mol	M_w/ g/mol	M_z/ g/mol	M_n/ g/mol	M_w/ g/mol	M_z/ g/mol
116.0	0.508	0.982				2818	8803	18920
161.0	0.711	0.951				9188	31060	73720
178.5	0.789	0.918	42880	176300	510700	13170	52620	156100
188.0	0.845	(cloud-point)	37500	183000	601000	(feed sample)		

copolymer (B): E-AA/03w 92WIN
solvent (A): ethene C_2H_4 74-85-1

Type of data: coexistence data

T/K = 433.15

continue

continue

The total feed concentration of the copolymer in the homogeneous system is: $w_B = 0.164$

Demixing pressure	w_A bottom phase	w_A top phase	Fractionation during demixing bottom phase			top phase		
P/ MPa			M_n/ g/mol	M_w/ g/mol	M_z/ g/mol	M_n/ g/mol	M_w/ g/mol	M_z/ g/mol
94.0	0.438	0.993				1520	6745	13920
113.0	0.536	0.982				2500	8300	17620
129.0	0.633	0.966				8200	29070	69600
147.7	0.766	0.939	41500	173700	472800	11600	47030	146200
157.5	0.836	(cloud-point)	37500	183000	601000	(feed sample)		

copolymer (B): **E-AA/06w** **92WIN**
solvent (A): **ethene** **C_2H_4** **74-85-1**

Type of data: cloud-points

w_B	0.038	0.038	0.038	0.038	0.057	0.057	0.057	0.057	0.076
T/K	433.15	453.15	473.15	493.15	433.15	453.15	473.15	493.15	433.15
P/MPa	203.4	176.8	157.3	142.0	195.7	170.5	152.3	138.9	191.2
w_B	0.076	0.076	0.076	0.116	0.116	0.116	0.116	0.150	0.150
T/K	453.15	473.15	493.15	433.15	453.15	473.15	493.15	433.15	453.15
P/MPa	167.2	148.9	134.0	189.5	164.2	146.2	132.5	183.9	161.1
w_B	0.150	0.150	0.215	0.215	0.215	0.215	0.260	0.260	0.260
T/K	473.15	493.15	433.15	453.15	473.15	493.15	433.15	453.15	473.15
P/MPa	145.2	130.5	172.4	152.2	135.4	125.4	165.3	146.3	132.6
w_B	0.260	0.295	0.295	0.295	0.295	0.335	0.335	0.335	0.335
T/K	493.15	433.15	453.15	473.15	493.15	433.15	453.15	473.15	493.15
P/MPa	121.8	158.3	142.0	129.8	119.0	147.5	135.9	125.6	115.5
w_B	0.383	0.383	0.383	0.383	0.430	0.430	0.430	0.430	0.505
T/K	433.15	453.15	473.15	493.15	433.15	453.15	473.15	493.15	433.15
P/MPa	137.5	126.9	118.5	112.2	130.1	120.2	112.0	106.3	114.4
w_B	0.505	0.505	0.505						
T/K	453.15	473.15	493.15						
P/MPa	108.0	103.1	99.2						

copolymer (B):	**E-AA/06w**		**92WIN**
solvent (A):	**ethene**	C_2H_4	**74-85-1**

Type of data: coexistence data

T/K = 433.15

The total feed concentration of the copolymer in the homogeneous system is: w_B = 0.1565

P/ MPa	w_A bottom phase	top phase
124.9	0.5550	0.9870
138.1	0.6130	0.9735
158.6	0.7130	0.9540
172.0	0.7855	0.9300
182.0	0.8435	(cloud-point)

copolymer (B):	**E-AA/07w**		**92WIN**
solvent (A):	**ethene**	C_2H_4	**74-85-1**

Type of data: cloud-points

w_B	0.043	0.043	0.043	0.102	0.102	0.102	0.102	0.160	0.160
T/K	453.15	473.15	493.15	443.15	453.15	473.15	493.15	443.15	453.15
P/MPa	192.4	165.6	143.9	196.2	181.4	157.7	140.2	186.0	171.4
w_B	0.160	0.160	0.203	0.203	0.203	0.203	0.272	0.272	0.272
T/K	473.15	493.15	443.15	453.15	473.15	493.15	443.15	453.15	473.15
P/MPa	150.8	135.3	178.2	165.3	146.3	132.5	166.0	156.4	140.6
w_B	0.272	0.330	0.330	0.330	0.330	0.410	0.410	0.410	0.410
T/K	493.15	443.15	453.15	473.15	493.15	443.15	453.15	473.15	493.15
P/MPa	126.8	155.5	147.0	134.0	122.5	141.0	134.3	123.6	114.9
w_B	0.500	0.500	0.500	0.500					
T/K	443.15	453.15	473.15	493.15					
P/MPa	121.4	117.0	110.8	104.8					

Copolymers from ethylene and butyl acrylate

Average chemical composition of the copolymers, acronyms and references:

Copolymer (B)	Acronym	Ref.
Ethylene/butyl acrylate copolymer		
15.0 wt% butyl acrylate	E-BA/15w	96MUE
19.0 wt% butyl acrylate	E-BA/19w	96MUE
35.0 wt% butyl acrylate	E-BA/35w	96MUE
67.0 wt% butyl acrylate	E-BA/67w	96MUE
75.0 wt% butyl acrylate	E-BA/75w	96MUE

Characterization:

Copolymer (B)	M_n/ g/mol	M_w/ g/mol	M_η/ g/mol	Further information
E-BA/15w	6980	34900		sample from BASF AG, Ger.
E-BA/19w	7740	38700		sample from BASF AG, Ger.
E-BA/35w	7960	39800		sample from BASF AG, Ger.
E-BA/67w	11500	37000		synthesized in the laboratory
E-BA/75w	13900	35200		synthesized in the laboratory

Experimental HPPE data:

copolymer (B): **E-BA/15w** **96MUE**
solvent (A): **ethene** **C_2H_4** **74-85-1**

Type of data: cloud-points

w_B	0.025	0.025	0.025	0.025	0.025	0.050	0.050	0.050	0.050
T/K	522.85	498.05	473.05	448.55	423.85	524.05	498.35	473.85	449.15
P/MPa	117.7	121.5	126.3	132.3	139.6	116.9	120.8	125.5	131.3

continue

continue

w_B	0.050	0.100	0.100	0.100	0.100	0.100	0.150	0.150	0.150
T/K	423.75	522.85	498.05	473.15	448.95	424.15	524.15	498.25	473.55
P/MPa	138.8	121.2	124.1	127.5	132.1	139.4	120.8	122.5	125.1
w_B	0.150	0.150	0.200	0.200	0.200	0.200	0.200		
T/K	448.95	424.35	523.35	498.05	473.35	449.15	423.35		
P/MPa	128.5	133.6	117.3	118.8	121.5	124.8	128.7		

copolymer (B): **E-BA/19w** **96MUE**
solvent (A): **ethene** **C_2H_4** **74-85-1**

Type of data: cloud-points

w_B	0.034	0.034	0.034	0.034	0.034	0.050	0.050	0.050	0.050
T/K	523.05	498.55	472.65	448.15	423.15	523.95	497.85	473.05	449.25
P/MPa	116.8	120.6	125.4	131.6	138.4	118.4	121.3	125.3	130.8
w_B	0.050	0.100	0.100	0.100	0.100	0.100	0.150	0.150	0.150
T/K	423.95	523.55	498.85	473.45	449.05	424.25	524.85	498.15	472.65
P/MPa	139.0	121.7	123.6	128.2	131.6	139.4	120.3	122.1	128.4
w_B	0.150	0.150	0.199	0.199	0.199	0.199			
T/K	448.95	424.25	523.25	498.55	473.35	448.95			
P/MPa	130.6	134.7	120.7	123.0	126.1	129.8			

copolymer (B): **E-BA/35w** **96MUE**
solvent (A): **ethene** **C_2H_4** **74-85-1**

Type of data: cloud-points

w_B	0.025	0.025	0.025	0.025	0.025	0.050	0.050	0.050	0.050
T/K	523.65	498.55	473.05	448.55	423.45	523.75	498.05	472.95	448.05
P/MPa	111.5	115.9	120.4	127.2	134.4	109.4	113.1	117.2	122.3
w_B	0.050	0.100	0.100	0.100	0.100	0.100	0.150	0.150	0.150
T/K	423.25	523.45	498.05	473.15	448.45	423.45	523.85	498.35	472.95
P/MPa	129.7	107.1	109.8	113.1	116.9	121.9	105.9	108.1	111.5
w_B	0.150	0.150	0.199	0.199	0.199	0.199	0.199		
T/K	448.05	423.85	523.75	498.05	473.25	449.15	424.05		
P/MPa	115.1	119.8	104.9	106.7	109.5	115.4	120.5		

copolymer (B): **E-BA/67w** **96MUE**
solvent (A): **ethene** **C_2H_4** **74-85-1**

Type of data: cloud-points

w_B	0.168	0.168	0.168	0.168	0.168
T/K	522.35	498.15	473.65	448.35	424.65
P/MPa	96.6	98.4	100.1	102.3	104.0

copolymer (B):	**E-BA/75w**		**96MUE**
solvent (A):	**ethene**	C_2H_4	**74-85-1**

Type of data: cloud-points

w_B	0.069	0.069	0.069	0.069	0.069	0.069
T/K	524.05	499.05	474.45	448.65	423.75	398.75
P/MPa	89.4	90.3	92.9	96.5	99.8	102.1

Copolymers from ethylene and 1-hexene

Average chemical composition of the copolymers, acronyms and references:

Copolymer (B)	Acronym	Ref.
Ethylene/1-hexene copolymer		
10.6 wt% 1-hexene	E-H/11w	2000CH3
16.1 wt% 1-hexene	E-H/16w	99KIN
20.6 wt% 1-hexene	E-H/21w	2000CH2
35.0 wt% 1-hexene	E-H/35w(1)	2000CH3
35.0 wt% 1-hexene	E-H/35w(2)	2000CH3

Characterization:

Copolymer (B)	M_n/ g/mol	M_w/ g/mol	M_η/ g/mol	Further information
E-H/11w	52600	80000		
E-H/16w	60000	129000		
E-H/21w	51300	126000		
E-H/35w(1)	48100	103000		
E-H/35w(2)	112100	139000		

Experimental HPPE data:

copolymer (B): **E-H/11w** **2000CH3**
solvent (A): **ethene** C_2H_4 **74-85-1**

Type of data: cloud-points

w_B	0.148	0.148	0.148	0.148
T/K	393.25	413.15	433.25	453.15
P/MPa	168.0	154.5	146.0	138.9

copolymer (B): **E-H/16w** **99KIN**
solvent (A): **ethene** C_2H_4 **74-85-1**

Type of data: cloud-points

w_B	0.150	0.150	0.150	0.150	0.150	0.150
T/K	393.15	413.15	433.15	453.15	473.15	493.15
P/MPa	156.0	144.3	137.3	129.1	123.1	118.6

copolymer (B): **E-H/35w(1)** **2000CH3**
solvent (A): **ethene** C_2H_4 **74-85-1**

Type of data: cloud-points

w_B	0.050	0.050	0.050	0.050	0.050	0.050	0.050	0.100	0.100
T/K	393.15	403.15	413.15	423.15	433.15	443.15	453.15	393.15	403.15
P/MPa	136.3	132.6	129.6	127.0	124.9	122.9	121.0	134.8	131.4
w_B	0.100	0.100	0.100	0.100	0.100	0.150	0.150	0.150	0.150
T/K	413.15	423.15	433.15	443.15	453.15	393.15	403.15	413.15	423.15
P/MPa	128.3	125.8	123.7	121.8	120.8	129.4	126.1	123.6	121.5
w_B	0.150	0.150	0.150						
T/K	433.15	443.15	453.15						
P/MPa	119.8	118.7	119.4						

copolymer (B): **E-H/11w** **2000CH3**
solvent (A): **propane** C_3H_8 **74-98-6**

Type of data: cloud-points

w_B	0.010	0.010	0.010	0.010	0.051	0.051	0.051	0.051	0.148
T/K	363.15	373.25	413.15	453.15	363.15	373.15	413.15	453.15	373.25
P/MPa	50.3	50.0	49.9	50.7	51.0	50.4	50.4	50.4	50.6
w_B	0.148	0.148							
T/K	413.15	453.15							
P/MPa	50.4	51.1							

copolymer (B): **E-H/21w** **2000CH2**
solvent (A): **propane** **C_3H_8** **74-98-6**

Type of data: cloud-points

T/K = 453.15

w_B	0.018	0.031	0.055	0.066	0.075	0.079	0.089	0.098	0.100
P/bar	446.0	454.0	456.0	459.0	468.0	475.0	478.0	470.0	474.0
w_B	0.108	0.114	0.148						
P/bar	471.0	466.0	459.0						

copolymer (B): **E-H/35w(1)** **2000CH3**
solvent (A): **propane** **C_3H_8** **74-98-6**

Type of data: cloud-points

w_B	0.150	0.150	0.150	0.150	0.150	0.150
T/K	313.15	343.15	363.05	373.05	413.15	453.15
P/MPa	37.3	38.1	39.1	39.7	42.3	44.6

copolymer (B): **E-H/35w(2)** **2000CH3**
solvent (A): **propane** **C_3H_8** **74-98-6**

Type of data: cloud-points

w_B	0.009	0.009	0.009	0.009	0.009	0.009	0.152	0.152	0.152
T/K	313.25	353.25	368.25	373.15	413.15	453.15	353.25	373.15	413.15
P/MPa	39.0	36.8	37.6	37.9	40.6	42.9	37.8	39.2	41.5
w_B	0.152								
T/K	453.15								
P/MPa	43.7								

Copolymers from ethylene and methyl acrylate

Average chemical composition of the copolymers, acronyms and references:

Copolymer (B)	Acronym	Ref.
Ethylene/methyl acrylate copolymer		
25.0 wt% methyl acrylate	E-MA/25w	96MUE
58.0 wt% methyl acrylate	E-MA/58w	96MUE
68.0 wt% methyl acrylate	E-MA/68w	96MUE

Characterization:

Copolymer (B)	M_n/ g/mol	M_w/ g/mol	M_η/ g/mol	Further information
E-MA/25w	17000	75400		synthesized in the laboratory
E-MA/58w	33000	108900		synthesized in the laboratory
E-MA/68w	42000	110000		synthesized in the laboratory

Experimental HPPE data:

copolymer (B): E-MA/25w 96MUE
solvent (A): ethene C_2H_4 74-85-1

Type of data: cloud-points

w_B	0.010	0.010	0.010	0.010	0.010	0.050	0.050	0.050	0.050
T/K	522.65	498.15	473.85	448.55	424.05	523.95	498.85	472.75	449.15
P/MPa	113.7	119.8	123.6	128.8	133.0	115.9	119.4	123.2	127.5
w_B	0.050	0.050	0.100	0.100	0.100	0.100	0.100	0.113	0.113
T/K	423.95	404.15	525.05	498.35	474.35	448.65	424.95	523.15	498.35
P/MPa	134.2	140.0	117.3	117.4	120.2	124.6	129.2	117.2	119.0
w_B	0.113	0.113	0.113	0.149	0.149	0.149	0.149	0.149	0.207
T/K	473.55	448.85	423.45	522.85	498.55	473.25	448.45	423.05	523.95
P/MPa	122.7	125.5	131.2	116.7	118.8	121.2	125.0	130.2	109.4
w_B	0.207	0.207	0.207	0.207					
T/K	497.85	473.05	448.85	424.15					
P/MPa	112.2	115.5	119.5	124.1					

copolymer (B): E-MA/58w 96MUE
solvent (A): ethene C_2H_4 74-85-1

Type of data: cloud-points

w_B	0.010	0.010	0.010	0.010	0.010	0.050	0.050	0.050	0.050
T/K	523.75	498.35	473.75	448.05	423.55	522.95	498.65	473.95	449.05
P/MPa	125.8	133.7	140.6	149.7	161.3	124.9	130.8	138.2	148.5
w_B	0.050	0.050	0.101	0.101	0.101	0.101	0.101	0.150	0.150
T/K	423.95	404.05	524.65	499.15	473.25	448.45	424.05	524.05	498.95
P/MPa	163.1	178.8	123.0	128.4	135.6	145.0	162.0	122.7	127.8
w_B	0.150	0.150	0.150	0.200	0.200	0.200	0.200		
T/K	473.45	448.95	423.55	523.75	498.45	473.85	448.45		
P/MPa	132.3	140.6	150.6	117.3	122.1	128.2	136.2		

copolymer (B):	**E-MA/68w**		**96MUE**
solvent (A):	**ethene**	C_2H_4	**74-85-1**

Type of data: cloud-points

w_B	0.011	0.011	0.011	0.011	0.011	0.050	0.050	0.050	0.050
T/K	523.85	498.95	473.55	448.45	423.75	523.25	498.35	473.15	448.85
P/MPa	136.5	145.5	157.8	165.2	206.4	136.1	145.2	157.7	174.4
w_B	0.050	0.050	0.101	0.101	0.101	0.101	0.101	0.150	0.150
T/K	423.85	403.65	523.65	498.25	473.25	448.25	423.15	523.35	498.15
P/MPa	201.3	234.3	133.6	142.0	153.0	167.4	189.3	133.9	141.0
w_B	0.150	0.150	0.150	0.192	0.192	0.192	0.192	0.192	
T/K	473.05	448.95	424.35	523.45	497.95	472.65	448.85	423.45	
P/MPa	151.8	164.9	184.1	132.0	139.9	150.0	163.2	181.9	

Copolymers from ethylene and 1-octene

Average chemical composition of the copolymers, acronyms and references:

Copolymer (B)	Acronym	Ref.
Ethylene/1-octene copolymer		
13.9 wt% 1-octene	E-O/14w	2000CH1
25.8 wt% 1-octene	E-O/26w	2000CH1
38.7 wt% 1-octene	E-O/39w	2000CH1

Characterization:

Copolymer (B)	M_n/ g/mol	M_w/ g/mol	M_η/ g/mol	Further information
E-O/14w	51200	83000		4 hexyl branches/100 ethyl units
E-O/26w	54800	115000		8 hexyl branches/100 ethyl units
E-O/39w	83300	120000		13.6 hexyl branches/100 ethyl units

Experimental HPPE data:

copolymer (B): **E-O/14w** **2000CH1**
solvent (A): **propane** C_3H_8 **74-98-6**

Type of data: cloud-points

w_B	0.050	0.050	0.050	0.050	0.050	0.050
T/K	453.15	413.15	393.15	373.15	368.15	363.15
P/MPa	53.4	52.8	52.8	53.2	53.5	53.7

copolymer (B): **E-O/26w** **2000CH1**
solvent (A): **propane** C_3H_8 **74-98-6**

Type of data: cloud-points

w_B	0.049	0.049	0.049
T/K	453.15	413.15	373.15
P/MPa	48.5	47.1	46.3

copolymer (B): **E-O/39w** **2000CH1**
solvent (A): **propane** C_3H_8 **74-98-6**

Type of data: cloud-points

w_B	0.010	0.010	0.010	0.010	0.010	0.049	0.049	0.049	0.049
T/K	453.15	413.15	373.15	343.15	333.15	453.15	413.15	373.15	343.15
P/MPa	38.6	36.0	33.2	34.8	37.8	40.8	38.4	35.6	33.7

w_B	0.049	0.100	0.100	0.100
T/K	333.15	453.15	413.15	373.15
P/MPa	33.6	40.4	37.9	35.5

Copolymers from ethylene and propylene

Average chemical composition of the copolymers, acronyms and references:

Copolymer (B)	Acronym	Ref.
Ethylene/propylene copolymer		
50.0 mol% propene	E-P/50x(1)	92CH1
50.0 mol% propene	E-P/50x(1)	93GRE
50.0 mol% propene	E-P/50x(1)	94GR1
50.0 mol% propene	E-P/50x(2)	92CH1
50.0 mol% propene	E-P/50x(2)	93GRE
50.0 mol% propene	E-P/50x(2)	94GR1
50.0 mol% propene	E-P/50x(2)	94GR2
50.0 mol% propene	E-P/50x(3)	92CH1
50.0 mol% propene	E-P/50x(3)	92CH2
50.0 mol% propene	E-P/50x(4)	92CH1
50.0 mol% propene	E-P/50x(4)	93GRE
50.0 mol% propene	E-P/50x(4)	94GR1
Ethylene/propylene/isoprene terpolymer		
28.2 mol% propene, 43.6 mol% isoprene	E-P-IP/28x	96ALB
36.7 mol% propene, 26.6 mol% isoprene	E-P-IP/37x	96ALB

Characterization:

Copolymer (B)	M_n/ g/mol	M_w/ g/mol	M_η/ g/mol	Further information
E-P/50x(1)	782	790		alternating ethylene/propylene units from complete hydrogenation of polyisoprene
E-P/50x(2)	5460	5900		alternating ethylene/propylene units from complete hydrogenation of polyisoprene
E-P/50x(3)	25200	26000		alternating ethylene/propylene units from complete hydrogenation of polyisoprene
E-P/50x(4)	87700	96400		alternating ethylene/propylene units from complete hydrogenation of polyisoprene
E-P-IP/28x	7650	13620		alternating ethylene/propylene units from partial hydrogenation of polyisoprene
E-P-IP/37x	7690	13690		alternating ethylene/propylene units from partial hydrogenation of polyisoprene

Experimental HPPE data:

copolymer (B): **E-P/50x(1)** **92CH1**
solvent (A): **1-butene** **C_4H_8** **106-98-9**

Type of data: cloud-points

$w_B = 0.151$

T/K	355.05	373.15	398.15	421.65	424.05	426.35	428.15	432.95	444.45
P/bar	11.5	16.4	27.1	39.8	42.5	44.8	46.5	51.7	63.4
T/K	473.15								
P/bar	90.8								

copolymer (B): **E-P/50x(2)** **92CH1**
solvent (A): **1-butene** **C_4H_8** **106-98-9**

Type of data: cloud-points

$w_B = 0.157$

T/K	376.15	378.85	382.65	397.95	423.15	447.95	473.15
P/bar	21.3	27.2	35.5	66.7	110.0	146.0	172.5

Comments: a lower critical endpoint is found at 373.15 K

copolymer (B): **E-P/50x(3)** **92CH1**
solvent (A): **1-butene** **C_4H_8** **106-98-9**

Type of data: cloud-points

$w_B = 0.160$

T/K	346.35	348.35	350.35	355.15	361.35	361.85	367.45	373.15	398.15
P/bar	12.3	15.2	20.5	29.0	40.2	38.2	50.3	61.2	108.0
T/K	416.35	423.15	448.35	473.35					
P/bar	137.3	149.0	183.0	213.4					

Comments: a lower critical endpoint is found at 345.15 K

copolymer (B):	**E-P/50x(4)**		**92CH1**
solvent (A):	**1-butene**	C_4H_8	**106-98-9**

Type of data: cloud-points

$w_B = 0.152$

T/K	334.45	336.05	339.55	343.75	355.15	373.15	398.15	423.75	448.15
P/bar	8.4	13.1	19.5	28.4	54.0	90.8	137.3	177.3	210.1

T/K	473.15
P/bar	240.5

Comments: a lower critical endpoint is found at 333.15 K

copolymer (B):	**E-P/50x(1)**		**93GRE, 94GR1**
solvent (A):	**ethene**	C_2H_4	**74-85-1**

Type of data: cloud-points

w_B	0.156	0.156	0.156	0.152	0.152
T/K	293.15	323.15	363.15	423.15	475.15
P/bar	571.5	555.0	550.0	600.9	689.6

copolymer (B):	**E-P/50x(2)**		**93GRE, 94GR1**
solvent (A):	**ethene**	C_2H_4	**74-85-1**

Type of data: cloud-points

w_B	0.156	0.156	0.156	0.158	0.158	0.157	0.158	0.158	0.158
T/K	308.15	323.15	325.65	348.15	373.15	398.15	448.15	468.15	473.15
P/bar	1080.1	1018.6	1001.1	941.9	897.9	899.9	899.1	892.8	900.0

copolymer (B):	**E-P/50x(4)**		**93GRE, 94GR1**
solvent (A):	**ethene**	C_2H_4	**74-85-1**

Type of data: cloud-points

w_B	0.038	0.038	0.038	0.038	0.038
T/K	353.15	373.15	398.15	423.15	473.15
P/bar	1722.5	1569.0	1389.0	1235.5	1132.7

copolymer (B):	**E-P/50x(3)**		**92CH1, 92CH2**
solvent (A):	**1-hexene**	C_6H_{12}	**592-41-6**

Type of data: cloud-points

continue

continue

w_B = 0.160

T/K	427.15	447.95	467.85	473.15	477.95	478.65	473.15	477.95	478.55
P/bar	7.2	12.7	17.2	23.5	29.4	31.6	19.3	20.0	20.4
	(VLE	VLE	VLE	LLE	LLE	LLE	VLLE	VLLE	VLLE)

Comments: a lower critical endpoint is found at 472.15 K

copolymer (B): **E-P/50x(4)** **92CH1**
solvent (A): **1-hexene** C_6H_{12} **592-41-6**

Type of data: cloud-points

w_B = 0.150

T/K	416.95	453.15	459.35	454.95	474.45	463.95
P/bar	8.3	14.0	18.3	28.5	43.3	17.3
	(VLE	VLE	LLE	LLE	LLE	VLLE)

Comments: a lower critical endpoint is found at 457.15 K

copolymer (B): **E-P/50x(1)** **92CH1**
solvent (A): **propene** C_3H_6 **115-07-1**

Type of data: cloud-points

w_B = 0.172

T/K	319.45	343.35	350.55	352.15	355.05	373.35	397.95	423.15	447.95
P/bar	19.0	30.2	34.4	37.7	43.0	73.0	113.0	144.0	170.0
	(VLE	VLE	VLE	LLE	LLE	LLE	LLE	LLE	LLE)
T/K	473.35	352.15	353.75	355.05					
P/bar	180.0	35.9	36.9	38.3					
	(LLE	VLLE	VLLE	VLLE)					

Comments: a lower critical endpoint is found at 351.1 K

copolymer (B): **E-P/50x(2)** **92CH1**
solvent (A): **propene** C_3H_6 **115-07-1**

Type of data: cloud-points

w_B = 0.155

T/K	272.65	298.15	325.15	356.15	373.15	398.95	423.35	447.95	472.75
P/bar	5.5	64.0	126.0	190.5	224.4	267.0	296.0	325.0	344.0

Comments: a lower critical endpoint is found at 272.6 K

copolymer (B): **E-P/50x(2)** **93GRE, 94GR1, 94GR2**
solvent (A): **propene** **C_3H_6** **115-07-1**

Type of data: cloud-points

w_B	0.018	0.073	0.150	0.216	0.281	0.392	0.018	0.073	0.150
T/K	373.15	373.75	373.15	373.35	373.45	373.15	423.15	423.15	423.15
P/bar	179.5	202.8	212.5	209.8	200.2	189.1	248.6	273.7	295.0
w_B	0.216	0.329							
T/K	423.15	423.15							
P/bar	293.0	267.1							

copolymer (B): **E-P/50x(3)** **92CH1**
solvent (A): **propene** **C_3H_6** **115-07-1**

Type of data: cloud-points

$w_B = 0.151$

T/K	323.55	355.55	373.55	398.05	423.55	448.15
P/bar	273.0	312.0	332.0	361.2	388.0	412.0

copolymer (B): **E-P/50x(4)** **92CH1**
solvent (A): **propene** **C_3H_6** **115-07-1**

Type of data: cloud-points

$w_B = 0.158$

T/K	264.15	265.85	271.15	274.15	277.95	280.35	285.25	287.05	293.55
P/bar	354.0	347.0	337.0	331.0	325.5	322.5	316.7	315.0	311.0
T/K	303.05	317.15	322.55	331.15	354.65	373.05	397.75	423.35	447.95
P/bar	310.6	319.5	322.3	330.0	352.0	371.0	396.4	423.0	442.3
T/K	473.55								
P/bar	461.0								

copolymer (B): **E-P/50x(4)** **93GRE, 94GR1, 94GR2**
solvent (A): **propene** **C_3H_6** **115-07-1**

Type of data: cloud-points

w_B	0.008	0.045	0.088	0.158	0.231	0.008	0.045	0.088	0.158
T/K	373.15	373.15	373.15	373.15	373.15	423.15	423.15	423.15	423.15
P/bar	393.1	363.0	371.3	371.2	343.0	402.1	413.8	423.0	423.1
w_B	0.231								
T/K	423.15								
P/bar	393.0								

copolymer (B): **E-P-IP/28x** **96ALB**
solvent (A): **dimethyl ether** C_2H_6O **115-10-6**

Type of data: cloud-points

w_B = 0.0536

T/K	424.75	383.25	347.55	329.95
P/bar	216.6	133.0	46.2	12.2

Comments: a lower critical endpoint is found at 329.95 K

copolymer (B): **E-P-IP/37x** **96ALB**
solvent (A): **dimethyl ether** C_2H_6O **115-10-6**

Type of data: cloud-points

w_B = 0.0524

T/K	423.45	411.95	410.65	384.35	362.35	348.65	328.15
P/bar	208.9	186.1	184.4	132.9	85.5	53.5	11.9

Comments: a lower critical endpoint is found at 328.15 K

copolymer (B): **E-P-IP/37x** **96ALB**
solvent (A): **ethane** C_2H_6 **74-84-0**

Type of data: cloud-points

w_B = 0.0722

T/K	423.65	414.15	393.75	392.85	383.05	382.25	373.75	364.85	362.25
P/bar	818.1	824.2	869.4	874.3	918.7	923.0	964.9	1005.4	1038.7

copolymer (B): **E-P-IP/37x** **96ALB**
solvent (A): **ethene** C_2H_4 **74-85-1**

Type of data: cloud-points

w_B = 0.0843

T/K	423.75	423.15	421.65	409.15	394.35
P/bar	1249.1	1250.5	1256.2	1293.2	1352.9

copolymer (B): **E-P-IP/37x** **96ALB**
solvent (A): **propane** **C_3H_8** **74-98-6**

Type of data: cloud-points

w_B = 0.0528

T/K	422.75	422.05	390.05	389.85	388.75	388.45	363.95	347.75	333.65
P/bar	377.8	378.3	370.2	370.2	366.6	369.5	398.0	423.0	453.0
T/K	327.25								
P/bar	472.3								

copolymer (B): **E-P-IP/37x** **96ALB**
solvent (A): **propene** **C_3H_6** **115-07-1**

Type of data: cloud-points

w_B = 0.0478

T/K	424.25	390.45	363.65	339.05	335.85	310.95
P/bar	357.5	320.7	301.1	304.5	305.0	337.5

Copolymers from ethylene and vinyl acetate

Average chemical composition of the copolymers, acronyms and references:

Copolymer (B)	Acronym	Ref.
Ethylene/vinyl acetate copolymer		
7.5 wt% vinyl acetate	E-VA/08w	82RAE
7.5 wt% vinyl acetate	E-VA/08w	84WAG
10.9 wt% vinyl acetate	E-VA/11w	91NIE
12.7 wt% vinyl acetate	E-VA/13w	82RAE
12.7 wt% vinyl acetate	E-VA/13w	84WAG
17.5 wt% vinyl acetate	E-VA/18w	91NIE
26.0 wt% vinyl acetate	E-VA/26w	91NIE
27.3 wt% vinyl acetate	E-VA/27w	82RAE
27.3 wt% vinyl acetate	E-VA/27w	84WAG
27.5 wt% vinyl acetate	E-VA/28w(1)	91NIE
27.5 wt% vinyl acetate	E-VA/28w(2)	99KIN
29.0 wt% vinyl acetate	E-VA/29w	91FIN

continue

continue

Copolymer (B)	Acronym	Ref.
31.4 wt% vinyl acetate	E-VA/31w	83RAE
31.4 wt% vinyl acetate	E-VA/31w	84WAG
31.8 wt% vinyl acetate	E-VA/32w(1)	82RAE
31.8 wt% vinyl acetate	E-VA/32w(1)	84WAG
32.1 wt% vinyl acetate	E-VA/32w(2)	84WAG
32.3 wt% vinyl acetate	E-VA/32w(3)	83RAE
32.3 wt% vinyl acetate	E-VA/32w(3)	84WAG
32.5 wt% vinyl acetate	E-VA/33w(1)	91NIE
33.0 wt% vinyl acetate	E-VA/33w(2)	83RAE
33.0 wt% vinyl acetate	E-VA/33w(2)	84WAG
33.2 wt% vinyl acetate	E-VA/33w(3)	83RAE
33.2 wt% vinyl acetate	E-VA/33w(3)	84WAG
33.4 wt% vinyl acetate	E-VA/33w(4)	84WAG
34.2 wt% vinyl acetate	E-VA/34w	91FIN
42.7 wt% vinyl acetate	E-VA/43w	82RAE
42.7 wt% vinyl acetate	E-VA/43w	84WAG
58.0 wt% vinyl acetate	E-VA/58w	91NIE

Characterization:

Copolymer (B)	M_n/ g/mol	M_w/ g/mol	M_η/ g/mol	Further information
E-VA/08w	20400	145000		Leuna-Werke, Germany
E-VA/11w	29300	161600		Exxon Chemical Company
E-VA/13w	19700	120000		Leuna-Werke, Germany
E-VA/18w	33400	137700		Exxon Chemical Company
E-VA/26w	33600	231000		Exxon Chemical Company
E-VA/27w	13700	60800		Leuna-Werke, Germany
E-VA/28w(1)	35000	126500		Exxon Chemical Company
E-VA/28w(2)	61900	167000		
E-VA/29w	2180	5680		Leuna-Werke, Germany
E-VA/31w	27500	60300		Leuna-Werke, Germany
E-VA/32w(1)	16200	45800		Leuna-Werke, Germany
E-VA/32w(2)	21800	46400		Leuna-Werke, Germany
E-VA/32w(3)	15500	40800		Leuna-Werke, Germany
E-VA/33w(1)	13000	22100		Exxon Chemical Company
E-VA/33w(2)	5430	8220		Leuna-Werke, Germany
E-VA/33w(3)	15000	28300		Leuna-Werke, Germany
E-VA/33w(4)	9590	16900		Leuna-Werke, Germany
E-VA/34w	13100	41300		Leuna-Werke, Germany
E-VA/43w	19800	44000		Leuna-Werke, Germany
E-VA/58w	35000			Exxon Chemical Company

Experimental HPPE data:

copolymer (B): **E-VA/08w** **82RAE**
solvent (A): **ethene** C_2H_4 **74-85-1**

Type of data: cloud-points

T/K = 433.15

w_B	0.05	0.10	0.15	0.20	0.30
P/MPa	140.3	139.3	137.3	131.2	121.0

copolymer (B): **E-VA/08w** **84WAG**
solvent (A): **ethene** C_2H_4 **74-85-1**

Type of data: cloud-points

w_B	0.05	0.10	0.15	0.20	0.30	0.05	0.10	0.15	0.20
T/K	393.15	393.15	393.15	393.15	393.15	413.15	413.15	413.15	413.15
P/MPa	155.0	154.5	149.5	144.9	133.2	146.4	144.9	141.9	136.8
w_B	0.30	0.05	0.10	0.15	0.20	0.30	0.05	0.10	0.15
T/K	413.15	433.15	433.15	433.15	433.15	433.15	453.15	453.15	453.15
P/MPa	126.1	140.7	139.3	137.9	131.2	121.0	135.3	134.8	133.2
w_B	0.20	0.30	0.05	0.10	0.15	0.20	0.30		
T/K	453.15	453.15	473.15	473.15	473.15	473.15	473.15		
P/MPa	127.2	116.0	130.7	129.7	126.7	122.6	113.0		

copolymer (B): **E-VA/11w** **91NIE**
solvent (A): **ethene** C_2H_4 **74-85-1**

Type of data: coexistence data

T/K = 403.15

The total feed concentration of the copolymer in the homogeneous system is: w_B = 0.151

P/ MPa	w_A bottom phase	top phase
88.6	0.449	0.991
108.8	0.567	0.983
125.5	0.674	0.968
136.2	0.768	0.949
141.0	0.849	(cloud point)

copolymer (B): **E-VA/11w** **91NIE**
solvent (A): **ethene** C_2H_4 **74-85-1**

Type of data: coexistence data

T/K = 433.15

The total feed concentration of the copolymer in the homogeneous system is: w_B = 0.155

P/ MPa	w_A bottom phase	top phase
94.8	0.501	0.992
104.9	0.568	0.985
115.2	0.652	0.975
128.2	0.782	0.952
131.0	0.845	(cloud point)

copolymer (B): **E-VA/13w** **82RAE**
solvent (A): **ethene** C_2H_4 **74-85-1**

Type of data: cloud-points

T/K = 433.15

w_B	0.05	0.10	0.15	0.20	0.30
P/MPa	132.4	128.5	125.0	121.6	114.7

copolymer (B): **E-VA/13w** **84WAG**
solvent (A): **ethene** C_2H_4 **74-85-1**

Type of data: cloud-points

w_B	0.05	0.075	0.10	0.15	0.20	0.30	0.05	0.075	0.10
T/K	393.15	393.15	393.15	393.15	393.15	393.15	413.15	413.15	413.15
P/MPa	146.6	145.1	143.3	138.9	133.9	125.5	137.8	137.3	134.8
w_B	0.15	0.20	0.30	0.05	0.075	0.10	0.15	0.20	0.30
T/K	413.15	413.15	413.15	433.15	433.15	433.15	433.15	433.15	433.15
P/MPa	130.4	127.0	119.6	132.4	130.4	128.5	125.0	121.6	114.7
w_B	0.05	0.075	0.10	0.15	0.20	0.30	0.05	0.075	0.10
T/K	453.15	453.15	453.15	453.15	453.15	453.15	473.15	473.15	473.15
P/MPa	127.5	125.5	124.0	120.6	117.7	109.8	122.6	121.1	119.6
w_B	0.15	0.20	0.30						
T/K	473.15	473.15	473.15						
P/MPa	116.7	113.8	105.5						

copolymer (B): **E-VA/18w** **91NIE**
solvent (A): **ethene** C_2H_4 **74-85-1**

Type of data: cloud-points

w_B	0.030	0.030	0.030	0.081	0.081	0.081	0.081	0.211	0.211
T/K	373.65	403.15	433.15	363.15	393.15	423.15	453.15	388.15	403.15
P/MPa	158.6	143.5	133.0	156.2	141.4	131.8	124.5	136.5	130.5
w_B	0.211	0.211	0.344	0.344	0.344	0.344	0.462	0.462	0.462
T/K	433.15	448.15	398.65	403.15	433.15	453.15	373.15	433.15	463.15
P/MPa	122.8	119.6	125.5	117.5	111.4	108.2	104.2	95.0	93.5
w_B	0.585	0.585	0.585						
T/K	393.15	433.15	453.15						
P/MPa	78.2	77.2	76.8						

copolymer (B): **E-VA/18w** **91NIE**
solvent (A): **ethene** C_2H_4 **74-85-1**

Type of data: coexistence data

T/K = 403.15

The total feed concentration of the copolymer in the homogeneous system is: w_B = 0.145

Demixing pressure	w_A bottom phase	w_A top phase	Fractionation during demixing bottom phase		top phase		
P/MPa			M_n/g/mol	M_w/g/mol	M_n/g/mol	M_w/g/mol	M_z/g/mol
48.7	0.276	0.9992					
100.1	0.543	0.989			7833	15338	30553
118.7	0.680	0.980			11676	23830	42441
134.1	0.855	(cloud point)	33400	137700	(feed sample)		

copolymer (B): **E-VA/18w** **91NIE**
solvent (A): **ethene** C_2H_4 **74-85-1**

Type of data: coexistence data

T/K = 433.15

continue

continue

The total feed concentration of the copolymer in the homogeneous system is: $w_B = 0.133$

P/ MPa	w_A bottom phase	top phase
43.1	0.280	
47.8		0.999
75.5	0.410	
77.5		0.995
89.7	0.497	0.993
103.5	0.584	0.984
107.9	0.623	0.985
113.8	0.679	0.972
121.0	0.757	0.948
123.8	0.792	0.940
126.1	0.867	(cloud point)

copolymer (B): **E-VA/18w** **91NIE**
solvent (A): **ethene** C_2H_4 **74-85-1**

Type of data: coexistence data

T/K = 463.15

The total feed concentration of the copolymer in the homogeneous system is: $w_B = 0.138$

P/ MPa	w_A bottom phase	top phase
48.1	0.252	0.9994
97.4	0.542	0.9905
108.7	0.650	0.9792
122.4	0.862	(cloud point)

copolymer (B): **E-VA/26w** **91NIE**
solvent (A): **ethene** C_2H_4 **74-85-1**

Type of data: cloud-points

w_B	0.040	0.040	0.040	0.067	0.067	0.101	0.101	0.101	0.157
T/K	373.15	403.15	463.15	393.15	433.15	393.15	433.15	453.15	373.15
P/MPa	146.6	139.8	125.5	139.7	129.0	138.5	128.0	123.8	143.5

continue

continue

w_B	0.157	0.157	0.157	0.262	0.262	0.262	0.262	0.360	0.360
T/K	393.15	433.15	453.15	393.15	413.15	433.15	453.15	403.15	418.15
P/MPa	136.5	126.2	122.5	130.2	125.4	121.2	117.0	115.5	113.0
w_B	0.360	0.408	0.408	0.408	0.408	0.480	0.480	0.480	0.480
T/K	453.15	393.15	413.15	433.15	453.15	373.15	393.15	433.15	449.15
P/MPa	108.2	110.6	107.5	105.2	102.8	99.2	96.5	94.0	93.8

copolymer (B): **E-VA/27w** **82RAE**
solvent (A): **ethene** **C_2H_4** **74-85-1**

Type of data: cloud-points

T/K = 433.15

w_B	0.05	0.10	0.15	0.20	0.30
P/MPa	121.0	120.0	118.0	115.5	109.0

copolymer (B): **E-VA/27w** **84WAG**
solvent (A): **ethene** **C_2H_4** **74-85-1**

Type of data: cloud-points

w_B	0.05	0.10	0.15	0.20	0.30	0.05	0.10	0.15	0.20
T/K	393.15	393.15	393.15	393.15	393.15	413.15	413.15	413.15	413.15
P/MPa	128.5	127.5	125.5	123.0	116.5	125.5	124.0	122.0	119.0
w_B	0.30	0.05	0.10	0.15	0.20	0.30	0.05	0.10	0.15
T/K	413.15	433.15	433.15	433.15	433.15	433.15	453.15	453.15	453.15
P/MPa	112.0	121.0	120.0	118.0	115.5	109.0	118.5	117.0	115.0
w_B	0.20	0.30	0.05	0.10	0.15	0.20	0.30		
T/K	453.15	453.15	473.15	473.15	473.15	473.15	473.15		
P/MPa	112.0	105.0	115.5	113.5	111.5	109.0	102.5		

copolymer (B): **E-VA/28w(1)** **91NIE**
solvent (A): **ethene** **C_2H_4** **74-85-1**

Type of data: coexistence data

T/K = 433.15

The total feed concentration of the copolymer in the homogeneous system is: w_B = 0.138

continue

continue

P/ MPa	w_A bottom phase	top phase
70.8		0.998
80.0		0.996
80.8	0.450	
92.8	0.551	0.987
104.0		0.981
104.2	0.600	
113.8	0.693	
114.2		0.964
119.0	0.745	0.945
120.5		0.936
123.8	0.792	
124.7	0.850	(cloud point)

copolymer (B): **E-VA/28w(2)** **99KIN**
solvent (A): **ethene** **C_2H_4** **74-85-1**

Type of data: cloud-points

w_B	0.030	0.060	0.100	0.150	0.200	0.230	0.240	0.250	0.270
T/K	393.15	393.15	393.15	393.15	393.15	393.15	393.15	393.15	393.15
P/MPa	121.6	128.7	128.3	125.9	121.4	118.0	116.9	115.7	113.4
w_B	0.300	0.350	0.030	0.060	0.100	0.150	0.200	0.230	0.240
T/K	393.15	393.15	413.15	413.15	413.15	413.15	413.15	413.15	413.15
P/MPa	111.4	110.4	117.5	124.0	123.6	121.5	117.4	114.5	113.4
w_B	0.250	0.270	0.300	0.350	0.030	0.060	0.100	0.150	0.200
T/K	413.15	413.15	413.15	413.15	433.15	433.15	433.15	433.15	433.15
P/MPa	112.3	110.2	108.0	107.0	113.7	119.5	119.2	117.4	113.6
w_B	0.230	0.240	0.250	0.270	0.300	0.350	0.030	0.060	0.100
T/K	433.15	433.15	433.15	433.15	433.15	433.15	453.15	453.15	453.15
P/MPa	110.7	109.8	109.2	107.3	105.4	104.4	110.5	116.3	116.0
w_B	0.150	0.200	0.230	0.240	0.250	0.270	0.300	0.350	0.030
T/K	453.15	453.15	453.15	453.15	453.15	453.15	453.15	453.15	473.15
P/MPa	113.9	110.5	108.1	107.1	106.4	104.7	102.9	101.9	108.1
w_B	0.060	0.100	0.150	0.200	0.230	0.240	0.250	0.270	0.300
T/K	473.15	473.15	473.15	473.15	473.15	473.15	473.15	473.15	473.15
P/MPa	113.0	112.9	111.2	108.0	105.9	104.9	104.2	102.1	100.8

continue

continue

w_B	0.350	0.030	0.060	0.100	0.150	0.200	0.230	0.240	0.250
T/K	473.15	493.15	493.15	493.15	493.15	493.15	493.15	493.15	493.15
P/MPa	99.8	105.7	110.4	110.3	108.5	105.7	103.7	102.8	101.9

w_B	0.270	0.300	0.350
T/K	493.15	493.15	493.15
P/MPa	100.0	98.8	97.8

copolymer (B): **E-VA/29w** **91FIN**
solvent (A): **ethene** **C_2H_4** **74-85-1**

Type of data: gas solubility

T/K = 443.15

w_A	0.0116	0.0184	0.0657	0.0538	0.0777	0.0884	0.1005	0.1207	0.1430
P/MPa	1.991	3.041	5.882	8.972	12.864	14.415	16.768	20.788	24.318

w_A	0.1619
P/MPa	28.143

copolymer (B): **E-VA/29w** **91FIN**
solvent (A): **ethene** **C_2H_4** **74-85-1**

Type of data: coexistence data

T/K = 443.15

The total feed concentration of the copolymer in the homogeneous system is: w_B = 0.250

P/ MPa	w_A bottom phase	w_A top phase
48.06	0.2719	
52.08		0.9710
55.12	0.3199	
67.77	0.4045	0.9364
80.71	0.5468	0.8985
81.49	0.5211	0.8930
90.12	0.7401	0.7584

copolymer (B): **E-VA/29w** **91FIN**
solvent (A): **vinyl acetate** **$C_4H_6O_2$** **108-05-4**

Type of data: gas solubility

T/K = 413.15

w_A	0.042	0.077	0.079	0.114	0.115	0.205	0.245	0.237	0.318
P/MPa	0.152	0.169	0.186	0.273	0.288	0.371	0.405	0.435	0.444
w_A	0.299								
P/MPa	0.464								

copolymer (B): **E-VA/29w** **91FIN**
solvent (A): **vinyl acetate** **$C_4H_6O_2$** **108-05-4**

Type of data: gas solubility

T/K = 443.15

w_A	0.028	0.030	0.047	0.050	0.065	0.071	0.116	0.127	0.129
P/MPa	0.125	0.144	0.165	0.166	0.185	0.265	0.494	0.557	0.567
w_A	0.322	0.456	0.684						
P/MPa	0.886	1.02	1.14						

copolymer (B): **E-VA/29w** **91FIN**
solvent (A): **vinyl acetate** **$C_4H_6O_2$** **108-05-4**

Type of data: gas solubility

T/K = 473.15

w_A	0.024	0.037	0.039	0.055	0.056	0.086	0.122	0.155
P/MPa	0.179	0.211	0.234	0.425	0.495	0.543	0.788	0.935

copolymer (B): **E-VA/31w** **83RAE**
solvent (A): **ethene** **C_2H_4** **74-85-1**

Type of data: cloud-points

w_B	0.014	0.04	0.07	0.10	0.20	0.25	0.30	0.05	0.10
T/K	393.15	393.15	393.15	393.15	393.15	393.15	393.15	433.15	433.15
P/MPa	122.6	124.0	123.1	121.6	118.7	115.2	114.2	117.9	115.1
w_B	0.20	0.30	0.014	0.04	0.07	0.10	0.20	0.25	0.30
T/K	433.15	433.15	473.15	473.15	473.15	473.15	473.15	473.15	473.15
P/MPa	112.0	108.3	111.3	113.3	110.3	107.4	106.4	103.9	102.5

copolymer (B): **E-VA/31w** **84WAG**
solvent (A): **ethene** C_2H_4 **74-85-1**

Type of data: cloud-points

w_B	0.014	0.04	0.07	0.10	0.15	0.20	0.25	0.30	0.014
T/K	393.15	393.15	393.15	393.15	393.15	393.15	393.15	393.15	413.15
P/MPa	122.6	124.0	123.1	121.6	122.1	119.7	115.2	114.2	119.2
w_B	0.04	0.07	0.10	0.15	0.20	0.25	0.30	0.014	0.04
T/K	413.15	413.15	413.15	413.15	413.15	413.15	413.15	433.15	433.15
P/MPa	121.6	118.7	118.1	118.7	115.2	112.8	110.3	116.2	118.2
w_B	0.07	0.10	0.15	0.20	0.25	0.30	0.014	0.04	0.07
T/K	433.15	433.15	433.15	433.15	433.15	433.15	453.15	453.15	453.15
P/MPa	115.7	114.5	114.5	110.8	110.3	108.4	114.2	115.2	112.8
w_B	0.10	0.15	0.20	0.25	0.30	0.014	0.04	0.07	0.10
T/K	453.15	453.15	453.15	453.15	453.15	473.15	473.15	473.15	473.15
P/MPa	111.8	111.8	108.8	108.8	106.4	111.3	113.3	110.3	107.4
w_B	0.15	0.20	0.25	0.30					
T/K	473.15	473.15	473.15	473.15					
P/MPa	108.4	106.4	105.9	102.5					

copolymer (B): **E-VA/32w(1)** **82RAE**
solvent (A): **ethene** C_2H_4 **74-85-1**

Type of data: cloud-points

T/K = 433.15

w_B	0.05	0.10	0.15	0.20	0.30
P/MPa	119.7	119.5	117.5	115.5	111.5

copolymer (B): **E-VA/32w(1)** **84WAG**
solvent (A): **ethene** C_2H_4 **74-85-1**

Type of data: cloud-points

w_B	0.05	0.10	0.15	0.20	0.30	0.05	0.10	0.15	0.20
T/K	393.15	393.15	393.15	393.15	393.15	413.15	413.15	413.15	413.15
P/MPa	126.8	126.0	125.8	124.5	119.5	123.0	122.2	121.0	120.0
w_B	0.30	0.05	0.10	0.15	0.20	0.30	0.05	0.10	0.15
T/K	413.15	433.15	433.15	433.15	433.15	433.15	453.15	453.15	453.15
P/MPa	115.5	119.7	119.5	117.5	115.5	111.5	116.0	115.7	114.0
w_B	0.20	0.30	0.05	0.10	0.15	0.20	0.30		
T/K	453.15	453.15	473.15	473.15	473.15	473.15	473.15		
P/MPa	112.0	108.0	113.5	112.5	111.0	109.2	105.5		

copolymer (B): **E-VA/32w(2)** **84WAG**
solvent (A): **ethene** **C_2H_4** **74-85-1**

Type of data: cloud-points

w_B	0.12	0.16	0.20	0.25	0.30	0.0075	0.12	0.16	0.20
T/K	393.15	393.15	393.15	393.15	393.15	413.15	413.15	413.15	413.15
P/MPa	121.1	120.6	118.7	116.2	112.3	114.2	117.2	117.2	114.7
w_B	0.25	0.30	0.0075	0.04	0.08	0.12	0.16	0.20	0.25
T/K	413.15	413.15	433.15	433.15	433.15	433.15	433.15	433.15	433.15
P/MPa	113.7	108.8	111.3	114.2	113.7	114.7	113.7	111.8	110.3
w_B	0.30	0.0075	0.04	0.08	0.12	0.16	0.20	0.25	0.30
T/K	433.15	453.15	453.15	453.15	453.15	453.15	453.15	453.15	453.15
P/MPa	106.9	110.9	112.3	110.8	111.8	110.8	109.8	107.4	103.9
w_B	0.0075	0.04	0.08	0.12	0.16	0.20	0.25	0.30	
T/K	473.15	473.15	473.15	473.15	473.15	473.15	473.15	473.15	
P/MPa	108.4	109.8	108.4	108.8	107.9	105.9	104.9	102.0	

copolymer (B): **E-VA/32w(3)** **83RAE**
solvent (A): **ethene** **C_2H_4** **74-85-1**

Type of data: cloud-points

w_B	0.009	0.04	0.08	0.12	0.20	0.25	0.30	0.009	0.04
T/K	393.15	393.15	393.15	393.15	393.15	393.15	393.15	473.15	473.15
P/MPa	118.7	120.1	118.2	117.4	114.9	111.8	105.9	107.9	109.8
w_B	0.08	0.12	0.20	0.25	0.30				
T/K	473.15	473.15	473.15	473.15	473.15				
P/MPa	107.9	106.9	105.9	102.7	99.1				

copolymer (B): **E-VA/32w(3)** **84WAG**
solvent (A): **ethene** **C_2H_4** **74-85-1**

Type of data: cloud-points

w_B	0.009	0.04	0.08	0.12	0.20	0.25	0.304	0.009	0.04
T/K	393.15	393.15	393.15	393.15	393.15	393.15	393.15	413.15	413.15
P/MPa	118.7	120.1	118.2	117.4	114.9	111.8	105.9	116.2	118.2
w_B	0.08	0.12	0.20	0.25	0.304	0.009	0.04	0.08	0.12
T/K	413.15	413.15	413.15	413.15	413.15	433.15	433.15	433.15	433.15
P/MPa	115.2	114.9	112.3	108.4	104.5	112.3	115.2	112.8	112.5
w_B	0.20	0.25	0.304	0.009	0.04	0.08	0.12	0.20	0.25
T/K	433.15	433.15	433.15	453.15	453.15	453.15	453.15	453.15	453.15
P/MPa	110.3	106.4	103.2	110.8	112.8	110.3	108.9	107.9	104.5
w_B	0.304	0.009	0.04	0.08	0.12	0.20	0.25	0.30	
T/K	453.15	473.15	473.15	473.15	473.15	473.15	473.15	473.15	
P/MPa	100.2	107.9	109.8	107.9	106.9	105.9	102.7	99.1	

copolymer (B): **E-VA/33w(1)** **91NIE**
solvent (A): **ethene** C_2H_4 **74-85-1**

Type of data: cloud-points

w_B	0.068	0.068	0.068	0.068	0.157	0.157	0.178	0.178	0.178
T/K	383.15	393.15	433.15	453.15	383.15	433.15	373.15	403.15	453.15
P/MPa	118.2	115.8	108.5	105.7	116.5	107.0	116.0	110.0	103.2
w_B	0.300	0.300	0.300	0.393	0.393	0.393	0.393	0.393	0.439
T/K	393.15	403.15	433.15	353.15	373.15	403.15	433.15	453.15	373.15
P/MPa	107.5	106.0	102.3	107.4	103.5	99.2	96.4	95.0	98.0
w_B	0.439								
T/K	433.15								
P/MPa	92.2								

copolymer (B): **E-VA/33w(2)** **83RAE**
solvent (A): **ethene** C_2H_4 **74-85-1**

Type of data: cloud-points

w_B	0.014	0.04	0.08	0.12	0.20	0.25	0.32	0.014	0.04
T/K	393.15	393.15	393.15	393.15	393.15	393.15	393.15	473.15	473.15
P/MPa	92.7	98.6	100.0	99.0	98.6	96.6	94.3	86.8	93.6
w_B	0.08	0.12	0.20	0.25	0.32				
T/K	473.15	473.15	473.15	473.15	473.15				
P/MPa	93.2	91.2	90.7	88.7	86.8				

copolymer (B): **E-VA/33w(2)** **84WAG**
solvent (A): **ethene** C_2H_4 **74-85-1**

Type of data: cloud-points

w_B	0.014	0.04	0.08	0.12	0.16	0.20	0.25	0.32	0.014
T/K	393.15	393.15	393.15	393.15	393.15	393.15	393.15	393.15	413.15
P/MPa	92.7	98.6	100.0	99.0	98.6	99.0	96.6	94.2	90.2
w_B	0.04	0.08	0.12	0.16	0.20	0.25	0.32	0.014	0.04
T/K	413.15	413.15	413.15	413.15	413.15	413.15	413.15	433.15	433.15
P/MPa	97.1	98.1	96.8	97.1	96.9	95.1	93.2	88.7	96.1
w_B	0.08	0.12	0.16	0.20	0.25	0.32	0.014	0.04	0.08
T/K	433.15	433.15	433.15	433.15	433.15	433.15	453.15	453.15	453.15
P/MPa	96.1	94.6	95.1	94.3	92.7	90.2	87.8	95.1	94.6
w_B	0.12	0.16	0.20	0.25	0.32	0.014	0.04	0.08	0.12
T/K	453.15	453.15	453.15	453.15	453.15	473.15	473.15	473.15	473.15
P/MPa	92.7	93.2	92.7	90.4	88.7	86.8	93.6	93.2	91.2
w_B	0.16	0.20	0.25	0.32					
T/K	473.15	473.15	473.15	473.15					
P/MPa	91.2	90.7	88.7	80.8					

copolymer (B): **E-VA/33w(3)** **83RAE**
solvent (A): **ethene** C_2H_4 **74-85-1**

Type of data: cloud-points

w_B	0.05	0.10	0.20	0.30
T/K	433.15	433.15	433.15	433.15
P/MPa	111.2	111.0	108.4	104.4

copolymer (B): **E-VA/33w(3)** **84WAG**
solvent (A): **ethene** C_2H_4 **74-85-1**

Type of data: cloud-points

w_B	0.007	0.03	0.08	0.12	0.16	0.20	0.25	0.30	0.007
T/K	393.15	393.15	393.15	393.15	393.15	393.15	393.15	393.15	413.15
P/MPa	114.2	116.2	118.2	116.7	116.7	115.2	113.3	110.8	110.8
w_B	0.03	0.08	0.12	0.16	0.20	0.25	0.30	0.007	0.03
T/K	413.15	413.15	413.15	413.15	413.15	413.15	413.15	433.15	433.15
P/MPa	113.3	114.9	112.8	112.8	112.3	109.8	106.7	108.4	110.8
w_B	0.08	0.12	0.16	0.20	0.25	0.30	0.007	0.03	0.08
T/K	433.15	433.15	433.15	433.15	433.15	433.15	453.15	453.15	453.15
P/MPa	111.8	109.3	108.8	108.4	107.4	104.1	105.9	108.4	108.4
w_B	0.12	0.16	0.20	0.25	0.30	0.007	0.03	0.08	0.12
T/K	453.15	453.15	453.15	453.15	453.15	473.15	473.15	473.15	473.15
P/MPa	106.9	106.9	105.9	103.9	100.8	104.4	105.9	106.4	104.4
w_B	0.16	0.20	0.25	0.30					
T/K	473.15	473.15	473.15	473.15					
P/MPa	104.4	103.0	101.0	98.6					

copolymer (B): **E-VA/33w(4)** **84WAG**
solvent (A): **ethene** C_2H_4 **74-85-1**

Type of data: cloud-points

w_B	0.0105	0.04	0.08	0.12	0.16	0.20	0.25	0.30	0.0105
T/K	393.15	393.15	393.15	393.15	393.15	393.15	393.15	393.15	413.15
P/MPa	114.2	114.7	110.8	110.3	110.3	107.8	107.8	106.4	111.0
w_B	0.04	0.08	0.12	0.16	0.20	0.25	0.30	0.0105	0.04
T/K	413.15	413.15	413.15	413.15	413.15	413.15	413.15	433.15	433.15
P/MPa	111.8	108.4	107.4	105.9	103.9	104.2	103.5	108.4	108.8
w_B	0.08	0.12	0.16	0.20	0.25	0.30	0.0105	0.04	0.08
T/K	433.15	433.15	433.15	433.15	433.15	433.15	453.15	453.15	453.15
P/MPa	105.4	104.4	102.5	102.0	102.0	100.5	106.9	106.4	103.0

continue

continue

w_B	0.12	0.16	0.20	0.25	0.30	0.0105	0.04	0.08	0.12
T/K	453.15	453.15	453.15	453.15	453.15	473.15	473.15	473.15	473.15
P/MPa	101.5	100.5	100.1	100.1	98.5	104.4	104.4	101.0	99.5
w_B	0.16	0.20	0.25	0.30					
T/K	473.15	473.15	473.15	473.15					
P/MPa	98.5	98.0	97.6	96.6					

copolymer (B): **E-VA/34w** **91FIN**
solvent (A): **ethene** **C_2H_4** **74-85-1**

Type of data: gas solubility

T/K = 473.15

w_A	0.0135	0.0264	0.0336	0.0337	0.0373	0.0462	0.0577	0.0652	0.0752
P/MPa	2.695	4.683	6.464	6.551	6.779	8.847	11.356	14.006	16.135
w_A	0.1091	0.1323							
P/MPa	22.361	29.520							

copolymer (B): **E-VA/43w** **83RAE**
solvent (A): **ethene** **C_2H_4** **74-85-1**

Type of data: cloud-points

w_B	0.05	0.10	0.15	0.20	0.30
T/K	433.15	433.15	433.15	433.15	433.15
P/MPa	114.5	112.0	109.3	107.5	106.2

copolymer (B): **E-VA/43w** **84WAG**
solvent (A): **ethene** **C_2H_4** **74-85-1**

Type of data: cloud-points

w_B	0.05	0.10	0.15	0.20	0.30	0.05	0.10	0.15	0.20
T/K	393.15	393.15	393.15	393.15	393.15	413.15	413.15	413.15	413.15
P/MPa	121.3	118.6	116.5	114.1	112.4	117.2	115.5	112.7	110.3
w_B	0.30	0.05	0.10	0.15	0.20	0.30	0.05	0.10	0.15
T/K	413.15	433.15	433.15	433.15	433.15	433.15	453.15	453.15	453.15
P/MPa	109.6	114.5	112.0	109.8	107.5	106.2	111.3	108.9	106.9
w_B	0.20	0.30	0.05	0.10	0.15	0.20	0.30		
T/K	453.15	453.15	473.15	473.15	473.15	473.15	473.15		
P/MPa	105.5	103.4	108.6	106.8	103.4	102.7	101.0		

copolymer (B): **E-VA/58w** **91NIE**
solvent (A): **ethene** C_2H_4 **74-85-1**

Type of data: coexistence data

T/K = 463.15

The total feed concentration of the copolymer in the homogeneous system is: w_B = 0.157

P/ MPa	w_A bottom phase	top phase
93.5	0.376	
133.0		0.980
139.5	0.560	
151.6	0.602	
156.2		0.972
160.2	0.630	
164.5		0.964
170.0		0.942
171.5	0.718	
174.5		0.920
175.7	0.843	(cloud point)

Copolymers from ethylene oxide and propylene oxide

Average chemical composition of the copolymers, acronyms and references:

Copolymer (B)	Acronym	Ref.
Poly(ethylene oxide)-b-poly(propylene oxide) diblock copolymer		
8.6 mol% ethylene oxide	EO-b-PO/09x	98ONE
10.0 mol% ethylene oxide	EO-b-PO/10x	98ONE

continue

continue

Copolymer (B)	Acronym	Ref.
Poly(ethylene oxide)-b-poly(propylene oxide)-b-poly(ethylene oxide) triblock copolymer		
11.8 mol% ethylene oxide	EO-PO-EO/11x	98ONE
12.6 mol% ethylene oxide	EO-PO-EO/13x	98ONE
13.7 mol% ethylene oxide	EO-PO-EO/14x	98ONE
24.8 mol% ethylene oxide	EO-PO-EO/25x(1)	98ONE
24.8 mol% ethylene oxide	EO-PO-EO/25x(2)	98ONE
36.2 mol% ethylene oxide	EO-PO-EO/36x	98ONE
Poly(propylene oxide)-b-poly(ethylene oxide)-b-poly(propylene oxide) triblock copolymer		
24.7 mol% ethylene oxide	PO-EO-PO/25x(1)	98ONE
24.8 mol% ethylene oxide	PO-EO-PO/25x(2)	98ONE
46.7 mol% ethylene oxide	PO-EO-PO/47x	98ONE
56.9 mol% ethylene oxide	PO-EO-PO/57x	98ONE

Characterization:

Copolymer (B)	M_n/ g/mol	Further information
EO-b-PO/09x	1988 (= 132 + 1856)	Jeffamine M2000, hydroxyl terminated EO, NH_2 terminated PO
EO-b-PO/10x	566 (= 44 + 522)	Jeffamine M600, hydroxyl terminated EO, NH_2 terminated PO
EO-PO-EO/11x	1050 (= 50 + 950 + 50)	Pluronic L31, hydroxyl terminated
EO-PO-EO/13x	2944 (= 147 + 2650 + 147)	Pluronic L81, hydroxyl terminated
EO-PO-EO/14x	2210 (= 120 + 1970 + 120)	Pluronic L61, hydroxyl terminated
EO-PO-EO/25x(1)	2460 (= 245 + 1970 + 245)	Pluronic L62, hydroxyl terminated
EO-PO-EO/25x(2)	3738 (= 374 + 2990 + 374)	Pluronic L92, hydroxyl terminated
EO-PO-EO/36x	1844 (= 277 + 1290 + 277)	Pluronic L43, hydroxyl terminated
PO-EO-PO/25x(1)	3125 (= 1250 + 625 + 1250)	Pluronic 25R2, hydroxyl terminated
PO-EO-PO/25x(2)	2125 (= 850 + 425 + 850)	Pluronic 17R2, hydroxyl terminated
PO-EO-PO/47x	2833 (= 850 + 1133 + 850)	Pluronic 17R4, hydroxyl terminated
PO-EO-PO/57x	2000 (= 500 + 1000 + 500)	Pluronic 10R5, hydroxyl terminated

Experimental HPPE data:

---

copolymer (B): **EO-b-PO/09x** **98ONE**
solvent (A): **carbon dioxide** **CO_2** **124-38-9**

Type of data: cloud-points

w_B	0.001
T/K	308.15
P/MPa	27.58

copolymer (B): **EO-b-PO/10x** **98ONE**
solvent (A): **carbon dioxide** **CO_2** **124-38-9**

Type of data: cloud-points

w_B	0.007
T/K	308.15
P/MPa	27.58

copolymer (B): **EO-PO-EO/11x** **98ONE**
solvent (A): **carbon dioxide** **CO_2** **124-38-9**

Type of data: cloud-points

w_B	0.030
T/K	308.15
P/MPa	27.58

copolymer (B): **EO-PO-EO/13x** **98ONE**
solvent (A): **carbon dioxide** **CO_2** **124-38-9**

Type of data: cloud-points

w_B	0.002
T/K	308.15
P/MPa	27.58

copolymer (B): **EO-PO-EO/14x** **98ONE**
solvent (A): **carbon dioxide** **CO_2** **124-38-9**

Type of data: cloud-points

w_B	0.008
T/K	308.15
P/MPa	27.58

copolymer (B):	**EO-PO-EO/25x(1)**		**98ONE**
solvent (A):	**carbon dioxide**	CO_2	**124-38-9**

Type of data: cloud-points

w_B	0.002
T/K	308.15
P/MPa	27.58

copolymer (B):	**EO-PO-EO/25x(2)**		**98ONE**
solvent (A):	**carbon dioxide**	CO_2	**124-38-9**

Type of data: cloud-points

w_B	0.001
T/K	308.15
P/MPa	27.58

copolymer (B):	**EO-PO-EO/36x**		**98ONE**
solvent (A):	**carbon dioxide**	CO_2	**124-38-9**

Type of data: cloud-points

w_B	0.003
T/K	308.15
P/MPa	27.58

copolymer (B):	**PO-EO-PO/25x(1)**		**98ONE**
solvent (A):	**carbon dioxide**	CO_2	**124-38-9**

Type of data: cloud-points

w_B	0.001
T/K	308.15
P/MPa	27.58

copolymer (B):	**PO-EO-PO/25x(2)**		**98ONE**
solvent (A):	**carbon dioxide**	CO_2	**124-38-9**

Type of data: cloud-points

w_B	0.007
T/K	308.15
P/MPa	27.58

copolymer (B): **PO-EO-PO/47x** **98ONE**
solvent (A): **carbon dioxide** **CO_2** **124-38-9**

Type of data: cloud-points

w_B	0.001
T/K	308.15
P/MPa	27.58

copolymer (B): **PO-EO-PO/57x** **98ONE**
solvent (A): **carbon dioxide** **CO_2** **124-38-9**

Type of data: cloud-points

w_B	0.003
T/K	308.15
P/MPa	27.58

4.2. Table of systems where binary HPPE data were published only in graphical form as phase diagrams or related figures

Copolymer (B)	Solvent/gas (A)	Ref.
Ethylene/acrylic acid copolymer		
	n-butane	94LE1
	n-butane	96LE3
	1-butene	94LE1
	1-butene	93HA2
	1-butene	94LE2
	1-butene	96LE3
	dimethyl ether	94LE1
	dimethyl ether	96LE1
	dimethyl ether	96LE3
	ethene	92LUF
	ethene	93HA2
	ethene	97LEE
	propane	94LE1
	propane	96LE1
	propene	93HA2
	propene	94LE1
	propene	96LE1

Copolymer (B)	Solvent/gas (A)	Ref.
Ethylene/1-butene copolymer		
	n-heptane	96DEL
	propane	95CHE
	propane	98HAN
Ethylene/butyl acrylate copolymer		
	ethene	96BYU
Ethylene/1-hexene copolymer		
	n-heptane	96DEL
	2-methylpropane	99PAN
	propane	98HAN
	propane	99PAN
Ethylene/methacrylic acid copolymer		
	n-butane	97LEE
	1-butene	97LEE
	dimethyl ether	97LEE
	ethene	97LEE
	propane	97LEE
Ethylene/methyl acrylate copolymer		
	n-butane	93PRA
	n-butane	94LOS
	n-butane	94LE2
	n-butane	96LE1
	n-butane	96LE3
	1-butene	93PRA
	1-butene	96LE1
	carbon dioxide	99LOR
	chlorodifluoromethane	91MEI
	chlorodifluoromethane	93HA1
	chlorodifluoromethane	93PRA
	dimethyl ether	96LE3
	ethane	92HAS
	ethane	93HA2
	ethane	94LOS
	ethane	96LE3
	ethene	92HAS
	ethene	93HA2
	ethene	96BYU
	ethene	96LE3

Copolymer (B)	Solvent/gas (A)	Ref.
Ethylene/methyl acrylate copolymer		
	n-hexane	94LOS
	methanol	96LE3
	propane	91MEI
	propane	92HAS
	propane	93HA1
	propane	93HA2
	propane	94LOS
	propene	92HAS
	propene	93HA2
	propene	96LE3
Ethylene/4-methyl-1-pentene copolymer		
	n-heptane	96DEL
Ethylene/1-octene copolymer		
	cyclohexane	96DEL
	n-heptane	96DEL
	n-hexane	96DEL
	2-methylpentane	96DEL
	propane	97WHA
	propane	98HAN
Ethylene/propylene copolymer		
	1-butene	93CHE
	ethene	97HAN
	n-heptane	96DEL
	propane	97WHA
	propane	98HAN
	propene	93CHE
	propene	97HAN
Ethylene/vinyl acetate copolymer		
	ethene	82WOH
	ethene	96FOL
	ethene	2000KIN
Methyl methacrylate/2-hydroxyethyl methacrylate copolymer		
	carbon dioxide	97LEP

Copolymer (B)	Solvent/gas (A)	Ref.
Poly(ethylene oxide)-b-poly(propylene oxide)-b-poly(ethylene oxide) triblock copolymer		
	carbon dioxide	98ONE
Tetrafluoroethylene/hexafluoropropylene copolymer		
	carbon dioxide	96MER
	carbon dioxide	96RIN
	carbon dioxide	98MCH
	carbon dioxide	99MER
	chlorotrifluoromethane	96MER
	hexafluoroethane	96MER
	hexafluoropropene	96MER
	octafluoropropane	96MER
	sulfur hexafluoride	96MER
	sulfur hexafluoride	98MCH
	sulfur hexafluoride	99MER
	tetrafluoromethane	96MER
	tetrafluoromethane	98MCH
	trifluoromethane	99MER
Tetrafluoroethylene/2,2-bis(trifluoromethyl)-4,5-difluoro-1,3-dioxole copolymer		
	carbon dioxide	2000DIN
Vinylidene fluoride/hexafluoropropylene copolymer		
	carbon dioxide	96RIN
	carbon dioxide	97MER
	carbon dioxide	2000DIN
	1-chloro-1,1-difluoroethane	2000DIN
	chlorodifluoromethane	2000DIN
	chlorotrifluoromethane	97MER
	chlorotrifluoromethane	2000DIN
	1,1-difluoroethane	2000DIN
	difluoromethane	2000DIN
	hexafluoropropene	97MER
	pentafluoroethane	2000DIN
	1,1,1,2-tetrafluoroethane	2000DIN
	trifluoromethane	97MER
	trifluoromethane	2000DIN

4.3. Experimental data of quasiternary solutions containing at least one copolymer

Copolymers from ethylene and acrylic acid

Average chemical composition of the copolymers, acronyms and references:

Copolymer (B)	Acronym	Ref.
Ethylene/acrylic acid copolymer		
3.0 wt% acrylic acid	E-AA/03w	92WIN
6.0 wt% acrylic acid	E-AA/06w	92WIN
7.3 wt% acrylic acid	E-AA/07w	92WIN

Characterization:

Copolymer (B)	M_n/ g/mol	M_w/ g/mol	M_η/ g/mol	Further information
E-AA/03w	37500	183000		synthesized in the laboratory
E-AA/06w	30000	150000		synthesized in the laboratory
E-AA/07w	25000	126000		synthesized in the laboratory

Experimental HPPE data:

copolymer (B):	**E-AA/03w**		**92WIN**
solvent (A):	**acrylic acid**	$C_3H_4O_2$	**79-10-7**
solvent (C):	**ethene**	C_2H_4	**74-85-1**

Type of data: cloud-points

T/K = 413.15

w_A	0.007	0.014	0.026	0.043	0.068	0.095
w_B	0.135	0.135	0.135	0.135	0.135	0.135
w_C	0.858	0.851	0.839	0.822	0.797	0.770
P/MPa	164.4	157.1	152.9	148.0	142.7	138.2

copolymer (B):	**E-AA/03w**		**92WIN**
solvent (A):	**acrylic acid**	$C_3H_4O_2$	**79-10-7**
solvent (C):	**ethene**	C_2H_4	**74-85-1**

Type of data: cloud-points

T/K = 433.15

w_A	0.007	0.014	0.026	0.043	0.068	0.095
w_B	0.135	0.135	0.135	0.135	0.135	0.135
w_C	0.858	0.851	0.839	0.822	0.797	0.770
P/MPa	150.6	146.2	143.2	139.4	134.1	130.5

copolymer (B):	**E-AA/03w**		**92WIN**
solvent (A):	**acrylic acid**	$C_3H_4O_2$	**79-10-7**
solvent (C):	**ethene**	C_2H_4	**74-85-1**

Type of data: cloud-points

T/K = 453.15

w_A	0.007	0.014	0.026	0.043	0.068	0.095
w_B	0.135	0.135	0.135	0.135	0.135	0.135
w_C	0.858	0.851	0.839	0.822	0.797	0.770
P/MPa	141.3	138.2	135.8	132.7	128.4	124.7

copolymer (B):	**E-AA/03w**		**92WIN**
solvent (A):	**acrylic acid**	$C_3H_4O_2$	**79-10-7**
solvent (C):	**ethene**	C_2H_4	**74-85-1**

Type of data: cloud-points

T/K = 473.15

w_A	0.007	0.014	0.026	0.043	0.068	0.095
w_B	0.135	0.135	0.135	0.135	0.135	0.135
w_C	0.858	0.851	0.839	0.822	0.797	0.770
P/MPa	133.8	131.7	129.7	127.5	123.7	120.5

copolymer (B):	**E-AA/06w**		**92WIN**
solvent (A):	**acrylic acid**	$C_3H_4O_2$	**79-10-7**
solvent (C):	**ethene**	C_2H_4	**74-85-1**

Type of data: cloud-points

T/K = 413.15

continue

continue

w_A	0.005	0.017	0.026	0.043	0.068	0.104
w_B	0.135	0.135	0.135	0.135	0.135	0.135
w_C	0.860	0.848	0.839	0.822	0.797	0.761
P/MPa	195.1	175.5	166.0	157.5	150.4	144.0

copolymer (B): **E-AA/06w** **92WIN**
solvent (A): **acrylic acid** **$C_3H_4O_2$** **79-10-7**
solvent (C): **ethene** **C_2H_4** **74-85-1**

Type of data: cloud-points

T/K = 433.15

w_A	0.005	0.014	0.017	0.026	0.043	0.068	0.104
w_B	0.135	0.135	0.135	0.135	0.135	0.135	0.135
w_C	0.860	0.851	0.848	0.839	0.822	0.797	0.761
P/MPa	173.4	163.2	169.4	154.5	147.2	141.2	135.1

copolymer (B): **E-AA/06w** **92WIN**
solvent (A): **acrylic acid** **$C_3H_4O_2$** **79-10-7**
solvent (C): **ethene** **C_2H_4** **74-85-1**

Type of data: cloud-points

T/K = 453.15

w_A	0.005	0.014	0.017	0.026	0.043	0.068	0.104
w_B	0.135	0.135	0.135	0.135	0.135	0.135	0.135
w_C	0.860	0.851	0.848	0.839	0.822	0.797	0.761
P/MPa	155.2	149.1	147.5	144.5	139.1	134.1	130.0

copolymer (B): **E-AA/06w** **92WIN**
solvent (A): **acrylic acid** **$C_3H_4O_2$** **79-10-7**
solvent (C): **ethene** **C_2H_4** **74-85-1**

Type of data: cloud-points

T/K = 473.15

w_A	0.005	0.014	0.017	0.026	0.043	0.068	0.104
w_B	0.135	0.135	0.135	0.135	0.135	0.135	0.135
w_C	0.860	0.851	0.848	0.839	0.822	0.797	0.761
P/MPa	143.1	140.5	139.2	137.5	133.3	128.5	124.6

copolymer (B):	**E-AA/07w**		**92WIN**
solvent (A):	**acrylic acid**	$C_3H_4O_2$	**79-10-7**
solvent (C):	**ethene**	C_2H_4	**74-85-1**

Type of data: cloud-points

T/K = 413.15

w_A	0.007	0.014	0.023	0.051	0.074	0.113
w_B	0.135	0.135	0.135	0.135	0.135	0.135
w_C	0.858	0.851	0.842	0.814	0.791	0.752
P/MPa	210.3	189.5	174.4	157.6	152.8	150.3

copolymer (B):	**E-AA/07w**		**92WIN**
solvent (A):	**acrylic acid**	$C_3H_4O_2$	**79-10-7**
solvent (C):	**ethene**	C_2H_4	**74-85-1**

Type of data: cloud-points

T/K = 433.15

w_A	0.007	0.014	0.023	0.051	0.074	0.113
w_B	0.135	0.135	0.135	0.135	0.135	0.135
w_C	0.858	0.851	0.842	0.814	0.791	0.752
P/MPa	180.1	169.3	160.2	147.3	144.0	142.4

copolymer (B):	**E-AA/07w**		**92WIN**
solvent (A):	**acrylic acid**	$C_3H_4O_2$	**79-10-7**
solvent (C):	**ethene**	C_2H_4	**74-85-1**

Type of data: cloud-points

T/K = 453.15

w_A	0.007	0.014	0.023	0.051	0.074	0.113
w_B	0.135	0.135	0.135	0.135	0.135	0.135
w_C	0.858	0.851	0.842	0.814	0.791	0.752
P/MPa	162.4	155.3	149.0	139.5	137.3	136.5

copolymer (B):	**E-AA/07w**		**92WIN**
solvent (A):	**acrylic acid**	$C_3H_4O_2$	**79-10-7**
solvent (C):	**ethene**	C_2H_4	**74-85-1**

Type of data: cloud-points

T/K = 473.15

continue

continue

w_A	0.007	0.014	0.023	0.051	0.074	0.113
w_B	0.135	0.135	0.135	0.135	0.135	0.135
w_C	0.858	0.851	0.842	0.814	0.791	0.752
P/MPa	148.5	144.7	140.5	134.1	131.9	130.5

Copolymers from ethylene and 1-hexene

Average chemical composition of the copolymers, acronyms and references:

Copolymer (B)	Acronym	Ref.
Ethylene/1-hexene copolymer		
10.6 wt% 1-hexene	E-H/11w	2000CH3
16.1 wt% 1-hexene	E-H/16w	99KIN
35.0 wt% 1-hexene	E-H/35w	2000CH3

Characterization:

Copolymer (B)	M_n/ g/mol	M_w/ g/mol	M_η/ g/mol	Further information
E-H/11w	52600	80000		
E-H/16w	60000	129000		
E-H/35w	48100	103000		

Experimental HPPE data:

copolymer (B):	**E-H/16w**		**99KIN**
solvent (A):	**ethene**	C_2H_4	**74-85-1**
solvent (C):	**n-butane**	C_4H_{10}	**106-97-8**

Type of data: cloud-points

continue

continue

w_A	0.800	0.800	0.800	0.800	0.800	0.800	0.750	0.750	0.750
w_B	0.150	0.150	0.150	0.150	0.150	0.150	0.150	0.150	0.150
w_C	0.050	0.050	0.050	0.050	0.050	0.050	0.100	0.100	0.100
T/K	393.15	413.15	433.15	453.15	473.15	493.15	393.15	413.15	433.15
P/MPa	145.0	135.3	127.3	121.2	116.6	113.0	134.0	125.9	119.1
w_A	0.750	0.750	0.750	0.700	0.700	0.700	0.700	0.700	0.700
w_B	0.150	0.150	0.150	0.150	0.150	0.150	0.150	0.150	0.150
w_C	0.100	0.100	0.100	0.150	0.150	0.150	0.150	0.150	0.150
T/K	453.15	473.15	493.15	393.15	413.15	433.15	453.15	473.15	493.15
P/MPa	113.6	109.6	106.2	125.4	117.8	111.7	107.1	103.7	101.0

copolymer (B): **E-H/16w** **99KIN**
solvent (A): **ethene** **C_2H_4** **74-85-1**
solvent (C): **ethane** **C_2H_6** **74-84-0**

Type of data: cloud-points

w_A	0.800	0.800	0.800	0.800	0.800	0.800	0.750	0.750	0.750
w_B	0.150	0.150	0.150	0.150	0.150	0.150	0.150	0.150	0.150
w_C	0.050	0.050	0.050	0.050	0.050	0.050	0.100	0.100	0.100
T/K	393.15	413.15	433.15	453.15	473.15	493.15	393.15	413.15	433.15
P/MPa	151.7	140.7	133.4	126.3	120.8	116.5	147.1	136.8	129.1
w_A	0.750	0.750	0.750	0.700	0.700	0.700	0.700	0.700	0.700
w_B	0.150	0.150	0.150	0.150	0.150	0.150	0.150	0.150	0.150
w_C	0.100	0.100	0.100	0.150	0.150	0.150	0.150	0.150	0.150
T/K	453.15	473.15	493.15	393.15	413.15	433.15	453.15	473.15	493.15
P/MPa	123.0	118.1	114.3	143.2	133.5	125.5	120.0	116.0	112.0

copolymer (B): **E-H/16w** **99KIN**
solvent (A): **ethene** **C_2H_4** **74-85-1**
solvent (C): **helium** **He** **7440-59-7**

Type of data: cloud-points

w_A	0.840	0.840	0.840	0.840	0.840	0.840	0.830	0.830	0.830
w_B	0.150	0.150	0.150	0.150	0.150	0.150	0.150	0.150	0.150
w_C	0.010	0.010	0.010	0.010	0.010	0.010	0.020	0.020	0.020
T/K	393.15	413.15	433.15	453.15	473.15	493.15	433.15	453.15	473.15
P/MPa	195.0	178.6	165.8	155.4	148.2	143.0	214.6	200.7	188.3
w_A	0.830								
w_B	0.150								
w_C	0.020								
T/K	493.15								
P/MPa	176.8								

copolymer (B): **E-H/11w** **2000CH3**
solvent (A): **ethene** C_2H_4 **74-85-1**
solvent (C): **1-hexene** C_6H_{12} **592-41-6**

Type of data: cloud-points

w_A	0.4255	0.4255	0.4255	0.2755	0.2755	0.2755
w_B	0.1490	0.1490	0.1490	0.1390	0.1390	0.1390
w_C	0.4255	0.4255	0.4255	0.5855	0.5855	0.5855
T/K	393.15	423.15	453.35	393.25	423.15	453.25
P/MPa	65.4	66.5	65.1	44.2	45.1	44.6

copolymer (B): **E-H/16w** **99KIN**
solvent (A): **ethene** C_2H_4 **74-85-1**
solvent (C): **1-hexene** C_6H_{12} **592-41-6**

Type of data: cloud-points

w_A	0.6375	0.6375	0.6375	0.6375	0.6375	0.6375	0.4250	0.4250	0.4250
w_B	0.1500	0.1500	0.1500	0.1500	0.1500	0.1500	0.1500	0.1500	0.1500
w_C	0.2125	0.2125	0.2125	0.2125	0.2125	0.2125	0.4250	0.4250	0.4250
T/K	393.15	413.15	433.15	453.15	473.15	493.15	393.15	413.15	433.15
P/MPa	118.2	103.2	99.2	95.7	93.3	91.4	59.0	59.0	59.1
w_A	0.4250	0.4250	0.4250	0.2125	0.2125	0.2125	0.2125	0.2125	0.2125
w_B	0.1500	0.1500	0.1500	0.1500	0.1500	0.1500	0.1500	0.1500	0.1500
w_C	0.4250	0.4250	0.4250	0.6375	0.6375	0.6375	0.6375	0.6375	0.6375
T/K	453.15	473.15	493.15	393.15	413.15	433.15	453.15	473.15	493.15
P/MPa	59.0	59.1	59.1	25.6	27.0	28.3	30.3	32.3	34.6

copolymer (B): **E-H/35w** **2000CH3**
solvent (A): **ethene** C_2H_4 **74-85-1**
solvent (C): **1-hexene** C_6H_{12} **592-41-6**

Type of data: cloud-points

w_A	0.4558	0.4558	0.4558	0.4558
w_B	0.1400	0.1400	0.1400	0.1400
w_C	0.4042	0.4042	0.4042	0.4042
T/K	393.15	413.15	433.15	453.15
P/MPa	55.7	55.7	56.2	57.7

copolymer (B): **E-H/16w** **99KIN**
solvent (A): **ethene** C_2H_4 **74-85-1**
solvent (C): **methane** CH_4 **74-82-8**

Type of data: cloud-points

continue

continue

w_A	0.800	0.800	0.800	0.800	0.800	0.800	0.750	0.750	0.750
w_B	0.150	0.150	0.150	0.150	0.150	0.150	0.150	0.150	0.150
w_C	0.050	0.050	0.050	0.050	0.050	0.050	0.100	0.100	0.100
T/K	393.15	413.15	433.15	453.15	473.15	493.15	393.15	413.15	433.15
P/MPa	169.0	156.7	147.1	139.4	132.4	126.4	186.5	170.5	157.8
w_A	0.750	0.750	0.750	0.700	0.700	0.700	0.700	0.700	0.700
w_B	0.150	0.150	0.150	0.150	0.150	0.150	0.150	0.150	0.150
w_C	0.100	0.100	0.100	0.150	0.150	0.150	0.150	0.150	0.150
T/K	453.15	473.15	493.15	393.15	413.15	433.15	453.15	473.15	493.15
P/MPa	149.0	142.0	137.5	202.0	186.9	174.0	162.8	153.7	149.1

copolymer (B): **E-H/16w** **99KIN**
solvent (A): **ethene** **C_2H_4** **74-85-1**
solvent (C): **propane** **C_3H_8** **74-98-6**

Type of data: cloud-points

w_A	0.800	0.800	0.800	0.800	0.800	0.800	0.750	0.750	0.750
w_B	0.150	0.150	0.150	0.150	0.150	0.150	0.150	0.150	0.150
w_C	0.050	0.050	0.050	0.050	0.050	0.050	0.100	0.100	0.100
T/K	393.15	413.15	433.15	453.15	473.15	493.15	393.15	413.15	433.15
P/MPa	148.5	136.0	128.1	122.1	117.4	113.6	138.3	129.3	123.3
w_A	0.750	0.750	0.750	0.700	0.700	0.700	0.700	0.700	0.700
w_B	0.150	0.150	0.150	0.150	0.150	0.150	0.150	0.150	0.150
w_C	0.100	0.100	0.100	0.150	0.150	0.150	0.150	0.150	0.150
T/K	453.15	473.15	493.15	393.15	413.15	433.15	453.15	473.15	493.15
P/MPa	117.1	113.3	110.0	130.3	122.1	116.0	111.8	108.9	106.6

copolymer (B): **E-H/16w** **99KIN**
solvent (A): **ethene** **C_2H_4** **74-85-1**
solvent (C): **1-hexene** **C_6H_{12}** **592-41-6**
solvent (D): **n-butane** **C_4H_{10}** **106-97-8**

Type of data: cloud-points

w_A	0.6000	0.6000	0.6000	0.6000	0.6000	0.6000	0.5625	0.5625	0.5625
w_B	0.1500	0.1500	0.1500	0.1500	0.1500	0.1500	0.1500	0.1500	0.1500
w_C	0.2000	0.2000	0.2000	0.2000	0.2000	0.2000	0.1875	0.1875	0.1875
w_D	0.0500	0.0500	0.0500	0.0500	0.0500	0.0500	0.1000	0.1000	0.1000
T/K	393.15	413.15	433.15	453.15	473.15	493.15	393.15	413.15	433.15
P/MPa	99.9	95.8	91.6	89.7	87.7	86.0	90.5	87.2	84.6
w_A	0.5625	0.5625	0.5625	0.5250	0.5250	0.5250	0.5250	0.5250	0.5250
w_B	0.1500	0.1500	0.1500	0.1500	0.1500	0.1500	0.1500	0.1500	0.1500

continue

continue

w_C	0.1875	0.1875	0.1875	0.1750	0.1750	0.1750	0.1750	0.1750	0.1750
w_D	0.1000	0.1000	0.1000	0.1500	0.1500	0.1500	0.1500	0.1500	0.1500
T/K	453.15	473.15	493.15	393.15	413.15	433.15	453.15	473.15	493.15
P/MPa	82.6	81.1	79.9	85.5	83.1	81.1	79.5	78.2	77.2

copolymer (B): **E-H/16w** **99KIN**
solvent (A): **ethene** **C_2H_4** **74-85-1**
solvent (C): **1-hexene** **C_6H_{12}** **592-41-6**
solvent (D): **n-butane** **C_4H_{10}** **106-97-8**

Type of data: cloud-points

w_A	0.4000	0.4000	0.4000	0.4000	0.4000	0.4000	0.3750	0.3750	0.3750
w_B	0.1500	0.1500	0.1500	0.1500	0.1500	0.1500	0.1500	0.1500	0.1500
w_C	0.4000	0.4000	0.4000	0.4000	0.4000	0.4000	0.3750	0.3750	0.3750
w_D	0.0500	0.0500	0.0500	0.0500	0.0500	0.0500	0.1000	0.1000	0.1000
T/K	393.15	413.15	433.15	453.15	473.15	493.15	393.15	413.15	433.15
P/MPa	57.0	57.1	57.2	57.3	57.3	57.4	55.6	55.6	55.7
w_A	0.3750	0.3750	0.3750	0.3500	0.3500	0.3500	0.3500	0.3500	0.3500
w_B	0.1500	0.1500	0.1500	0.1500	0.1500	0.1500	0.1500	0.1500	0.1500
w_C	0.3750	0.3750	0.3750	0.3500	0.3500	0.3500	0.3500	0.3500	0.3500
w_D	0.1000	0.1000	0.1000	0.1500	0.1500	0.1500	0.1500	0.1500	0.1500
T/K	453.15	473.15	493.15	393.15	413.15	433.15	453.15	473.15	493.15
P/MPa	55.8	55.9	56.1	54.9	54.9	55.1	55.4	55.6	55.6

copolymer (B): **E-H/16w** **99KIN**
solvent (A): **ethene** **C_2H_4** **74-85-1**
solvent (C): **1-hexene** **C_6H_{12}** **592-41-6**
solvent (D): **n-butane** **C_4H_{10}** **106-97-8**

Type of data: cloud-points

w_A	0.2000	0.2000	0.2000	0.2000	0.2000	0.2000	0.1875	0.1875	0.1785
w_B	0.1500	0.1500	0.1500	0.1500	0.1500	0.1500	0.1500	0.1500	0.1500
w_C	0.6000	0.6000	0.6000	0.6000	0.6000	0.6000	0.5625	0.5625	0.5625
w_D	0.0500	0.0500	0.0500	0.0500	0.0500	0.0500	0.1000	0.1000	0.1000
T/K	393.15	413.15	433.15	453.15	473.15	493.15	393.15	413.15	433.15
P/MPa	25.1	26.6	27.7	29.7	32.0	34.0	23.7	25.1	26.6
w_A	0.1875	0.1875	0.1875	0.1750	0.1750	0.1750	0.1750	0.1750	0.1750
w_B	0.1500	0.1500	0.1500	0.1500	0.1500	0.1500	0.1500	0.1500	0.1500
w_C	0.5625	0.5625	0.5625	0.5250	0.5250	0.5250	0.5250	0.5250	0.5250
w_D	0.1000	0.1000	0.1000	0.1500	0.1500	0.1500	0.1500	0.1500	0.1500
T/K	453.15	473.15	493.15	393.15	413.15	433.15	453.15	473.15	493.15
P/MPa	28.6	30.8	33.1	23.4	24.9	26.5	28.4	30.6	32.8

copolymer (B):	**E-H/16w**		**99KIN**
solvent (A):	**ethene**	C_2H_4	**74-85-1**
solvent (C):	**1-hexene**	C_6H_{12}	**592-41-6**
solvent (D):	**helium**	**He**	**7440-59-7**

Type of data: cloud-points

w_A	0.6300	0.6300	0.6300	0.6300	0.6300	0.6300	0.6225	0.6225	0.6225
w_B	0.1500	0.1500	0.1500	0.1500	0.1500	0.1500	0.1500	0.1500	0.1500
w_C	0.2100	0.2100	0.2100	0.2100	0.2100	0.2100	0.2075	0.2075	0.2075
w_D	0.0100	0.0100	0.0100	0.0100	0.0100	0.0100	0.0200	0.0200	0.0200
T/K	393.15	413.15	433.15	453.15	473.15	493.15	393.15	413.15	433.15
P/MPa	150.3	139.9	131.5	125.1	119.9	115.5	200.5	185.5	172.6
w_A	0.6225	0.6225	0.6225						
w_B	0.1500	0.1500	0.1500						
w_C	0.2075	0.2075	0.2075						
w_D	0.0200	0.0200	0.0200						
T/K	453.15	473.15	493.15						
P/MPa	161.4	152.5	147.7						

copolymer (B):	**E-H/16w**		**99KIN**
solvent (A):	**ethene**	C_2H_4	**74-85-1**
solvent (C):	**1-hexene**	C_6H_{12}	**592-41-6**
solvent (D):	**helium**	**He**	**7440-59-7**

Type of data: cloud-points

w_A	0.4200	0.4200	0.4200	0.4200	0.4200	0.4200	0.4150	0.4150	0.4150
w_B	0.1500	0.1500	0.1500	0.1500	0.1500	0.1500	0.1500	0.1500	0.1500
w_C	0.4200	0.4200	0.4200	0.4200	0.4200	0.4200	0.4150	0.4150	0.4150
w_D	0.0100	0.0100	0.0100	0.0100	0.0100	0.0100	0.0200	0.0200	0.0200
T/K	393.15	413.15	433.15	453.15	473.15	493.15	393.15	413.15	433.15
P/MPa	100.0	95.1	91.2	88.2	86.1	84.4	149.7	139.0	130.5
w_A	0.4150	0.4150	0.4150	0.4100	0.4100	0.4100	0.4100	0.4100	0.4100
w_B	0.1500	0.1500	0.1500	0.1500	0.1500	0.1500	0.1500	0.1500	0.1500
w_C	0.4150	0.4150	0.4150	0.4100	0.4100	0.4100	0.4100	0.4100	0.4100
w_D	0.0200	0.0200	0.0200	0.0300	0.0300	0.0300	0.0300	0.0300	0.0300
T/K	453.15	473.15	493.15	393.15	413.15	433.15	453.15	473.15	493.15
P/MPa	123.6	118.2	113.6	209.9	189.9	172.9	161.6	153.8	148.4

copolymer (B):	**E-H/16w**		**99KIN**
solvent (A):	**ethene**	C_2H_4	**74-85-1**
solvent (C):	**1-hexene**	C_6H_{12}	**592-41-6**
solvent (D):	**helium**	**He**	**7440-59-7**

Type of data: cloud-points

continue

continue

w_A	0.2100	0.2100	0.2100	0.2100	0.2100	0.2100	0.2075	0.2075	0.2075
w_B	0.1500	0.1500	0.1500	0.1500	0.1500	0.1500	0.1500	0.1500	0.1500
w_C	0.6300	0.6300	0.6300	0.6300	0.6300	0.6300	0.6225	0.6225	0.6225
w_D	0.0100	0.0100	0.0100	0.0100	0.0100	0.0100	0.0200	0.0200	0.0200
T/K	393.15	413.15	433.15	453.15	473.15	493.15	393.15	413.15	433.15
P/MPa	54.4	54.6	54.5	54.3	54.0	54.0	93.3	88.8	85.1
w_A	0.2075	0.2075	0.2075	0.2050	0.2050	0.2050	0.2050	0.2050	0.2050
w_B	0.1500	0.1500	0.1500	0.1500	0.1500	0.1500	0.1500	0.1500	0.1500
w_C	0.6225	0.6225	0.6225	0.6150	0.6150	0.6150	0.6150	0.6150	0.6150
w_D	0.0200	0.0200	0.0200	0.0300	0.0300	0.0300	0.0300	0.0300	0.0300
T/K	453.15	473.15	493.15	393.15	413.15	433.15	453.15	473.15	493.15
P/MPa	82.1	79.8	78.0	134.5	127.1	121.7	117.1	113.3	110.2

copolymer (B):	**E-H/16w**		**99KIN**
solvent (A):	**ethene**	C_2H_4	**74-85-1**
solvent (C):	**1-hexene**	C_6H_{12}	**592-41-6**
solvent (D):	**methane**	CH_4	**74-82-8**

Type of data: cloud-points

w_A	0.6000	0.6000	0.6000	0.6000	0.6000	0.6000	0.5625	0.5625	0.5625
w_B	0.1500	0.1500	0.1500	0.1500	0.1500	0.1500	0.1500	0.1500	0.1500
w_C	0.2000	0.2000	0.2000	0.2000	0.2000	0.2000	0.1875	0.1875	0.1875
w_D	0.0500	0.0500	0.0500	0.0500	0.0500	0.0500	0.1000	0.1000	0.1000
T/K	393.15	413.15	433.15	453.15	473.15	493.15	393.15	413.15	433.15
P/MPa	123.0	117.5	112.4	108.1	104.7	102.0	137.2	129.7	124.2
w_A	0.5625	0.5625	0.5625	0.5250	0.5250	0.5250	0.5250	0.5250	0.5250
w_B	0.1500	0.1500	0.1500	0.1500	0.1500	0.1500	0.1500	0.1500	0.1500
w_C	0.1875	0.1875	0.1875	0.1750	0.1750	0.1750	0.1750	0.1750	0.1750
w_D	0.1000	0.1000	0.1000	0.1500	0.1500	0.1500	0.1500	0.1500	0.1500
T/K	453.15	473.15	493.15	393.15	413.15	433.15	453.15	473.15	493.15
P/MPa	119.6	115.8	112.8	151.9	144.0	137.1	131.4	127.0	123.4

copolymer (B):	**E-H/16w**		**99KIN**
solvent (A):	**ethene**	C_2H_4	**74-85-1**
solvent (C):	**1-hexene**	C_6H_{12}	**592-41-6**
solvent (D):	**methane**	CH_4	**74-82-8**

Type of data: cloud-points

w_A	0.4000	0.4000	0.4000	0.4000	0.4000	0.4000	0.3750	0.3750	0.3750
w_B	0.1500	0.1500	0.1500	0.1500	0.1500	0.1500	0.1500	0.1500	0.1500
w_C	0.4000	0.4000	0.4000	0.4000	0.4000	0.4000	0.3750	0.3750	0.3750
w_D	0.0500	0.0500	0.0500	0.0500	0.0500	0.0500	0.1000	0.1000	0.1000
T/K	393.15	413.15	433.15	453.15	473.15	493.15	393.15	413.15	433.15
P/MPa	83.0	80.9	79.0	77.4	76.0	74.8	100.5	97.1	94.0

continue

continue

w_A	0.3750	0.3750	0.3750	0.3500	0.3500	0.3500	0.3500	0.3500	0.3500
w_B	0.1500	0.1500	0.1500	0.1500	0.1500	0.1500	0.1500	0.1500	0.1500
w_C	0.3750	0.3750	0.3750	0.3500	0.3500	0.3500	0.3500	0.3500	0.3500
w_D	0.1000	0.1000	0.1000	0.1500	0.1500	0.1500	0.1500	0.1500	0.1500
T/K	453.15	473.15	493.15	393.15	413.15	433.15	453.15	473.15	493.15
P/MPa	91.3	89.1	87.2	121.6	115.2	109.6	105.4	101.9	99.0

copolymer (B):	**E-H/16w**		**99KIN**
solvent (A):	**ethene**	**C_2H_4**	**74-85-1**
solvent (C):	**1-hexene**	**C_6H_{12}**	**592-41-6**
solvent (D):	**methane**	**CH_4**	**74-82-8**

Type of data: cloud-points

w_A	0.2000	0.2000	0.2000	0.2000	0.2000	0.2000	0.1875	0.1875	0.1785
w_B	0.1500	0.1500	0.1500	0.1500	0.1500	0.1500	0.1500	0.1500	0.1500
w_C	0.6000	0.6000	0.6000	0.6000	0.6000	0.6000	0.5625	0.5625	0.5625
w_D	0.0500	0.0500	0.0500	0.0500	0.0500	0.0500	0.1000	0.1000	0.1000
T/K	393.15	413.15	433.15	453.15	473.15	493.15	393.15	413.15	433.15
P/MPa	40.9	42.5	44.0	45.5	46.9	48.3	58.6	58.8	59.1

w_A	0.1875	0.1875	0.1875	0.1750	0.1750	0.1750	0.1750	0.1750	0.1750
w_B	0.1500	0.1500	0.1500	0.1500	0.1500	0.1500	0.1500	0.1500	0.1500
w_C	0.5625	0.5625	0.5625	0.5250	0.5250	0.5250	0.5250	0.5250	0.5250
w_D	0.1000	0.1000	0.1000	0.1500	0.1500	0.1500	0.1500	0.1500	0.1500
T/K	453.15	473.15	493.15	393.15	413.15	433.15	453.15	473.15	493.15
P/MPa	59.2	59.4	59.5	82.5	81.0	79.6	78.5	77.4	76.6

Copolymers from ethylene and propylene

Average chemical composition of the copolymers, acronyms and references:

Copolymer (B)	Acronym	Ref.
Ethylene/propylene copolymer 50.0 mol% ethylene	E-P/50x	92CH2

Characterization:

Copolymer (B)	M_n/ g/mol	M_w/ g/mol	M_η/ g/mol	Further information
E-P/50x	25200	26000		alternating ethylene/propylene units from complete hydrogenation of polyisoprene

Experimental HPPE data:

copolymer (B): E-P/50x 92CH2
solvent (A): ethene C_2H_4 74-85-1
solvent (C): 1-butene C_4H_8 106-98-9

Type of data: cloud-points

w_A	0.074	0.074	0.074	0.074	0.074	0.074	0.117	0.117	0.117
w_B	0.187	0.187	0.187	0.187	0.187	0.187	0.162	0.162	0.162
w_C	0.739	0.739	0.739	0.739	0.739	0.739	0.721	0.721	0.721
T/K	355.35	373.45	398.15	423.15	448.15	473.55	355.35	373.35	398.25
P/bar	144.0	177.0	215.3	250.3	278.0	303.0	200.4	232.0	266.4
w_A	0.117	0.117	0.117	0.234	0.234	0.234	0.234		
w_B	0.162	0.162	0.162	0.154	0.154	0.154	0.154		
w_C	0.721	0.721	0.721	0.612	0.612	0.612	0.612		
T/K	423.15	447.95	473.45	355.35	373.55	398.15	423.25		
P/bar	296.8	322.0	344.5	318.5	340.2	365.0	389.0		

copolymer (B): E-P/50x 92CH2
solvent (A): ethene C_2H_4 74-85-1
solvent (C): 1-hexene C_6H_{12} 592-41-6

Type of data: cloud-points

$w_A = 0.122$ $w_B = 0.158$ $w_C = 0.820$

T/K	299.65	323.15	353.45	373.65	384.05	397.15	405.75	413.15	422.15
P/bar	16.9	23.0	32.1	38.5	41.2	44.0	58.0	68.0	85.0
	(VLE	VLE	VLE	VLE	VLE	VLE	LLE	LLE	LLE)
T/K	424.75	433.45	447.65	463.65	468.55	473.35	424.75	433.45	447.65
P/bar	89.2	100.0	124.0	138.0	144.0	148.0	53.1	54.1	59.8
	(LLE	LLE	LLE	LLE	LLE	LLE	VLLE	VLLE	VLLE)

continue

continue

T/K	468.15
P/bar	63.2
	(VLLE)

copolymer (B): **E-P/50x** **92CH2**
solvent (A): **ethene** **C_2H_4** **74-85-1**
solvent (C): **1-hexene** **C_6H_{12}** **592-41-6**

Type of data: cloud-points

$w_A = 0.243$ $w_B = 0.150$ $w_C = 0.607$

T/K	300.65	306.15	314.15	321.85	353.75	371.95	395.85	425.55	447.85
P/bar	28.0	31.7	43.0	56.5	114.0	144.0	179.0	222.0	246.0
	(VLE	VLE	LLE	LLE	LLE	LLE	LLE	LLE	LLE)
T/K	472.55	315.95	323.35	341.15	353.75	371.95	395.85	424.15	
P/bar	266.0	35.0	39.2	47.0	55.6	63.4	73.0	81.2	
	(LLE	VLLE	VLLE	VLLE	VLLE	VLLE	VLLE	VLLE)	

copolymer (B): **E-P/50x** **92CH2**
solvent (A): **ethene** **C_2H_4** **74-85-1**
solvent (C): **1-hexene** **C_6H_{12}** **592-41-6**

Type of data: cloud-points

$w_A = 0.355$ $w_B = 0.153$ $w_C = 0.492$

T/K	251.15	264.65	271.15	282.75	295.45	311.65	324.65	355.45	373.45
P/bar	187.0	176.0	175.0	182.5	194.7	216.5	233.0	269.0	290.0
	(LLE	LLE	LLE	LLE	LLE	LLE	LLE	LLE	LLE)
T/K	398.15	423.95	449.15	473.45	258.85	282.75	295.45	311.65	324.65
P/bar	320.5	348.0	370.0	384.0	17.2	30.0	36.5	45.8	55.0
	(LLE	LLE	LLE	LLE	VLLE	VLLE	VLLE	VLLE	VLLE)
T/K	355.45	373.45	398.15						
P/bar	75.5	85.0	91.8						
	(VLLE	VLLE	VLLE)						

copolymer (B): **E-P/50x** **92CH2**
solvent (A): **ethene** **C_2H_4** **74-85-1**
solvent (C): **1-hexene** **C_6H_{12}** **592-41-6**

Type of data: cloud-points

continue

continue

w_A = 0.504 w_B = 0.149 w_C = 0.347

T/K	296.05	310.85	325.75	343.25	358.85	374.85	397.25	424.15	446.55
P/bar	488.0	464.0	456.0	452.0	459.0	468.0	479.0	492.0	502.0
	(LLE	LLE	LLE	LLE	LLE	LLE	LLE	LLE	LLE)
T/K	469.65	295.65	311.15	325.95	343.25	358.85	374.85		
P/bar	511.0	48.6	59.4	70.0	79.0	88.6	92.0		
	(LLE	VLLE	VLLE	VLLE	VLLE	VLLE	VLLE)		

Copolymers from ethylene and vinyl acetate

Average chemical composition of the copolymers, acronyms and references:

Copolymer (B)	Acronym	Ref.
Ethylene/vinyl acetate copolymer		
10.9 wt% vinyl acetate	E-VA/11w	91NIE
17.5 wt% vinyl acetate	E-VA/18w	91NIE
26.0 wt% vinyl acetate	E-VA/26w	91NIE
27.5 wt% vinyl acetate	E-VA/28w(1)	91NIE
27.5 wt% vinyl acetate	E-VA/28w(2)	99KIN
32.3 wt% vinyl acetate	E-VA/32w	84WOH
32.5 wt% vinyl acetate	E-VA/33w	91NIE
34.2 wt% vinyl acetate	E-VA/34w	91FIN
58.0 wt% vinyl acetate	E-VA/58w	91NIE

Characterization:

Copolymer (B)	M_n/ g/mol	M_w/ g/mol	M_η/ g/mol	Further information
E-VA/11w	29300	161600		Exxon Chemical Company
E-VA/18w	33400	137700		Exxon Chemical Company
E-VA/26w	33600	231000		Exxon Chemical Company
E-VA/28w(1)	35000	126500		Exxon Chemical Company

continue

continue

Copolymer (B)	M_n/ g/mol	M_w/ g/mol	M_η/ g/mol	Further information
E-VA/28w(2)	61900	167000		
E-VA/32w	15500	40800		Leuna-Werke, Germany
E-VA/33w	15000	28300		Leuna-Werke, Germany
E-VA/34w	13100	41300		Leuna-Werke, Germany
E-VA/58w	35000			Exxon Chemical Company

Experimental HPPE data:

copolymer (B):	**E-VA/11w**		**91NIE**
solvent (A):	**ethene**	C_2H_4	**74-85-1**
solvent (C):	**vinyl acetate**	$C_4H_6O_2$	**108-05-4**

Type of data: coexistence data

T/K = 403.15

The total feed concentration of the homogeneous system before demixing is:

$w_A = 0.590$; $w_B = 0.153$; $w_C = 0.257$

P/ MPa	w_A	w_B	w_C	w_A	w_B	w_C
	bottom phase			top phase		
56.5	0.311	0.542	0.147			
75.5				0.683	0.016	0.301
87.8	0.450	0.352	0.198	0.678	0.023	0.299
100.2	0.546	0.212	0.242	0.660	0.048	0.292
104.1	0.590	0.153	0.257 (cloud point)			

copolymer (B):	**E-VA/11w**		**91NIE**
solvent (A):	**ethene**	C_2H_4	**74-85-1**
solvent (C):	**vinyl acetate**	$C_4H_6O_2$	**108-05-4**

Type of data: coexistence data

T/K = 403.15

continue

continue

The total feed concentration of the homogeneous system before demixing is:

$w_A = 0.347$; $w_B = 0.157$; $w_C = 0.496$

P/ MPa	w_A	w_B	w_C	w_A	w_B	w_C
	bottom phase			top phase		
58.5				0.387	0.037	0.576
68.0	0.322	0.215	0.463	0.380	0.056	0.564
72.7	0.347	0.157	0.496	(cloud point)		

copolymer (B): **E-VA/11w** **91NIE**
solvent (A): **ethene** **C_2H_4** **74-85-1**
solvent (C): **vinyl acetate** **$C_4H_6O_2$** **108-05-4**

Type of data: coexistence data

T/K = 433.15

The total feed concentration of the homogeneous system before demixing is:

$w_A = 0.254$; $w_B = 0.155$; $w_C = 0.591$

P/ MPa	w_A	w_B	w_C	w_A	w_B	w_C
	bottom phase			top phase		
48.7	0.231	0.230	0.539	0.277	0.050	0.673
53.4	0.254	0.155	0.591	(cloud point)		

copolymer (B): **E-VA/18w** **91NIE**
solvent (A): **ethene** **C_2H_4** **74-85-1**
solvent (C): **vinyl acetate** **$C_4H_6O_2$** **108-05-4**

Type of data: cloud-points

w_A	0.422	0.422	0.422	0.217	0.217	0.217	0.120	0.120	0.120
w_B	0.159	0.159	0.159	0.156	0.156	0.156	0.298	0.298	0.298
w_C	0.419	0.419	0.419	0.627	0.627	0.627	0.582	0.582	0.582
T/K	373.15	403.15	433.15	373.15	403.15	433.15	378.15	388.15	403.15
P/MPa	75.3	69.8	68.0	45.8	38.3	39.0	21.4	19.5	18.0

continue

continue

w_A	0.120	0.120	0.120
w_B	0298	0.298	0.298
w_C	0.582	0.582	0.582
T/K	423.15	433.15	443.15
P/MPa	19.4	20.5	22.2

copolymer (B):	**E-VA/18w**		**91NIE**
solvent (A):	**ethene**	**C_2H_4**	**74-85-1**
solvent (C):	**vinyl acetate**	**$C_4H_6O_2$**	**108-05-4**

Type of data: coexistence data

T/K = 433.15

The total feed concentration of the homogeneous system before demixing is:

w_A = 0.217; w_B = 0.156; w_C = 0.627

P/ MPa	w_A bottom phase	w_B	w_C	w_A top phase	w_B	w_C
27.5	0.160	0.360	0.480	0.251	0.041	0.708
33.5	0.188	0.276	0.536	0.242	0.061	0.697
39.0	0.217	0.156	0.627 (cloud point)			

copolymer (B):	**E-VA/18w**		**91NIE**
solvent (A):	**ethene**	**C_2H_4**	**74-85-1**
solvent (C):	**vinyl acetate**	**$C_4H_6O_2$**	**108-05-4**

Type of data: coexistence data

T/K = 433.15

The total feed concentration of the homogeneous system before demixing is:

w_A = 0.343; w_B = 0.136; w_C = 0.521

P/ MPa	w_A bottom phase	w_B	w_C	w_A top phase	w_B	w_C
19.9				0.4042	0.0004	0.5954
43.0	0.256	0.361	0.383	0.3936	0.0127	0.5937
57.3	0.343	0.136	0.521 (cloud point)			

copolymer (B):	**E-VA/18w**		**91NIE**
solvent (A):	**ethene**	C_2H_4	**74-85-1**
solvent (C):	**vinyl acetate**	$C_4H_6O_2$	**108-05-4**

Type of data: coexistence data

T/K = 433.15

The total feed concentration of the homogeneous system before demixing is:

$w_A = 0.442$; $w_B = 0.159$; $w_C = 0.419$

P/ MPa	w_A	w_B	w_C	w_A	w_B	w_C
	bottom phase			top phase		
37.7	0.221	0.532	0.247	0.498	0.025	0.477
50.0	0.288	0.404	0.308	0.487	0.038	0.475
61.9	0.365	0.272	0.363	0.476	0.062	0.462
68.0	0.422	0.159	0.419	(cloud point)		

copolymer (B):	**E-VA/18w**		**91NIE**
solvent (A):	**ethene**	C_2H_4	**74-85-1**
solvent (C):	**vinyl acetate**	$C_4H_6O_2$	**108-05-4**

Type of data: coexistence data

T/K = 433.15

The total feed concentration of the homogeneous system before demixing is:

$w_A = 0.645$; $w_B = 0.158$; $w_C = 0.197$

P/ MPa	w_A	w_B	w_C	w_A	w_B	w_C
	bottom phase			top phase		
30.4	0.176	0.760	0.064	0.779	0.0008	0.221
43.7	0.244	0.670	0.086	0.766	0.0009	0.233
66.6	0.370	0.506	0.124	0.754	0.0130	0.233
86.0	0.513	0.325	0.162	0.736	0.0290	0.235
97.2	0.645	0.158	0.197	(cloud point)		

copolymer (B):	**E-VA/18w**		**91NIE**
solvent (A):	**ethene**	C_2H_4	**74-85-1**
solvent (C):	**vinyl acetate**	$C_4H_6O_2$	**108-05-4**

Type of data: coexistence data

T/K = 463.15

The total feed concentration of the homogeneous system before demixing is:

$w_A = 0.460$; $w_B = 0.135$; $w_C = 0.405$

P/ MPa	w_A	w_B bottom phase	w_C	w_A	w_B top phase	w_C
47.3	0.273	0.506	0.221	0.582	0.012	0.406
58.8	0.346	0.363	0.291	0.546	0.051	0.403
70.7	0.460	0.135	0.405 (cloud point)			

Fractionation during demixing

Demixing pressure	molar mass averages bottom phase			top phase		
P/ MPa	M_n/ g/mol	M_w/ g/mol	M_z/ g/mol	M_n/ g/mol	M_w/ g/mol	M_z/ g/mol
47.3				5611	13634	27447
58.8				11958	26057	44198
70.7	33400	137700 (feed sample)				

copolymer (B):	**E-VA/26w**		**91NIE**
solvent (A):	**ethene**	C_2H_4	**74-85-1**
solvent (C):	**vinyl acetate**	$C_4H_6O_2$	**108-05-4**

Type of data: cloud-points

w_A	0.317	0.317	0.317	0.317	0.317	0.317	0.340	0.340	0.413
w_B	0.360	0.360	0.360	0.360	0.360	0.360	0.322	0.322	0.207
w_C	0.323	0.323	0.323	0.323	0.323	0.323	0.338	0.338	0.380
T/K	363.15	373.15	393.15	413.15	438.15	453.15	383.15	433.15	403.15
P/MPa	52.3	53.1	54.0	55.2	57.2	58.5	62.4	64.5	75.4

continue

continue

w_A	0.413	0.505	0.505	0.505	0.505	0.505	0.631	0.631	0.631
w_B	0.207	0.257	0.257	0.257	0.257	0.257	0.187	0.187	0.187
w_C	0.380	0.238	0.238	0.238	0.238	0.238	0.182	0.182	0.182
T/K	433.15	363.15	383.15	403.15	438.15	453.15	363.15	383.15	403.15
P/MPa	75.1	88.3	86.1	83.7	82.0	81.8	106.8	102.8	100.0

w_A	0.631	0.795	0.795	0.795
w_B	0.187	0.105	0.105	0.105
w_C	0.182	0.100	0.100	0.100
T/K	443.15	383.15	403.15	433.15
P/MPa	96.5	122.0	118.5	113.0

copolymer (B): **E-VA/28w(1)** **91NIE**
solvent (A): **ethene** C_2H_4 **74-85-1**
solvent (C): **vinyl acetate** $C_4H_6O_2$ **108-05-4**

Type of data: cloud-points

w_A	0.594	0.594	0.594	0.200	0.200	0.200
w_B	0.148	0.148	0.148	0.152	0.152	0.152
w_C	0.258	0.258	0.258	0.648	0.648	0.648
T/K	398.15	403.15	433.15	401.15	413.15	433.15
P/MPa	90.2	86.4	84.6	26.1	27.2	29.0

copolymer (B): **E-VA/28w(1)** **91NIE**
solvent (A): **ethene** C_2H_4 **74-85-1**
solvent (C): **vinyl acetate** $C_4H_6O_2$ **108-05-4**

Type of data: coexistence data

T/K = 403.15

The total feed concentration of the homogeneous system before demixing is:

w_A = 0.353; w_B = 0.159; w_C = 0.488

P/ MPa	w_A	w_B bottom phase	w_C	w_A	w_B top phase	w_C
30.7				0.415	0.016	0.569
31.2	0.220	0.446	0.334			
39.1				0.409	0.025	0.563
39.7	0.255	0.349	0.396			
46.7	0.319	0.224	0.457	0.403	0.044	0.553
49.5	0.353	0.159	0.488 (cloud point)			

copolymer (B):	**E-VA/28w(1)**		**91NIE**
solvent (A):	**ethene**	C_2H_4	**74-85-1**
solvent (C):	**vinyl acetate**	$C_4H_6O_2$	**108-05-4**

Type of data: coexistence data

T/K = 433.15

The total feed concentration of the homogeneous system before demixing is:

$w_A = 0.594$; $w_B = 0.148$; $w_C = 0.258$

P/ MPa	w_A	w_B	w_C	w_A	w_B	w_C
	bottom phase			top phase		
35.5	0.204	0.702	0.094			
48.2	0.273	0.613	0.114	0.709	0.008	0.283
56.6	0.322	0.551	0.127	0.705	0.013	0.282
67.0	0.392	0.447	0.161	0.692	0.020	0.288
78.6	0.500	0.290	0.210	0.677	0.040	0.283
84.6	0.594	0.148	0.258 (cloud point)			

copolymer (B):	**E-VA/28w(1)**		**91NIE**
solvent (A):	**ethene**	C_2H_4	**74-85-1**
solvent (C):	**vinyl acetate**	$C_4H_6O_2$	**108-05-4**

Type of data: coexistence data

T/K = 433.15

The total feed concentration of the homogeneous system before demixing is:

$w_A = 0.349$; $w_B = 0.152$; $w_C = 0.499$

P/ MPa	w_A	w_B	w_C	w_A	w_B	w_C
	bottom phase			top phase		
27.8	0.176	0.529	0.295	0.419	0.012	0.567
35.8	0.221	0.439	0.340	0.413	0.021	0.566
46.2	0.289	0.308	0.403	0.402	0.051	0.547
51.0	0.349	0.152	0.499 (cloud point)			

copolymer (B): **E-VA/28w(1)** **91NIE**
solvent (A): **ethene** **C_2H_4** **74-85-1**
solvent (C): **vinyl acetate** **$C_4H_6O_2$** **108-05-4**

Type of data: coexistence data

T/K = 433.15

The total feed concentration of the homogeneous system before demixing is:
w_A = 0.200; w_B = 0.152; w_C = 0.648

P/ MPa	w_A bottom phase	w_B bottom phase	w_C bottom phase	w_A top phase	w_B top phase	w_C top phase
14.5	0.140	0.400	0.460	0.238	0.014	0.748
19.5	0.154	0.356	0.490	0.231	0.031	0.738
26.0	0.181	0.233	0.586	0.228	0.051	0.721
29.0	0.200	0.152	0.648 (cloud point)			

copolymer (B): **E-VA/28w(2)** **99KIN**
solvent (A): **ethene** **C_2H_4** **74-85-1**
solvent (C): **carbon dioxide** **CO_2** **124-38-9**

Type of data: cloud-points

w_A	0.800	0.800	0.800	0.800	0.800	0.800	0.750	0.750	0.750
w_B	0.150	0.150	0.150	0.150	0.150	0.150	0.150	0.150	0.150
w_C	0.050	0.050	0.050	0.050	0.050	0.050	0.100	0.100	0.100
T/K	393.15	413.15	433.15	453.15	473.15	493.15	393.15	413.15	433.15
P/MPa	127.7	123.0	119.2	115.9	113.0	110.2	128.5	123.8	120.1
w_A	0.750	0.750	0.750	0.700	0.700	0.700	0.700	0.700	0.700
w_B	0.150	0.150	0.150	0.150	0.150	0.150	0.150	0.150	0.150
w_C	0.100	0.100	0.100	0.150	0.150	0.150	0.150	0.150	0.150
T/K	453.15	473.15	493.15	393.15	413.15	433.15	453.15	473.15	493.15
P/MPa	117.0	113.9	111.2	130.3	125.7	121.8	118.9	115.8	113.1

copolymer (B): **E-VA/28w(2)** **99KIN**
solvent (A): **ethene** **C_2H_4** **74-85-1**
solvent (C): **ethane** **C_2H_6** **74-84-0**

Type of data: cloud-points

w_A	0.800	0.800	0.800	0.800	0.800	0.800	0.750	0.750	0.750
w_B	0.150	0.150	0.150	0.150	0.150	0.150	0.150	0.150	0.150
w_C	0.050	0.050	0.050	0.050	0.050	0.050	0.100	0.100	0.100
T/K	393.15	413.15	433.15	453.15	473.15	493.15	393.15	413.15	433.15
P/MPa	125.6	121.2	117.2	113.9	110.8	108.0	126.0	121.5	117.3

continue

continue

w_A	0.750	0.750	0.750	0.700	0.700	0.700	0.700	0.700	0.700
w_B	0.150	0.150	0.150	0.150	0.150	0.150	0.150	0.150	0.150
w_C	0.100	0.100	0.100	0.150	0.150	0.150	0.150	0.150	0.150
T/K	453.15	473.15	493.15	393.15	413.15	433.15	453.15	473.15	493.15
P/MPa	114.0	111.3	108.4	124.8	120.0	115.8	112.4	109.7	107.5

copolymer (B): **E-VA/28w(2)** **99KIN**
solvent (A): **ethene** **C_2H_4** **74-85-1**
solvent (C): **helium** **He** **7440-59-7**

Type of data: cloud-points

w_A	0.955	0.955	0.955	0.955	0.955	0.955	0.920	0.920	0.920
w_B	0.035	0.035	0.035	0.035	0.035	0.035	0.070	0.070	0.070
w_C	0.010	0.010	0.010	0.010	0.010	0.010	0.010	0.010	0.010
T/K	393.15	413.15	433.15	453.15	473.15	493.15	393.15	413.15	433.15
P/MPa	163.6	156.1	149.2	142.8	137.5	132.7	168.1	159.3	151.5
w_A	0.920	0.920	0.920	0.870	0.870	0.870	0.870	0.870	0.870
w_B	0.070	0.070	0.070	0.120	0.120	0.120	0.120	0.120	0.120
w_C	0.010	0.010	0.010	0.010	0.010	0.010	0.010	0.010	0.010
T/K	453.15	473.15	493.15	393.15	413.15	433.15	453.15	473.15	493.15
P/MPa	145.4	139.9	135.3	165.2	157.6	150.7	144.6	139.2	134.5
w_A	0.840	0.840	0.840	0.840	0.840	0.840	0.790	0.790	0.790
w_B	0.150	0.150	0.150	0.150	0.150	0.150	0.200	0.200	0.200
w_C	0.010	0.010	0.010	0.010	0.010	0.010	0.010	0.010	0.010
T/K	393.15	413.15	433.15	453.15	473.15	493.15	393.15	413.15	433.15
P/MPa	163.8	156.2	149.4	143.3	137.9	133.3	163.2	155.7	148.8
w_A	0.790	0.790	0.790	0.640	0.640	0.640	0.640	0.640	0.640
w_B	0.200	0.200	0.200	0.350	0.350	0.350	0.350	0.350	0.350
w_C	0.010	0.010	0.010	0.010	0.010	0.010	0.010	0.010	0.010
T/K	453.15	473.15	493.15	393.15	413.15	433.15	453.15	473.15	493.15
P/MPa	142.5	137.5	132.6	158.7	150.1	142.7	137.4	132.5	128.4

copolymer (B): **E-VA/28w(2)** **99KIN**
solvent (A): **ethene** **C_2H_4** **74-85-1**
solvent (C): **nitrogen** **N_2** **7727-37-9**

Type of data: cloud-points

w_A	0.800	0.800	0.800	0.800	0.800	0.800	0.750	0.750	0.750
w_B	0.150	0.150	0.150	0.150	0.150	0.150	0.150	0.150	0.150
w_C	0.050	0.050	0.050	0.050	0.050	0.050	0.100	0.100	0.100
T/K	393.15	413.15	433.15	453.15	473.15	493.15	393.15	413.15	433.15
P/MPa	146.5	140.5	135.8	130.5	125.4	120.5	172.5	164.7	155.6

continue

continue

w_A	0.750	0.750	0.750	0.700	0.700	0.700	0.700	0.700	0.700
w_B	0.150	0.150	0.150	0.150	0.150	0.150	0.150	0.150	0.150
w_C	0.100	0.100	0.100	0.150	0.150	0.150	0.150	0.150	0.150
T/K	453.15	473.15	493.15	393.15	413.15	433.15	453.15	473.15	493.15
P/MPa	148.8	142.5	137.5	193.0	184.8	176.8	168.9	161.4	154.4

copolymer (B): **E-VA/28w(2)** **99KIN**
solvent (A): **ethene** **C_2H_4** **74-85-1**
solvent (C): **propane** **C_3H_8** **74-98-6**

Type of data: cloud-points

w_A	0.800	0.800	0.800	0.800	0.800	0.800	0.750	0.750	0.750
w_B	0.150	0.150	0.150	0.150	0.150	0.150	0.150	0.150	0.150
w_C	0.050	0.050	0.050	0.050	0.050	0.050	0.100	0.100	0.100
T/K	393.15	413.15	433.15	453.15	473.15	493.15	393.15	413.15	433.15
P/MPa	122.0	117.7	113.6	110.5	107.6	104.9	116.3	112.4	109.0
w_A	0.750	0.750	0.750	0.700	0.700	0.700	0.700	0.700	0.700
w_B	0.150	0.150	0.150	0.150	0.150	0.150	0.150	0.150	0.150
w_C	0.100	0.100	0.100	0.150	0.150	0.150	0.150	0.150	0.150
T/K	453.15	473.15	493.15	393.15	413.15	433.15	453.15	473.15	493.15
P/MPa	105.8	103.4	100.8	112.9	109.2	105.9	103.0	100.4	98.0

copolymer (B): **E-VA/28w(2)** **99KIN**
solvent (A): **ethene** **C_2H_4** **74-85-1**
solvent (C): **vinyl acetate** **$C_4H_6O_2$** **108-05-4**

Type of data: cloud-points

w_A	0.6715	0.6715	0.6715	0.6715	0.6715	0.6715	0.6375	0.6375	0.6375
w_B	0.1500	0.1500	0.1500	0.1500	0.1500	0.1500	0.1500	0.1500	0.1500
w_C	0.1785	0.1785	0.1785	0.1785	0.1785	0.1785	0.2125	0.2125	0.2125
T/K	393.15	413.15	433.15	453.15	473.15	493.15	393.15	413.15	433.15
P/MPa	99.5	97.5	95.5	94.1	92.7	91.5	94.3	92.6	91.1
w_A	0.6375	0.6375	0.6375	0.4250	0.4250	0.4250	0.4250	0.4250	0.4250
w_B	0.1500	0.1500	0.1500	0.1500	0.1500	0.1500	0.1500	0.1500	0.1500
w_C	0.2125	0.2125	0.2125	0.4250	0.4250	0.4250	0.4250	0.4250	0.4250
T/K	453.15	473.15	493.15	393.15	413.15	433.15	453.15	473.15	493.15
P/MPa	89.8	88.6	87.5	60.7	61.1	61.6	62.0	62.4	62.7
w_A	0.2125	0.2125	0.2125	0.2125	0.2125	0.2125	0.0850	0.0850	0.0850
w_B	0.1500	0.1500	0.1500	0.1500	0.1500	0.1500	0.1500	0.1500	0.1500
w_C	0.6375	0.6375	0.6375	0.6375	0.6375	0.6375	0.7650	0.7650	0.7650
T/K	393.15	413.15	433.15	453.15	473.15	493.15	393.15	413.15	433.15
P/MPa	26.7	28.9	30.9	33.1	34.8	37.6	4.8	7.1	9.8

continue

continue

w_A	0.0850	0.0850	0.0850
w_B	0.1500	0.1500	0.1500
w_C	0.7650	0.7650	0.7650
T/K	453.15	473.15	493.15
P/MPa	13.2	17.5	19.5

copolymer (B): **E-VA/28w(2)** **99KIN**
solvent (A): **ethene** C_2H_4 **74-85-1**
solvent (C): **vinyl acetate** $C_4H_6O_2$ **108-05-4**

Type of data: cloud-points

w_A	0.4250	0.4250	0.4250	0.4250	0.4250	0.4250	0.4250	0.3060	0.3060
w_B	0.1500	0.1500	0.1500	0.1500	0.1500	0.1500	0.1500	0.1500	0.1500
w_C	0.4250	0.4250	0.4250	0.4250	0.4250	0.4250	0.4250	0.5440	0.5440
T/K	348.15	353.15	358.15	368.15	383.15	393.15	423.15	348.15	353.15
P/MPa	48.7	47.5	46.8	46.7	47.5	48.0	50.8	42.2	41.4
w_A	0.3060	0.3060	0.3060	0.3060	0.3060	0.3060	0.3060	0.3060	0.3060
w_B	0.1500	0.1500	0.1500	0.1500	0.1500	0.1500	0.1500	0.1500	0.1500
w_C	0.5440	0.5440	0.5440	0.5440	0.5440	0.5440	0.5440	0.5440	0.5440
T/K	358.15	368.15	373.15	383.15	393.15	403.15	423.15	443.15	463.15
P/MPa	41.0	40.9	41.0	41.1	41.5	42.0	43.9	45.3	46.9
w_A	0.0850	0.0850	0.0850	0.0850	0.0850	0.0850	0.0850	0.0850	0.0850
w_B	0.1500	0.1500	0.1500	0.1500	0.1500	0.1500	0.1500	0.1500	0.1500
w_C	0.7650	0.7650	0.7650	0.7650	0.7650	0.7650	0.7650	0.7650	0.7650
T/K	341.15	343.15	348.15	353.15	358.15	368.15	373.15	383.15	403.15
P/MPa	4.8	3.5	2.9	2.8	2.7	3.0	3.4	4.0	5.7
w_A	0.0850	0.0850	0.0850						
w_B	0.1500	0.1500	0.1500						
w_C	0.7650	0.7650	0.7650						
T/K	423.15	443.15	463.15						
P/MPa	8.6	11.0	15.5						

copolymer (B): **E-VA/28w(2)** **99KIN**
solvent (A): **ethene** C_2H_4 **74-85-1**
solvent (C): **vinyl acetate** $C_4H_6O_2$ **108-05-4**

Type of data: cloud-points

w_A	0.4825	0.4825	0.4825	0.4825	0.4825	0.4825	0.4600	0.4600	0.4600
w_B	0.0350	0.0350	0.0350	0.0350	0.0350	0.0350	0.0800	0.0800	0.0800
w_C	0.4825	0.4825	0.4825	0.4825	0.4825	0.4825	0.4600	0.4600	0.4600
T/K	393.15	413.15	433.15	453.15	473.15	493.15	393.15	413.15	433.15
P/MPa	59.0	59.9	60.6	61.2	61.6	61.9	61.9	62.3	62.9

continue

continue

w_A	0.4600	0.4600	0.4600	0.4250	0.4250	0.4250	0.4250	0.4250	0.4250
w_B	0.0800	0.0800	0.0800	0.1500	0.1500	0.1500	0.1500	0.1500	0.1500
w_C	0.4600	0.4600	0.4600	0.4250	0.4250	0.4250	0.4250	0.4250	0.4250
T/K	453.15	473.15	493.15	393.15	413.15	433.15	453.15	473.15	493.15
P/MPa	63.3	63.8	64.1	60.7	61.1	61.6	62.0	62.4	62.7
w_A	0.3750	0.3750	0.3750	0.3750	0.3750	0.3750			
w_B	0.2500	0.2500	0.2500	0.2500	0.2500	0.2500			
w_C	0.3750	0.3750	0.3750	0.3750	0.3750	0.3750			
T/K	393.15	413.15	433.15	453.15	473.15	493.15			
P/MPa	53.8	54.9	56.0	56.9	58.0	58.9			

copolymer (B):	**E-VA/28w(2)**		**99KIN**
solvent (A):	**ethene**	**C_2H_4**	**74-85-1**
solvent (C):	**vinyl acetate**	**$C_4H_6O_2$**	**108-05-4**

Type of data: cloud-points

w_A	0.3450	0.3450	0.3450	0.3450	0.3450	0.3450	0.3400	0.3400	0.3400
w_B	0.3100	0.3100	0.3100	0.3100	0.3100	0.3100	0.3200	0.3200	0.3200
w_C	0.3450	0.3450	0.3450	0.3450	0.3450	0.3450	0.3400	0.3400	0.3400
T/K	393.15	413.15	433.15	453.15	473.15	493.15	393.15	413.15	433.15
P/MPa	47.5	49.2	50.7	52.1	53.5	54.7	49.0	50.4	51.8
w_A	0.3400	0.3400	0.3400	0.3250	0.3250	0.3250	0.3250	0.3250	0.3250
w_B	0.3200	0.3200	0.3200	0.3500	0.3500	0.3500	0.3500	0.3500	0.3500
w_C	0.3400	0.3400	0.3400	0.3250	0.3250	0.3250	0.3250	0.3250	0.3250
T/K	453.15	473.15	493.15	393.15	413.15	433.15	453.15	473.15	493.15
P/MPa	53.2	54.5	55.7	48.6	50.1	51.6	53.0	54.3	55.5
w_A	0.2750	0.2750	0.2750	0.2750	0.2750	0.2750			
w_B	0.4500	0.4500	0.4500	0.4500	0.4500	0.4500			
w_C	0.2750	0.2750	0.2750	0.2750	0.2750	0.2750			
T/K	393.15	413.15	433.15	453.15	473.15	493.15			
P/MPa	42.8	44.6	46.2	47.7	49.1	50.3			

copolymer (B):	**E-VA/28w(2)**		**99KIN**
solvent (A):	**ethene**	**C_2H_4**	**74-85-1**
solvent (C):	**vinyl acetate**	**$C_4H_6O_2$**	**108-05-4**
solvent (D):	**carbon dioxide**	**CO_2**	**124-38-9**

Type of data: cloud-points

w_A	0.6000	0.6000	0.6000	0.6000	0.6000	0.6000	0.5625	0.5625	0.5625
w_B	0.1500	0.1500	0.1500	0.1500	0.1500	0.1500	0.1500	0.1500	0.1500
w_C	0.2000	0.2000	0.2000	0.2000	0.2000	0.2000	0.1875	0.1875	0.1875
w_D	0.0500	0.0500	0.0500	0.0500	0.0500	0.0500	0.1000	0.1000	0.1000
T/K	393.15	413.15	433.15	453.15	473.15	493.15	393.15	413.15	433.15
P/MPa	97.9	96.0	94.3	92.6	91.1	90.0	101.5	99.4	97.3

continue

continue

w_A	0.5625	0.5625	0.5625	0.5250	0.5250	0.5250	0.5250	0.5250	0.5250
w_B	0.1500	0.1500	0.1500	0.1500	0.1500	0.1500	0.1500	0.1500	0.1500
w_C	0.1875	0.1875	0.1875	0.1750	0.1750	0.1750	0.1750	0.1750	0.1750
w_D	0.1000	0.1000	0.1000	0.1500	0.1500	0.1500	0.1500	0.1500	0.1500
T/K	453.15	473.15	493.15	393.15	413.15	433.15	453.15	473.15	493.15
P/MPa	95.7	94.2	92.9	105.1	102.4	100.1	97.7	95.9	94.3

copolymer (B): **E-VA/28w(2)** **99KIN**
solvent (A): **ethene** **C_2H_4** **74-85-1**
solvent (C): **vinyl acetate** **$C_4H_6O_2$** **108-05-4**
solvent (D): **carbon dioxide** **CO_2** **124-38-9**

Type of data: cloud-points

w_A	0.400	0.400	0.400	0.400	0.400	0.400	0.375	0.375	0.375
w_B	0.150	0.150	0.150	0.150	0.150	0.150	0.150	0.150	0.150
w_C	0.400	0.400	0.400	0.400	0.400	0.400	0.375	0.375	0.375
w_D	0.050	0.050	0.050	0.050	0.050	0.050	0.100	0.100	0.100
T/K	393.15	413.15	433.15	453.15	473.15	493.15	393.15	413.15	433.15
P/MPa	64.9	65.1	65.3	65.5	65.7	65.8	69.6	69.6	69.5
w_A	0.375	0.375	0.375	0.350	0.350	0.350	0.350	0.350	0.350
w_B	0.150	0.150	0.150	0.150	0.150	0.150	0.150	0.150	0.150
w_C	0.375	0.375	0.375	0.350	0.350	0.350	0.350	0.350	0.350
w_D	0.100	0.100	0.100	0.150	0.150	0.150	0.150	0.150	0.150
T/K	453.15	473.15	493.15	393.15	413.15	433.15	453.15	473.15	493.15
P/MPa	69.5	69.4	69.3	74.9	74.6	74.3	74.1	73.7	73.5

copolymer (B): **E-VA/28w(2)** **99KIN**
solvent (A): **ethene** **C_2H_4** **74-85-1**
solvent (C): **vinyl acetate** **$C_4H_6O_2$** **108-05-4**
solvent (D): **carbon dioxide** **CO_2** **124-38-9**

Type of data: cloud-points

w_A	0.2000	0.2000	0.2000	0.2000	0.2000	0.2000	0.1875	0.1875	0.1875
w_B	0.1500	0.1500	0.1500	0.1500	0.1500	0.1500	0.1500	0.1500	0.1500
w_C	0.6000	0.6000	0.6000	0.6000	0.6000	0.6000	0.5625	0.5625	0.5625
w_D	0.0500	0.0500	0.0500	0.0500	0.0500	0.0500	0.1000	0.1000	0.1000
T/K	393.15	413.15	433.15	453.15	473.15	493.15	393.15	413.15	433.15
P/MPa	32.5	34.4	36.1	38.0	39.4	40.9	38.9	40.3	41.8
w_A	0.1875	0.1875	0.1875	0.1750	0.1750	0.1750	0.1750	0.1750	0.1750
w_B	0.1500	0.1500	0.1500	0.1500	0.1500	0.1500	0.1500	0.1500	0.1500
w_C	0.5625	0.5625	0.5625	0.5250	0.5250	0.5250	0.5250	0.5250	0.5250
w_D	0.1000	0.1000	0.1000	0.1500	0.1500	0.1500	0.1500	0.1500	0.1500
T/K	453.15	473.15	493.15	393.15	413.15	433.15	453.15	473.15	493.15
P/MPa	43.1	44.2	45.2	45.2	46.1	47.2	48.3	48.9	49.6

copolymer (B):	**E-VA/28w(2)**		**99KIN**
solvent (A):	**ethene**	C_2H_4	**74-85-1**
solvent (C):	**vinyl acetate**	$C_4H_6O_2$	**108-05-4**
solvent (D):	**helium**	He	**7440-59-7**

Type of data: cloud-points

w_A	0.8300	0.8300	0.8300	0.8300	0.8300	0.8300	0.6300	0.6300	0.6300
w_B	0.1500	0.1500	0.1500	0.1500	0.1500	0.1500	0.1500	0.1500	0.1500
w_C	0.0000	0.0000	0.0000	0.0000	0.0000	0.0000	0.2100	0.2100	0.2100
w_D	0.0200	0.0200	0.0200	0.0200	0.0200	0.0200	0.0100	0.0100	0.0100
T/K	393.15	413.15	433.15	453.15	473.15	493.15	393.15	413.15	433.15
P/MPa	209.4	197.5	185.4	174.8	165.9	158.5	139.1	133.4	127.4
w_A	0.6300	0.6300	0.6300	0.6225	0.6225	0.6225	0.6225	0.6225	0.6225
w_B	0.1500	0.1500	0.1500	0.1500	0.1500	0.1500	0.1500	0.1500	0.1500
w_C	0.2100	0.2100	0.2100	0.2075	0.2075	0.2075	0.2075	0.2075	0.0750
w_D	0.0100	0.0100	0.0100	0.0200	0.0200	0.0200	0.0200	0.0200	0.0200
T/K	453.15	473.15	493.15	393.15	413.15	433.15	453.15	473.15	493.15
P/MPa	122.6	118.7	116.3	186.5	176.6	166.4	157.7	151.3	147.1

copolymer (B):	**E-VA/28w(2)**		**99KIN**
solvent (A):	**ethene**	C_2H_4	**74-85-1**
solvent (C):	**vinyl acetate**	$C_4H_6O_2$	**108-05-4**
solvent (D):	**helium**	He	**7440-59-7**

Type of data: cloud-points

w_A	0.420	0.420	0.420	0.420	0.420	0.420	0.415	0.415	0.415
w_B	0.150	0.150	0.150	0.150	0.150	0.150	0.150	0.150	0.150
w_C	0.420	0.420	0.420	0.420	0.420	0.420	0.415	0.415	0.415
w_D	0.010	0.010	0.010	0.010	0.010	0.010	0.020	0.020	0.020
T/K	393.15	413.15	433.15	453.15	473.15	493.15	393.15	413.15	433.15
P/MPa	99.2	96.8	95.0	93.0	91.3	89.9	148.0	139.6	132.0
w_A	0.415	0.415	0.415	0.410	0.410	0.410	0.410	0.410	0.410
w_B	0.150	0.150	0.150	0.150	0.150	0.150	0.150	0.150	0.150
w_C	0.415	0.415	0.415	0.410	0.410	0.410	0.410	0.410	0.410
w_D	0.020	0.020	0.020	0.030	0.030	0.030	0.030	0.030	0.030
T/K	453.15	473.15	493.15	393.15	413.15	433.15	453.15	473.15	493.15
P/MPa	126.8	121.7	117.3	215.6	201.1	185.4	173.2	163.1	153.5

copolymer (B):	**E-VA/28w(2)**		**99KIN**
solvent (A):	**ethene**	C_2H_4	**74-85-1**
solvent (C):	**vinyl acetate**	$C_4H_6O_2$	**108-05-4**
solvent (D):	**helium**	He	**7440-59-7**

Type of data: cloud-points

continue

continue

w_A	0.2100	0.2100	0.2100	0.2100	0.2100	0.2100	0.2075	0.2075	0.2075
w_B	0.1500	0.1500	0.1500	0.1500	0.1500	0.1500	0.1500	0.1500	0.1500
w_C	0.6300	0.6300	0.6300	0.6300	0.6300	0.6300	0.6225	0.6225	0.6225
w_D	0.0100	0.0100	0.0100	0.0100	0.0100	0.0100	0.0200	0.0200	0.0200
T/K	393.15	413.15	433.15	453.15	473.15	493.15	393.15	413.15	433.15
P/MPa	69.4	67.1	65.1	63.7	62.9	62.8	117.3	109.4	102.8
w_A	0.2075	0.2075	0.2075	0.2050	0.2050	0.2050	0.2050	0.2050	0.2050
w_B	0.1500	0.1500	0.1500	0.1500	0.1500	0.1500	0.1500	0.1500	0.1500
w_C	0.6225	0.6225	0.6225	0.6150	0.6150	0.6150	0.6150	0.6150	0.6150
w_D	0.0200	0.0200	0.0200	0.0300	0.0300	0.0300	0.0300	0.0300	0.0300
T/K	453.15	473.15	493.15	393.15	413.15	433.15	453.15	473.15	493.15
P/MPa	97.6	94.0	91.8	165.0	151.9	140.6	131.5	124.9	120.6

copolymer (B):	**E-VA/28w(2)**		**99KIN**
solvent (A):	**ethene**	C_2H_4	**74-85-1**
solvent (C):	**vinyl acetate**	$C_4H_6O_2$	**108-05-4**
solvent (C):	**nitrogen**	N_2	**7727-37-9**

Type of data: cloud-points

w_A	0.6000	0.6000	0.6000	0.6000	0.6000	0.6000	0.5625	0.5625	0.5625
w_B	0.1500	0.1500	0.1500	0.1500	0.1500	0.1500	0.1500	0.1500	0.1500
w_C	0.2000	0.2000	0.2000	0.2000	0.2000	0.2000	0.1875	0.1875	0.1875
w_D	0.0500	0.0500	0.0500	0.0500	0.0500	0.0500	0.1000	0.1000	0.1000
T/K	393.15	413.15	433.15	453.15	473.15	493.15	393.15	413.15	433.15
P/MPa	115.4	112.5	110.0	108.0	105.4	103.0	142.2	136.6	131.1
w_A	0.5625	0.5625	0.5625	0.5250	0.5250	0.5250	0.5250	0.5250	0.5250
w_B	0.1500	0.1500	0.1500	0.1500	0.1500	0.1500	0.1500	0.1500	0.1500
w_C	0.1875	0.1875	0.1875	0.1750	0.1750	0.1750	0.1750	0.1750	0.1750
w_D	0.1000	0.1000	0.1000	0.1500	0.1500	0.1500	0.1500	0.1500	0.1500
T/K	453.15	473.15	493.15	393.15	413.15	433.15	453.15	473.15	493.15
P/MPa	125.5	120.6	116.1	170.0	160.9	153.2	146.2	139.8	133.9

copolymer (B):	**E-VA/28w(2)**		**99KIN**
solvent (A):	**ethene**	C_2H_4	**74-85-1**
solvent (C):	**vinyl acetate**	$C_4H_6O_2$	**108-05-4**
solvent (C):	**nitrogen**	N_2	**7727-37-9**

Type of data: cloud-points

w_A	0.400	0.400	0.400	0.400	0.400	0.400	0.375	0.375	0.375
w_B	0.150	0.150	0.150	0.150	0.150	0.150	0.150	0.150	0.150
w_C	0.400	0.400	0.400	0.400	0.400	0.400	0.375	0.375	0.375
w_D	0.050	0.050	0.050	0.050	0.050	0.050	0.100	0.100	0.100
T/K	393.15	413.15	433.15	453.15	473.15	493.15	393.15	413.15	433.15
P/MPa	84.1	82.5	81.3	79.7	78.5	77.3	107.0	104.0	101.1

continue

continue

w_A	0.375	0.375	0.375	0.350	0.350	0.350	0.350	0.350	0.350
w_B	0.150	0.150	0.150	0.150	0.150	0.150	0.150	0.150	0.150
w_C	0.375	0.375	0.375	0.350	0.350	0.350	0.350	0.350	0.350
w_D	0.100	0.100	0.100	0.150	0.150	0.150	0.150	0.150	0.150
T/K	453.15	473.15	493.15	393.15	413.15	433.15	453.15	473.15	493.15
P/MPa	98.4	96.1	93.9	130.8	126.1	122.0	118.6	115.5	112.5

copolymer (B): **E-VA/28w(2)** **99KIN**
solvent (A): **ethene** **C_2H_4** **74-85-1**
solvent (C): **vinyl acetate** **$C_4H_6O_2$** **108-05-4**
solvent (C): **nitrogen** **N_2** **7727-37-9**

Type of data: cloud-points

w_A	0.2000	0.2000	0.2000	0.2000	0.2000	0.2000	0.1875	0.1875	0.1875
w_B	0.1500	0.1500	0.1500	0.1500	0.1500	0.1500	0.1500	0.1500	0.1500
w_C	0.6000	0.6000	0.6000	0.6000	0.6000	0.6000	0.5625	0.5625	0.5625
w_D	0.0500	0.0500	0.0500	0.0500	0.0500	0.0500	0.1000	0.1000	0.1000
T/K	393.15	413.15	433.15	453.15	473.15	493.15	393.15	413.15	433.15
P/MPa	48.4	49.0	49.4	49.6	50.2	50.7	73.6	72.2	70.9
w_A	0.1875	0.1875	0.1875	0.1750	0.1750	0.1750	0.1750	0.1750	0.1750
w_B	0.1500	0.1500	0.1500	0.1500	0.1500	0.1500	0.1500	0.1500	0.1500
w_C	0.5625	0.5625	0.5625	0.5250	0.5250	0.5250	0.5250	0.5250	0.5250
w_D	0.1000	0.1000	0.1000	0.1500	0.1500	0.1500	0.1500	0.1500	0.1500
T/K	453.15	473.15	493.15	393.15	413.15	433.15	453.15	473.15	493.15
P/MPa	70.0	69.3	68.7	99.8	97.7	95.8	94.1	92.6	91.3

copolymer (B): **E-VA/32w** **84WOH**
solvent (A): **ethene** **C_2H_4** **74-85-1**
solvent (C): **vinyl acetate** **$C_4H_6O_2$** **108-05-4**

Type of data: cloud-points

Comments: the feed (total) monomer weight fraction ratio w_A/w_C = 64/36 was kept constant

w_B	0.05	0.10	0.15	0.20	0.30	0.05	0.10	0.15	0.20
T/K	393.15	393.15	393.15	393.15	393.15	413.15	413.15	413.15	413.15
P/MPa	80.4	77.8	75.8	73.3	71.3	78.3	76.8	75.0	72.3
w_B	0.30	0.05	0.10	0.15	0.20	0.30	0.05	0.10	0.15
T/K	413.15	433.15	433.15	433.15	433.15	433.15	453.15	453.15	453.15
P/MPa	70.3	76.8	76.0	74.0	70.3	68.8	76.3	74.8	73.3
w_B	0.20	0.30							
T/K	453.15	453.15							
P/MPa	68.8	67.8							

copolymer (B):	**E-VA/33w**		**91NIE**
solvent (A):	**ethene**	C_2H_4	**74-85-1**
solvent (C):	**vinyl acetate**	$C_4H_6O_2$	**108-05-4**

Type of data: cloud-points

w_A	0.761	0.761	0.761	0.761	0.761	0.657	0.657	0.657	0.657
w_B	0.073	0.073	0.073	0.073	0.073	0.136	0.136	0.136	0.136
w_C	0.166	0.166	0.166	0.166	0.166	0.207	0.207	0.207	0.207
T/K	348.15	373.15	403.15	433.15	453.15	373.15	403.15	433.15	453.15
P/MPa	65.8	65.4	65.8	66.8	67.8	84.9	82.9	81.4	80.6

w_A	0.505	0.505	0.505	0.361	0.361	0.361	0.430	0.430	0.521
w_B	0.184	0.184	0.184	0.246	0.246	0.246	0.273	0.273	0.084
w_C	0.311	0.311	0.311	0.393	0.393	0.393	0.297	0.297	0.395
T/K	358.15	383.15	438.15	373.15	403.15	433.15	373.15	433.15	373.15
P/MPa	98.1	94.6	90.2	53.4	55.6	58.5	63.1	64.3	61.7

w_A	0.521
w_B	0.084
w_C	0.395
T/K	433.15
P/MPa	64.1

copolymer (B):	**E-VA/34w**		**91FIN**
solvent (A):	**ethene**	C_2H_4	**74-85-1**
solvent (C):	**vinyl acetate**	$C_4H_6O_2$	**108-05-4**

Type of data: coexistence data

T/K = 473.15

The total feed concentration of the homogeneous system before demixing is:

w_A = 0.4395; w_B = 0.2500; w_C = 0.2565

P/ MPa	w_A bottom phase	w_B	w_C	w_A top phase	w_B	w_C
28.15	0.1554	0.7016	0.1489			
29.32	0.1640	0.6923	0.1515	0.6095	0.0700	0.3299
31.28	0.1720	0.7010	0.1498	0.6801	0.0090	0.3305
32.54	0.1863	0.6719	0.1600	0.6573	0.0307	0.3314
36.19	0.1968	0.6550	0.1507			
38.05	0.2118	0.6400	0.1557	0.6211	0.0314	0.3279
50.12	0.2594	0.6017	0.1591	0.6653	0.0381	0.3331
51.40	0.2736	0.5415	0.1870	0.6590	0.0266	0.3318

continue

continue

P/ MPa	w_A	w_B bottom phase	w_C	w_A	w_B top phase	w_C
52.54	0.3404	0.4725	0.2066	0.6541	0.0300	0.3441
59.01	0.3348	0.4848	0.2051	0.6361	0.0552	0.3302
64.90	0.3782	0.4188	0.2262	0.6550	0.0692	0.3210
66.94	0.4114	0.3811	0.2295	0.6427	0.0483	0.3296
71.00	0.4675	0.3011	0.2425	0.6232	0.0734	0.3230

copolymer (B): **E-VA/34w** **91FIN**
solvent (A): **ethene** **C_2H_4** **74-85-1**
solvent (C): **vinyl acetate** **$C_4H_6O_2$** **108-05-4**

Type of data: coexistence data

T/K = 473.15

The total feed concentration of the homogeneous system before demixing is:
$w_A = 0.4125$; $w_B = 0.2500$; $w_C = 0.3375$

P/ MPa	w_A	w_B bottom phase	w_C	w_A	w_B top phase	w_C
20.11	0.1116	0.7008	0.1876	0.5426	0.0392	0.4182
20.70	0.1193	0.6861	0.1947	0.5672	0.0000	0.4328
20.99	0.1171	0.6957	0.1872	0.5395	0.0169	0.4436
21.09	0.1166	0.6876	0.1959	0.5397	0.0303	0.4299
21.68	0.1191	0.6864	0.1944	0.5416	0.0299	0.4284
21.68	0.1204	0.6815	0.1981	0.5366	0.0214	0.4420
29.32	0.1650	0.6410	0.2009	0.5621	0.0232	0.4292
30.70	0.1742	0.6255	0.2047	0.5571	0.0330	0.4217
36.58	0.2132	0.5672	0.2132	0.5274	0.0362	0.4432
38.35	0.2273	0.5445	0.2405	0.5283	0.0414	0.4431
42.77	0.2487	0.6113	0.2400	0.5705	0.0215	0.4070
44.63	0.2630	0.4983	0.2387	0.5659	0.0281	0.4060
48.16	0.2846	0.4441	0.2714	0.5340	0.0547	0.4112
49.15	0.3009	0.4163	0.2828	0.5317	0.0330	0.4350
50.81	0.3188	0.3966	0.2846	0.5334	0.0512	0.4154
51.60	0.3308	0.3757	0.2935	0.4975	0.0870	0.4155
56.99	0.3618	0.3357	0.3025	0.5201	0.0696	0.4102
60.43	0.3881	0.2553	0.3567	0.5047	0.0974	0.3979

copolymer (B):	**E-VA/58w**		**91NIE**
solvent (A):	**ethene**	C_2H_4	**74-85-1**
solvent (C):	**vinyl acetate**	$C_4H_6O_2$	**108-05-4**

Type of data: coexistence data

T/K = 393.15

The total feed concentration of the homogeneous system before demixing is:

w_A = 0.167; w_B = 0.219; w_C = 0.614

P/ MPa	w_A	w_B	w_C	w_A	w_B	w_C
	bottom phase			top phase		
8.3				0.231	0.013	0.756
8.65	0.128	0.350	0.522			
9.6	0.130	0.337	0.533	0.227	0.021	0.752
11.0	0.151	0.286	0.563			
11.65				0.217	0.050	0.733
12.1	0.167	0.219	0.614 (cloud point)			

Comments: some three-phase demixing data below 8 MPa can be found in the original source 91NIE

copolymer (B):	**E-VA/58w**		**91NIE**
solvent (A):	**ethene**	C_2H_4	**74-85-1**
solvent (C):	**vinyl acetate**	$C_4H_6O_2$	**108-05-4**

Type of data: coexistence data

T/K = 393.15

The total feed concentration of the homogeneous system before demixing is:

w_A = 0.251; w_B = 0.200; w_C = 0.549

P/ MPa	w_A	w_B	w_C	w_A	w_B	w_C
	bottom phase			top phase		
9.85	0.112	0.489	0.399	0.330	0.004	0.666

continue

continue

P/ MPa	w_A	w_B	w_C		w_A	w_B	w_C
	bottom phase				top phase		
13.9	0.132	0.487	0.381		0.346	0.004	0.645
17.4	0.150	0.441	0.409		0.343	0.007	0.650
20.0	0.164	0.415	0.421		0.336	0.009	0.650
24.9	0.200	0.328	0.472				
25.2					0.319	0.037	0.644
27.1	0.224	0.272	0.504				
27.7					0.311	0.061	0.628
28.1	0.251	0.200	0.549	(cloud point)			

Comments: some three-phase demixing data below 10 MPa can be found in the original source 91NIE

copolymer (B):	**E-VA/58w**		**91NIE**
solvent (A):	**ethene**	C_2H_4	**74-85-1**
solvent (C):	**vinyl acetate**	$C_4H_6O_2$	**108-05-4**

Type of data: cloud-points

w_A	0.096	0.096	0.096	0.148	0.152	0.152	0.160	0.196	0.196
w_B	0.200	0.200	0.200	0.214	0.214	0.214	0.205	0.198	0.198
w_C	0.704	0.704	0.704	0.638	0.642	0.642	0.635	0.606	0.606
T/K	363.15	393.15	423.15	393.15	363.15	393.15	393.15	363.15	393.15
P/MPa	4.2	5.3	6.8	7.9	6.3	8.4	9.8	12.0	18.0

w_A	0.196	0.299	0.401	0.401	0.401
w_B	0.198	0.203	0.204	0.204	0.204
w_C	0.606	0.498	0.405	0.405	0.405
T/K	423.15	363.15	363.15	393.15	423.15
P/MPa	23.3	36.0	65.0	64.5	64.2

4.4. Table of systems where ternary HPPE data were published only in graphical form as phase diagrams or related figures

Copolymer (B)	Second and third component	Ref.
Ethylene/acrylic acid copolymer		
	n-butane and dimethyl ether	96LE2
	n-butane and ethanol	96LE1
	n-butane and ethanol	96LE2
Ethylene/ethyl acrylate copolymer		
	1-butene and polyethylene	98LEE
Ethylene/2-ethylhexyl acrylate copolymer		
	ethene and 2-ethylhexyl acrylate	96BUB
Ethylene/1-hexene copolymer		
	ethene and n-butane	2000KIN
	ethene and carbon dioxide	2000KIN
	ethene and 1-hexene	2000KIN
	ethene and nitrogen	2000KIN
	2-methylpropane and 1-hexene	99PAN
Ethylene/methacrylic acid copolymer		
	n-butane and dimethyl ether	97LEE
	n-butane and ethanol	97LEE
Ethylene/methyl acrylate copolymer		
	n-butane and ethylene/ methyl acrylate copolymer	98LEE
	n-butane and ethylene/ vinyl alcohol copolymer	98LEE
	1-butene and ethylene/ methyl acrylate copolymer	98LEE
	chlorodifluoromethane and ethanol	92MEI
	chlorodifluoromethane and ethanol	93HA1
	chlorodifluoromethane and 2-propanone	92MEI
	chlorodifluoromethane and 2-propanone	93HA1
	ethene and methyl acrylate	87LUF
	propane and 1-butanol	94LOS

Copolymer (B)	Second and third component	Ref.
Ethylene/methyl acrylate copolymer		
	propane and ethanol	92MEI
	propane and ethanol	93HA1
	propane and ethanol	93HA2
	propane and ethanol	94LOS
	propane and n-hexane	94LOS
	propane and 1-hexene	94LOS
	propane and methanol	94LOS
	propane and 1-propanol	94LOS
	propane and 2-propanone	92MEI
	propane and 2-propanone	93HA1
	propane and 2-propanone	93HA2
Ethylene/propylene copolymer		
	carbon dioxide and C6-fraction solvent	85MCH
	ethene and C6-fraction solvent	85MCH
	ethene and 1-butene	93CHE
	ethene and n-hexane	85MCC
	ethene and 1-hexene	93CHE
	methane and C6-fraction solvent	85MCH
	propene and C6-fraction solvent	85MCH
	propene and C6-fraction solvent	86IRA
	propene and 2-methylbutane	86IRA
Ethylene/vinyl acetate copolymer		
	ethene and n-butane	2000KIN
	ethene and carbon dioxide	2000KIN
	ethene and nitrogen	2000KIN
	ethene and vinyl acetate	80RAE
	ethene and vinyl acetate	83WOH
	ethene and vinyl acetate	92FIN
	ethene and vinyl acetate	96FOL
	ethene and vinyl acetate	2000KIN
Ethylene/vinyl alcohol copolymer		
	n-butane and ethylene/ methyl acrylate copolymer	98LEE
Tetrafluoroethylene/hexafluoropropylene copolymer		
	carbon dioxide and sulfur hexafluoride	99MER
	carbon dioxide and trifluoromethane	99MER
	trifluoromethane and sulfur hexafluoride	99MER

4.5. References

80RAE Raetzsch, M., Findeisen, R., and Sernov, V.S., Untersuchungen zum Phasenverhalten von Monomer-Polymer-Systemen unter hohem Druck, *Z. Phys. Chemie, Leipzig*, 261, 995, 1980.

82RAE Raetzsch, M.T., Wagner, P., Wohlfarth, C., and Heise, D., High-pressure phase equilibrium studies in mixtures of ethylene and (ethylene-vinyl acetate) copolymers. Part I. Dependence on vinyl acetate content of copolymers (Ger.), *Acta Polym.*, 33, 463, 1982.

82WOH Wohlfarth, C., Wagner, P., Raetzsch, M.T., and Westmeier, S., High-pressure phase equilibrium studies in mixtures of ethylene and (ethylene-vinyl acetate) copolymers. Part II. Temperature effect (Ger.), *Acta Polym.*, 33, 468, 1982.

83RAE Raetzsch, M.T., Wagner, P., Wohlfarth, C., and Gleditzsch, S., Studies of phase equilibriums in mixtures of ethylene and ethylene-vinyl acetate copolymers at high pressures. Part III. Dependence on the molar weight distribution (Ger.), *Acta Polym.*, 34, 340, 1983.

83WOH Wohlfarth, C. and Rätzsch, M.T., Calculation of high-pressure phase equilibria in mixtures of ethylene, vinyl acetate, and ethylene-vinyl acetate copolymers, *Acta Polym.*, 34, 255, 1983.

84WAG Wagner, P., Zum Phasengleichgewicht im System Ethylen + Ethylen-Vinylacetat-Copolymer unter hohem Druck, *PhD-Thesis*, TH Leuna-Merseburg, 1984.

84WOH Wohlfarth, Ch., Wagner, P., Glindemann, D., Voelkner, M., and Rätzsch, M.T., High-pressure phase equilibrium in the system ethylene + vinyl acetate + (ethylene-vinyl acetate) copolymers (Ger.), *Acta Polym.*, 35, 498, 1984.

85MCC McClellan, A.K. and McHugh, M.A., Separating polymer solutions using high pressure lower critical solution temperature (LCST) phenomena, *Polym. Eng. Sci.*, 25, 1088, 1985.

85MCH McHugh, M.A. and Guckes, T.L., Separating polymer solutions with supercritical fluids, *Macromolecules*, 18, 674, 1985.

86IRA Irani, C.A. and Cozewith, C., Lower critical solution temperature behavior of ethylene propylene copolymers in multicomponent solvents, *J. Appl. Polym. Sci.*, 31, 1879, 1986.

87LUF Luft, G. and Subramanian, N.S., Phase behavior of mixtures of ethylene, methyl acrylate, and copolymers under high pressures, *Ind. Eng. Chem. Res.*, 26, 750, 1987.

91FIN Finck, U., Phasengleichgewichte in Monomer + Copolymer - Systemen, *Diploma paper*, TH Leuna-Merseburg, 1991.

91MEI Meilchen, M.A., Hasch, B.M., and McHugh, M.A., Effect of copolymer composition on the phase behavior of mixtures of poly(ethylene-co-methyl acrylate) with propane and chlorodifluoromethane, *Macromolecules*, 24, 4874, 1991.

91NIE Nieszporek, B., Untersuchungen zum Phasenverhalten quasibinärer und quasiternärer Mischungen aus Polyethylen, Ethylen-Vinylacetat-Copolymeren, Ethylen und Vinylacetat unter Druck, *PhD-Thesis*, TH Darmstadt, 1991.

92CH1 Chen, S.-J. and Radosz, M., Density-tuned polyolefin phase equilibria 1, *Macromolecules*, 25, 3089, 1992.

92CH2 Chen, S.-J., Economou, I.G., Radosz, M., Density-tuned polyolefin phase equilibria 2, *Macromolecules*, 25, 4987, 1992.

92FIN Finck, U., Wohlfarth, Ch., and Heuer, T., Calculation of high pressure phase equilibria of mixtures of ethylene, vinyl acetate and an (ethylene-vinyl acetate) copolymer, *Ber. Bunsenges. Phys. Chem.*, 96, 179, 1992.

92HAS Hasch, B.A., Meilchen, M.A., Lee, S.-H., and McHugh, M.A., High-pressure phase behavior of mixtures of poly(ethylene-co-methyl acrylate) with low-molecular weight hydrocarbons, *J. Polym. Sci., Part B: Polym. Phys.*, 30, 1365, 1992.

92LUF Luft, G. and Wind, R.W., Phasenverhalten von Mischungen aus Ethylen und Ethylen-Acrylsäure-Copolymeren unter hohem Druck, *Chem.-Ing.-Techn.*, 64, 1114, 1992.

92MEI Meilchen, M.A., Hasch, B.M., Lee, S.-H., and McHugh, M.A., Poly(ethylene-co-methyl acrylate)-solvent-cosolvent phase behaviour at high pressures, *Polymer*, 33, 1922, 1992.

92WIN Wind, R.W., Untersuchungen zum Phasenverhalten von Mischungen aus Ethylen, Acrylsäure und Ethylen-Acrylsäure-Copolymer unter hohem Druck, *PhD-Thesis*, TH Darmstadt, 1992.

93CHE Chen, S.-J., Economou, I.G., and Radosz, M., Phase behavior of LCST and UCST solutions of branchy copolymers. Experiment and SAFT modelling, *Fluid Phase Equil.*, 83, 391, 1993

93GRE Gregg, C.J., Chen, S.-J., Stein, F.P., and Radosz, M., Phase behavior of binary ethylene-propylene copolymer solutions in sub- and supercritical ethylene and propylene, *Fluid Phase Equil.*, 83, 375, 1993.

93HA1 Hasch, B.M., Meilchen, M.A., Lee, S.-H., and McHugh, M.A., Cosolvency effects on copolymer solutions at high pressure, *J. Polym. Sci., Part B: Polym. Phys.*, 31, 429, 1993.

93HA2 Hasch, B.M., Lee, S.-H., and McHugh, M.A., The effect of copolymer architecture on solution behavior, *Fluid Phase Equil.*, 83, 341, 1993.

93PRA Pratt, J.A., Lee, S.-H., and McHugh, M.A., Supercritical fluid fractionation of copolymers based on chemical composition and molecular weight, *J. Appl. Polym. Sci.*, 49, 953, 1993.

94GR1 Gregg, C.J., Phase equilibria of supercritical fluid solutions of associating polymers in nonpolar and polar fluids, *PhD-Thesis*, Lehigh University, Bethlehem, 1994.

94GR2 Gregg, C.J., Stein, F.P., Morgan, C.K., and Radosz, M., A variable-volume optical pressure-volume-temperature cell for high-pressure cloud points, densities, and IR-spectra, applicable to supercritical fluid solutions of polymers up to 2 kbar, *J. Chem. Eng. Data*, 39, 219, 1994.

94LE1 Lee, S.-H., LoStracco, M.A., Hasch, B.M., and McHugh, M.A., Solubility of poly(ethylene-co-acrylic acid) in low molecular weight hydrocarbons, *J. Phys. Chem.*, 98, 4055, 1994.

94LE2 Lee, S.-H., LoStracco, A., and McHugh, M.A., High pressure, molecular weight-dependent behavior of (co)polymer-solvent mixtures, *Macromolecules*, 27, 4652, 1994.

94LOS LoStracco, M.A., Lee, S.-H., and McHugh, M.A., Comparison of the effect of density and hydrogen bonding on the cloud point behavior of poly(ethylene-co-methyl acrylate)-propane-cosolvent mixtures, *Polymer*, 35, 3272, 1994.

95CHE Chen, S., Banaszak, M., and Radosz, M., Phase behavior of poly(ethylene-1-butene) in subcritical and supercritical propane, *Macromolecules*, 28, 1812, 1995.

96ALB Albrecht, K.L., Stein, F.P., Han, S.J., Gregg, C.J., and Radosz, M., Phase equilibria of saturated and unsaturated polyisoprene in sub- and supercritical ethane, ethylene, propane, propylene, and dimethyl ether, *Fluid Phase Equil.*, 117, 84, 1996.

96BUB Buback, M., Busch, M., Dietzsch, H., Droege, T., and Lovis, K., Cloud-point curves in ethylene-acrylate-poly(ethylene-co-acrylate) systems, in *High Pressure Chemical Engineering*, von Rohr, Ph.R., Trepp, Ch., Eds., Elsevier Sci. B.V., 1996, 175.

96BYU Byun, H.-S., Hasch, B.M., McHugh, M.A., Maehling, F.-O., Busch, M., and Buback, M., Poly(ethylene-co-butyl acrylate). Phase behavior in ethylene compared to the poly(ethylene-co-methyl acrylate)-ethylene system and aspects of copolymerization kinetics at high pressures, *Macromolecules*, 29, 1625, 1996.

96DEL DeLoos, Th.W., de Graaf, L.J., and de Swaan Arons, J., Liquid-liquid phase separation in linear low density polyethylene - solvent systems, *Fluid Phase Equil.*, 117, 40, 1996.

96FOL Folie, B., Gregg, C., Luft, G., and Radosz, M., Phase equilibria of poly(ethylene-co-vinyl acetate) copolymers in subcritical and supercritical ethylene and ethylene - vinyl acetate mixtures, *Fluid Phase Equil.*, 120, 11, 1996.

96LE1 Lee, S.-H., Hasch, B.M., and McHugh, M.A., Calculating copolymer solution behavior with statistical associating fluid theory, *Fluid Phase Equil.*, 117, 61, 1996.

96LE2 Lee, S.-H., LoStracco, M.A., and McHugh, M.A., Cosolvent effect on the phase behavior of poly(ethylene-co-acrylic acid) - butane mixtures, *Macromolecules*, 29, 1349, 1996.

96LE3 Lee, S.-H. and McHugh, M.A., Influence of chain architecture on high-pressure co-polymer solution behavior: Experiments and modeling, in *High Pressure Chemical Engineering*, von Rohr, Ph.R., Trepp, Ch., Eds., Elsevier Sci. B.V., 1996, 11.

96MER Mertdogan, C.A., Byun, H.-S., McHugh, M.A., and Tuminello, W.H., Solubility of poly (tetrafluoroethylene-co-19 mol% hexafluoropropylene) in supercritical CO_2 and halogenated supercritical solvents, *Macromolecules*, 29, 6548, 1996.

96MUE Mueller, C., Untersuchungen zum Phasenverhalten von quasibinären Gemischen aus Ethylen und Ethylen-Copolymeren, *PhD-Thesis*, Univ. Karlsruhe (TH), 1996.

96RIN Rindfleisch, F., DiNoia, T.P., and McHugh, M.A., Solubility of polymers and copolymers in supercritical CO_2, *J. Phys. Chem.*, 100, 15581, 1996.

97HAN Han, S.J., Gregg, C.J., and Radosz, M., How the solute polydispersity affects the cloud-point and coexistence pressures in propylene and ethylene solutions of alternating poly(ethylene-co-propylene), *Ind. Eng. Chem. Res.*, 36, 5520, 1997.

97LEE Lee, S.-H. and McHugh, M.A., Phase behaviour studies with poly(ethylene-co-methacrylic acid) at high pressures, *Polymer*, 38, 1317, 1997.

97LEP Lepilleur, C., Beckman, E.J., Schonemann, H., and Krukonis, V.J., Effect of molecular architecture on the phase behavior of fluoroether-functional graft copolymers in supercritical CO_2, *Fluid Phase Equil.*, 134, 285, 1997.

97MER Mertdogan, C.A., DiNoia, T.P., and McHugh, M.A., Impact of backbone architecture on the solubility of fluoropolymers in supercritical CO_2 and halogenated supercritical solvents, *Macromolecules*, 30, 7511, 1997.

97WHA Whaley, P.D., Winter, H.H., and Ehrlich, P., Phase equilibria of polypropylene with compressed propane and related systems. 2, *Macromolecules*, 30, 4887, 1997.

98HAN Han, S.J., Lohse, D.J., Radosz, M., and Sperling, L.H., Short chain branching effect on the cloud-point pressures of ethylene copolymers in subcritical and supercritical propane, *Macromolecules*, 31, 2533, 1998.

98LEE Lee, S.-H. and McHugh, M.A., Phase behavior of copolymer-copolymer-solvent mixtures at high pressures, *Polymer*, 39, 5447, 1998.

98MCH McHugh, M.A., Mertdogan, C.A., DiNoia, T.P., Anolick, C., Tuminello, W.H., and Wheland, R., Impact of melting temperature on poly(tetrafluoroethylene-co-hexafluoropropylene) solubility in supercritical fluid solvents, *Macromolecules*, 31, 2252, 1998.

98ONE O'Neill, M.L., Cao, Q., Fang, C.M., Johnston, K.P., Wilkinson, S.P., Smith, C.D., Kerschner, J.L., and Jureller, S.H., Solubility of homopolymers and copolymers in carbon dioxide, *Ind. Eng. Chem. Res.*, 37, 3067, 1998.

99KIN Kinzl, M., Einfluss der Zugabe von Inertkomponenten auf die Phasengleichgewichte von Copolymerlösungen in überkritischem Ethen, *PhD-Thesis*, TH Darmstadt, 1999.

99LOR Lora, M., Rindfleisch, F., and McHugh, M.A., Influence of the alkyl tail on the solubility of poly(alkyl acrylates) in ethylene and CO_2 at high pressures: Experiments and modeling, *J. Appl. Polym. Sci.*, 73, 1979, 1999.

99MER Mertdogan, C.A., McHugh, M.A., and Tuminello, W.H., Cosolvency effect of SF_6 on the solubility of poly(tetrafluoroethylene-co-19 mol% hexafluoropropylene) in supercritical CO_2 and CHF_3, *J. Appl. Polym. Sci.*, 74, 2039, 1999.

99PAN Pan, C. and Radosz, M., Phase behavior of poly(ethylene-co-hexene-1) solutions in isobutane and propane, *Ind. Eng. Chem. Res.*, 38, 2842, 1999.

2000CH1 Chan, K.C., Adidharma, H., and Radosz, M., Fluid-liquid and fluid-solid transitions of poly(ethylene-co-octene-1) in sub- and supercritical propane solutions, *Ind. Eng. Chem. Res.*, 39, 3069, 2000.

2000CH2 Chan, A.K.C., Russo, P.S., and Radosz, M., Fluid-liquid equilibria in poly(ethylene-co-hexene-1) + propane: a light-scattering probe of cloud-point pressure and critical polymer concentration, *Fluid Phase Equil.*, 173, 149, 2000.

2000CH3 Chan, A.K.C. and Radosz, M., Fluid-liquid and fluid-solid phase behavior of poly(ethylene-co-hexene-1) solutions in sub- and supercritical propane, ethylene, and ethylene + hexene-1, *Macromolecules*, 33, 6800, 2000.

2000DIN DiNoia, T.P., Conway, S.E., Lim, J.S., McHugh, M.A., Solubility of vinylidene fluoride polymers in supercritical CO_2 and halogenated solvents, *J. Polym. Sci., Part B: Polym. Phys.*, 38, 2832, 2000

2000KIN Kinzl, M., Luft, G., Adidharma, H., and Radosz, M., SAFT modeling of inert-gas effects on the cloud-point pressures in ethylene copolymerization systems: Poly(ethylene-co-vinyl acetate) + vinyl acetate + ethylene and poly(ethylene-co-hex-1-ene) + 1-hexene + ethylene with carbon dioxide, nitrogen, or butane, *Ind. Eng. Chem. Res.*, 39, 541, 2000.

5. ENTHALPY CHANGES FOR BINARY COPOLYMER SOLUTIONS

5.1. Enthalpies of mixing or intermediary enthalpies of dilution, and copolymer partial enthalpies of mixing (at infinite dilution), or copolymer (first) integral enthalpies of solution

Copolymers from acrylonitrile and butadiene

Average chemical composition of the copolymers, acronyms and references:

Copolymer (B)	Acronym	Ref.
Acrylonitrile/butadiene copolymer		
18.0 wt% acrylonitrile	AN-B/18w	55TA1
26.0 wt% acrylonitrile	AN-B/26w(1)	50TAG
26.0 wt% acrylonitrile	AN-B/26w(1)	51TAG
26.0 wt% acrylonitrile	AN-B/26w(2)	55TA1
40.0 wt% acrylonitrile	AN-B/40w(1)	50TAG
40.0 wt% acrylonitrile	AN-B/40w(1)	51TAG
40.0 wt% acrylonitrile	AN-B/40w(2)	55TA1

Characterization:

Copolymer (B)	M_n/ g/mol	M_w/ g/mol	M_η/ g/mol	Further information
AN-B/18w				no data
AN-B/26w(1)				Buna No. 26
AN-B/26w(2)				no data
AN-B/40w(1)				Buna No. 40
AN-B/40w(2)				no data

Experimental enthalpy data:

copolymer (B):	**AN-B/18w**		**55TA1**
solvent (A):	**benzene**	C_6H_6	**71-43-2**

T/K = 298.15

$\Delta_{sol}H_B^{\infty}$ = 0.0 J/(g copolymer)

Comments: the final concentration is about 2 g copolymer/132 g solvent

copolymer (B):	**AN-B/26w(1)**		**50TAG, 51TAG**
solvent (A):	**benzene**	C_6H_6	**71-43-2**

T/K = 298.15

$\Delta_{sol}H_B^{\infty}$ = 5.82 J/(g copolymer)

Comments: the final concentration is about 0.55 g copolymer/100 g solution

copolymer (B):	**AN-B/26w(2)**		**55TA1**
solvent (A):	**benzene**	C_6H_6	**71-43-2**

T/K = 298.15

$\Delta_{sol}H_B^{\infty}$ = −1.88 J/(g copolymer)

Comments: the final concentration is about 3 g copolymer/132 g solvent

copolymer (B):	**AN-B/40w(1)**		**50TAG, 51TAG**
solvent (A):	**benzene**	C_6H_6	**71-43-2**

T/K = 298.15

$\Delta_{sol}H_B^{\infty}$ = −7.0 J/(g copolymer)

Comments: the final concentration is about 0.85 g copolymer/100 g solution

copolymer (B):	**AN-B/40w(2)**		**55TA1**
solvent (A):	**benzene**	C_6H_6	**71-43-2**

T/K = 298.15

$\Delta_{sol}H_B^{\infty}$ = −2.93 J/(g copolymer)

Comments: the final concentration is about 3 g copolymer/88 g solvent

Copolymers from acrylonitrile and isoprene

Average chemical composition of the copolymers, acronyms and references:

Copolymer (B)	Acronym	Ref.
Acrylonitrile/isoprene copolymer 85.0 mol% acrylonitrile	AN-IP/85x	76PET

Characterization:

Copolymer (B)	M_n/ g/mol	M_w/ g/mol	M_η/ g/mol	Further information
AN-IP/85x				no data

Experimental enthalpy data:

copolymer (B):	**AN-IP/85x**		**76PET**
solvent (A):	**N,N-dimethylformamide**	**C_3H_7NO**	**68-12-2**

T/K = 323.15

$\Delta_{sol}H_B^{\infty} = -31.8$ J/(g copolymer)

Comments: additional information is given for varying isoprene contents between 2 and 22 mol% in the copolymer in Fig. 1 of the original source 76PET

Copolymers from acrylonitrile and vinyl chloride

Average chemical composition of the copolymers, acronyms and references:

Copolymer (B)	Acronym	Ref.
Acrylonitrile/vinyl chloride copolymer		
13.0 wt% acrylonitrile	AN-VC/13w	59ZEL
29.0 wt% acrylonitrile	AN-VC/29w	59ZEL
40.0 wt% acrylonitrile	AN-VC/40w	59ZEL

Characterization:

Copolymer (B)	M_n/ g/mol	M_w/ g/mol	M_η/ g/mol	Further information
AN-VC/13w				no data
AN-VC/29w				no data
AN-VC/40w				no data

Experimental enthalpy data:

copolymer (B):	**AN-VC/13w**		**59ZEL**
solvent (A):	**N,N-dimethylformamide**	**C_3H_7NO**	**68-12-2**

T/K	295.15	308.15	308.15	323.15	323.15	338.15	338.15
$\Delta_{sol}H_B^\infty$/(g copolymer)	−37.95	−22.18	−21.34	−19.25	−16.95	−15.48	−14.64
T/K	353.15	353.15					
$\Delta_{sol}H_B^\infty$/(g copolymer)	−12.97	−11.30					

Comments: the final concentration is about 1 g copolymer/165 g solvent

copolymer (B): **AN-VC/29w** **59ZEL**
solvent (A): **N,N-dimethylformamide** **C_3H_7NO** **68-12-2**

T/K	295.15	295.15	308.15	308.15	323.15	323.15	338.15
$\Delta_{sol}H_B^{\infty}$/(g copolymer)	–43.93	–40.21	–27.61	–26.78	–22.18	–20.08	–20.50
T/K	338.15	353.15	353.15				
$\Delta_{sol}H_B^{\infty}$/(g copolymer)	–17.99	–16.32	–15.90				

Comments: the final concentration is about 1 g copolymer/165 g solvent

copolymer (B): **AN-VC/40w** **59ZEL**
solvent (A): **N,N-dimethylformamide** **C_3H_7NO** **68-12-2**

T/K	295.15	295.15	308.15	323.15	323.15	338.15	338.15
$\Delta_{sol}H_B^{\infty}$/(g copolymer)	–47.28	–46.65	–30.54	–28.87	–26.36	–20.08	–19.66
T/K	353.15						
$\Delta_{sol}H_B^{\infty}$/(g copolymer)	–17.15						

Comments: the final concentration is about 1 g copolymer/165 g solvent

Copolymers from acrylonitrile and vinylidene chloride

Average chemical composition of the copolymers, acronyms and references:

Copolymer (B)	Acronym	Ref.
Acrylonitrile/vinylidene chloride copolymer unspecified contents of acrylonitrile	AN-VdC/?	55ZAZ

Characterization:

Copolymer (B)	M_n/ g/mol	M_w/ g/mol	M_η/ g/mol	Further information
AN-VdC/?				fibre "Saniv", USSR

Experimental enthalpy data:

copolymer (B):	**AN-VdC/?**		**55ZAZ**
solvent (A):	**2-propanone**	**C_3H_6O**	**67-64-1**

T/K = 298.15 $\Delta_{sol}H_B^{\infty}$ = −28.03 J/g copolymer

T/K = 298.15 $\Delta_{sol}H_B^{\infty}$ = −19.25 J/g copolymer (fibre drawn 156%)

T/K = 298.15 $\Delta_{sol}H_B^{\infty}$ = −15.92 J/g copolymer (fibre drawn 300%)

T/K = 298.15 $\Delta_{sol}H_B^{\infty}$ = −14.23 J/g copolymer (fibre drawn 400%)

Copolymers from bisphenol A-isophthaloyl chloride and terephthaloyl chloride

Average chemical composition of the copolymers, acronyms and references:

Copolymer (B)	Acronym	Ref.
Bisphenol A-isophthaloyl chloride/terephthaloyl chloride copolymer 50 mol% terephthaloyl chloride	BAIPC-TPC/50x	78SOK

Characterization:

Copolymer (B)	M_n/ g/mol	M_w/ g/mol	M_η/ g/mol	Further information
BAIPC-TPC/50x				synthesized in the laboratory

Experimental enthalpy data:

copolymer (B):	**BAIPC-TPC/50x**		**78SOK**
solvent (A):	**N,N-dimethylacetamide**	C_4H_9NO	**127-19-5**

T/K = 298.15

$\Delta_{sol}H_B^{\infty} = -56.5$ J/(g copolymer)

Comments: amorphous sample

copolymer (B):	**BAIPC-TPC/50x**		**78SOK**
solvent (A):	**1,1,2,2-tetrachloroethane**	$C_2H_2Cl_4$	**79-34-5**

T/K = 298.15

$\Delta_{sol}H_B^{\infty} = 72.5$ J/(g copolymer)

Comments: amorphous sample

copolymer (B):	**BAIPC-TPC/50x**		**78SOK**
solvent (A):	**1,1,2,2-tetrachloroethane**	$C_2H_2Cl_4$	**79-34-5**

T/K = 298.15

$\Delta_{sol}H_B^{\infty} = 41.9$ J/(g copolymer)

Comments: partially crystalline sample

Copolymers from butadiene and styrene

Average chemical composition of the copolymers, acronyms and references:

Copolymer (B)	Acronym	Ref.
Butadiene/styrene copolymer		
10.0 wt% styrene	S-BR/10w	55TA2
10.0 wt% styrene	S-BR/10w	58TA2
30.0 wt% styrene	S-BR/30w(1)	55TA2
30.0 wt% styrene	S-BR/30w(1)	58TA2

continue

continue

Copolymer (B)	Acronym	Ref.
Butadiene/styrene copolymer		
30.0 wt% styrene	S-BR/30w(2)	56STR
30.0 wt% styrene	S-BR/30w(2)	58SLO
50.0 wt% styrene	S-BR/50w	55TA2
50.0 wt% styrene	S-BR/50w	58TA2
60.0 wt% styrene	S-BR/60w	55TA2
60.0 wt% styrene	S-BR/60w	58TA2
70.0 wt% styrene	S-BR/70w	55TA2
70.0 wt% styrene	S-BR/70w	58TA2
75.0 wt% styrene	S-BR/75w	50TAG
75.0 wt% styrene	S-BR/75w	51TAG
80.0 wt% styrene	S-BR/80w	55TA2
80.0 wt% styrene	S-BR/80w	58TA2
90.0 wt% styrene	S-BR/90w	55TA2
90.0 wt% styrene	S-BR/90w	58TA2

Characterization:

Copolymer (B)	M_n/ g/mol	M_w/ g/mol	M_η/ g/mol	Further information
S-BR/10w				synthesized in the laboratory
S-BR/30w(1)				synthesized in the laboratory
S-BR/30w(2)				intrinsic viscosity = 1.50
S-BR/50w				synthesized in the laboratory
S-BR/60w				synthesized in the laboratory
S-BR/70w				synthesized in the laboratory
S-BR/75w				Buna-S
S-BR/80w				synthesized in the laboratory
S-BR/90w				synthesized in the laboratory

Experimental enthalpy data:

copolymer (B): **S-BR/10w** **55TA2**
solvent (A): **benzene** **C_6H_6** **71-43-2**

T/K = 293.65

w_B	0.1	0.2	0.3	0.4	0.5	0.7	0.8	0.99
$\Delta_M H$/(J/g)	0.33	0.63	0.92	1.26	1.59	2.43	2.55	0.42

copolymer (B):	**S-BR/10w**		**55TA2**
solvent (A):	**benzene**	**C_6H_6**	**71-43-2**

T/K = 293.65

$\Delta_{sol}H_B^{\infty}$ = 4.94 J/(g copolymer)

Comments: the final concentration is about 2.7 g copolymer/100 ml solvent

copolymer (B):	**S-BR/10w**		**55TA2**
solvent (A):	**ethylbenzene**	**C_8H_{10}**	**100-41-4**
solvent (C):	**2,2,4-trimethylpentane**	**C_8H_{18}**	**540-84-1**

T/K = 298.15

$\Delta_{sol}H_B^{\infty}$ = 1.46 J/(g copolymer)

Comments: solvent mixture concentration 10 wt% C_8H_{10} / 90 wt% C_8H_{18}

copolymer (B):	**S-BR/30w(1)**		**55TA2**
solvent (A):	**benzene**	**C_6H_6**	**71-43-2**

T/K = 293.65

w_B	0.1	0.2	0.3	0.4	0.5	0.685	0.943	0.96
$\Delta_M H$/(J/g)	0.17	0.38	0.54	0.71	0.88	1.21	0.92	0.67

copolymer (B):	**S-BR/30w(1)**		**55TA2**
solvent (A):	**benzene**	**C_6H_6**	**71-43-2**

T/K = 293.65

$\Delta_{sol}H_B^{\infty}$ = 3.01 J/(g copolymer)

Comments: the final concentration is about 2.7 g copolymer/100 ml solvent

copolymer (B):	**S-BR/30w(2)**		**56STR, 58SLO**
solvent (A):	**benzene**	**C_6H_6**	**71-43-2**

T/K = 298.15

$\Delta_{sol}H_B^{\infty}$ = 3.05 J/(g copolymer)

Comments: final concentration 1 g copolymer/100 g solvent

copolymer (B):	**S-BR/30w(1)**		**55TA2**
solvent (A):	**ethylbenzene**	C_8H_{10}	**100-41-4**
solvent (C):	**2,2,4-trimethylpentane**	C_8H_{18}	**540-84-1**

T/K = 298.15

$\Delta_{sol}H_B^{\infty}$ = 0.75 J/(g copolymer)

Comments: solvent mixture concentration 30 wt% C_8H_{10} / 70 wt% C_8H_{18}

copolymer (B):	**S-BR/50w**		**55TA2**
solvent (A):	**benzene**	C_6H_6	**71-43-2**

T/K = 293.65

$\Delta_{sol}H_B^{\infty}$ = 1.80 J/(g copolymer)

Comments: the final concentration is about 4 g copolymer/100 ml solvent

copolymer (B):	**S-BR/50w**		**55TA2**
solvent (A):	**ethylbenzene**	C_8H_{10}	**100-41-4**
solvent (C):	**2,2,4-trimethylpentane**	C_8H_{18}	**540-84-1**

T/K = 298.15

$\Delta_{sol}H_B^{\infty}$ = 0.54 J/(g copolymer)

Comments: solvent mixture concentration 50 wt% C_8H_{10} / 50 wt% C_8H_{18}

copolymer (B):	**S-BR/60w**		**55TA2**
solvent (A):	**benzene**	C_6H_6	**71-43-2**

T/K = 293.65

w_B	0.1	0.2	0.3	0.4	0.5	0.6	0.7	0.8	0.9
$\Delta_M H$/(J/g)	0.0	0.0	0.0	0.0	0.0	0.0	0.0	0.0	0.0

copolymer (B):	**S-BR/60w**		**55TA2**
solvent (A):	**benzene**	C_6H_6	**71-43-2**

T/K = 293.65

$\Delta_{sol}H_B^{\infty}$ = 0.0 J/(g copolymer)

Comments: the final concentration is about 4 g copolymer/100 ml solvent

copolymer (B):	**S-BR/60w**		**55TA2**
solvent (A):	**ethylbenzene**	C_8H_{10}	**100-41-4**
solvent (C):	**2,2,4-trimethylpentane**	C_8H_{18}	**540-84-1**

T/K = 298.15

$\Delta_{sol}H_B^{\infty}$ = 0.42 J/(g copolymer)

Comments: solvent mixture concentration 60 wt% C_8H_{10} / 40 wt% C_8H_{18}

copolymer (B):	**S-BR/70w**		**55TA2**
solvent (A):	**benzene**	C_6H_6	**71-43-2**

T/K = 293.65

$\Delta_{sol}H_B^{\infty}$ = 0.0 J/(g copolymer)

Comments: the final concentration is about 4.4 g copolymer/100 ml solvent

copolymer (B):	**S-BR/70w**		**55TA2**
solvent (A):	**ethylbenzene**	C_8H_{10}	**100-41-4**
solvent (C):	**2,2,4-trimethylpentane**	C_8H_{18}	**540-84-1**

T/K = 298.15

$\Delta_{sol}H_B^{\infty}$ = 0.33 J/(g copolymer)

Comments: solvent mixture concentration 70 wt% C_8H_{10} / 30 wt% C_8H_{18}

copolymer (B):	**S-BR/75w**		**50TAG, 51TAG**
solvent (A):	**benzene**	C_6H_6	**71-43-2**

T/K = 298.15

$\Delta_{sol}H_B^{\infty}$ = 1.50 J/(g copolymer)

Comments: the final concentration is about 3 g copolymer/100 ml solvent

copolymer (B):	**S-BR/80w**		**55TA2**
solvent (A):	**benzene**	C_6H_6	**71-43-2**

T/K = 293.65

$\Delta_{sol}H_B^{\infty}$ = –0.59 J/(g copolymer)

Comments: the final concentration is about 3.5 g copolymer/150 ml solvent

copolymer (B):	**S-BR/80w**		**55TA2**
solvent (A):	**ethylbenzene**	C_8H_{10}	**100-41-4**
solvent (C):	**2,2,4-trimethylpentane**	C_8H_{18}	**540-84-1**

T/K = 298.15

$\Delta_{sol}H_B^\infty$ = 0.21 J/(g copolymer)

Comments: solvent mixture concentration 80 wt% C_8H_{10} / 20 wt% C_8H_{18}

copolymer (B):	**S-BR/90w**		**55TA2**
solvent (A):	**benzene**	C_6H_6	**71-43-2**

T/K = 293.65

w_B	0.10	0.20	0.30	0.44	0.55	0.65	0.79	0.81	0.86	0.93
$\Delta_M H$/(J/g)	–0.50	–1.00	–1.46	–2.13	–2.72	–3.18	–3.89	–4.02	–4.22	–3.39
w_B	0.95	0.95	0.96							
$\Delta_M H$/(J/g)	–2.80	–2.59	–1.88							

copolymer (B):	**S-BR/90w**		**55TA2**
solvent (A):	**benzene**	C_6H_6	**71-43-2**

T/K = 293.65

$\Delta_{sol}H_B^\infty$ = –4.94 J/(g copolymer)

Comments: the final concentration is about 2 g copolymer/150 ml solvent

copolymer (B):	**S-BR/90w**		**55TA2**
solvent (A):	**ethylbenzene**	C_8H_{10}	**100-41-4**
solvent (C):	**2,2,4-trimethylpentane**	C_8H_{18}	**540-84-1**

T/K = 298.15

$\Delta_{sol}H_B^\infty$ = 0.13 J/(g copolymer)

Comments: solvent mixture concentration 90 wt% C_8H_{10} / 10 wt% C_8H_{18}

Copolymers from butyl methacrylate and isobutyl methacrylate

Average chemical composition of the copolymers, acronyms and references:

Copolymer (B)	Acronym	Ref.
Butyl methacrylate/isobutyl methacrylate copolymer 50 wt% isobutyl methacrylate	BMA-IBMA/50w	96SAT

Characterization:

Copolymer (B)	M_n/ g/mol	M_w/ g/mol	M_η/ g/mol	Further information
BMA-IBMA/50w		150000		synthesized in the laboratory

Experimental enthalpy data:

copolymer (B):	**BMA-IBMA/50w**		**96SAT**
solvent (A):	**cyclohexanone**	$\mathbf{C_6H_{10}O}$	**108-94-1**

T/K = 298.15

$\Delta_{sol}H_B^\infty$ = 5.9 J/(g copolymer)

and

T/K = 298.15

$\Delta_M H_B^\infty$ = 14.0 J/(g copolymer)

Comments: $\Delta_M H_B^\infty$ is corrected for the enthalpy from the glass transition

Copolymers from butyl methacrylate and methyl methacrylate

Average chemical composition of the copolymers, acronyms and references:

Copolymer (B)	Acronym	Ref.
Butyl methacrylate/methyl methacrylate copolymer 55 wt% methyl methacrylate	BMA-MMA/55w	96SHI

Characterization:

Copolymer (B)	M_n/ g/mol	M_w/ g/mol	M_η/ g/mol	Further information
BMA-MMA/55w	109000	250000		synthesized in the laboratory

Experimental enthalpy data:

copolymer (B): **BMA-MMA/55w** **96SHI**
solvent (A): **cyclohexanone** **$C_6H_{10}O$** **108-94-1**

T/K = 298.15

$\Delta_{sol}H_B^\infty$ = –5.4 J/(g copolymer)

and

T/K = 298.15

$\Delta_M H_B^\infty$ = +9.1 J/(g copolymer)

Comments: $\Delta_M H_B^\infty$ is corrected for the enthalpy from the glass transition

Copolymers from ε-carbobenzoxy-L-lysine and L-phenylalanine

Average chemical composition of the copolymers, acronyms and references:

Copolymer (B)	Acronym	Ref.
ε-Carbobenzoxy-L-lysine/L-phenylalanine copolymer		
25 mol% L-phenylalanine	CBL-PHA/25x	70GIA
50 mol% L-phenylalanine	CBL-PHA/50x	70GIA
75 mol% L-phenylalanine	CBL-PHA/75x	70GIA

Characterization:

Copolymer (B)	M_n/ g/mol	M_w/ g/mol	M_η/ g/mol	Further information
CBL-PHA/25x				no data
CBL-PHA/50x				no data
CBL-PHA/75x				no data

Experimental enthalpy data:

copolymer (B): **CBL-PHA/25x** **70GIA**
solvent (A): **1,2-dichloroethane** $C_2H_4Cl_2$ **107-06-2**

T/K = 303.15

$\Delta_{sol}H_B^{\infty}$ = –6900 J/(base mol copolymer)

Comments: the final concentration is about 0.1 wt%

A graph is given in the original source for $\Delta_{sol}H_B^{\infty}$ for the binary mixture of 1,2-dichloroethane with dichloroacetic acid

copolymer (B):	**CBL-PHA/50x**		**70GIA**
solvent (A):	**1,2-dichloroethane**	**$C_2H_4Cl_2$**	**107-06-2**

T/K = 303.15

$\Delta_{sol}H_B^\infty = -6500$ J/(base mol copolymer)

Comments: the final concentration is about 0.1 wt%

A graph is given in the original source for $\Delta_{sol}H_B^\infty$ for the binary mixture of 1,2-dichloroethane with dichloroacetic acid

copolymer (B):	**CBL-PHA/75x**		**70GIA**
solvent (A):	**1,2-dichloroethane**	**$C_2H_4Cl_2$**	**107-06-2**

T/K = 303.15

$\Delta_{sol}H_B^\infty = -5860$ J/(base mol copolymer)

Comments: the final concentration is about 0.1 wt%

A graph is given in the original source for $\Delta_{sol}H_B^\infty$ for the binary mixture of 1,2-dichloroethane with dichloroacetic acid

Copolymers from ethylene and propylene

Average chemical composition of the copolymers, acronyms and references:

Copolymer (B)	Acronym	Ref.
Ethylene/propylene copolymer		
33 mol% ethylene	E-P/33x	79PH1
33 mol% ethylene	E-P/33x	79PH2
63 mol% ethylene	E-P/63x	79PH1
63 mol% ethylene	E-P/63x	79PH2
75 mol% ethylene	E-P/75x	79PH1
75 mol% ethylene	E-P/75x	79PH2

Characterization:

Copolymer (B)	M_n/ g/mol	M_w/ g/mol	M_η/ g/mol	Further information
E-P/33x			145000	
E-P/63x			236000	
E-P/75x			109000	

Experimental enthalpy data:

copolymer (B): **E-P/33x** **79PH1, 79PH2**
solvent (A): **cyclohexane** **C_6H_{12}** **110-82-7**

T/K = 298.15

$\Delta_M H_B^\infty$ = +1.4 J/(g copolymer)

copolymer (B): **E-P/63x** **79PH1, 79PH2**
solvent (A): **cyclohexane** **C_6H_{12}** **110-82-7**

T/K = 298.15

$\Delta_M H_B^\infty$ = +8.1 J/(g copolymer)

copolymer (B): **E-P/75x** **79PH1, 79PH2**
solvent (A): **cyclohexane** **C_6H_{12}** **110-82-7**

T/K = 298.15

$\Delta_M H_B^\infty$ = +11.8 J/(g copolymer)

copolymer (B): **E-P/33x** **79PH1, 79PH2**
solvent (A): **cyclooctane** **C_8H_{16}** **292-64-8**

T/K = 298.15

$\Delta_M H_B^\infty$ = +1.2 J/(g copolymer)

copolymer (B): **E-P/63x** **79PH1, 79PH2**
solvent (A): **cyclooctane** **C_8H_{16}** **292-64-8**

T/K = 298.15

$\Delta_M H_B^\infty$ = +6.9 J/(g copolymer)

copolymer (B): **E-P/75x** **79PH1, 79PH2**
solvent (A): **cyclooctane** C_8H_{16} **292-64-8**

T/K = 298.15

$\Delta_M H_B^\infty$ = +8.6 J/(g copolymer)

copolymer (B): **E-P/33x** **79PH1, 79PH2**
solvent (A): **cyclopentane** C_5H_{10} **287-92-3**

T/K = 298.15

$\Delta_M H_B^\infty$ = –3.5 J/(g copolymer)

copolymer (B): **E-P/63x** **79PH1, 79PH2**
solvent (A): **cyclopentane** C_5H_{10} **287-92-3**

T/K = 298.15

$\Delta_M H_B^\infty$ = +1.1 J/(g copolymer)

copolymer (B): **E-P/33x** **79PH1, 79PH2**
solvent (A): ***cis*-decahydronaphthalene** $C_{10}H_{18}$ **493-01-6**

T/K = 298.15

$\Delta_M H_B^\infty$ = –2.4 J/(g copolymer)

copolymer (B): **E-P/63x** **79PH1, 79PH2**
solvent (A): ***cis*-decahydronaphthalene** $C_{10}H_{18}$ **493-01-6**

T/K = 298.15

$\Delta_M H_B^\infty$ = +2.4 J/(g copolymer)

copolymer (B): **E-P/75x** **79PH1, 79PH2**
solvent (A): ***cis*-decahydronaphthalene** $C_{10}H_{18}$ **493-01-6**

T/K = 298.15

$\Delta_M H_B^\infty$ = +3.9 J/(g copolymer)

copolymer (B): **E-P/33x** **79PH1, 79PH2**
solvent (A): ***trans*-decahydronaphthalene** $C_{10}H_{18}$ **493-02-7**

T/K = 298.15

$\Delta_M H_B^\infty$ = –4.8 J/(g copolymer)

copolymer (B): **E-P/63x** **79PH1, 79PH2**
solvent (A): ***trans*-decahydronaphthalene** **$C_{10}H_{18}$** **493-02-7**

T/K = 298.15

$\Delta_M H_B^\infty$ = −1.3 J/(g copolymer)

copolymer (B): **E-P/75x** **79PH1, 79PH2**
solvent (A): ***trans*-decahydronaphthalene** **$C_{10}H_{18}$** **493-02-7**

T/K = 298.15

$\Delta_M H_B^\infty$ = −0.3 J/(g copolymer)

copolymer (B): **E-P/63x** **79PH1, 79PH2**
solvent (A): **3,3-diethylpentane** **C_9H_{20}** **4032-86-4**

T/K = 298.15

$\Delta_M H_B^\infty$ = −1.4 J/(g copolymer)

copolymer (B): **E-P/75x** **79PH1, 79PH2**
solvent (A): **3,3-diethylpentane** **C_9H_{20}** **4032-86-4**

T/K = 298.15

$\Delta_M H_B^\infty$ = < 0.1 J/(g copolymer)

copolymer (B): **E-P/63x** **79PH1, 79PH2**
solvent (A): **2,2-dimethylpentane** **C_7H_{16}** **590-35-2**

T/K = 298.15

$\Delta_M H_B^\infty$ = +5.3 J/(g copolymer)

copolymer (B): **E-P/75x** **79PH1, 79PH2**
solvent (A): **2,2-dimethylpentane** **C_7H_{16}** **590-35-2**

T/K = 298.15

$\Delta_M H_B^\infty$ = +2.3 J/(g copolymer)

copolymer (B): **E-P/63x** **79PH1, 79PH2**
solvent (A): **2,3-dimethylpentane** **C_7H_{16}** **565-59-3**

T/K = 298.15

$\Delta_M H_B^\infty$ = +0.7 J/(g copolymer)

copolymer (B): **E-P/75x** **79PH1, 79PH2**
solvent (A): **2,3-dimethylpentane** **C_7H_{16}** **565-59-3**

T/K = 298.15

$\Delta_M H_B^\infty$ = +0.4 J/(g copolymer)

copolymer (B): **E-P/33x** **79PH1, 79PH2**
solvent (A): **2,4-dimethylpentane** **C_7H_{16}** **108-08-7**

T/K = 298.15

$\Delta_M H_B^\infty$ = –1.2 J/(g copolymer)

copolymer (B): **E-P/63x** **79PH1, 79PH2**
solvent (A): **2,4-dimethylpentane** **C_7H_{16}** **108-08-7**

T/K = 298.15

$\Delta_M H_B^\infty$ = +3.0 J/(g copolymer)

copolymer (B): **E-P/75x** **79PH1, 79PH2**
solvent (A): **2,4-dimethylpentane** **C_7H_{16}** **108-08-7**

T/K = 298.15

$\Delta_M H_B^\infty$ = +0.2 J/(g copolymer)

copolymer (B): **E-P/33x** **79PH1, 79PH2**
solvent (A): **3,3-dimethylpentane** **C_7H_{16}** **562-49-2**

T/K = 298.15

$\Delta_M H_B^\infty$ = –2.7 J/(g copolymer)

copolymer (B): **E-P/63x** **79PH1, 79PH2**
solvent (A): **3,3-dimethylpentane** **C_7H_{16}** **562-49-2**

T/K = 298.15

$\Delta_M H_B^\infty$ = +0.3 J/(g copolymer)

copolymer (B): **E-P/33x** **79PH1, 79PH2**
solvent (A): **n-dodecane** **$C_{12}H_{26}$** **112-40-3**

T/K = 298.15

$\Delta_M H_B^\infty$ = –0.1 J/(g copolymer)

copolymer (B):	**E-P/63x**		**79PH1, 79PH2**
solvent (A):	**n-dodecane**	$C_{12}H_{26}$	**112-40-3**

T/K = 298.15

$\Delta_M H_B^\infty$ = +0.8 J/(g copolymer)

copolymer (B):	**E-P/75x**		**79PH1, 79PH2**
solvent (A):	**n-dodecane**	$C_{12}H_{26}$	**112-40-3**

T/K = 298.15

$\Delta_M H_B^\infty$ = –4.0 J/(g copolymer)

copolymer (B):	**E-P/63x**		**79PH1, 79PH2**
solvent (A):	**3-ethylpentane**	C_7H_{16}	**617-78-7**

T/K = 298.15

$\Delta_M H_B^\infty$ = +2.6 J/(g copolymer)

copolymer (B):	**E-P/75x**		**79PH1, 79PH2**
solvent (A):	**3-ethylpentane**	C_7H_{16}	**617-78-7**

T/K = 298.15

$\Delta_M H_B^\infty$ = –0.6 J/(g copolymer)

copolymer (B):	**E-P/33x**		**79PH1, 79PH2**
solvent (A):	**2,2,4,4,6,8,8-heptamethylnonane**	$C_{16}H_{34}$	**4390-04-9**

T/K = 298.15

$\Delta_M H_B^\infty$ = –0.5 J/(g copolymer)

copolymer (B):	**E-P/63x**		**79PH1, 79PH2**
solvent (A):	**2,2,4,4,6,8,8-heptamethylnonane**	$C_{16}H_{34}$	**4390-04-9**

T/K = 298.15

$\Delta_M H_B^\infty$ = +2.2 J/(g copolymer)

copolymer (B):	**E-P/75x**		**79PH1, 79PH2**
solvent (A):	**2,2,4,4,6,8,8-heptamethylnonane**	$C_{16}H_{34}$	**4390-04-9**

T/K = 298.15

$\Delta_M H_B^\infty$ = –0.9 J/(g copolymer)

copolymer (B):	**E-P/33x**		**79PH1, 79PH2**
solvent (A):	**n-hexadecane**	$C_{16}H_{34}$	**544-76-3**

T/K = 298.15

$\Delta_M H_B^\infty$ = +0.7 J/(g copolymer)

copolymer (B):	**E-P/63x**		**79PH1, 79PH2**
solvent (A):	**n-hexadecane**	$C_{16}H_{34}$	**544-76-3**

T/K = 298.15

$\Delta_M H_B^\infty$ = −1.1 J/(g copolymer)

copolymer (B):	**E-P/75x**		**79PH1, 79PH2**
solvent (A):	**n-hexadecane**	$C_{16}H_{34}$	**544-76-3**

T/K = 298.15

$\Delta_M H_B^\infty$ = −4.6 J/(g copolymer)

Comments: after subtracting the small heat of fusion of this copolymer

copolymer (B):	**E-P/63x**		**79PH1, 79PH2**
solvent (A):	**3-methylhexane**	C_7H_{16}	**589-34-4**

T/K = 298.15

$\Delta_M H_B^\infty$ = +0.7 J/(g copolymer)

copolymer (B):	**E-P/75x**		**79PH1, 79PH2**
solvent (A):	**3-methylhexane**	C_7H_{16}	**589-34-4**

T/K = 298.15

$\Delta_M H_B^\infty$ = +1.7 J/(g copolymer)

copolymer (B):	**E-P/33x**		**79PH1, 79PH2**
solvent (A):	**n-octane**	C_8H_{18}	**111-65-9**

T/K = 298.15

$\Delta_M H_B^\infty$ = −1.6 J/(g copolymer)

copolymer (B): **E-P/63x** **79PH1, 79PH2**
solvent (A): **n-octane** C_8H_{18} **111-65-9**

T/K = 298.15

$\Delta_M H_B^\infty$ = +3.6 J/(g copolymer)

copolymer (B): **E-P/75x** **79PH1, 79PH2**
solvent (A): **n-octane** C_8H_{18} **111-65-9**

T/K = 298.15

$\Delta_M H_B^\infty$ = +0.3 J/(g copolymer)

Comments: after subtracting the small heat of fusion of this copolymer

copolymer (B): **E-P/33x** **79PH1, 79PH2**
solvent (A): **2,2,4,6,6-pentamethylheptane** $C_{12}H_{26}$ **13475-82-6**

T/K = 298.15

$\Delta_M H_B^\infty$ = –0.3 J/(g copolymer)

copolymer (B): **E-P/63x** **79PH1, 79PH2**
solvent (A): **2,2,4,6,6-pentamethylheptane** $C_{12}H_{26}$ **13475-82-6**

T/K = 298.15

$\Delta_M H_B^\infty$ = +3.6 J/(g copolymer)

copolymer (B): **E-P/75x** **79PH1, 79PH2**
solvent (A): **2,2,4,6,6-pentamethylheptane** $C_{12}H_{26}$ **13475-82-6**

T/K = 298.15

$\Delta_M H_B^\infty$ = +0.0 J/(g copolymer)

Comments: after subtracting the small heat of fusion of this copolymer

copolymer (B): **E-P/33x** **79PH1, 79PH2**
solvent (A): **2,2,4,4-tetramethylpentane** C_9H_{20} **1070-87-7**

T/K = 298.15

$\Delta_M H_B^\infty$ = +3.2 J/(g copolymer)

copolymer (B):	**E-P/63x**		**79PH1, 79PH2**
solvent (A):	**2,2,4,4-tetramethylpentane**	**C_9H_{20}**	**1070-87-7**

T/K = 298.15

$\Delta_M H_B^\infty$ = +2.7 J/(g copolymer)

copolymer (B):	**E-P/75x**		**79PH1, 79PH2**
solvent (A):	**2,2,4,4-tetramethylpentane**	**C_9H_{20}**	**1070-87-7**

T/K = 298.15

$\Delta_M H_B^\infty$ = +3.1 J/(g copolymer)

copolymer (B):	**E-P/33x**		**79PH1, 79PH2**
solvent (A):	**2,2,4-trimethylpentane**	**C_8H_{18}**	**540-84-1**

T/K = 298.15

$\Delta_M H_B^\infty$ = –0.2 J/(g copolymer)

copolymer (B):	**E-P/63x**		**79PH1, 79PH2**
solvent (A):	**2,2,4-trimethylpentane**	**C_8H_{18}**	**540-84-1**

T/K = 298.15

$\Delta_M H_B^\infty$ = +1.9 J/(g copolymer)

copolymer (B):	**E-P/75x**		**79PH1, 79PH2**
solvent (A):	**2,2,4-trimethylpentane**	**C_8H_{18}**	**540-84-1**

T/K = 298.15

$\Delta_M H_B^\infty$ = +3.5 J/(g copolymer)

Copolymers from ethylene and vinyl acetate

Average chemical composition of the copolymers, acronyms and references:

Copolymer (B)	Acronym	Ref.
Ethylene/vinyl acetate copolymer 70.0 wt% vinyl acetate	E-VA/70w	90SHI

Characterization:

Copolymer (B)	M_n/ g/mol	M_w/ g/mol	M_η/ g/mol	Further information
E-VA/70w		220000		Bayer, Japan

Experimental enthalpy data:

copolymer (B):	**E-VA/70w**		**90SHI**
solvent (A):	**tetrahydrofuran**	C_4H_8O	**109-99-9**

T/K = 304.65

$\Delta_{sol}H_B^\infty = -1.33$ J/(g copolymer)

Copolymers from ethylene oxide and propylene oxide

Average chemical composition of the copolymers, acronyms and references:

Copolymer (B)	Acronym	Ref.
Poly(ethylene oxide)-b-poly(propylene oxide) diblock copolymer		
58.0 wt% propylene oxide	EO-b-PO/58w	89MOE
58.0 wt% propylene oxide	EO-b-PO/58w	90KIL
Poly(ethylene oxide)-b-poly(propylene oxide)-b-poly(ethylene oxide) triblock copolymer		
60.6 mol% propylene oxide	EO-PO-EO/61x	85COR
60.6 mol% propylene oxide	EO-PO-EO/61x	95KIL
83.3 mol% propylene oxide	EO-PO-EO/83x	85COR
83.3 mol% propylene oxide	EO-PO-EO/83x	95KIL
91.1 mol% propylene oxide	EO-PO-EO/91x	85COR
91.1 mol% propylene oxide	EO-PO-EO/91x	95KIL

Characterization:

Copolymer (B)	M_n/ g/mol	M_w/ g/mol	M_η/ g/mol	Further information
EO-b-PO/58w	1700			
EO-PO-EO/61x	1900			PE 4300, BASF, Germany
EO-PO-EO/83x	950			PE 3100, BASF, Germany
EO-PO-EO/91x	1800			PE 6100, BASF, Germany

Experimental enthalpy data:

copolymer (B): **EO-b-PO/58w** **89MOE, 90KIL**
solvent (A): **tetrachloromethane** **CCl_4** **56-23-5**

T/K = 303.15

w_B	0.0180	0.0408	0.0628	0.0838	0.1235	0.1602	0.1942	0.2370	0.2402
$\Delta_M H$/(J/g)	−0.285	−0.618	−0.925	−1.205	−1.685	−2.096	−2.437	−2.856	−2.897
w_B	0.2763	0.2936	0.3121	0.3449	0.3804	0.4241	0.4792	0.5306	0.5942
$\Delta_M H$/(J/g)	−3.178	−3.287	−3.436	−3.578	−3.728	−3.844	−3.902	−3.860	−3.690
w_B	0.6751	0.7246	0.7818	0.8487	0.9283	0.9739	0.9935		
$\Delta_M H$/(J/g)	−3.324	−3.010	−2.564	−1.917	−0.950	−0.298	−0.078		

copolymer (B): **EO-PO-EO/61x** **85COR, 95KIL**
solvent (A): **tetrachloromethane** **CCl_4** **56-23-5**

T/K = 303.15

w_B	0.0076	0.0147	0.0251	0.0353	0.0453	0.0551	0.0658	0.0763	0.0866
$\Delta_M H$/(J/g)	−0.067	−0.134	−0.236	−0.340	−0.443	−0.546	−0.663	−0.776	−0.889
w_B	0.0966	0.1064	0.1160	0.1265	0.1368	0.1564	0.1687	0.1806	0.1935
$\Delta_M H$/(J/g)	−0.998	−1.104	−1.209	−1.321	−1.429	−1.612	−1.718	−1.822	−1.932
w_B	0.2059	0.2180	0.2297	0.2411	0.2567	0.2852	0.3131	0.3506	0.3587
$\Delta_M H$/(J/g)	−2.037	−2.139	−2.235	−2.327	−2.425	−2.653	−2.818	−2.990	−3.021
w_B	0.3750	0.3865	0.3912	0.4101	0.4215	0.4524	0.4923	0.5305	0.5581
$\Delta_M H$/(J/g)	−3.092	−3.153	−3.162	−3.226	−3.245	−3.329	−3.352	−3.331	−3.296
w_B	0.5930	0.6263	0.6578	0.7062	0.8022	0.8708	0.9475	0.9821	
$\Delta_M H$/(J/g)	−3.194	−3.043	−2.871	−2.592	−1.982	−1.381	−0.584	−0.198	

copolymer (B): **EO-PO-EO/83x** **85COR, 95KIL**
solvent (A): **tetrachloromethane** CCl_4 **56-23-5**

T/K = 303.15

w_B	0.0040	0.0110	0.0215	0.0316	0.0449	0.0577	0.0850	0.1107	0.1339
$\Delta_M H$/(J/g)	−0.021	−0.065	−0.136	−0.209	−0.309	−0.407	−0.624	−0.837	−0.967
w_B	0.1573	0.1936	0.2284	0.2729	0.3432	0.4134	0.4892	0.5528	0.6162
$\Delta_M H$/(J/g)	−1.123	−1.331	−1.503	−1.694	−1.924	−2.094	−2.138	−2.158	−2.045
w_B	0.7140	0.8178	0.9076	0.9389	0.9723				
$\Delta_M H$/(J/g)	−1.817	−1.455	−0.829	−0.566	−0.254				

copolymer (B): **EO-PO-EO/91x** **85COR, 95KIL**
solvent (A): **tetrachloromethane** CCl_4 **56-23-5**

T/K = 303.15

w_B	0.0027	0.0043	0.0389	0.0613	0.0896	0.1207	0.1498	0.1770	0.2272
$\Delta_M H$/(J/g)	−0.024	−0.038	−0.361	−0.564	−0.834	−1.130	−1.394	−1.617	−2.048
w_B	0.2726	0.3285	0.3974	0.4479	0.5449	0.6677	0.7389	0.8270	0.9053
$\Delta_M H$/(J/g)	−2.367	−2.668	−2.907	−2.999	−2.945	−2.535	−2.130	−1.496	−0.837

Copolymers from isobutyl methacrylate and methyl methacrylate

Average chemical composition of the copolymers, acronyms and references:

Copolymer (B)	Acronym	Ref.
Isobutyl methacrylate/methyl methacrylate copolymer 49 wt% methyl methacrylate	IBMA-MMA/49w	96SAT

Characterization:

Copolymer (B)	M_n/ g/mol	M_w/ g/mol	M_η/ g/mol	Further information
IBMA-MMA/49w		240000		synthesized in the laboratory

Experimental enthalpy data:

copolymer (B):	**IBMA-MMA/49w**		**96SAT**
solvent (A):	**cyclohexanone**	**$C_6H_{10}O$**	**108-94-1**

T/K = 298.15

$\Delta_{sol}H_B^{\infty}$ = −10.8 J/(g copolymer)

and

T/K = 298.15

$\Delta_M H_B^{\infty}$ = +14.5 J/(g copolymer)

Comments: $\Delta_M H_B^{\infty}$ is corrected for the enthalpy from the glass transition

Copolymers from styrene and butyl methacrylate

Average chemical composition of the copolymers, acronyms and references:

Copolymer (B)	Acronym	Ref.
Styrene/butyl methacrylate copolymer		
20.6 wt% styrene	S-BMA/21w	87KYO
67.7 wt% styrene	S-BMA/68w	87KYO
80.0 wt% styrene	S-BMA/80w	87KYO
85.0 wt% styrene	S-BMA/85w	87KYO

Characterization:

Copolymer (B)	M_n/ g/mol	M_w/ g/mol	M_η/ g/mol	Further information
S-BMA/21w	185000	311000		Scientific Polymer Products
S-BMA/68w	121000	395000		Scientific Polymer Products
S-BMA/80w	193000	249000		Scientific Polymer Products
S-BMA/85w	176000	308000		Scientific Polymer Products

Experimental enthalpy data:

copolymer (B): **S-BMA/21w** **87KYO**
solvent (A): **2-butanone** **C_4H_8O** **78-93-3**

T/K = 298.15

$\varphi_B^{(1)}$	0.0908	0.0908	0.0908	0.0908	0.0908	0.0908
$\varphi_B^{(2)}$	0.0272	0.0454	0.0545	0.0636	0.0724	0.0814
$\Delta_{dil}H^{12}$/(J/mol)	0.564	0.778	0.860	1.056	1.390	1.748

Comments: the table provides the ratio of $\Delta_{dil}H^{12}/\Delta n_A$, i.e., the enthalpy change caused by diluting the primary solution by 1 mol solvent, where $\varphi_B^{(1)}$ denotes the volume fraction of the copolymer in the starting solution and $\varphi_B^{(2)}$ denotes the volume fraction after the dilution process.

copolymer (B): **S-BMA/68w** **87KYO**
solvent (A): **2-butanone** **C_4H_8O** **78-93-3**

T/K = 298.15

$\varphi_B^{(1)}$	0.0788	0.0788	0.0788	0.0788	0.0788	0.0788	0.0788
$\varphi_B^{(2)}$	0.0221	0.0323	0.0355	0.0497	0.0551	0.0654	0.0607
$\Delta_{dil}H^{12}$/(J/mol)	0.170	0.282	0.261	0.351	0.393	0.399	0.569

Comments: the table provides the ratio of $\Delta_{dil}H^{12}/\Delta n_A$, i.e., the enthalpy change caused by diluting the primary solution by 1 mol solvent, where $\varphi_B^{(1)}$ denotes the volume fraction of the copolymer in the starting solution and $\varphi_B^{(2)}$ denotes the volume fraction after the dilution process.

copolymer (B): **S-BMA/80w** **87KYO**
solvent (A): **2-butanone** **C_4H_8O** **78-93-3**

T/K = 298.15

$\varphi_B^{(1)}$	0.0465	0.0465	0.0465	0.0465	0.0465	0.0776	0.0776	0.0766
$\varphi_B^{(2)}$	0.0189	0.0279	0.0326	0.0372	0.0419	0.0380	0.0388	0.0465
$\Delta_{dil}H^{12}$/(J/mol)	0.104	0.164	0.121	0.228	0.232	0.304	0.326	0.432
$\varphi_B^{(1)}$	0.0766	0.0766	0.0766					
$\varphi_B^{(2)}$	0.0543	0.0621	0.0698					
$\Delta_{dil}H^{12}$/(J/mol)	0.555	0.706	0.867					

Comments: the table provides the ratio of $\Delta_{dil}H^{12}/\Delta n_A$, i.e., the enthalpy change caused by diluting the primary solution by 1 mol solvent, where $\varphi_B^{(1)}$ denotes the volume fraction of the copolymer in the starting solution and $\varphi_B^{(2)}$ denotes the volume fraction after the dilution process.

copolymer (B):	**S-BMA/85w**		**87KYO**
solvent (A):	**2-butanone**	**C_4H_8O**	**78-93-3**

T/K = 298.15

$\varphi_B^{(1)}$	0.0777	0.0777	0.0777	0.0777	0.0777	0.0777
$\varphi_B^{(2)}$	0.0583	0.0505	0.0428	0.0350	0.0272	0.0192
$\Delta_{dil}H^{12}$/(J/mol)	−0.453	−0.328	−0.255	−0.195	−0.169	−0.105

Comments: the table provides the ratio of $\Delta_{dil}H^{12}/\Delta n_A$, i.e., the enthalpy change caused by diluting the primary solution by 1 mol solvent, where $\varphi_B^{(1)}$ denotes the volume fraction of the copolymer in the starting solution and $\varphi_B^{(2)}$ denotes the volume fraction after the dilution process.

Copolymers from vinyl acetate and vinyl alcohol

Average chemical composition of the copolymers, acronyms and references:

Copolymer (B)	Acronym	Ref.
Vinyl acetate/vinyl alcohol copolymer		
0.4 mol% vinyl acetate	VA-VAL/01x	56AMA
4.2 mol% vinyl acetate	VA-VAL/04x(1)	55OYA
4.3 mol% vinyl acetate	VA-VAL/04x(2)	55OYA
7.8 mol% vinyl acetate	VA-VAL/08x	56AMA
9.0 mol% vinyl acetate	VA-VAL/09x	55OYA
10.3 mol% vinyl acetate	VA-VAL/10x	55OYA
15.3 mol% vinyl acetate	VA-VAL/15x(1)	55OYA
15.4 mol% vinyl acetate	VA-VAL/15x(2)	55OYA
15.6 mol% vinyl acetate	VA-VAL/16x	56AMA
19.4 mol% vinyl acetate	VA-VAL/19x	56AMA
19.5 mol% vinyl acetate	VA-VAL/20x	55OYA
22.1 mol% vinyl acetate	VA-VAL/22x	55OYA
26.2 mol% vinyl acetate	VA-VAL/26x	55OYA
30.6 mol% vinyl acetate	VA-VAL/31x	55OYA
34.0 mol% vinyl acetate	VA-VAL/34x	55OYA
34.7 mol% vinyl acetate	VA-VAL/35x	55OYA
9.0 wt% vinyl acetate	VA-VAL/09w	58TA1
9.0 wt% vinyl acetate	VA-VAL/09w	58TA2
44.0 wt% vinyl acetate	VA-VAL/44w	58TA1
44.0 wt% vinyl acetate	VA-VAL/44w	58TA2
57.0 wt% vinyl acetate	VA-VAL/57w	58TA1
57.0 wt% vinyl acetate	VA-VAL/57w	58TA2
67.0 wt% vinyl acetate	VA-VAL/67w	58TA1
67.0 wt% vinyl acetate	VA-VAL/67w	58TA2
68.6 wt% vinyl acetate	VA-VAL/69w	58TA2

Characterization:

Copolymer (B)	M_n/ g/mol	M_w/ g/mol	M_η/ g/mol	Further information
VA-VAL/01x	53000			Mitsubishi Rayon Co., Japan
VA-VAL/04x(1)	7560			
VA-VAL/04x(2)	64300			
VA-VAL/08x	53000			Mitsubishi Rayon Co., Japan
VA-VAL/09x	66900			
VA-VAL/10x	7970			
VA-VAL/15x(1)	8300			
VA-VAL/15x(2)	70700			
VA-VAL/16x	53000			Mitsubishi Rayon Co., Japan
VA-VAL/19x	53000			Mitsubishi Rayon Co., Japan
VA-VAL/20x	73100			
VA-VAL/22x	8800			
VA-VAL/26x	77000			
VA-VAL/31x	9370			
VA-VAL/34x	81600			
VA-VAL/35x	9670			
VA-VAL/09w				synthesized in the laboratory
VA-VAL/44w				synthesized in the laboratory
VA-VAL/57w				synthesized in the laboratory
VA-VAL/67w				synthesized in the laboratory
VA-VAL/69w				synthesized in the laboratory

Experimental enthalpy data:

copolymer (B): **VA-VAL/01x** **56AMA**
solvent (A): **water** **H_2O** **7732-18-5**

T/K = 343.15

$V^{(1)}$/cm^3	9.00	9.51
$\varphi_B^{(1)}$	0.099	0.109
$V^{(2)}$/cm^3	19.00	19.51
$\varphi_B^{(2)}$	0.047	0.053
$\Delta_{dil}h^{12}$/J	−0.607	−0.741

Comments: $\Delta_{dil}h^{12}$ is the extensive quantity obtained for a given total volume change from $V^{(1)}$ to $V^{(2)}$, where $\varphi_B^{(1)}$ denotes the volume fraction of the copolymer in the starting solution and $\varphi_B^{(2)}$ in denotes the volume fraction after the dilution process.

copolymer (B):	**VA-VAL/04x(1)**		**55OYA**
solvent (A):	**water**	**H_2O**	**7732-18-5**

T/K = 303.15

$\Delta_{sol}H_B^{\infty} = -41.3$ J/(g copolymer)

Comments: the final concentration is < 2 g copolymer/200 cm^3 solvent

copolymer (B):	**VA-VAL/04x(2)**		**55OYA**
solvent (A):	**water**	**H_2O**	**7732-18-5**

T/K = 303.15

$\Delta_{sol}H_B^{\infty} = -48.6$ J/(g copolymer)

Comments: the final concentration is < 2 g copolymer/200 cm^3 solvent

copolymer (B):	**VA-VAL/09x**		**55OYA**
solvent (A):	**water**	**H_2O**	**7732-18-5**

T/K = 303.15

$\Delta_{sol}H_B^{\infty} = -54.6$ J/(g copolymer)

Comments: the final concentration is < 2 g copolymer/200 cm^3 solvent

copolymer (B):	**VA-VAL/10x**		**55OYA**
solvent (A):	**water**	**H_2O**	**7732-18-5**

T/K = 303.15

$\Delta_{sol}H_B^{\infty} = -46.8$ J/(g copolymer)

Comments: the final concentration is < 2 g copolymer/200 cm^3 solvent

copolymer (B):	**VA-VAL/08x**		**56AMA**
solvent (A):	**water**	**H_2O**	**7732-18-5**

T/K = 343.15

$V^{(1)}$/cm^3	9.57	8.57
$\varphi_B^{(1)}$	0.077	0.077
$V^{(2)}$/cm^3	19.57	18.57
$\varphi_B^{(2)}$	0.038	0.0355
$\Delta_{dil}h^{12}$/J	−0.699	−0.632

Comments: $\Delta_{dil}h^{12}$ is the extensive quantity obtained for a given total volume change from $V^{(1)}$ to $V^{(2)}$, where $\varphi_B^{(1)}$ denotes the volume fraction of the copolymer in the starting solution and $\varphi_B^{(2)}$ in denotes the volume fraction after the dilution process.

copolymer (B):	**VA-VAL/15x(1)**		**55OYA**
solvent (A):	**water**	H_2O	**7732-18-5**

T/K = 303.15

$\Delta_{sol}H_B^{\infty}$ = –60.2 J/(g copolymer)

Comments: the final concentration is < 2 g copolymer/200 cm^3 solvent

copolymer (B):	**VA-VAL/15x(2)**		**55OYA**
solvent (A):	**water**	H_2O	**7732-18-5**

T/K = 303.15

$\Delta_{sol}H_B^{\infty}$ = –64.7 J/(g copolymer)

Comments: the final concentration is < 2 g copolymer/200 cm^3 solvent

copolymer (B):	**VA-VAL/16x**		**56AMA**
solvent (A):	**water**	H_2O	**7732-18-5**

T/K = 343.15

$V^{(1)}$/cm^3	9.91	9.94
$\varphi_B^{(1)}$	0.052	0.052
$V^{(2)}$/cm^3	19.91	19.94
$\varphi_B^{(2)}$	0.026	0.026
$\Delta_{dil}h^{12}$/J	–0.167	–0.172

Comments: $\Delta_{dil}h^{12}$ is the extensive quantity obtained for a given total volume change from $V^{(1)}$ to $V^{(2)}$, where $\varphi_B^{(1)}$ denotes the volume fraction of the copolymer in the starting solution and $\varphi_B^{(2)}$ in denotes the volume fraction after the dilution process.

copolymer (B):	**VA-VAL/19x**		**56AMA**
solvent (A):	**water**	H_2O	**7732-18-5**

T/K = 343.15

$V^{(1)}$/cm^3	9.86	9.94
$\varphi_B^{(1)}$	0.038	0.038
$V^{(2)}$/cm^3	19.86	19.94
$\varphi_B^{(2)}$	0.019	0.019
$\Delta_{dil}h^{12}$/J	–0.071	–0.063

Comments: $\Delta_{dil}h^{12}$ is the extensive quantity obtained for a given total volume change from $V^{(1)}$ to $V^{(2)}$, where $\varphi_B^{(1)}$ denotes the volume fraction of the copolymer in the starting solution and $\varphi_B^{(2)}$ in denotes the volume fraction after the dilution process.

copolymer (B):	**VA-VAL/20x**		**55OYA**
solvent (A):	**water**	H_2O	**7732-18-5**

T/K = 303.15

$\Delta_{sol}H_B^\infty = -66.4$ J/(g copolymer)

Comments: the final concentration is < 2 g copolymer/200 cm^3 solvent

copolymer (B):	**VA-VAL/22x**		**55OYA**
solvent (A):	**water**	H_2O	**7732-18-5**

T/K = 303.15

$\Delta_{sol}H_B^\infty = -60.1$ J/(g copolymer)

Comments: the final concentration is < 2 g copolymer/200 cm^3 solvent

copolymer (B):	**VA-VAL/26x**		**55OYA**
solvent (A):	**water**	H_2O	**7732-18-5**

T/K = 303.15

$\Delta_{sol}H_B^\infty = -64.0$ J/(g copolymer)

Comments: the final concentration is < 2 g copolymer/200 cm^3 solvent

copolymer (B):	**VA-VAL/31x**		**55OYA**
solvent (A):	**water**	H_2O	**7732-18-5**

T/K = 303.15

$\Delta_{sol}H_B^\infty = -53.2$ J/(g copolymer)

Comments: the final concentration is < 2 g copolymer/200 cm^3 solvent

copolymer (B):	**VA-VAL/34x**		**55OYA**
solvent (A):	**water**	H_2O	**7732-18-5**

T/K = 303.15

$\Delta_{sol}H_B^\infty = -60.4$ J/(g copolymer)

Comments: the final concentration is < 2 g copolymer/200 cm^3 solvent

copolymer (B):	**VA-VAL/35x**		**55OYA**
solvent (A):	**water**	**H_2O**	**7732-18-5**

T/K = 303.15

$\Delta_{sol}H_B^{\infty}$ = –44.4 J/(g copolymer)

Comments: the final concentration is < 2 g copolymer/200 cm^3 solvent

copolymer (B):	**VA-VAL/09w**		**58TA2**
solvent (A):	**ethyl acetate**	**$C_4H_8O_2$**	**141-78-6**
solvent (C):	**ethanol**	**C_2H_6O**	**64-17-5**

T/K = 298.15

$\Delta_{sol}H_B^{\infty}$ = 1.67 J/(g copolymer)

Comments: the solvent mixture concentration is equal to the comonomer contents, i.e., 9.0 wt% ethyl acetate/91.0 wt% ethanol

copolymer (B):	**VA-VAL/09w**		**58TA1**
solvent (A):	**2-propanone**	**C_3H_6O**	**67-64-1**

T/K = 298.15

$\Delta_{sol}H_B^{\infty}$ = 6.3 J/(g copolymer)

copolymer (B):	**VA-VAL/44w**		**58TA2**
solvent (A):	**ethyl acetate**	**$C_4H_8O_2$**	**141-78-6**
solvent (C):	**ethanol**	**C_2H_6O**	**64-17-5**

T/K = 298.15

$\Delta_{sol}H_B^{\infty}$ = 0.84 J/(g copolymer)

Comments: the solvent mixture concentration is equal to the comonomer contents, i.e., 44.0 wt% ethyl acetate/56.0 wt% ethanol

copolymer (B):	**VA-VAL/44w**		**58TA1**
solvent (A):	**2-propanone**	**C_3H_6O**	**67-64-1**

T/K = 298.15

$\Delta_{sol}H_B^{\infty}$ = 4.6 J/(g copolymer)

copolymer (B):	**VA-VAL/57w**		**58TA2**
solvent (A):	**ethyl acetate**	$C_4H_8O_2$	**141-78-6**
solvent (C):	**ethanol**	C_2H_6O	**64-17-5**

T/K = 298.15

$\Delta_{sol}H_B^{\infty}$ = 0.84 J/(g copolymer)

Comments: the solvent mixture concentration is equal to the comonomer contents, i.e., 57.0 wt% ethyl acetate/43.0 wt% ethanol

copolymer (B):	**VA-VAL/57w**		**58TA1**
solvent (A):	**2-propanone**	C_3H_6O	**67-64-1**

T/K = 298.15

$\Delta_{sol}H_B^{\infty}$ = 0.0 J/(g copolymer)

copolymer (B):	**VA-VAL/67w**		**58TA1**
solvent (A):	**ethyl acetate**	$C_4H_8O_2$	**141-78-6**
solvent (C):	**ethanol**	C_2H_6O	**64-17-5**

T/K = 298.15

$\Delta_{sol}H_B^{\infty}$ = 0.0 J/(g copolymer)

Comments: the solvent mixture concentration is equal to the comonomer contents, i.e., 67.0 wt% ethyl acetate/33.0 wt% ethanol

copolymer (B):	**VA-VAL/67w**		**58TA1**
solvent (A):	**2-propanone**	C_3H_6O	**67-64-1**

T/K = 298.15

$\Delta_{sol}H_B^{\infty}$ = −1.26 J/(g copolymer)

copolymer (B):	**VA-VAL/69w**		**58TA1**
solvent (A):	**ethyl acetate**	$C_4H_8O_2$	**141-78-6**
solvent (C):	**ethanol**	C_2H_6O	**64-17-5**

T/K = 298.15

$\Delta_{sol}H_B^{\infty}$ = 0.0 J/(g copolymer)

Comments: the solvent mixture concentration is equal to the comonomer contents, i.e., 68.6 wt% ethyl acetate/31.4 wt% ethanol

Copolymers from vinyl acetate and vinyl chloride

Average chemical composition of the copolymers, acronyms and references:

Copolymer (B)	Acronym	Ref.
Vinyl acetate/vinyl chloride copolymer 10.0 wt% vinyl acetate	VA-VC/10w	97SAT

Characterization:

Copolymer (B)	M_n/ g/mol	M_w/ g/mol	M_η/ g/mol	Further information
VA-VC/10w	12400	26000		Scientific Polymer Products

Experimental enthalpy data:

copolymer (B):	**VA-VC/10w**		**97SAT**
solvent (A):	**cyclohexanone**	**$C_6H_{10}O$**	**108-94-1**

T/K = 304.15

$\Delta_{sol}H_B^\infty = -37.2$ J/(g copolymer)

and

T/K = 304.15

$\Delta_M H_B^\infty = -16.5$ J/(g copolymer)

Comments: $\Delta_M H_B^\infty$ is corrected for the enthalpy from the glass transition

Copolymers from vinyl chloride and vinylidene chloride

Average chemical composition of the copolymers, acronyms and references:

Copolymer (B)	Acronym	Ref.
Vinyl chloride/vinylidene chloride copolymer unspecified contents of vinyl chloride	VC-VdC/?	56LIP

Characterization:

Copolymer (B)	M_n/ g/mol	M_w/ g/mol	M_η/ g/mol	Further information
VC-VdC/?				no data

Experimental enthalpy data:

copolymer (B):	**VC-VdC/?**		**56LIP**
solvent (A):	**trichloromethane**	**$CHCl_3$**	**67-66-3**

T/K = 297.15

$\Delta_{sol}H_B^\infty$ = −17.15 J/(g copolymer)

Comments: non-oriented sample

and

T/K = 297.15

$\Delta_{sol}H_B^\infty$ = −15.9 J/(g copolymer)

Comments: oriented sample, drawn 100%

5.2. Partial molar enthalpies of mixing at infinite dilution of solvents and enthalpies of solution of gases/vapors of solvents in molten copolymers from inverse gas-liquid chromatography (IGC)

Copolymers from acrylonitrile and butadiene

Average chemical composition of the copolymers, acronyms and references:

Copolymer (B)	Acronym	Ref.
Acrylonitrile/butadiene copolymer		
34.0 wt% acrylonitrile	AN-B/34w	93ISS
34.0 wt% acrylonitrile	AN-B/34w	97SCH

Characterization:

Copolymer (B)	M_n/ g/mol	M_w/ g/mol	M_η/ g/mol	Further information
AN-B/34w	60000			Perbunan 3307, Bayer AG, Ger.

Experimental enthalpy data:

copolymer (B): AN-B/34w 93ISS

solvent (A)	T-range/ K	$\Delta_{sol}H_{A(vap)}^{\infty}$/ kJ/mol	solvent (A)	T-range/ K	$\Delta_{sol}H_{A(vap)}^{\infty}$/ kJ/mol
benzene	333.15-393.15	−31.80	n-octane	333.15-393.15	−30.23
n-heptane	333.15-393.15	−25.94	toluene	333.15-393.15	−35.30
n-nonane	333.15-393.15	−35.62			

copolymer (B): AN-B/34w **97SCH**

solvent (A)	T-range/ K	$\Delta_{sol}H_{A(vap)}{}^{\infty}$/ kJ/mol	solvent (A)	T-range/ K	$\Delta_{sol}H_{A(vap)}{}^{\infty}$/ kJ/mol
benzene	323.15-393.15	−31.82	n-hexane	323.15-393.15	−30.50
cyclohexane	323.15-393.15	−31.65	toluene	323.15-393.15	−35.60
n-heptane	323.15-393.15	−35.41			

Copolymers from acrylonitrile and methyl acrylate

Average chemical composition of the copolymers, acronyms and references:

Copolymer (B)	Acronym	Ref.
Acrylonitrile/methyl acrylate copolymer 91.9 wt% acrylonitrile	AN-MA/92w	94COS

Characterization:

Copolymer (B)	M_n/ g/mol	M_w/ g/mol	M_η/ g/mol	Further information
AN-MA/92w				Yalova Fibre Co., Turkey

Experimental enthalpy data:

copolymer (B): AN-MA/92w 94COS

solvent (A)	T-range/ K	$\Delta_M H_A^\infty$/ kJ/mol	solvent (A)	T-range/ K	$\Delta_M H_A^\infty$/ kJ/mol
1-butanol	383.15-413.15	31.74	1-pentanol	383.15-413.15	31.11
ethanol	383.15-413.15	33.24	1-propanol	383.15-413.15	32.41
1-hexanol	383.15-413.15	45.05			

Copolymers from acrylonitrile and 2-(3-methyl-3-phenylcyclobutyl)-2-hydroxyethyl methacrylate

Average chemical composition of the copolymers, acronyms and references:

Copolymer (B)	Acronym	Ref.
Acrylonitrile/2-(3-methyl-3-phenylcyclobutyl)-2-hydroxyethyl methacrylate copolymer 45.0 mol% acrylonitrile	AN-MPCHEMA/45x	2000KAY

Characterization:

Copolymer (B)	M_n/ g/mol	M_w/ g/mol	M_η/ g/mol	Further information
AN-MPCHEMA/45x				ρ_B = 1.075 g/cm^3, synthesized in the laboratory

Experimental enthalpy data:

copolymer (B): AN-MPCHEMA/45x 2000KAY

solvent (A)	T-range/ K	$\Delta_M H_A^\infty$/ kJ/mol	solvent (A)	T-range/ K	$\Delta_M H_A^\infty$/ kJ/mol
benzene	443.15-453.15	24.95	methyl acetate	443.15-453.15	16.62
2-butanone	443.15-453.15	8.33	n-nonane	443.15-453.15	23.28
n-decane	443.15-453.15	23.28	n-octane	443.15-453.15	16.62
n-dodecane	443.15-453.15	26.63	2-propanone	443.15-453.15	18.30
ethanol	443.15-453.15	38.27	toluene	443.15-453.15	24.95
ethyl acetate	443.15-453.15	14.99	n-undecane	443.15-453.15	19.97
methanol	443.15-453.15	24.95	o-xylene	443.15-453.15	23.28

Copolymers from acrylonitrile and α-methylstyrene

Average chemical composition of the copolymers, acronyms and references:

Copolymer (B)	Acronym	Ref.
Acrylonitrile/α-methylstyrene copolymer		
30.0 wt% acrylonitrile	AN-αMS/30w	84SIO
30.0 wt% acrylonitrile	AN-αMS/30w	86SIO

Characterization:

Copolymer (B)	M_n/ g/mol	M_w/ g/mol	M_η/ g/mol	Further information
AN-αMS/30w		160000		synthesized in the laboratory

Experimental enthalpy data:

copolymer (B): AN-αMS/30w 84SIO, 86SIO

solvent (A)	T-range/ K	$\Delta_M H_A^\infty$/ kJ/mol	$\Delta_{sol} H_{A(vap)}^\infty$/ kJ/mol
benzene	438.15-468.15	8.46	−18.46
1-butanol	438.15-468.15	14.57	−24.12
2-butanone	438.15-468.15	6.95	−20.10
butyl acetate	438.15-468.15	11.35	−21.94
butylbenzene	438.15-468.15	10.59	−30.60
chlorobenzene	438.15-468.15	4.31	−29.14
cyclohexanol	438.15-468.15	15.78	−25.79
dichloromethane	438.15-468.15	−2.51	−24.91
1,4-dioxane	438.15-468.15	4.48	−25.87
n-dodecane	438.15-468.15	31.36	−16.66
n-tetradecane	438.15-468.15	26.08	−29.98
tetrahydrofuran	438.15-468.15	11.68	−13.19
toluene	438.15-468.15	6.82	−23.57
trichloromethane	438.15-468.15	7.41	−17.08

Copolymers from butadiene and styrene

Average chemical composition of the copolymers, acronyms and references:

Copolymer (B)	Acronym	Ref.
Butadiene/styrene copolymer		
15.0 wt% styrene	S-BR/15w	97SCH
23.0 wt% styrene	S-BR/23w	97SCH
40.0 wt% styrene	S-BR/40w	97SCH
Polystyrene-b-polybutadiene-b-polystyrene triblock copolymer		
17.0 wt% styrene	S-B-S/17w	92ROM
31.0 wt% styrene	S-B-S/31w	92ROM

Characterization:

Copolymer (B)	M_n/ g/mol	M_w/ g/mol	M_η/ g/mol	Further information
S-BR/15w	240000			Buna EM BT98, Buna Werke Huels/Marl
S-BR/23w	210000	400000		Buna EM 1500, Buna Werke Huels/Marl
S-BR/40w	190000			Buna EM 1516, Buna Werke Huels/Marl
S-B-S/17w	114800	188300		Kraton D-1301X
S-B-S/31w	110800	139600		Kraton D-1101

Experimental enthalpy data:

copolymer (B): S-BR/15w **97SCH**

solvent (A)	T-range/ K	$\Delta_{sol}H_{A(vap)}{}^{\infty}$/ kJ/mol	solvent (A)	T-range/ K	$\Delta_{sol}H_{A(vap)}{}^{\infty}$/ kJ/mol
benzene	323.15-393.15	−32.25	n-hexane	323.15-393.15	−29.44
cyclohexane	323.15-393.15	−30.75	toluene	323.15-393.15	−35.96
n-heptane	323.15-393.15	−33.67			

copolymer (B): S-BR/23w **97SCH**

solvent (A)	T-range/ K	$\Delta_{sol}H_{A(vap)}{}^{\infty}$/ kJ/mol	solvent (A)	T-range/ K	$\Delta_{sol}H_{A(vap)}{}^{\infty}$/ kJ/mol
benzene	323.15-393.15	−31.84	n-hexane	323.15-393.15	−29.00
cyclohexane	323.15-393.15	−30.58	toluene	323.15-393.15	−36.41
n-heptane	323.15-393.15	−34.93			

copolymer (B): **S-BR/40w** **97SCH**

solvent (A)	T-range/ K	$\Delta_{sol}H_{A(vap)}{}^{\infty}$/ kJ/mol	solvent (A)	T-range/ K	$\Delta_{sol}H_{A(vap)}{}^{\infty}$/ kJ/mol
benzene	323.15-393.15	−32.47	n-hexane	323.15-393.15	−29.22
cyclohexane	323.15-393.15	−30.87	toluene	323.15-393.15	−35.55
n-heptane	323.15-393.15	−33.48			

copolymer (B): **S-B-S/17w** **92ROM**

solvent (A)	T-range/ K	$\Delta_M H_A{}^{\infty}$/ kJ/mol	$\Delta_{sol}H_{A(vap)}{}^{\infty}$/ kJ/mol
benzene	308.15-348.15	−0.252	−32.7
2-butanone	308.15-348.15	0.302	−32.8
cyclohexane	308.15-348.15	0.727	−30.8
ethylbenzene	308.15-348.15	0.397	−40.2
n-heptane	308.15-348.15	0.464	−34.4
n-hexane	308.15-348.15	−0.228	−30.0
toluene	308.15-348.15	−0.106	−36.7
trichloromethane	308.15-348.15	−2.467	−32.4
p-xylene	308.15-348.15	0.270	−40.4

copolymer (B): **S-B-S/31w** **92ROM**

solvent (A)	T-range/ K	$\Delta_M H_A{}^{\infty}$/ kJ/mol	$\Delta_{sol}H_{A(vap)}{}^{\infty}$/ kJ/mol
benzene	308.15-348.15	−0.745	−33.2
2-butanone	308.15-348.15	−1.11	−34.2
cyclohexane	308.15-348.15	1.09	−30.4
ethylbenzene	308.15-348.15	−0.130	−40.7
n-heptane	308.15-348.15	0.459	−34.4
n-hexane	308.15-348.15	−0.765	−30.6
toluene	308.15-348.15	−0.134	−36.7
trichloromethane	308.15-348.15	−4.49	−34.4
p-xylene	308.15-348.15	0.158	−40.5

Copolymers from chloroprene and methyl methacrylate

Average chemical composition of the copolymers, acronyms and references:

Copolymer (B)	Acronym	Ref.
Chloroprene/methyl methacrylate copolymer		
15.0 mol% methyl methacrylate	CP-MMA/15x	98YAM
Chloroprene/methyl methacrylate/methacrylic acid terpolymer		
13.8 mol% methyl methacrylate and 78.2 mol% chloroprene	CP-MMA-MA/14x	98YAM
14.8 mol% methyl methacrylate and 83.7 mol% chloroprene	CP-MMA-MA/15x	98YAM

Characterization:

Copolymer (B)	M_n/ g/mol	M_w/ g/mol	M_η/ g/mol	Further information
CP-MMA/15x				synthesized in the laboratory
CP-MMA-MA/14x				synthesized in the laboratory
CP-MMA-MA/15x				synthesized in the laboratory

Experimental enthalpy data:

copolymer (B): CP-MMA/15x 98YAM

solvent (A)	T-range/ K	$\Delta_M H_A^\infty$/ kJ/mol	$\Delta_{sol} H_{A(vap)}^\infty$/ kJ/mol
dichloromethane	298.15-383.15	−1.68	−22.2
tetrachloromethane	298.15-383.15	−4.64	−20.1
trichloromethane	298.15-383.15	−1.47	−22.2

copolymer (B): CP-MMA-MA/14x 98YAM

solvent (A)	T-range/ K	$\Delta_M H_A^\infty$/ kJ/mol	$\Delta_{sol} H_{A(vap)}^\infty$/ kJ/mol
dichloromethane	298.15-383.15	−1.30	−18.8
tetrachloromethane	298.15-383.15	−3.75	−29.7
trichloromethane	298.15-383.15	−0.52	−30.6

copolymer (B): CP-MMA-MA/15x 98YAM

solvent (A)	T-range/ K	$\Delta_M H_A^\infty$/ kJ/mol	$\Delta_{sol} H_{A(vap)}^\infty$/ kJ/mol
dichloromethane	298.15-383.15	−0.77	−28.9
tetrachloromethane	298.15-383.15	−3.36	−27.6
trichloromethane	298.15-383.15	−1.67	−29.7

Copolymers from diethylmaleate and vinyl acetate

Average chemical composition of the copolymers, acronyms and references:

Copolymer (B)	Acronym	Ref.
Diethylmaleate/vinyl acetate copolymer 50 mol% vinyl acetate	DEM-VA/50x	95NEM

Characterization:

Copolymer (B)	M_n/ g/mol	M_w/ g/mol	M_η/ g/mol	Further information
DEM-VA/50x	51200			synthesized in the laboratory

Experimental enthalpy data:

copolymer (B): DEM-VA/50x **95NEM**

solvent (A)	T-range/ K	$\Delta_M H_A^\infty$/ kJ/mol	$\Delta_{sol} H_{A(vap)}^\infty$/ kJ/mol
n-heptane	359.96-378.36	12.748	−19.147
n-hexane	359.96-378.36	3.949	−24.071
n-nonane	359.96-378.36	29.693	−11.010
n-octane	359.96-378.36	21.209	−14.892

Copolymers from ethylene and carbon monoxide

Average chemical composition of the copolymers, acronyms and references:

Copolymer (B)	Acronym	Ref.
Ethylene/carbon monoxide copolymer 10.5 wt% carbon monoxide	E-CO/11w	78DIP

Characterization:

Copolymer (B)	M_n/ g/mol	M_w/ g/mol	M_η/ g/mol	Further information
E-CO/11w				no data

Experimental enthalpy data:

copolymer (B): **E-CO/11w** **78DIP**

solvent (A)	T-range/ K	$\Delta_M H_A^{\infty}$/ kJ/mol
n-octane	381.23-411.23	1.08

Copolymers from ethylene and propylene

Average chemical composition of the copolymers, acronyms and references:

Copolymer (B)	Acronym	Ref.
Ethylene/propylene copolymer		
40 wt% propylene	E-P/40w	79ITO
Ethylene/propylene/diene terpolymer		
50 mol% propylene	E-P-D/50x	93ISS
50 mol% propylene	E-P-D/50x	97SCH

Characterization:

Copolymer (B)	M_n/ g/mol	M_w/ g/mol	M_η/ g/mol	Further information
E-P/40w				no data
E-P-D/50x	90000	200000		Buna AP341, Buna-Werke Huels/Marl

Experimental enthalpy data:

copolymer (B): E-P/40w **79ITO**

solvent (A)	T-range/ K	$\Delta_M H_A^\infty$/ kJ/mol	solvent (A)	T-range/ K	$\Delta_M H_A^\infty$/ kJ/mol
benzene	303.15-346.15	2.39	n-decane	303.15-346.15	1.00
cyclohexane	303.15-346.15	0.92	n-hexane	303.15-346.15	1.05
ethylbenzene	303.15-346.15	2.13	n-octane	303.15-346.15	1.37
tert-butyl-benzene	303.15-346.15	1.93	2,2,4-trimethyl-pentane	303.15-346.15	1.30

copolymer (B): E-P-D/50x **93ISS**

solvent (A)	T-range/ K	$\Delta_{sol} H_{A(vap)}^\infty$/ kJ/mol	solvent (A)	T-range/ K	$\Delta_{sol} H_{A(vap)}^\infty$/ kJ/mol
benzene	333.15-393.15	–29.38	n-nonane	333.15-393.15	–43.25
n-heptane	333.15-393.15	–33.91	n-octane	333.15-393.15	–38.71
n-hexane	333.15-393.15	–29.05	toluene	333.15-393.15	–34.51

copolymer (B): E-P-D/50x **97SCH**

solvent (A)	T-range/ K	$\Delta_{sol} H_{A(vap)}^\infty$/ kJ/mol	solvent (A)	T-range/ K	$\Delta_{sol} H_{A(vap)}^\infty$/ kJ/mol
benzene	323.15-393.15	–29.35	n-hexane	323.15-393.15	–29.41
n-heptane	323.15-393.15	–33.89	toluene	323.15-393.15	–34.50

Copolymers from ethylene and vinyl acetate

Average chemical composition of the copolymers, acronyms and references:

Copolymer (B)	Acronym	Ref.
Ethylene/vinyl acetate copolymer		
18.0 wt% vinyl acetate	E-VA/18w	80DIP
20.0 wt% vinyl acetate	E-VA/20w	98KIM
29.0 wt% vinyl acetate	E-VA/29w	78DIN

Characterization:

Copolymer (B)	M_n/ g/mol	M_w/ g/mol	M_η/ g/mol	Further information
E-VA/18w				DuPont Elvax460, $MI = 2.5$
E-VA/20w	37000	49000		Polymer Laboratories
E-VA/29w	43200			Union Carbide, DX-31034, $MI = 15.0$

Experimental enthalpy data:

copolymer (B): E-VA/18w 80DIP

solvent (A)	T-range/ K	$\Delta_M H_A^\infty$/ kJ/mol	$\Delta_{sol} H_{A(vap)}^\infty$/ kJ/mol
benzene	388.15-408.15	0.08	−28.93
1-butanol	398.15-418.15	10.47	−32.07
chlorobenzene	388.15-408.15	0.88	−35.13
1-chlorobutane	388.15-408.15	2.01	−26.08
chlorocyclohexane	388.15-408.15		−36.13
cyclohexane	388.15-408.15	1.72	−26.59
cyclohexanol	388.15-408.15	9.59	−39.94
cyclohexanone	388.15-408.15		−37.30
n-hexane	388.15-408.15	1.09	−25.67
2-pentanone	388.15-408.15	3.01	−30.02
phenol	398.15-418.15	−1.42	−51.16
tetrachloromethane	388.15-408.15	2.60	−25.50
trichloromethane	388.15-408.15	−2.47	−28.81

copolymer (B): **E-VA/20w** **98KIM**

solvent (A)	T-range/ K	$\Delta_M H_A^\infty$/ kJ/mol	$\Delta_{sol} H_{A(vap)}^\infty$/ kJ/mol
benzene	423.15-463.15	−14.07	−41.88
1-butanol	423.15-463.15	−11.56	−48.85
2-butanone	423.15-463.15	−11.79	−26.23
butyl acetate	423.15-463.15	−11.55	−45.30
cyclohexane	423.15-463.15	−12.97	−38.91
n-decane	423.15-463.15	−12.61	−51.39
ethanol	423.15-463.15	− 9.72	−41.50
ethyl acetate	423.15-463.15	−10.42	−37.10
methanol	423.15-463.15	−12.01	−41.78
n-octane	423.15-463.15	−12.29	−44.55
2-propanone	423.15-463.15	−12.19	−36.52
tetrachloromethane	423.15-463.15	−15.11	−46.88
toluene	423.15-463.15	−11.02	−38.79
trichloromethane	423.15-463.15	−14.62	−39.19
p-xylene	423.15-463.15	−13.14	−47.60

copolymer (B): **E-VA/29w** **78DIN**

solvent (A)	T-range/ K	$\Delta_{sol} H_{A(vap)}^\infty$/ kJ/mol	solvent (A)	T-range/ K	$\Delta_{sol} H_{A(vap)}^\infty$/ kJ/mol
acetaldehyde	423.61-433.68	−23.08	n-hexane	423.61-433.68	−24.63
acetic acid	423.61-433.68	−24.91	methanol	423.61-433.68	−14.48
acetonitrile	423.61-433.68	−28.21	methylcyclohexane	423.61-433.68	−30.94
acrylonitrile	423.61-433.68	−30.46	nitroethane	423.61-433.68	−32.98
benzene	423.61-433.68	−31.12	nitromethane	423.61-433.68	−31.78
1-bromobutane	423.61-433.68	−37.10	1-nitropropane	423.61-433.68	−33.52
2-bromobutane	423.61-433.68	−33.76	2-nitropropane	423.61-433.68	−35.36
1-butanol	423.61-433.68	−36.23	n-octane	423.61-433.68	−38.49
2-butanol	423.61-433.68	−26.99	1-octene	423.61-433.68	−37.04
2-butanone	423.61-433.68	−27.27	n-pentane	423.61-433.68	−16.92
chlorobenzene	423.61-433.68	−41.02	3-pentanone	423.61-433.68	−36.17
cyclohexane	423.61-433.68	−26.03	2-propanol	423.61-433.68	−17.65
1,2-dichloro-ethane	423.61-433.68	−41.30	2-propanone	423.61-433.68	−25.48
			propionitrile	423.61-433.68	−30.06

continue

continue

solvent (A)	T-range/ K	$\Delta_{sol}H_{A(vap)}{}^{\infty}$/ kJ/mol	solvent (A)	T-range/ K	$\Delta_{sol}H_{A(vap)}{}^{\infty}$/ kJ/mol
dichloromethane	423.61-433.68	−22.64	propyl acetate	423.61-433.68	−35.46
diethyl ether	423.61-433.68	−34.21	tetrachloro-methane	423.61-433.68	−29.65
1,4-dioxane	423.61-433.68	−33.08			
dipropyl ether	423.61-433.68	−34.44	tetrahydrofuran	423.61-433.68	−29.94
ethanol	423.61-433.68	−21.32	trichloromethane	423.61-433.68	−29.83
ethyl acetate	423.61-433.68	−35.34	2,2,2-trifluoro-ethanol	423.61-433.68	−35.11
formic acid	423.61-433.68	−30.09			
furan	423.61-433.68	−21.97	water	423.61-433.68	−28.00
n-heptane	423.61-433.68	−31.79	m-xylene	423.61-433.68	−44.36

Copolymers from ethylene and vinyl alcohol

Average chemical composition of the copolymers, acronyms and references:

Copolymer (B)	Acronym	Ref.
Ethylene/vinyl alcohol copolymer 71.0 wt% vinyl alcohol	E-VAL/71w	96GAV

Characterization:

Copolymer (B)	M_n/ g/mol	M_w/ g/mol	M_η/ g/mol	Further information
E-VAL/71w				CERDATO, Elf Atochem, France

Experimental enthalpy data:

copolymer (B): E-VAL/71w 96GAV

solvent (A)	T-range/ K	$\Delta_{sol}H_{A(vap)}{}^{\infty}$/ kJ/mol	solvent (A)	T-range/ K	$\Delta_{sol}H_{A(vap)}{}^{\infty}$/ kJ/mol
1-butanol	333.15-363.15	−71.18	methanol	333.15-363.15	−15.49
2-butanol	333.15-363.15	−19.26	1-propanol	333.15-363.15	−81.22
ethanol	333.15-363.15	−74.52	2-propanol	333.15-363.15	−67.41
2-methyl-1-propanol	333.15-363.15	−17.58	2-methyl-2-propanol	333.15-363.15	−25.12

Copolymers from styrene and butyl methacrylate

Average chemical composition of the copolymers, acronyms and references:

Copolymer (B)	Acronym	Ref.
Styrene/butyl methacrylate copolymer 58.0 wt% styrene	S-BMA/58w	81DIP

Characterization:

Copolymer (B)	M_n/ g/mol	M_w/ g/mol	M_η/ g/mol	Further information
S-BMA/58w	32500	72500		Xerox Corp., Can.

Experimental enthalpy data:

copolymer (B): S-BMA/58w **81DIP**

solvent (A)	T-range/ K	$\Delta_M H_A^\infty$/ kJ/mol	solvent (A)	T-range/ K	$\Delta_M H_A^\infty$/ kJ/mol
benzene	393.15-423.15	−0.25	ethylbenzene	393.15-423.15	0.13
1-butanol	393.15-423.15	5.11	methylcyclohexane	393.15-423.15	3.22
butyl acetate	393.15-423.15	−0.46	*tert*-butyl-benzene	393.15-423.15	0.84
butylbenzene	393.15-423.15	0.25	n-octane	393.15-423.15	3.81
butylcyclohexane	393.15-423.15	3.85	2-pentanone	393.15-423.15	−1.17
chlorobenzene	393.15-423.15	−0.75	tetrachloromethane	393.15-423.15	0.08
1-chlorobutane	393.15-423.15	−1.38	trichloromethane	393.15-423.15	−3.94
cyclohexane	393.15-423.15	2.39	2,2,4-trimethyl-pentane	393.15-423.15	5.73
cyclohexanol	393.15-423.15	5.74			
n-decane	393.15-423.15	4.73	3,4,5-trimethyl-heptane	393.15-423.15	5.31
dichloromethane	393.15-423.15	−2.22			
n-dodecane	393.15-423.15	5.15			

Copolymers from styrene and divinylbenzene

Average chemical composition of the copolymers, acronyms and references:

Copolymer (B)	Acronym	Ref.
Styrene/divinylbenzene copolymer		
5 wt% divinylbenzene	S-DVB/05w	87SAN, 89SAN
10 wt% divinylbenzene	S-DVB/10w	87SAN, 89SAN
20 wt% divinylbenzene	S-DVB/20w	87SAN, 89SAN

Characterization:

Copolymer (B)	M_n/ g/mol	M_w/ g/mol	M_η/ g/mol	Further information
S-DVB/05w				network, synthesized in the laboratory
S-DVB/10w				network, synthesized in the laboratory
S-DVB/20w				network, synthesized in the laboratory

Experimental enthalpy data:

copolymer (B): **S-DVB/05w** **87SAN, 89SAN**

solvent (A)	T-range/ K	$\Delta_M H_A^\infty$/ kJ/mol	$\Delta_{sol} H_{A(vap)}^\infty$/ kJ/mol
1-butanol	445.-465.	22.0	− 5.7
ethanol	445.-465.	17.0	−16.0
methanol	445.-465.	11.9	−12.7
1-propanol	445.-465.	22.6	− 8.8

copolymer (B): **S-DVB/10w** **87SAN, 89SAN**

solvent (A)	T-range/ K	$\Delta_M H_A^\infty$/ kJ/mol	$\Delta_{sol} H_{A(vap)}^\infty$/ kJ/mol
1-butanol	455.-480.	17.0	−16.0
methanol	455.-480.	10.3	−17.9
1-pentanol	455.-480.	16.0	−15.1
1-propanol	455.-480.	10.2	−16.9

copolymer (B): **S-DVB/20w** **87SAN, 89SAN**

solvent (A)	T-range/ K	$\Delta_M H_A^\infty$/ kJ/mol	$\Delta_{sol} H_{A(vap)}^\infty$/ kJ/mol
1-butanol	460.-480.	25.9	−11.7
ethanol	460.-480.	13.8	−16.8
methanol	460.-480.	14.6	−13.4
1-pentanol	460.-480.	22.2	− 8.5
1-propanol	460.-480.	17.2	−11.9

Copolymers from styrene and ethyl acrylate

Average chemical composition of the copolymers, acronyms and references:

Copolymer (B)	Acronym	Ref.
Styrene/ethyl acrylate copolymer		
18.5 mol% ethyl acrylate	S-EA/19x	92COS
37.8 mol% ethyl acrylate	S-EA/38x	92COS
63.4 mol% ethyl acrylate	S-EA/63x	92COS
81.0 mol% ethyl acrylate	S-EA/81x	92COS

Characterization:

Copolymer (B)	M_n/ g/mol	M_w/ g/mol	M_η/ g/mol	Further information
S-EA/19x				synthesized in the laboratory
S-EA/38x				synthesized in the laboratory
S-EA/63x				synthesized in the laboratory
S-EA/81x				synthesized in the laboratory

Experimental enthalpy data:

copolymer (B): S-EA/19x 92COS

solvent (A)	T-range/ K	$\Delta_{sol}H_{A(vap)}{}^{\infty}$/ kJ/mol
n-decane	373.15-423.15	−21.20
n-hexane	373.15-423.15	− 2.66
n-nonane	373.15-423.15	−21.28

copolymer (B): **S-EA/38x** **92COS**

solvent (A)	T-range/ K	$\Delta_{sol}H_{A(vap)}^{\infty}$/ kJ/mol
n-decane	373.15-423.15	−23.82
n-hexane	373.15-423.15	− 6.47
n-nonane	373.15-423.15	−15.49

copolymer (B): **S-EA/63x** **92COS**

solvent (A)	T-range/ K	$\Delta_{sol}H_{A(vap)}^{\infty}$/ kJ/mol
n-decane	373.15-423.15	−13.63
n-hexane	373.15-423.15	−14.21
n-nonane	373.15-423.15	−28.50

copolymer (B): **S-EA/81x** **92COS**

solvent (A)	T-range/ K	$\Delta_{sol}H_{A(vap)}^{\infty}$/ kJ/mol
n-decane	373.15-423.15	−26.15
n-hexane	373.15-423.15	−19.05
n-nonane	373.15-423.15	−10.91

Copolymers from styrene and maleic anhydride

Average chemical composition of the copolymers, acronyms and references:

Copolymer (B)	Acronym	Ref.
Styrene/maleic anhydride/methacrylic acid terpolymer (1:1:1) 33 mol% styrene	S-MAH-MA/33x	95BOG

Characterization:

Copolymer (B)	M_n/ g/mol	M_w/ g/mol	M_η/ g/mol	Further information
S-MAH-MA/33x	28000			synthesized in the laboratory

Experimental enthalpy data:

copolymer (B): **S-MAH-MA/33x** **95BOG**

solvent (A)	T-range/ K	$\Delta_M H_A^\infty$/ kJ/mol	$\Delta_{sol} H_{A(vap)}^\infty$/ kJ/mol
1-butanol	526.6-556.6	−23.0	−11.0
n-decane	526.6-556.6	−13.0	−18.0
dibutyl ether	526.6-556.6	−16.0	−15.0
1,4-dioxane	526.6-556.6	−13.0	−15.0
n-heptane	526.6-556.6	−16.0	−15.0
n-nonane	526.6-556.6	−15.0	−15.0
n-octane	526.6-556.6	−13.0	−17.0

Copolymers from styrene and methyl methacrylate

Average chemical composition of the copolymers, acronyms and references:

Copolymer (B)	Acronym	Ref.
Styrene/methyl methacrylate copolymer 42 mol% styrene	S-MMA/42x	95NEM

Characterization:

Copolymer (B)	M_n/ g/mol	M_w/ g/mol	M_η/ g/mol	Further information
S-MMA/42x	57800			synthesized in the laboratory

Experimental enthalpy data:

copolymer (B): **S-MMA/42x** **95NEM**

solvent (A)	T-range/ K	$\Delta_M H_A{}^\infty$/ kJ/mol	$\Delta_{sol} H_{A(vap)}{}^\infty$/ kJ/mol
n-heptane	443.2-471.7	8.147	−19.147
n-hexane	443.2-471.7	10.096	−12.674
n-nonane	443.2-471.7	6.651	−25.529
n-octane	443.2-471.7	8.314	−28.720

Copolymers from styrene and 2-(3-methyl-3-phenylcyclobutyl)-2-hydroxyethyl methacrylate

Average chemical composition of the copolymers, acronyms and references:

Copolymer (B)	Acronym	Ref.
Styrene/2-(3-methyl-3-phenylcyclobutyl)-2-hydroxyethyl methacrylate copolymer 40.0 mol% styrene	S-MPCHEMA/40x	2000KAY

Characterization:

Copolymer (B)	M_n/ g/mol	M_w/ g/mol	M_η/ g/mol	Further information
S-MPCHEMA/40x				$\rho_B = 1.045$ g/cm^3, synthesized in the laboratory

Experimental enthalpy data:

copolymer (B): S-MPCHEMA/40x 2000KAY

solvent (A)	T-range/ K	$\Delta_M H_A^\infty$/ kJ/mol	solvent (A)	T-range/ K	$\Delta_M H_A^\infty$/ kJ/mol
benzene	443.15-453.15	11.64	methyl acetate	443.15-453.15	21.65
2-butanone	443.15-453.15	23.28	n-nonane	443.15-453.15	19.97
n-decane	443.15-453.15	21.65	n-octane	443.15-453.15	21.65
n-dodecane	443.15-453.15	16.62	2-propanone	443.15-453.15	16.62
ethanol	443.15-453.15	41.62	toluene	443.15-453.15	43.25
ethyl acetate	443.15-453.15	29.94	n-undecane	443.15-453.15	4.98
methanol	443.15-453.15	21.65	o-xylene	443.15-453.15	9.96

Copolymers from styrene and nonyl methacrylate

Average chemical composition of the copolymers, acronyms and references:

Copolymer (B)	Acronym	Ref.
Styrene/nonyl methacrylate copolymer 80 mol% styrene	S-NMA/80x	95BOG

Characterization:

Copolymer (B)	M_n/ g/mol	M_w/ g/mol	M_η/ g/mol	Further information
S-NMA/80x	29000			synthesized in the laboratory

Experimental enthalpy data:

copolymer (B): S-NMA/80x **95BOG**

solvent (A)	T-range/ K	$\Delta_M H_A^\infty$/ kJ/mol	$\Delta_{sol} H_{A(vap)}^\infty$/ kJ/mol
1-butanol	358.15-383.15	− 5.0	−38.0
n-decane	358.15-383.15	− 4.0	−41.0
dibutyl ether	358.15-383.15	8.0	−46.0
1,4-dioxane	358.15-383.15	0.0	−34.0
n-heptane	358.15-383.15	−14.0	−16.0
n-nonane	358.15-383.15	0.0	−40.0
n-octane	358.15-383.15	− 9.0	−25.0

Copolymers from tetrafluoroethylene and 2,2-bis(trifluoromethyl)-4,5-difluoro-1,3-dioxole

Average chemical composition of the copolymers, acronyms and references:

Copolymer (B)	Acronym	Ref.
Tetrafluoroethylene/2,2-bis(trifluoromethyl)-4,5-difluoro-1,3-dioxole copolymer 13 mol% tetrafluoroethylene	TFE-PFD/13x	99BON

Characterization:

Copolymer (B)	M_n/ g/mol	M_w/ g/mol	M_η/ g/mol	Further information
TFE-PFD/13x				AF2400, DuPont, Wilmington

Experimental enthalpy data:

copolymer (B): **TFE-PFD/13x** **99BON**

solvent (A)	T-range/ K	$\Delta_{sol}H_{A(vap)}^{\infty}$/ kJ/mol	solvent (A)	T-range/ K	$\Delta_{sol}H_{A(vap)}^{\infty}$/ kJ/mol
benzene	423.15	−29.98	n-hexane	423.15	−31.74
n-butane	423.15	−21.14	2-methylpropane	423.15	−24.95
n-heptane	423.15	−36.34	octafluorotoluene	423.15	−44.38
hexafluorobenzene	423.15	−37.68	n-pentane	423.15	−27.13

Copolymers from vinyl chloride and vinylidene chloride

Average chemical composition of the copolymers, acronyms and references:

Copolymer (B)	Acronym	Ref.
Vinyl chloride/vinylidene chloride copolymer 20 wt% vinyl chloride	VC-VdC/20w	92KAL

Characterization:

Copolymer (B)	M_n/ g/mol	M_w/ g/mol	M_η/ g/mol	Further information
VC-VdC/20w		90000		Polysciences Inc.

Experimental enthalpy data:

copolymer (B): VC-VdC/20w **92KAL**

solvent (A)	T-range/ K	$\Delta_{sol}H_{A(vap)}^{\infty}$/ kJ/mol
water	298.15-323.15	–42.70

5.3. Table of systems where additional information on enthalpy effects in copolymer solutions can be found

Copolymer (B)	Solvent (A)	Enthalpy	T-range	Ref.
Acrylic acid/methyl acrylate copolymer				
	water	$\Delta_M H$	298 K	93ADA
	water	$\Delta_M H$	298 K	93TAG
Acrylic acid/methyl methacrylate copolymer				
	water	$\Delta_M H$	298 K	93ADA
Acrylonitrile/butadiene copolymer				
	2-butanone	$\Delta_M H_A$	311 K	69POD
	benzene	$\Delta_M H_A^\infty$	298 K	55TA1
	benzene	$\Delta_M H$	298 K	55TA1
	cyclohexane	$\Delta_M H_A$	311 K	69POD
	hydrogen	$\Delta_{sol} H_{A(vap)}^\infty$	293-343 K	39BAR
	nitrogen	$\Delta_{sol} H_{A(vap)}^\infty$	293-343 K	39BAR
Acrylonitrile/isoprene copolymer				
	carbon dioxide	$\Delta_{sol} H_{A(vap)}^\infty$		50AME
	helium	$\Delta_{sol} H_{A(vap)}^\infty$		50AME
	hydrogen	$\Delta_{sol} H_{A(vap)}^\infty$		50AME
	nitrogen	$\Delta_{sol} H_{A(vap)}^\infty$		50AME
	oxygen	$\Delta_{sol} H_{A(vap)}^\infty$		50AME
Acrylonitrile/2-methylfuran copolymer				
	N,N-dimethylformamide	$\Delta_{sol} H_B^\infty$		64USM
Acrylonitrile/styrene copolymer				
	acrylonitrile	$\Delta_{sol} H_B^\infty$	298 K	89EGO
	N,N-dimethylformamide	$\Delta_{sol} H_B^\infty$	298 K	91ZVE
	styrene	$\Delta_{sol} H_B^\infty$	298 K	89EGO
	water	$\Delta_{sol} H_B^\infty$	298 K	91ZVE

Copolymer (B)	Solvent (A)	Enthalpy	T-range	Ref.
Acrylonitrile/styrenesulfonate copolymer				
	N,N-dimethyl-formamide	$\Delta_{sol}H_B^{\infty}$	298 K	91ZVE
	water	$\Delta_{sol}H_B^{\infty}$	298 K	91ZVE
Acrylonitrile/vinyl chloride copolymer				
	N,N-dimethyl-formamide	$\Delta_{sol}H_B^{\infty}$	298 K	91ZVE
	water	$\Delta_{sol}H_B^{\infty}$	298 K	91ZVE
Butadiene/methyl methacrylate copolymer				
	argon	$\Delta_{sol}H_{A(vap)}^{\infty}$	293-343 K	39BAR
	hydrogen	$\Delta_{sol}H_{A(vap)}^{\infty}$	293-343 K	39BAR
	nitrogen	$\Delta_{sol}H_{A(vap)}^{\infty}$	293-343 K	39BAR
	oxygen	$\Delta_{sol}H_{A(vap)}^{\infty}$	293-343 K	39BAR
Butadiene/styrene copolymer				
	argon	$\Delta_{sol}H_{A(vap)}^{\infty}$	293-343 K	39BAR
	argon	$\Delta_{sol}H_{A(vap)}^{\infty}$	298 K	60TIK
	benzene	$\Delta_M H_A$	293 K	55TA2
	hydrogen	$\Delta_{sol}H_{A(vap)}^{\infty}$	293-343 K	39BAR
	nitrogen	$\Delta_{sol}H_{A(vap)}^{\infty}$	293-343 K	39BAR
Cyclopentene/maleic acid copolymer				
	water	$\Delta_{dil}H^{12}$	298 K	81PAO
Ethyl acrylate/tetraethylene glycol dimethacrylate copolymer				
	argon	$\Delta_{sol}H_{A(vap)}^{\infty}$	333 K	68BAR
	carbon dioxide	$\Delta_{sol}H_{A(vap)}^{\infty}$	333 K	68BAR
	hydrogen	$\Delta_{sol}H_{A(vap)}^{\infty}$	333 K	68BAR
	krypton	$\Delta_{sol}H_{A(vap)}^{\infty}$	333 K	68BAR
	methane	$\Delta_{sol}H_{A(vap)}^{\infty}$	333 K	68BAR
	neon	$\Delta_{sol}H_{A(vap)}^{\infty}$	333 K	68BAR
	nitrogen	$\Delta_{sol}H_{A(vap)}^{\infty}$	333 K	68BAR
	oxygen	$\Delta_{sol}H_{A(vap)}^{\infty}$	333 K	68BAR
Ethylene/maleic acid copolymer				
	water	$\Delta_{dil}H^{12}$	298 K	81PAO

Copolymer (B)	Solvent (A)	Enthalpy	T-range	Ref.
Ethylene/propylene copolymer				
	benzene	$\Delta_{sol}H_{A(vap)}^{\infty}$	283-313 K	64FRE
	benzene	$\Delta_M H_A$	298-373 K	64HOL
	chlorobenzene	$\Delta_M H_A$	298-373 K	64HOL
	cyclohexane	$\Delta_{sol}H_{A(vap)}^{\infty}$	283-313 K	64FRE
	cyclohexane	$\Delta_M H_A$	298-373 K	64HOL
	dichloromethane	$\Delta_{sol}H_{A(vap)}^{\infty}$	283-313 K	64FRE
	n-heptane	$\Delta_M H_A$	298-373 K	64HOL
	n-hexane	$\Delta_{sol}H_{A(vap)}^{\infty}$	283-313 K	64FRE
	n-pentane	$\Delta_{sol}H_{A(vap)}^{\infty}$	283-313 K	64FRE
	tetrachloro-methane	$\Delta_{sol}H_{A(vap)}^{\infty}$	283-313 K	64FRE
	tetrachloro-methane	$\Delta_M H_A$	298-373 K	64HOL
	1,1,2-trichloro-trifluoroethane	$\Delta_{sol}H_{A(vap)}^{\infty}$	283-313 K	64FRE
Ethylene/vinyl acetate copolymer				
	n-dodecane	$\Delta_M H$	383-433 K	81DRU
	n-octadecane	$\Delta_M H$	383-433 K	81DRU
9,9-bis(4-Hydroxyphenyl)fluorene/terephthaloyl chloride copolymer				
	1,1,2,2-tetra-chloroethane	$\Delta_M H$	298K	77TA2
	trichloro-methane	$\Delta_M H$	298K	77TA2
Isobutylene/maleic acid copolymer				
	water	$\Delta_{dil}H^{12}$	298 K	81PAO
N-Isopropylacrylamide/N-octadecyl-acrylamide copolymer				
	water	$\Delta_{dil}H^{12}$	303 K	96FAE
N-Methacryloyl-L-alanine/N-phenyl-methacrylamide copolymer				
	water	$\Delta_{dil}H^{12}$	298 K	82MOR
4-Methyl-1-pentene/maleic acid copolymer				
	water	$\Delta_{dil}H^{12}$	298 K	81PAO

Copolymer (B)	Solvent (A)	Enthalpy	T-range	Ref.
Phenolphthalein/terephthaloyl chloride copolymer				
	1,1,2,2-tetra-chloroethane	$\Delta_M H$	298K	77TA2
	trichloro-methane	$\Delta_M H$	298K	77TA1
Polyisoprene-b-polystyrene diblock copolymer				
	cyclohexane	$\Delta_M H_A$	293-333 K	72GIR
	4-methyl-2-pentanone	$\Delta_M H_A$	293-333 K	72GIR
	toluene	$\Delta_M H_A$	293-333 K	72GIR
Propylene/maleic acid copolymer	water	$\Delta_{dil} H^{12}$	298 K	81PAO
Styrene/acrylonitrile copolymer				
	trichloro-methane	$\Delta_{sol} H_B^{\infty}$	303-323 K	93FRE
	trichloro-methane	$\Delta_{sol} H_B^{\infty}$	303-323 K	94FRE
Trimethyl-1-pentene/maleic acid copolymer	water	$\Delta_{dil} H^{12}$	298 K	81PAO
Vinyl acetate/1-vinyl-2-pyrrolidinone copolymer				
	vinyl acetate	$\Delta_{sol} H_B^{\infty}$	298 K	89EGO
	1-vinyl-2-pyr-rolidinone	$\Delta_{sol} H_B^{\infty}$	298 K	89EGO
Vinyl amine/vinyl caprolactam copolymer	water	$\Delta_M H$	298-308 K	94TAG
5-Vinyltetrazole/2-methyl-5-vinyltetrazole copolymer				
	N,N-dimethyl-formamide	$\Delta_{sol} H_B^{\infty}$	298 K	95KIZ
	water	$\Delta_{sol} H_B^{\infty}$, $\Delta_{dil} H^{12}$	298 K	97KIZ
	water	$\Delta_M H$	298 K	97KIZ

5.4. References

39BAR Barrer, R.M., Permeation, diffusion and solution of gases in organic polymers, *Trans. Faraday Soc.*, 35, 628, 1939.

50AME Amerongen, G.J. van, Influence of structure of elastomers on their permeability to gases, *J. Polym. Sci.*, 5, 307, 1950.

50TAG Tager, A. and Sanatina, V., Heats of solution and swelling of some synthetic high-molecular compounds (Russ.), *Kolloidn. Zh.*, 12, 474, 1950.

51TAG Tager, A. and Sanatina, V., Heats of solution and swelling of some synthetic high-molecular compounds, *Rubber Chem. Technol.*, 24, 773, 1951.

55OYA Oya, S., Heat of solution of synthetic high polymers. III. Heat of solution of partially acetylated poly(vinyl alcohol), *Chem. High Polym. Japan*, 12, 122, 1955.

55TA1 Tager, A.A. and Kosova, L.K., Thermodynamic studies of copolymer solutions 2. Thermodynamic study of butadiene-acrylonitrile copolymer solutions (Russ.), *Kolloidn. Zh.*, 17, 391, 1955.

55TA2 Tager, A.A., Kosova, L.K., Karlinskaya, D.Yu., and Yurina, I.A., Thermodynamic studies of copolymer solutions 1. Thermodynamic study of butadiene-styrene copolymer solutions (Russ.), *Kolloidn. Zh.*, 17, 315, 1955.

55ZAZ Zazulina, Z.A. and Rogovin, Z.A., Studies on properties of synthetic carbochain fibres. 2. Mechanism of the thermorelaxation behavior of carbochain fibres (Russ.), *Kolloidn. Zh.*, 17, 343, 1955.

56AMA Amaya, K. and Fujishiro, R., Heats of dilution of polyvinylalcohol solutions II, *Bull. Chem. Soc. Japan*, 29, 830, 1956.

56LIP Lipatov, Yu.S., Kargin, V.A., and Slonimskii, G.L., Study on orientation in high polymers II. Crystalline polymers (Russ.), *Zh. Fiz. Khim.*, 30, 1207, 1956.

56STR Struminskii, G.V. and Slonimskii, G.L., Mutual solubility of polymers III. Heats of mixing (Russ.), *Zh. Fiz. Khim.*, 30, 1941, 1956.

58SLO Slonimskii, G.L., Mutual solubility of polymers and properties of their mixtures, *J. Polym. Sci.*, 30, 625, 1958.

58TA1 Tager, A.A. and Iovleva, M., Thermodynamic studies of copolymer solutions III. Saponified poly(vinyl acetate)s (Russ.), *Zh. Fiz. Khim.*, 32, 1774, 1958.

58TA2 Tager, A.A. and Kargin, V.A., Thermodynamic studies of polymer-hydrogenated monomer systems II. Enthalpies of solution of copolymers in mixtures of hydrogenated monomers (Russ.), *Zh. Fiz. Khim.*, 32, 1362, 1958.

59ZEL Zelikman, S.G. and Mikhailov, N.V., Studies on the structure and properties of carbochain polymers in dilute solutions. IV. The integral and differential heats of solution and densities of the polymers (Russ.), *Vysokomol. Soedin.*, 1, 1077, 1959.

60TIK Tikhomirova, N.S., Malinksii, Yu.M., and Karpov, V.L., Studies on diffusion processes in polymers. I. Diffusion of monoatomic gases through polymer films of different structures (Russ.), *Vysokomol. Soedin.*, 2, 221, 1960.

64FRE Frensdorff, H.K., Diffusion and sorption of vapors in ethylene-propylene copolymers. I. Equilibrium sorption, *J. Polym. Sci., Pt. A*, 2, 333, 1964.

64HOL Holly, E.D., Interaction parameters and heats of dilution for ethylene-propylene rubber in various solvents, *J. Polym. Sci., Pt. A, 2*, 5267, 1964.

64USM Usmanov, Kh.U., Tillaev, R.S., and Musaev, Y.N., Radiation copolymerization of acrylonitrile with sylvane (Russ.), *Nauchn. Tr. Tashkent. Gos. Univ.*, 257, 3, 1964.

68BAR Barrer, R.M., Barrie, J.A., and Wong, P.S.-L., The diffusion and solution of gases in highly crosslinked copolymers, *Polymer*, 9, 609, 1968.

69POD Poddubnyi, I.Ya. and Podalinskii, A.V., Possible production of thermodynamically ideal polymer-solvent systems (Russ.), *Dokl. Akad. Nauk SSSR*, 185, 401, 1969.

70GIA Giacommeti, G., Turolla, A., and Boni, R., Enthalpy of helix-coil transitions from heats of solution. II. Poly-eps-carbobenzoxy-L-lysine and copolymers with L-phenylalanine, *Biopolymers*, 9, 979, 1970.

72GIR Girolamo, M. and Urwin, J.R., Thermodynamic parameters from osmotic studies on solutions of block copolymers of polyisoprene and polystyrene, *Eur. Polym. J.*, 8, 299, 1972.

76PET Petrosyan, V.A., Gabrielyan, G.A., and Rogovin, Z.A., Investigation of some properties of fiber-forming copolymers of acrylonitrile and isoprene, *Arm. Khim. Zh.*, 29, 516, 1976.

77TA1 Tager, A.A., Kolmakova, L.K., Bessonov, Yu.S., Salazkin, S.N., and Trofimova, N.M., Effect of the molecular mass and porous structure of cardo polyarylate on the thermodynamic parameters of dissolution (Russ.), *Vysokomol. Soedin., Ser. A*, 19, 1475, 1977.

77TA2 Tager, A.A., Kolmakova, L.K., Anufriev, V.A., Bessonov, Yu.S., Zhigunova, O.A., Vinogradova, S.V., Salazkin, S.N., and Tsilipotkina, M.V., Thermodynamics of the dissolution of cardo polyarylates in chloroform and tetrachloroethane (Russ.), *Vysokomol. Soedin., Ser. A*, 19, 2367, 1977.

78DIN Dincer, S. and Bonner, D.C., Thermodynamic analysis of ethylene and vinyl acetate copolymer with various solvents by gas chromatography, *Macromolecules*, 11, 107, 1978.

78DIP Dipaola-Baranyi, G., Braun, J.-M., and Guillet, J.E., Partial molar heats of mixing of small molecules with polymers by gas chromatography, *Macromolecules*, 11, 224, 1978.

78SOK Sokolova, D.F., Kudim, T.V., Sokolov, L.B., Zhegalova, N.I., and Zhuravlev, N.D., Thermochemical evaluation of the orderliness of copolymer polyarylates (Russ.), *Vysokomol. Soedin., Ser. B*, 20, 596, 1978.

79ITO Ito, K. and Guillet, J.E., Estimation of solubility parameters for some olefin polymers and copolymers by inverse gas chromatography, *Macromolecules*, 12, 1163, 1979.

79PH1 Phuong-Nguyen, H. and Delmas, G., Heats of mixing at infinite dilution of atactic polymers. 1. Effect of correlations of orientations in the solvents or in the polymers, *Macromolecules*, 12, 740, 1979.

79PH2 Phuong-Nguyen, H. and Delmas, G., Heats of mixing at infinite dilution of atactic polymers. 2. Effect of steric hindrance and of the shape of the solvent, *Macromolecules*, 12, 746, 1979.

80DIP DiPaola-Baranyi, G., Guillet, J.E., Jeberien, H.-E., and Klein, J., Thermodynamics of hydrogen bonding polymer-solute interactions by inverse gas chromatography, *Makromol. Chem.*, 181, 215, 1980.

81DIP DiPaola-Baranyi, G., Thermodynamic miscibility of various solutes with styrene-butyl methacrylate polymers and copolymers, *Macromolecules*, 14, 683, 1981.

81DRU Druz, N.I., Kreitus, A., and Chalykh, A.E., Thermodynamic parameters of mixing in the systems ethylene-vinyl acetate copolymers - hydrocarbons (Russ.), *Latv. PSR Zinat. Akad. Vestis. Kim. Ser.*, (3), 199, 1981.

81PAO Paoletti, S., Delben, F., and Crescenzi, V., Enthalpies of dilution of partially neutralized maleic acid copolymers in water. Correlation of experiments with theories, *J. Phys. Chem.*, 85, 1413, 1981.

82MOR Morcellet, M., Loucheux, C., and Daoust, H., Poly(methacrylic acid) derivatives 5. Microcalorimetric study of poly(N-methacryloyl-L-alanine) and poly(N-methacryloyl-L-alanine-co-N-phenylmethacrylamide) in aqueous solutions, *Macromolecules*, 15, 890, 1982.

84SIO Siow, K.S., Goh, S.H., and Yap, K.S., Solubility parameters of poly(α-methylstyrene-co-acrylnitrile) from gas-liquid chromatography, *Polym. Mater. Sci. Eng.*, 51, 532, 1984.

85COR Cordt, F., Die Energetik der Wechselwirkungen von Oligomeren des Ethylenoxids und dessen Copolymeren mit Propylenoxid in CCl_4-Lösung, *PhD-Thesis*, TU München, 1985.

86SIO Siow, K.S., Goh, S.H., and Yap, K.S., Solubility parameters of poly(α-methylstyrene-co-acrylonitrile) from gas-liquid chromatography, *J. Chromatogr.*, 354, 75, 1986.

87KYO Kyohmen, M., Inoue, K., Baba, Y., Kagemoto, A., and Beatty, Ch.L., Heats of dilution of poly[styrene-ran-(butyl methacrylate)] solutions measured with an automatic flow microcalorimeter, *Makromol. Chem.*, 188, 2721, 1987.

87SAN Sanetra, R., Kolarz, B., and Wlochowicz, A., Determination of thermodynamic data for the interaction of aliphatic alcohols with poly(styrene-co-divinylbenzene) using inverse gas chromatography, *Polymer*, 28, 1753, 1987.

89EGO Egorochkin, G.A., Semchikov, Yu.D., Smirnova, L.A., Knyazeva, T.E., Tokhonova, Z.A., Karyakin, N.V., and Sveshnikova, T.G., Thermodynamic analysis of the copolymerization of styrene with acrylonitrile and N-vinylpyrrolidone with vinyl acetate (Russ.), *Vysokomol. Soedin., Ser. B*, 31, 46, 1989.

89MOE Moeller, F., Energetik der Wechselwirkungen von Oligomeren des Ethylen- und Propylenoxids und deren Cooligomeren in Mischung mit CCl_4, *PhD-Thesis*, TU München, 1989.

89SAN Sanetra, R., Kolarz, B.N., and Wlochowicz, A., Badanie metoda inwesyjnej chromatografii gazowej struktury usieciowanych kopolimerow na przykladzie kopolimerow styren/ diwinylobenzen, *Polimery*, 34, 490, 1989.

90KIL Killmann, E., Cordt, F., and Moeller, F., Thermodynamics of mixing ethylene oxide oligomers with different end groups in tetrachloromethane, *Makromol. Chem.*, 191, 2929, 1990.

90SHI Shiomi, T., Ishimatsu, H., Eguchi, T., and Imai, K., Application of equation-of-state theory to random copolymer systems. 1, *Macromolecules*, 23, 4970, 1990.

91ZVE Zverev, M.P., Zenkov, I.D., Zakharova, N.N., Zashchenkina, E.S., and Bondarenko, O.A., Properties of polyacrylonitrile and its fibers containing chemically active groups (Russ.), *Khim. Volokna*, (5), 32, 1991.

92COS Coskun, M., Oezdemir, E., Benzen, R., and Pulat, E., Stiren-etilakrilat kopolimerlerinin termodinamik oezellikerinin invers gaz kromatografisi ile incelenmesi, *Doga - Turk Kimya Dergisi*, 16, 76, 1992.

92KAL Kalaouzis, P.J. and Demertzis, P.G., Water sorption and water vapour diffusion in food-grade plastics packaging materials: Effect of a polymeric plasticizer, *Packag. Technol. Sci.*, 5, 133, 1992.

92ROM Romdhane, I.H., Plana, A., Hwang, S., and Danner, R.P., Thermodynamic interactions of solvents with styrene-butadiene-styrene triblock copolymers, *J. Appl. Polym. Sci.*, 45, 2049, 1992.

93ADA Adamova, L.V., Klyuzhin, E.S., Safronov, A.P., Nerush, N.T., and Tager, A.A., Thermodynamics of interactions between water and copolymers of acrylate and acrylic acid (Russ.), *Vysokomol. Soedin., Ser. B*, 35, 893, 1993.

93FRE Frezzotti, D. and Ravanetti, G.P., Evaluation of the Flory-Huggins interaction parameter for poly(styrene-co-acrylonitrile) and poly(methyl methacrylate) blend from enthalpy of mixing measurements, *J. Calorim. Anal. Therm.*, 24, 135, 1993.

93ISS Issel, H.-M., Thermodynamische und rheologische Steuerung der Materialeigenschaften von Elastomeren durch trans-Poly(octenylen), *PhD-Thesis*, Univ. Hannover, 1993.

93TAG Tager, A.A., Klyuzhin, E.S., Adamova, L.V., and Safronov, A.P., Thermodynamics of dissolution of acrylic acid-methyl acrylate copolymers in water (Russ.), *Vysokomol. Soedin., Ser. B*, 35, 1357, 1993.

94COS Coskun, M., Oezdemir, E., Celik, S., and Cansiz, A., Inverse gas chromatography of n-alcohols - poly(acrylonitrile-co-methyl acrylate) systems, *J. Macromol. Sci., Macromol. Rep. A, (Suppl.1&2)*, 31, 63, 1994.

94FRE Frezzotti, D. and Ravanetti, G.P., Evaluation of the Flory-Huggins interaction parameter for poly(styrene-co-acrylonitrile) and poly(methyl methacrylate) blend from enthalpy of mixing measurements, *J. Therm. Anal.*, 41, 1237, 1994.

94TAG Tager, A.A., Safronov, A.P., Berezyuk, E.A., and Galaev, I.Yu., Lower critical solution temperature and hydrophobic hydration in aqueous polymer solutions, *Colloid Polym. Sci.*, 272, 1234, 1994.

95BOG Bogillo, V.I. and Voelkel, A., Solution properties of amorphous co- and terpolymers of styrene as examined by inverse gas chromatography, *J. Chromatogr. A*, 715, 127, 1995.

95KIL Killmann, E., Cordt, F., Moeller, F., and Zellner, H., Thermodynamics of mixing propylene oxide oligomers with different end groups of statistical and block cooligomers of ethylene oxide and propylene oxide in tetrachloromethane, *Macromol. Chem. Phys.*, 196, 47, 1995.

95KIZ Kizhnyaev, V.N., Gorkovenko, O.P., Bazhenov, D.N., and Smirnov, A.I., Interrelation between hydrodynamic and thermodynamic characteristics of solutions of poly(vinyl tetrazoles) in mixed solvents (Russ.), *Vysokomol. Soedin., Ser. B*, 37, 1948, 1995.

95NEM Nemtoi, G. and Beldie, C., Thermodynamic characterization of some copolymers by gas-chromatographic measurements, *Rev. Roum. Chim.*, 40, 335, 1995.

96FAE Faes, H., De Schryver, F.C., Sein, A., Bijma, K., Kevelam, J., and Engberts, J.B.F., Study of self-associating amphiphilic copolymers and their interaction with surfactants, *Macromolecules*, 29, 3875, 1996.

96GAV Gavara, R., Catala, R., Aucejo, S., Cabedo, D., and Hernandez, R., Solubility of alcohols in ethylene-vinyl alcohol copolymers by inverse gas chromatography, *J. Polym. Sci., Part B: Polym. Phys.*, 34, 1907, 1966.

96SAT Sato, T., Tohyama, M., Suzuki, M., Shiomi, T., and Imai, K., Application of equation-of-state theory to random copolymer blends with upper critical solution temperature type miscibility, *Macromolecules*, 29, 8231, 1996.

96SHI Shiomi, T., Tohyama, M., Endo, M., Sato, T., and Imai, K., Dependence of Flory-Huggins χ-parameters on the copolymer composition for solutions of poly(methyl methacrylate-ran-n-butyl methacrylate) in cyclohexanone, *J. Polym. Sci., Part B: Polym. Phys.*, 34, 2599, 1996.

97KIZ Kizhnyaev, V.N., Gorkovenko, O.P., Safronov, A.P., and Adamova, L.V., Thermodynamics of the interaction between tetrazole-containing polyelectrolytes and water (Russ.), *Vysokomol. Soedin., Ser. A*, 39, 527, 1997.

97SAT Sato, T., Suzuki, M., Tohyama, M., Endo, M., Shiomi, T., and Imai, K., Behavior of temperature dependence of χ-parameter in random copolymer blends showing an immiscibility window, *Polym. J.*, 29, 417, 1997.

97SCH Schuster, R.H., Issel, H.M., and Peterseim, V., Charakterisierung der Kautschuk-Lösungsmittel Wechselwirkung mittels inverser Gaschromatographie, *Kautschuk Gummi Kunststoffe*, 50, 890, 1997.

98KIM Kim, N.H., Won, Y.S., and Choi, J.S., Partial molar heat of mixing at infinite dilution in solvent/polymer (PEG, PMMA, EVA) solutions, *Fluid Phase Equil.*, 146, 223, 1998.

98YAM Yampolskii, Yu.P. and Bondarenko, G.N., Evidence of hydrogen bonding during sorption of chloromethanes in copolymers of chloroprene with methyl methacrylate and methacrylic acid, *Polymer*, 39, 2241, 1998.

99BON Bondar, V.I., Freeman, B.D., and Yampolskii, Yu.P., Sorption of gases and vapors in an amorphous glassy perfluorodioxol copolymer, *Macromolecules*, 32, 6163, 1999.

2000KAY Kaya, I. and Demirelli, K., Determination of thermodynamic properties of poly[2-(3-methyl-3-phenyl-cyclobutyl)-2-hydroxyethylmethacrylate] and its copolymers at infinite dilution using inverse gas chromatography, *Polymer*, 41, 2855, 2000.

6. PVT DATA OF MOLTEN COPOLYMERS

6.1. Experimental data and/or Tait equation parameters

Copolymers from acrylonitrile and styrene

Average chemical composition of the copolymers, acronyms, range of data, and references:

Copolymer	Acronym	Range of data T/K	P/MPa	Ref.
Acrylonitrile/styrene copolymer				
2.7 wt% acrylonitrile	S-AN/03w	438-539	0.1-200	92KIM
5.7 wt% acrylonitrile	S-AN/06w	437-540	0.1-200	92KIM
15.3 wt% acrylonitrile	S-AN/15w	453-543	0.1-200	92KIM
18.1 wt% acrylonitrile	S-AN/18w	423-528	0.1-200	92KIM
25.0 wt% acrylonitrile	S-AN/25w	433-573	0.1-200	95ZOL
40.0 wt% acrylonitrile	S-AN/40w	423-544	0.1-200	92KIM
69.7 wt% acrylonitrile	S-AN/70w	453-543	0.1-200	92KIM

Characterization:

Copolymer (B)	M_n/ g/mol	M_w/ g/mol	M_η/ g/mol	Further information
S-AN/03w	93500	211000		Asahi Chemical
S-AN/06w		270000		Asahi Chemical
S-AN/15w	56300	149000		Asahi Chemical
S-AN/18w		197000		Asahi Chemical
S-AN/25w	81900	506000		Polysciences Inc.
S-AN/40w	61000	122000		Asahi Chemical
S-AN/70w				Monsanto Co.

Experimental *PVT* data:

copolymer (B): **S-AN/03w** **92KIM**

P/MPa	438.25	450.65	456.65	469.75	494.45	505.85	516.55	529.55	539.25
	V_{spez}/cm^{3}g^{-1}								
0.1	1.0042	1.0105	1.0154	1.0218	1.0379	1.0465	1.0519	1.0620	1.0716
10	0.9963	1.0024	1.0068	1.0131	1.0279	1.0353	1.0406	1.0496	1.0585
20	0.9883	0.9945	0.9977	1.0045	1.0176	1.0237	1.0293	1.0370	1.0448
30	0.9820	0.9881	0.9913	0.9975	1.0102	1.0156	1.0211	1.0284	1.0359
40	0.9762	0.9820	0.9853	0.9914	1.0033	1.0085	1.0135	1.0204	1.0279
50	0.9710	0.9765	0.9797	0.9854	0.9970	1.0018	1.0070	1.0132	1.0204
60	0.9660	0.9713	0.9744	0.9800	0.9911	0.9958	1.0004	1.0067	1.0133
70	0.9611	0.9663	0.9695	0.9747	0.9855	0.9900	0.9947	1.0005	1.0071
80	0.9566	0.9616	0.9647	0.9698	0.9803	0.9846	0.9890	0.9948	1.0008
90	0.9524	0.9573	0.9603	0.9651	0.9753	0.9796	0.9839	0.9890	0.9953
100	0.9483	0.9533	0.9559	0.9608	0.9706	0.9750	0.9789	0.9840	0.9899
110	0.9446	0.9493	0.9520	0.9568	0.9662	0.9701	0.9741	0.9790	0.9849
120	0.9408	0.9455	0.9481	0.9527	0.9619	0.9658	0.9696	0.9746	0.9803
130			0.9444	0.9489	0.9578	0.9617	0.9654	0.9703	0.9755
140			0.9409	0.9452	0.9538	0.9577	0.9610	0.9656	0.9709
150			0.9373	0.9417	0.9501	0.9537	0.9571	0.9616	0.9666
160			0.9339	0.9380	0.9462	0.9497	0.9532	0.9574	0.9625
170			0.9306	0.9347	0.9425	0.9461	0.9495	0.9536	0.9585
180			0.9274	0.9314	0.9389	0.9425	0.9457	0.9498	0.9545
190			0.9241	0.9281	0.9356	0.9389	0.9418	0.9458	0.9506
200			0.9209	0.9248	0.9321	0.9354	0.9383	0.9421	0.9466

copolymer (B): **S-AN/06w** **92KIM**

P/MPa	436.75	449.85	463.15	475.75	495.65	506.45	517.25	527.55	540.35
	V_{spez}/cm^{3}g^{-1}								
0.1	1.0086	1.0169	1.0259	1.0336	1.0467	1.0535	1.0619	1.0701	1.0809
10	1.0009	1.0088	1.0171	1.0244	1.0364	1.0428	1.0507	1.0583	1.0678
20	0.9934	1.0009	1.0084	1.0153	1.0264	1.0325	1.0399	1.0467	1.0546
30	0.9871	0.9945	1.0015	1.0082	1.0186	1.0243	1.0315	1.0382	1.0455
40	0.9812	0.9884	0.9951	1.0019	1.0111	1.0170	1.0238	1.0302	1.0370
50	0.9758	0.9829	0.9891	0.9956	1.0046	1.0100	1.0168	1.0229	1.0292

continue

continue

P/MPa	T/K 436.75	449.85	463.15	475.75	495.65	506.45	517.25	527.55	540.35
	V_{spez}/cm^{3}g^{-1}								
60	0.9707	0.9777	0.9838	0.9895	0.9982	1.0037	1.0101	1.0158	1.0218
70	0.9660	0.9722	0.9782	0.9842	0.9925	0.9977	1.0040	1.0093	1.0150
80	0.9616	0.9676	0.9735	0.9791	0.9872	0.9922	0.9983	1.0036	1.0088
90	0.9573	0.9634	0.9690	0.9745	0.9821	0.9872	0.9928	0.9980	1.0034
100	0.9532	0.9591	0.9646	0.9701	0.9776	0.9820	0.9879	0.9927	0.9977
110	0.9492	0.9552	0.9605	0.9659	0.9728	0.9775	0.9831	0.9877	0.9926
120	0.9456	0.9514	0.9566	0.9616	0.9684	0.9730	0.9785	0.9831	0.9878
130		0.9474	0.9526	0.9576	0.9641	0.9686	0.9741	0.9786	0.9826
140		0.9439	0.9487	0.9538	0.9602	0.9644	0.9700	0.9740	0.9781
150		0.9403	0.9448	0.9499	0.9562	0.9606	0.9656	0.9696	0.9738
160		0.9370	0.9411	0.9462	0.9523	0.9565	0.9613	0.9653	0.9694
170			0.9376	0.9428	0.9486	0.9528	0.9573	0.9613	0.9651
180			0.9342	0.9393	0.9450	0.9489	0.9534	0.9572	0.9613
190			0.9307	0.9358	0.9412	0.9453	0.9494	0.9533	0.9569
200			0.9271	0.9324	0.9377	0.9416	0.9458	0.9494	0.9534

copolymer (B): **S-AN/18w** **92KIM**

P/MPa	T/K 423.25	436.55	450.05	463.05	475.95	488.55	501.55	515.05	528.05
	V_{spez}/cm^{3}g^{-1}								
0.1	0.9714	0.9788	0.9862	0.9936	1.0006	1.0080	1.0155	1.0240	1.0317
10	0.9650	0.9721	0.9791	0.9861	0.9928	0.9997	1.0067	1.0145	1.0217
20	0.9586	0.9654	0.9720	0.9786	0.9850	0.9916	0.9979	1.0049	1.0119
30	0.9533	0.9598	0.9662	0.9725	0.9786	0.9848	0.9910	0.9976	1.0041
40	0.9483	0.9547	0.9608	0.9667	0.9727	0.9786	0.9845	0.9908	0.9970
50	0.9436	0.9498	0.9557	0.9615	0.9673	0.9727	0.9784	0.9844	0.9905
60	0.9393	0.9452	0.9510	0.9565	0.9620	0.9676	0.9728	0.9789	0.9842
70	0.9349	0.9407	0.9463	0.9516	0.9569	0.9623	0.9675	0.9731	0.9783
80		0.9366	0.9421	0.9470	0.9522	0.9575	0.9624	0.9678	0.9729
90		0.9327	0.9382	0.9429	0.9479	0.9529	0.9577	0.9629	0.9680
100		0.9290	0.9340	0.9389	0.9437	0.9487	0.9534	0.9582	0.9631
110		0.9254	0.9301	0.9349	0.9397	0.9444	0.9492	0.9538	0.9586
120		0.9218	0.9265	0.9314	0.9358	0.9404	0.9453	0.9498	0.9542
130			0.9232	0.9279	0.9320	0.9364	0.9412	0.9454	0.9499

continue

continue

P/MPa	T/K 423.25	436.55	450.05	463.05	475.95	488.55	501.55	515.05	528.05
	V_{spez}/cm^{3}g^{-1}								
140			0.9197	0.9244	0.9286	0.9325	0.9373	0.9414	0.9458
150			0.9164	0.9207	0.9248	0.9292	0.9333	0.9377	0.9420
160			0.9132	0.9174	0.9215	0.9254	0.9298	0.9339	0.9378
170				0.9140	0.9181	0.9220	0.9260	0.9300	0.9341
180				0.9108	0.9146	0.9185	0.9228	0.9266	0.9301
190				0.9077	0.9113	0.9153	0.9191	0.9230	0.9265
200				0.9047	0.9082	0.9117	0.9159	0.9193	0.9229

copolymer (B): S-AN/25w **95ZOL**

P/MPa	T/K 433.05	442.95	452.75	462.55	472.75	482.25	493.65	503.85	514.45
	V_{spez}/cm^{3}g^{-1}								
0.1	0.9856	0.9912	0.9968	1.0026	1.0085	1.0142	1.0213	1.0274	1.0343
20	0.9742	0.9795	0.9847	0.9898	0.9953	1.0004	1.0065	1.0121	1.0181
40	0.9643	0.9694	0.9741	0.9789	0.9839	0.9887	0.9942	0.9992	1.0047
60	0.9556	0.9603	0.9648	0.9692	0.9737	0.9784	0.9835	0.9882	0.9929
80	0.9479	0.9525	0.9566	0.9607	0.9650	0.9693	0.9741	0.9784	0.9829
100	0.9406	0.9451	0.9489	0.9528	0.9569	0.9610	0.9655	0.9697	0.9738
120	0.9339	0.9382	0.9419	0.9457	0.9495	0.9535	0.9577	0.9616	0.9655
140	0.9278	0.9319	0.9354	0.9390	0.9428	0.9464	0.9506	0.9543	0.9579
160	0.9220	0.9259	0.9293	0.9328	0.9364	0.9400	0.9438	0.9474	0.9509
180	0.9166	0.9204	0.9235	0.9269	0.9303	0.9338	0.9375	0.9409	0.9443
200	0.9115	0.9151	0.9181	0.9214	0.9246	0.9280	0.9316	0.9348	0.9382

continue **S-AN/25w**

P/MPa	T/K 524.05	533.15	543.95	553.05	563.25	573.75
	V_{spez}/cm^{3}g^{-1}					
0.1	1.0403	1.0455	1.0522	1.0578	1.0642	1.0711
20	1.0231	1.0280	1.0337	1.0385	1.0442	1.0501

continue

continue

P/MPa	*T*/K 524.05	533.15	543.95	553.05	563.25	573.75
	V_{spez}/cm^{3}g^{-1}					
40	1.0092	1.0136	1.0187	1.0230	1.0281	1.0333
60	0.9971	1.0012	1.0060	1.0100	1.0146	1.0199
80	0.9868	0.9905	0.9949	0.9986	1.0030	1.0074
100	0.9774	0.9810	0.9850	0.9885	0.9927	0.9967
120	0.9689	0.9723	0.9761	0.9795	0.9832	0.9871
140	0.9613	0.9644	0.9680	0.9712	0.9747	0.9785
160	0.9540	0.9571	0.9605	0.9635	0.9669	0.9703
180	0.9472	0.9503	0.9536	0.9562	0.9595	0.9627
200	0.9410	0.9437	0.9468	0.9497	0.9526	0.9558

copolymer (B): **S-AN/40w** **92KIM**

P/MPa	*T*/K 423.25	440.65	455.65	475.45	492.95	509.85	522.75	535.55	544.35
	V_{spez}/cm^{3}g^{-1}								
0.1	0.9488	0.9578	0.9683	0.9755	0.9866	0.9954	1.0020	1.0098	1.0168
10	0.9431	0.9517	0.9612	0.9684	0.9783	0.9867	0.9931	1.0003	1.0062
20	0.9375	0.9455	0.9538	0.9614	0.9695	0.9777	0.9843	0.9909	0.9952
30	0.9328	0.9406	0.9483	0.9558	0.9635	0.9713	0.9774	0.9838	0.9877
40	0.9283	0.9359	0.9434	0.9504	0.9579	0.9654	0.9713	0.9772	0.9810
50	0.9241	0.9314	0.9386	0.9458	0.9527	0.9600	0.9656	0.9712	0.9748
60	0.9200	0.9271	0.9339	0.9409	0.9474	0.9545	0.9600	0.9653	0.9687
70	0.9161	0.9232	0.9298	0.9366	0.9428	0.9496	0.9546	0.9599	0.9632
80		0.9193	0.9261	0.9321	0.9384	0.9448	0.9499	0.9549	0.9580
90		0.9158	0.9221	0.9280	0.9342	0.9405	0.9453	0.9502	0.9532
100		0.9122	0.9183	0.9244	0.9304	0.9364	0.9410	0.9458	0.9487
110		0.9089	0.9147	0.9209	0.9265	0.9326	0.9370	0.9417	0.9446
120		0.9057	0.9114	0.9174	0.9229	0.9287	0.9331	0.9377	0.9406
130			0.9080	0.9139	0.9193	0.9250	0.9292	0.9335	0.9365
140			0.9048	0.9106	0.9159	0.9216	0.9257	0.9297	0.9325
150			0.9018	0.9075	0.9124	0.9180	0.9218	0.9261	0.9284
160			0.8987	0.9041	0.9091	0.9144	0.9185	0.9225	0.9249
170				0.9011	0.9060	0.9111	0.9152	0.9190	0.9214
180				0.8981	0.9027	0.9076	0.9116	0.9152	0.9177
190				0.8949	0.8996	0.9042	0.9080	0.9117	0.9140
200				0.8917	0.8963	0.9011	0.9045	0.9083	0.9105

Tait equation parameter functions:

Copolymer	$V(P/\text{MPa}, T/\text{K}) = V(0, T/\text{K})\{1 - C^*\ln[1 + (P/\text{MPa})/B(T/\text{K})]\}$ with C = 0.0894 and $\theta = T/\text{K} - 273.15$	
	$V(0,T/\text{K})/\text{cm}^3\ \text{g}^{-1}$	$B(T/\text{K})/\text{MPa}$
S-AN/03w	$0.9233 + 3.9355\ 10^{-4}\theta + 5.6848\ 10^{-7}\theta^2$	$239.8\ \exp(-4.3763\ 10^{-3}\theta)$
S-AN/15w	$0.9044 + 4.2068\ 10^{-4}\theta + 4.0772\ 10^{-7}\theta^2$	$238.4\ \exp(-3.9434\ 10^{-3}\theta)$
S-AN/18w	$0.9016 + 4.0365\ 10^{-4}\theta + 4.2061\ 10^{-7}\theta^2$	$240.4\ \exp(-3.8578\ 10^{-3}\theta)$
S-AN/40w	$0.8871 + 3.4057\ 10^{-4}\theta + 4.9378\ 10^{-7}\theta^2$	$289.3\ \exp(-4.4313\ 10^{-3}\theta)$
S-AN/70w	$0.8528 + 3.6159\ 10^{-4}\theta + 2.6336\ 10^{-7}\theta^2$	$335.4\ \exp(-3.9230\ 10^{-3}\theta)$

Copolymers from acrylonitrile and butadiene

Average chemical composition of the copolymers, acronyms, range of data, and references:

Copolymer	Acronym	Range of data *T*/K	*P*/MPa	Ref.
Acrylonitrile/butadiene copolymer 33.0 wt% acrylonitrile	AN-B/33w	303-513	0.1-200	95ZOL

Characterization:

Copolymer (B)	M_n/ g/mol	M_w/ g/mol	M_η/ g/mol	Further information
AN-B/33w				Polysciences Inc.

Experimental *PVT* data:

copolymer (B): AN-B/33w **95ZOL**

P/kg cm^{-2}	*T*/K 303.05	312.25	322.25	333.55	343.75	354.25	363.75	374.15	383.85
	V_{spez}/cm^{3}g^{-1}								
1	0.9985	1.0055	1.0166	1.0184	1.0246	1.0319	1.0380	1.0442	1.0499
200	0.9907	0.9969	1.0027	1.0090	1.0148	1.0213	1.0269	1.0329	1.0384
400	0.9834	0.9890	0.9946	1.0006	1.0062	1.0122	1.0174	1.0229	1.0281
600	0.9764	0.9818	0.9871	0.9927	0.9980	1.0038	1.0085	1.0138	1.0185
800	0.9702	0.9754	0.9802	0.9857	0.9906	0.9961	1.0006	1.0056	1.0102
1000	0.9641	0.9691	0.9739	0.9791	0.9838	0.9891	0.9932	0.9980	1.0023
1200	0.9586	0.9634	0.9680	0.9730	0.9774	0.9823	0.9864	0.9910	0.9953
1400	0.9535	0.9580	0.9624	0.9673	0.9716	0.9763	0.9801	0.9847	0.9885
1600	0.9486	0.9529	0.9570	0.9619	0.9660	0.9705	0.9743	0.9785	0.9824
1800	0.9441	0.9481	0.9523	0.9567	0.9607	0.9651	0.9688	0.9729	0.9765
2000	0.9393	0.9432	0.9474	0.9518	0.9556	0.9599	0.9634	0.9675	0.9710

continue **AN-B/33w**

P/kg cm^{-2}	*T*/K 394.05	404.55	413.35	423.55	432.55	442.65	453.05	462.75	472.75
	V_{spez}/cm^{3}g^{-1}								
1	1.0569	1.0634	1.0694	1.0768	1.0825	1.0892	1.0961	1.1023	1.1098
200	1.0445	1.0507	1.0560	1.0626	1.0679	1.0738	1.0801	1.0858	1.0919
400	1.0337	1.0394	1.0446	1.0504	1.0554	1.0606	1.0665	1.0716	1.0774
600	1.0238	1.0291	1.0341	1.0395	1.0441	1.0488	1.0542	1.0589	1.0643
800	1.0149	1.0202	1.0247	1.0297	1.0341	1.0386	1.0435	1.0479	1.0530
1000	1.0070	1.0118	1.0162	1.0208	1.0250	1.0291	1.0339	1.0379	1.0428
1200	0.9994	1.0041	1.0082	1.0128	1.0166	1.0206	1.0250	1.0289	1.0335
1400	0.9928	0.9972	1.0011	1.0054	1.0092	1.0129	1.0172	1.0209	1.0253
1600	0.9866	0.9906	0.9943	0.9987	1.0021	1.0057	1.0099	1.0134	1.0175
1800	0.9806	0.9845	0.9881	0.9927	0.9995	0.9991	1.0029	1.0063	1.0101
2000	0.9749	0.9786	0.9821	0.9861	0.9893	0.9928	0.9962	0.9996	1.0035

continue **AN-B/33w**

P/kg cm^{-2}	T/K 482.55	493.65	502.45	513.35
	V_{spez}/cm^{3}g^{-1}			
1	1.1167	1.1245	1.1309	1.1384
200	1.0983	1.1052	1.1106	1.1171
400	1.0830	1.0892	1.0938	1.0996
600	1.0694	1.0752	1.0793	1.0849
800	1.0577	1.0632	1.0670	1.0721
1000	1.0471	1.0521	1.0559	1.0603
1200	1.0378	1.0426	1.0459	1.0502
1400	1.0293	1.0337	1.0370	1.0409
1600	1.0213	1.0256	1.0287	1.0323
1800	1.0140	1.0178	1.0209	1.0245
2000	1.0071	1.0109	1.0135	1.0173

Copolymers from butadiene and styrene

Average chemical composition of the copolymers, acronyms, range of data, and references:

Copolymer	Acronym	Range of data T/K	P/MPa	Ref.
Butadiene/styrene copolymer				
10.0 wt% styrene	S-BR/10w	393-533	0.1-196	94WAN
23.5 wt% styrene	S-BR/23w	393-533	0.1-196	94WAN
25.0 % styrene	S-BR/25	348-373	0.1-10	77REN
60.0 wt% styrene	S-BR/60w	393-533	0.1-196	94WAN
85.0 wt% styrene	S-BR/85w	393-533	0.1-196	94WAN
Butadiene/styrene copolymer (block copolymer without specification)				
48.0 % styrene	S-?-B/48	348-373	0.1-10	77REN
70.0 % styrene	S-?-B/70	348-373	0.1-10	77REN
Polystyrene-b-polybutadiene-b-polystyrene triblock copolymer				
25.4 % styrene	S-B-S/25	348-373	0.1-10	77REN
30.0 % styrene	S-B-S/30	348-373	0.1-10	77REN

Characterization:

Copolymer (B)	M_n/ g/mol	M_w/ g/mol	M_η/ g/mol	Further information
S-BR/10w	112000	650000		35% 1,4-cis, 52% 1,4-trans, 13% 1,2-vinyl
S-BR/23w	126000	720000		30% 1,4-cis, 47% 1,4-trans, 23% 1,2-vinyl
S-BR/25	239000	481000		Europrene R130, ANIC Mailand
S-BR/60w	128000	820000		30% 1,4-cis, 47% 1,4-trans, 23% 1,2-vinyl
S-BR/85w	121000	730000		30% 1,4-cis, 47% 1,4-trans, 23% 1,2-vinyl
S-?-B/48	61800	76100		Europrene S141, ANIC Mailand (32% styrene in blocks)
S-?-B/70	97600	126400		Europrene S142, ANIC Mailand (45% styrene in blocks)
S-B-S/25	3800			(7Styr-43Buta-7Styr), 40% 1,4-cis, 50% 1,4-trans, 10% 1,2-vinyl
S-B-S/30	130500	229600		

Tait equation parameter functions:

Copolymer	$V(P/\text{MPa}, T/\text{K}) = V(0, T/\text{K})\{1 - \text{C*ln}[1 + (P/\text{MPa})/\text{B}(T/\text{K})]\}$ with C = 0.0894	
	$V(0,T/\text{K})/\text{cm}^3\text{g}^{-1}$	$\text{B}(T/\text{K})/\text{MPa}$
S-BR/10w	$0.9053 \exp(2.437\ 10^{-5}\ T^{1.5})$	$530.3 \exp[-3.99\ 10^{-3}\ (T\text{-}273.15)]$
S-BR/23w	$0.8986 \exp(2.317\ 10^{-5}\ T^{1.5})$	$551.6 \exp[-4.17\ 10^{-3}\ (T\text{-}273.15)]$
S-BR/60w	$0.8812 \exp(2.031\ 10^{-5}\ T^{1.5})$	$486.0 \exp[-4.34\ 10^{-3}\ (T\text{-}273.15)]$
S-BR/85w	$0.8704 \exp(1.846\ 10^{-5}\ T^{1.5})$	$356.7 \exp[-4.24\ 10^{-3}\ (T\text{-}273.15)]$

Copolymer	$V(P/\text{MPa}, T/\text{K}) = V(0, T/\text{K})\{1 - \text{C}(T/\text{K}) \ln[1 + (P/\text{MPa})/\text{B}(T/\text{K})]\}$	
S-BR/25	$T/\text{K} = 348.15;$	$V/\text{cm}^3\text{g}^{-1} = 1.1074\ [1 - 0.10290 \ln(1 + P/130.1)]$
	$T/\text{K} = 373.15;$	$V/\text{cm}^3\text{g}^{-1} = 1.1260\ [1 - 0.11820 \ln(1 + P/137.1)]$
S-?-B/48	$T/\text{K} = 348.15;$	$V/\text{cm}^3\text{g}^{-1} = 1.0607\ [1 - 0.10780 \ln(1 + P/168.1)]$
	$T/\text{K} = 373.15;$	$V/\text{cm}^3\text{g}^{-1} = 1.0770\ [1 - 0.08794 \ln(1 + P/121.1)]$
S-?-B/70	$T/\text{K} = 348.15;$	$V/\text{cm}^3\text{g}^{-1} = 1.0298\ [1 - 0.08723 \ln(1 + P/177.2)]$
	$T/\text{K} = 373.15;$	$V/\text{cm}^3\text{g}^{-1} = 1.0148\ [1 - 0.07611 \ln(1 + P/126.2)]$
S-B-S/25	$T/\text{K} = 348.15;$	$V/\text{cm}^3\text{g}^{-1} = 1.0989\ [1 - 0.07866 \ln(1 + P/117.1)]$
	$T/\text{K} = 373.15;$	$V/\text{cm}^3\text{g}^{-1} = 1.1177\ [1 - 0.08936 \ln(1 + P/118.8)]$
S-B-S/30	$T/\text{K} = 348.15;$	$V/\text{cm}^3\text{g}^{-1} = 1.0972\ [1 - 0.13030 \ln(1 + P/222.5)]$
	$T/\text{K} = 373.15;$	$V/\text{cm}^3\text{g}^{-1} = 1.1158\ [1 - 0.11420 \ln(1 + P/165.0)]$

Copolymers from ethylene and acrylic acid

Average chemical composition of the copolymers, acronyms, range of data, and references:

Copolymer	Acronym	Range of data T/K	Range of data P/MPa	Ref.
Ethylene/acrylic acid copolymer				
3.0 wt% acrylic acid	E-AA/03w	393-473	0.1-200	92WIN
6.0 wt% acrylic acid	E-AA/06w	393-473	0.1-200	92WIN
7.3 wt% acrylic acid	E-AA/07w	393-473	0.1-200	92WIN
9.0 wt% acrylic acid	E-AA/09w	410-515	0.1-200	95ZOL
10.0 wt% acrylic acid	E-AA/10w	410-525	0.1-200	95ZOL
12.4 wt% acrylic acid	E-AA/12w	393-473	3.4-200	92WIN
20.0 wt% acrylic acid	E-AA/20w	402-535	0.1-200	95ZOL

Characterization:

Copolymer (B)	M_n/ g/mol	M_w/ g/mol	M_η/ g/mol	Further information
E-AA/03w	37500	183000		synthesized in the laboratory
E-AA/06w	30000	150000		synthesized in the laboratory
E-AA/07w	25000	126000		synthesized in the laboratory
E-AA/09w				unspecified commercial product
E-AA/10w				Du Pont, Nucrel, MI = 1.3
E-AA/12w	26000	130000		synthesized in the laboratory
E-AA/20w				Sci. Polym. Prod., No. 10

Experimental *PVT* data:

copolymer (B): E-AA/03w **92WIN**

T/K	393.2		413.2		433.2		453.2		473.2
P/MPa	V_{spez}/cm^{3}g^{-1}	*P*/MPa	V_{spez}/cm^{3}g^{-1}	*P*/MPa	V_{spez}/cm^{3}g^{-1}	*P*/MPa	V_{spez}/cm^{3}g^{-1}	*P*/MPa	V_{spez}/cm^{3}g^{-1}
0.1	1.2366	0.1	1.2516	0.1	1.2686	0.1	1.2856	0.1	1.3035
0.9	1.2360	5.3	1.2455	1.2	1.2675	1.0	1.2861	4.3	1.2969
2.1	1.2343	10.2	1.2400	2.9	1.2649	3.1	1.2814	10.2	1.2878
5.0	1.2312	20.4	1.2295	6.9	1.2598	7.6	1.2751	18.4	1.2766
7.3	1.2290	29.4	1.2211	11.6	1.2541	13.4	1.2675	30.2	1.2617
10.7	1.2257	39.0	1.2127	21.2	1.2433	24.7	1.2540	43.9	1.2470
14.4	1.2221	46.7	1.2066	31.0	1.2335	44.9	1.2334	63.3	1.2292
17.8	1.2189	59.5	1.1970	44.7	1.2209	64.2	1.2166	83.4	1.2131
21.5	1.2156	73.4	1.1875	64.3	1.2051	85.2	1.2009	102.7	1.1996
32.6	1.2058	87.5	1.1786	83.7	1.1911	102.4	1.1894	122.2	1.1869
		101.2	1.1709	102.5	1.1792	121.6	1.1778	142.0	1.1759
		123.7	1.1587	121.6	1.1682	142.2	1.1665	160.0	1.1665
				141.7	1.1579	159.8	1.1578	179.8	1.1573
				160.4	1.1491	179.5	1.1488	200.9	1.1479
				179.0	1.1411	200.4	1.1401		

copolymer (B): E-AA/06w **92WIN**

T/K	393.2		413.2		433.2		453.2		473.2
P/MPa	V_{spez}/cm^{3}g^{-1}	*P*/MPa	V_{spez}/cm^{3}g^{-1}	*P*/MPa	V_{spez}/cm^{3}g^{-1}	*P*/MPa	V_{spez}/cm^{3}g^{-1}	*P*/MPa	V_{spez}/cm^{3}g^{-1}
0.1	1.2256	2.9	1.2364	2.7	1.2531	11.7	1.2574	23.3	1.2564
1.4	1.2243	6.5	1.2321	6.5	1.2483	18.6	1.2487	34.0	1.2442
3.7	1.2218	11.0	1.2271	11.4	1.2423	29.1	1.2370	47.6	1.2303
7.6	1.2179	18.5	1.2193	17.9	1.2349	44.5	1.2214	64.3	1.2151
14.3	1.2113	31.2	1.2073	32.9	1.2195	63.9	1.2046	83.0	1.2001
22.7	1.2038	44.4	1.1960	45.1	1.2082	83.4	1.1899	102.8	1.1863
32.7	1.1956	63.1	1.1819	64.2	1.1929	102.2	1.1775	121.8	1.1743
36.3	1.1918	82.2	1.1689	83.4	1.1791	121.0	1.1655	141.0	1.1634
41.9	1.1885	101.7	1.1571	102.7	1.1670	141.6	1.1551	160.3	1.1534
		121.2	1.1465	121.6	1.1563	160.3	1.1457	180.1	1.1440
		144.9	1.1335	141.1	1.1463	178.7	1.1374	201.8	1.1345
				160.2	1.1372	197.0	1.1295		

copolymer (B): **E-AA/07w** **92WIN**

T/K	393.2		413.2		433.2		453.2		473.2
P/ MPa	V_{spez}/ cm^3g^{-1}	P/ MPa	V_{spez}/ cm^3g^{-1}	P/ MPa	V_{spez}/ cm^3g^{-1}	P/ MPa	V_{spez}/ cm^3g^{-1}	P/ MPa	V_{spez}/ cm^3g^{-1}
18.4	1.1989	3.4	1.2271	3.6	1.2436	4.0	1.2593	3.7	1.2755
37.2	1.1832	6.6	1.2234	7.6	1.2384	8.5	1.2532	7.5	1.2699
56.4	1.1691	11.8	1.2179	15.6	1.2293	16.8	1.2128	12.2	1.2632
77.5	1.1557	20.4	1.2092	27.6	1.2169	27.7	1.2306	20.9	1.2515
		30.6	1.1998	43.7	1.2018	43.7	1.2145	31.5	1.2391
		44.1	1.1884	63.1	1.1866	63.3	1.1980	43.6	1.2263
		64.3	1.1736	82.2	1.1730	82.6	1.1837	62.8	1.2089
		83.1	1.1613	101.9	1.1607	102.4	1.1707	82.7	1.1931
		102.8	1.1500	121.3	1.1495	121.3	1.1595	102.5	1.1794
		121.6	1.1399	140.7	1.1395	140.9	1.1487	120.8	1.1679
		142.0	1.1292	159.8	1.1306	159.8	1.1393	140.6	1.1565
				177.5	1.1228	178.4	1.1308	160.2	1.1464
				193.4	1.1164	197.9	1.1225	179.2	1.1375
								199.4	1.1288

copolymer (B): **E-AA/09w** **95ZOL**

P/MPa	T/K 410.15	418.55	426.45	434.75	442.05	450.45	458.75	466.95	475.15
	V_{spez}/cm^3g^{-1}								
0.1	1.2221	1.2301	1.2365	1.2443	1.2506	1.2588	1.2655	1.2736	1.2819
20	1.2017	1.2081	1.2142	1.2205	1.2265	1.2330	1.2393	1.2458	1.2522
40	1.1855	1.1911	1.1967	1.2023	1.2078	1.2139	1.2192	1.2252	1.2306
60	1.1714	1.1766	1.1818	1.1869	1.1922	1.1973	1.2025	1.2077	1.2128
80	1.1591	1.1640	1.1688	1.1735	1.1785	1.1833	1.1879	1.1927	1.1974
100	1.1481	1.1526	1.1572	1.1615	1.1663	1.1707	1.1751	1.1795	1.1841
120	1.1381	1.1421	1.1464	1.1509	1.1553	1.1595	1.1635	1.1676	1.1719
140	1.1289	1.1327	1.1367	1.1407	1.1451	1.1492	1.1530	1.1569	1.1610
160	1.1202	1.1238	1.1277	1.1315	1.1358	1.1397	1.1433	1.1471	1.1507
180	1.1124	1.1157	1.1193	1.1230	1.1270	1.1308	1.1343	1.1377	1.1414
200	1.1046	1.1078	1.1116	1.1151	1.1193	1.1225	1.1259	1.1291	1.1327

continue **E-AA/09w**

P/MPa	T/K 483.05	490.85	499.15	507.35	515.55
	V_{spez}/cm^{3}g^{-1}				
0.1	1.2887	1.2963	1.3046	1.3129	1.3213
20	1.2588	1.2652	1.2719	1.2787	1.2858
40	1.2366	1.2424	1.2480	1.2541	1.2602
60	1.2181	1.2233	1.2287	1.2340	1.2395
80	1.2024	1.2073	1.2122	1.2171	1.2221
100	1.1888	1.1931	1.1977	1.2025	1.2070
120	1.1763	1.1805	1.1847	1.1892	1.1937
140	1.1650	1.1690	1.1730	1.1772	1.1814
160	1.1549	1.1586	1.1623	1.1664	1.1702
180	1.1452	1.1488	1.1525	1.1561	1.1601
200	1.1364	1.1400	1.1433	1.1469	1.1506

copolymer (B): E-AA/10w 95ZOL

P/MPa	T/K 409.45	417.85	426.35	434.65	442.65	450.75	458.95	467.25	475.45
	V_{spez}/cm^{3}g^{-1}								
0.1	1.2214	1.2286	1.2363	1.2436	1.2506	1.2581	1.2657	1.2732	1.2806
20	1.2027	1.2089	1.2156	1.2220	1.2282	1.2346	1.2409	1.2478	1.2541
40	1.1867	1.1926	1.1986	1.2044	1.2101	1.2158	1.2214	1.2275	1.2332
60	1.1729	1.1783	1.1839	1.1892	1.1945	1.1997	1.2053	1.2102	1.2157
80	1.1609	1.1658	1.1710	1.1759	1.1808	1.1856	1.1906	1.1956	1.2005
100	1.1498	1.1545	1.1595	1.1640	1.1686	1.1732	1.1777	1.1822	1.1870
120	1.1399	1.1442	1.1489	1.1532	1.1576	1.1621	1.1663	1.1704	1.1749
140	1.1306	1.1347	1.1393	1.1434	1.1476	1.1515	1.1556	1.1596	1.1640
160	1.1220	1.1260	1.1306	1.1341	1.1381	1.1422	1.1459	1.1498	1.1539
180	1.1142	1.1180	1.1221	1.1257	1.1297	1.1333	1.1372	1.1406	1.1446
200	1.1070	1.1105	1.1147	1.1179	1.1215	1.1252	1.1288	1.1323	1.1360

continue **E-AA/10w**

P/MPa	*T*/K					
	484.15	492.05	500.55	508.85	516.95	525.75
	V_{spez}/cm^{3}g^{-1}					
0.1	1.2889	1.2966	1.3049	1.3134	1.3215	1.3306
20	1.2613	1.2677	1.2747	1.2814	1.2883	1.2955
40	1.2392	1.2450	1.2512	1.2575	1.2634	1.2698
60	1.2210	1.2265	1.2319	1.2374	1.2429	1.2486
80	1.2055	1.2104	1.2154	1.2204	1.2256	1.2304
100	1.1917	1.1965	1.2009	1.2056	1.2103	1.2149
120	1.1792	1.1837	1.1880	1.1924	1.1969	1.2011
140	1.1679	1.1722	1.1763	1.1804	1.1847	1.1886
160	1.1577	1.1617	1.1657	1.1695	1.1737	1.1774
180	1.1482	1.1523	1.1558	1.1596	1.1634	1.1672
200	1.1397	1.1433	1.1469	1.1504	1.1540	1.1574

copolymer (B): **E-AA/12w** **92WIN**

T/K	393.2		413.2		433.2		453.2		473.2
P/ MPa	V_{spez}/ cm^{3}g^{-1}	*P*/ MPa	V_{spez}/ cm^{3}g^{-1}	*P*/ MPa	V_{spez}/ cm^{3}g^{-1}	*P*/ MPa	V_{spez}/ cm^{3}g^{-1}	*P*/ MPa	V_{spez}/ cm^{3}g^{-1}
8.6	1.1749	3.4	1.1930	6.1	1.2047	5.6	1.2208	5.3	1.2355
19.6	1.1653	7.6	1.1884	10.7	1.1994	9.8	1.2157	10.2	1.2288
30.4	1.1568	14.7	1.1814	19.2	1.1905	17.9	1.2056	18.2	1.2187
39.7	1.1498	27.9	1.1698	31.6	1.1788	31.1	1.1923	32.2	1.2035
49.6	1.1430	43.1	1.1575	44.2	1.1683	45.5	1.1792	44.4	1.1916
62.8	1.1348	62.8	1.1434	63.4	1.1538	63.5	1.1652	64.1	1.1748
78.9	1.1249	82.0	1.1316	83.3	1.1409	82.8	1.1518	83.1	1.1607
94.1	1.1167	99.0	1.1227	103.1	1.1296	102.7	1.1397	102.8	1.1481
99.3	1.1128	102.2	1.1205	122.3	1.1196	122.1	1.1287	121.4	1.1369
112.3	1.1076	121.0	1.1114	141.5	1.1105	141.2	1.1189	141.9	1.1263
129.6	1.0995	141.7	1.1020	160.4	1.1022	159.8	1.1101	161.0	1.1169
		159.9	1.0942	179.3	1.0946	178.6	1.1021	180.0	1.1085
		178.3	1.0863	198.3	1.0873	197.5	1.0944	200.1	1.1005

copolymer (B): **E-AA/20w** **95ZOL**

P/MPa	*T*/K 401.75	409.75	418.45	426.15	434.75	443.55	451.95	459.75	468.25
	V_{spez}/cm^{3}g^{-1}								
0.1	1.1543	1.1605	1.1673	1.1740	1.1807	1.1878	1.1952	1.2017	1.2090
20	1.1384	1.1441	1.1501	1.1562	1.1620	1.1680	1.1742	1.1804	1.1866
40	1.1248	1.1300	1.1354	1.1410	1.1465	1.1518	1.1571	1.1629	1.1686
60	1.1128	1.1176	1.1226	1.1279	1.1327	1.1377	1.1424	1.1479	1.1530
80	1.1018	1.1064	1.1111	1.1159	1.1206	1.1250	1.1295	1.1347	1.1393
100	1.0923	1.0967	1.1011	1.1056	1.1099	1.1143	1.1183	1.1232	1.1275
120	1.0833	1.0880	1.0919	1.0962	1.1002	1.1041	1.1080	1.1129	1.1166
140	1.0751	1.0791	1.0831	1.0871	1.0909	1.0947	1.0985	1.1029	1.1068
160	1.0675	1.0713	1.0751	1.0789	1.0826	1.0862	1.0898	1.0940	1.0978
180	1.0603	1.0641	1.0679	1.0713	1.0749	1.0749	1.0817	1.0858	1.0894
200	1.0536	1.0572	1.0607	1.0641	1.0677	1.0677	1.0742	1.0782	1.0815

continue **E-AA/20w**

P/MPa	*T*/K 476.55	485.15	493.25	501.95	510.25	518.65	526.75	535.35
	V_{spez}/cm^{3}g^{-1}							
0.1	1.2163	1.2237	1.2309	1.2387	1.2467	1.2543	1.2622	1.2707
20	1.1927	1.1993	1.2057	1.2121	1.2189	1.2254	1.2319	1.2388
40	1.1741	1.1798	1.1856	1.1914	1.1971	1.2031	1.2088	1.2146
60	1.1581	1.1634	1.1686	1.1736	1.1790	1.1844	1.1896	1.1949
80	1.1441	1.1490	1.1539	1.1585	1.1635	1.1683	1.1732	1.1781
100	1.1321	1.1367	1.1412	1.1454	1.1500	1.1546	1.1591	1.1637
120	1.1210	1.1254	1.1296	1.1336	1.1380	1.1423	1.1465	1.1507
140	1.1108	1.1149	1.1190	1.1227	1.1270	1.1309	1.1350	1.1389
160	1.1017	1.1054	1.1094	1.1130	1.1169	1.1207	1.1247	1.1283
180	1.0931	1.0967	1.1006	1.1039	1.1077	1.1114	1.1150	1.1186
200	1.0851	1.0886	1.0922	1.0955	1.0991	1.1026	1.1061	1.1095

Copolymers from ethylene and 1-butene

Average chemical composition of the copolymers, acronyms, range of data, and references:

Copolymer	Acronym	Range of data *T*/K	*P*/MPa	Ref.
Ethylene/1-butene copolymer				
20.2 wt% 1-butene	E-B/20w	425-545	0.1-200	99MAI
52.2 wt% 1-butene	E-B/52w	425-545	0.1-200	99MAI
69.6 wt% 1-butene	E-B/70w	425-545	0.1-200	99MAI
81.6 wt% 1-butene	E-B/82w	425-545	0.1-200	99MAI
87.5 wt% 1-butene	E-B/88w	425-545	0.1-200	99MAI

Characterization:

Copolymer (B)	M_n/ g/mol	M_w/ g/mol	M_η/ g/mol	Further information
E-B/20w	121900	305000		synthesized in the laboratory, $\rho_B = 0.8817$ g/cm^3
E-B/52w	82800	199000		synthesized in the laboratory, $\rho_B = 0.8662$ g/cm^3
E-B/70w	62400	150000		synthesized in the laboratory, $\rho_B = 0.8683$ g/cm^3
E-B/82w	63500	152000		synthesized in the laboratory, $\rho_B = 0.8701$ g/cm^3
E-B/88w	82100	197000		synthesized in the laboratory, $\rho_B = 0.8830$ g/cm^3

Experimental *PVT* data:

copolymer (B): E-B/20w 99MAI

P/MPa	T/K 425.24	434.75	444.75	455.35	465.22	475.34	485.06	495.15	505.89
	V_{spez}/cm^{3}g^{-1}								
0.1	1.2752	1.2841	1.2936	1.3046	1.3135	1.3221	1.3320	1.3418	1.3523
10	1.2636	1.2718	1.2806	1.2904	1.2988	1.3067	1.3157	1.3244	1.3339
20	1.2531	1.2605	1.2688	1.2773	1.2855	1.2929	1.3012	1.3090	1.3176
30	1.2437	1.2507	1.2585	1.2666	1.2738	1.2812	1.2888	1.2962	1.3041
40	1.2351	1.2417	1.2490	1.2568	1.2637	1.2702	1.2777	1.2846	1.2921
50	1.2272	1.2335	1.2403	1.2478	1.2543	1.2605	1.2672	1.2742	1.2813
60	1.2198	1.2258	1.2324	1.2395	1.2457	1.2517	1.2580	1.2648	1.2714
70	1.2129	1.2186	1.2250	1.2317	1.2378	1.2435	1.2497	1.2557	1.2624
80	1.2064	1.2119	1.2181	1.2245	1.2304	1.2359	1.2419	1.2477	1.2536
90	1.2003	1.2056	1.2116	1.2178	1.2233	1.2287	1.2344	1.2400	1.2458
100	1.1945	1.1996	1.2055	1.2114	1.2167	1.2220	1.2275	1.2330	1.2386
110	1.1890	1.1939	1.1996	1.2053	1.2106	1.2156	1.2209	1.2264	1.2317
120	1.1837	1.1884	1.1939	1.1996	1.2046	1.2094	1.2148	1.2199	1.2252
130	1.1787	1.1832	1.1887	1.1941	1.1989	1.2038	1.2088	1.2140	1.2190
140	1.1738	1.1782	1.1835	1.1888	1.1935	1.1982	1.2033	1.2081	1.2131
150	1.1692	1.1734	1.1786	1.1838	1.1882	1.1928	1.1978	1.2026	1.2075
160	1.1647	1.1687	1.1739	1.1789	1.1833	1.1877	1.1927	1.1974	1.2021
170	1.1604	1.1643	1.1693	1.1742	1.1785	1.1830	1.1877	1.1922	1.1968
180	1.1562	1.1599	1.1650	1.1697	1.1740	1.1782	1.1828	1.1875	1.1919
190	1.1522	1.1559	1.1608	1.1654	1.1695	1.1737	1.1782	1.1827	1.1870
200	1.1484	1.1518	1.1568	1.1613	1.1654	1.1694	1.1738	1.1783	1.1824

continue **E-B/20w**

P/MPa	T/K 515.09	524.94	535.01	545.00
	V_{spez}/cm^{3}g^{-1}			
0.1	1.3623	1.3729	1.3841	1.3950
10	1.3429	1.3524	1.3623	1.3719
20	1.3258	1.3346	1.3434	1.3521
30	1.3118	1.3200	1.3281	1.3360
40	1.2993	1.3069	1.3146	1.3219
50	1.2882	1.2953	1.3026	1.3096
60	1.2781	1.2849	1.2918	1.2983
70	1.2689	1.2753	1.2819	1.2881

continue

continue

P/MPa	*T*/K 515.09	524.94	535.01	545.00
	V_{spez}/cm^{3}g^{-1}			
80	1.2602	1.2665	1.2728	1.2789
90	1.2519	1.2583	1.2643	1.2700
100	1.2444	1.2501	1.2563	1.2619
110	1.2373	1.2429	1.2485	1.2543
120	1.2306	1.2361	1.2415	1.2467
130	1.2243	1.2295	1.2349	1.2399
140	1.2181	1.2234	1.2284	1.2335
150	1.2123	1.2174	1.2224	1.2273
160	1.2069	1.2118	1.2167	1.2214
170	1.2015	1.2063	1.2110	1.2157
180	1.1965	1.2012	1.2058	1.2103
190	1.1915	1.1962	1.2007	1.2051
200	1.1869	1.1914	1.1958	1.2001

copolymer (B): E-B/52w 99MAI

P/MPa	*T*/K 424.48	433.74	443.82	454.18	464.52	474.15	484.04	493.79	504.43
	V_{spez}/cm^{3}g^{-1}								
0.1	1.2558	1.2645	1.2734	1.2828	1.2924	1.3025	1.3119	1.3210	1.3314
10	1.2443	1.2522	1.2604	1.2692	1.2778	1.2868	1.2953	1.3038	1.3133
20	1.2337	1.2410	1.2487	1.2569	1.2647	1.2728	1.2804	1.2887	1.2974
30	1.2243	1.2313	1.2384	1.2462	1.2534	1.2610	1.2681	1.2756	1.2839
40	1.2158	1.2224	1.2290	1.2365	1.2433	1.2504	1.2571	1.2642	1.2716
50	1.2080	1.2142	1.2205	1.2276	1.2342	1.2409	1.2472	1.2539	1.2611
60	1.2007	1.2066	1.2127	1.2193	1.2257	1.2322	1.2382	1.2446	1.2514
70	1.1939	1.1995	1.2053	1.2118	1.2179	1.2240	1.2300	1.2360	1.2425
80	1.1874	1.1929	1.1985	1.2046	1.2107	1.2166	1.2222	1.2281	1.2342
90	1.1815	1.1867	1.1922	1.1980	1.2039	1.2096	1.2150	1.2207	1.2265
100	1.1757	1.1807	1.1861	1.1917	1.1974	1.2029	1.2081	1.2137	1.2193
110	1.1703	1.1752	1.1803	1.1857	1.1913	1.1966	1.2018	1.2072	1.2125
120	1.1651	1.1698	1.1748	1.1802	1.1856	1.1906	1.1957	1.2009	1.2061
130	1.1602	1.1646	1.1696	1.1748	1.1800	1.1850	1.1899	1.1949	1.2000
140	1.1554	1.1597	1.1646	1.1697	1.1747	1.1796	1.1844	1.1893	1.1942
150	1.1508	1.1551	1.1598	1.1647	1.1697	1.1745	1.1791	1.1839	1.1887
160	1.1464	1.1505	1.1550	1.1600	1.1648	1.1695	1.1740	1.1787	1.1834
170	1.1422	1.1462	1.1507	1.1554	1.1601	1.1648	1.1691	1.1738	1.1782
180	1.1382	1.1419	1.1463	1.1510	1.1556	1.1601	1.1645	1.1690	1.1733
190	1.1343	1.1379	1.1422	1.1468	1.1514	1.1558	1.1601	1.1645	1.1686
200	1.1304	1.1340	1.1383	1.1428	1.1472	1.1516	1.1558	1.1601	1.1642

continue **E-B/52w**

P/MPa	*T*/K 513.63	523.99	533.96	544.35
	V_{spez}/cm^3g^{-1}			
0.1	1.3413	1.3521	1.3626	1.3734
10	1.3221	1.3316	1.3409	1.3503
20	1.3052	1.3138	1.3221	1.3307
30	1.2913	1.2991	1.3068	1.3145
40	1.2788	1.2862	1.2934	1.3006
50	1.2673	1.2748	1.2815	1.2882
60	1.2574	1.2641	1.2708	1.2771
70	1.2482	1.2546	1.2607	1.2670
80	1.2398	1.2459	1.2517	1.2573
90	1.2319	1.2378	1.2434	1.2487
100	1.2246	1.2302	1.2356	1.2407
110	1.2176	1.2230	1.2283	1.2332
120	1.2111	1.2164	1.2213	1.2261
130	1.2048	1.2100	1.2148	1.2194
140	1.1989	1.2040	1.2087	1.2131
150	1.1933	1.1981	1.2027	1.2070
160	1.1879	1.1926	1.1970	1.2012
170	1.1827	1.1873	1.1917	1.1957
180	1.1779	1.1823	1.1864	1.1905
190	1.1731	1.1774	1.1815	1.1854
200	1.1685	1.1728	1.1767	1.1805

copolymer (B): E-B/70w 99MAI

P/MPa	*T*/K 424.88	435.03	445.15	455.26	465.29	475.12	484.94	494.74	504.78
	V_{spez}/cm^3g^{-1}								
0.1	1.2548	1.2640	1.2729	1.2822	1.2914	1.3016	1.3110	1.3203	1.3300
10	1.2423	1.2507	1.2589	1.2674	1.2758	1.2848	1.2933	1.3019	1.3106
20	1.2308	1.2385	1.2463	1.2539	1.2618	1.2697	1.2778	1.2856	1.2934
30	1.2211	1.2283	1.2358	1.2428	1.2501	1.2574	1.2648	1.2720	1.2797
40	1.2123	1.2191	1.2261	1.2327	1.2398	1.2467	1.2534	1.2602	1.2672
50	1.2044	1.2106	1.2174	1.2237	1.2304	1.2369	1.2433	1.2498	1.2562
60	1.1969	1.2028	1.2094	1.2154	1.2217	1.2281	1.2342	1.2403	1.2464
70	1.1900	1.1956	1.2019	1.2078	1.2137	1.2199	1.2257	1.2316	1.2374
80	1.1836	1.1888	1.1950	1.2006	1.2063	1.2123	1.2179	1.2235	1.2291

continue

continue

P/MPa	*T*/K 424.88	435.03	445.15	455.26	465.29	475.12	484.94	494.74	504.78
	V_{spez}/cm^{3}g^{-1}								
90	1.1776	1.1826	1.1884	1.1939	1.1994	1.2052	1.2106	1.2161	1.2214
100	1.1718	1.1766	1.1823	1.1876	1.1929	1.1985	1.2038	1.2090	1.2143
110	1.1664	1.1709	1.1764	1.1816	1.1866	1.1922	1.1973	1.2024	1.2072
120	1.1611	1.1656	1.1709	1.1760	1.1809	1.1862	1.1912	1.1961	1.2008
130	1.1562	1.1604	1.1655	1.1705	1.1752	1.1804	1.1854	1.1903	1.1948
140	1.1513	1.1555	1.1604	1.1654	1.1700	1.1750	1.1799	1.1846	1.1889
150	1.1468	1.1508	1.1555	1.1603	1.1648	1.1699	1.1746	1.1791	1.1835
160	1.1424	1.1461	1.1508	1.1557	1.1598	1.1649	1.1696	1.1740	1.1781
170	1.1381	1.1417	1.1464	1.1510	1.1552	1.1601	1.1647	1.1690	1.1731
180	1.1340	1.1376	1.1421	1.1467	1.1508	1.1555	1.1601	1.1642	1.1682
190	1.1301	1.1335	1.1379	1.1424	1.1463	1.1511	1.1555	1.1596	1.1636
200	1.1263	1.1297	1.1339	1.1383	1.1422	1.1468	1.1512	1.1553	1.1590

continue **E-B/70w**

P/MPa	*T*/K 514.68	525.18	535.17	545.27
	V_{spez}/cm^{3}g^{-1}			
0.1	1.3405	1.3515	1.3616	1.3724
10	1.3198	1.3293	1.3382	1.3475
20	1.3016	1.3100	1.3182	1.3262
30	1.2873	1.2949	1.3024	1.3097
40	1.2746	1.2818	1.2887	1.2954
50	1.2629	1.2700	1.2765	1.2829
60	1.2527	1.2591	1.2657	1.2716
70	1.2435	1.2495	1.2554	1.2616
80	1.2350	1.2407	1.2463	1.2518
90	1.2270	1.2325	1.2380	1.2430
100	1.2196	1.2249	1.2302	1.2350
110	1.2126	1.2177	1.2227	1.2274
120	1.2060	1.2109	1.2157	1.2204
130	1.1996	1.2046	1.2092	1.2137
140	1.1937	1.1984	1.2030	1.2074
150	1.1881	1.1926	1.1971	1.2013
160	1.1826	1.1871	1.1914	1.1956
170	1.1774	1.1818	1.1860	1.1900
180	1.1726	1.1767	1.1809	1.1847
190	1.1677	1.1719	1.1758	1.1797
200	1.1632	1.1673	1.1711	1.1748

copolymer (B): E-B/82w 99MAI

P/MPa	*T*/K 424.74	434.29	444.47	454.98	465.22	474.85	485.01	494.29	504.54
	V_{spez}/cm^{3}g^{-1}								
0.1	1.2589	1.2673	1.2763	1.2858	1.2952	1.3042	1.3134	1.3220	1.3322
10	1.2466	1.2545	1.2627	1.2714	1.2798	1.2878	1.2961	1.3042	1.3132
20	1.2355	1.2429	1.2505	1.2583	1.2661	1.2730	1.2809	1.2884	1.2966
30	1.2257	1.2327	1.2400	1.2472	1.2544	1.2609	1.2678	1.2752	1.2828
40	1.2169	1.2235	1.2302	1.2371	1.2438	1.2499	1.2564	1.2630	1.2705
50	1.2087	1.2149	1.2214	1.2281	1.2344	1.2401	1.2462	1.2525	1.2591
60	1.2012	1.2070	1.2133	1.2197	1.2255	1.2312	1.2370	1.2429	1.2491
70	1.1942	1.1999	1.2057	1.2119	1.2175	1.2228	1.2284	1.2340	1.2400
80	1.1877	1.1930	1.1988	1.2046	1.2100	1.2152	1.2206	1.2260	1.2317
90	1.1815	1.1866	1.1922	1.1978	1.2030	1.2080	1.2131	1.2183	1.2240
100	1.1758	1.1806	1.1861	1.1914	1.1964	1.2013	1.2064	1.2112	1.2166
110	1.1703	1.1748	1.1802	1.1853	1.1901	1.1949	1.1998	1.2046	1.2097
120	1.1650	1.1694	1.1746	1.1796	1.1843	1.1889	1.1937	1.1983	1.2033
130	1.1599	1.1642	1.1692	1.1741	1.1787	1.1832	1.1876	1.1924	1.1972
140	1.1552	1.1592	1.1641	1.1687	1.1734	1.1777	1.1820	1.1867	1.1913
150	1.1504	1.1545	1.1593	1.1639	1.1681	1.1725	1.1767	1.1813	1.1858
160	1.1461	1.1499	1.1545	1.1590	1.1633	1.1675	1.1716	1.1760	1.1805
170	1.1417	1.1456	1.1500	1.1544	1.1586	1.1627	1.1667	1.1711	1.1754
180	1.1376	1.1412	1.1458	1.1499	1.1541	1.1581	1.1619	1.1664	1.1706
190	1.1336	1.1372	1.1416	1.1458	1.1497	1.1538	1.1574	1.1618	1.1659
200	1.1298	1.1334	1.1376	1.1417	1.1456	1.1495	1.1530	1.1574	1.1614

continue **E-B/82w**

P/MPa	*T*/K 514.64	524.44	534.44	544.66
	V_{spez}/cm^{3}g^{-1}			
0.1	1.3416	1.3520	1.3622	1.3731
10	1.3214	1.3307	1.3396	1.3490
20	1.3038	1.3123	1.3203	1.3285
30	1.2893	1.2972	1.3046	1.3121
40	1.2765	1.2839	1.2908	1.2978
50	1.2652	1.2721	1.2786	1.2852
60	1.2546	1.2616	1.2678	1.2739
70	1.2453	1.2515	1.2580	1.2638
80	1.2368	1.2427	1.2485	1.2545
90	1.2287	1.2344	1.2400	1.2453

continue

continue

P/MPa	*T*/K 514.64	524.44	534.44	544.66
	V_{spez}/cm^{3}g^{-1}			
100	1.2213	1.2268	1.2322	1.2374
110	1.2144	1.2196	1.2248	1.2298
120	1.2078	1.2128	1.2178	1.2227
130	1.2015	1.2064	1.2114	1.2160
140	1.1956	1.2003	1.2050	1.2096
150	1.1899	1.1945	1.1992	1.2036
160	1.1847	1.1889	1.1935	1.1979
170	1.1795	1.1837	1.1881	1.1924
180	1.1745	1.1785	1.1829	1.1870
190	1.1698	1.1738	1.1780	1.1820
200	1.1652	1.1691	1.1733	1.1771

copolymer (B): **E-B/88w** **99MAI**

P/MPa	*T*/K 424.88	434.88	445.04	455.03	465.05	475.25	485.00	495.09	505.18
	V_{spez}/cm^{3}g^{-1}								
0.1	1.2609	1.2695	1.2784	1.2874	1.2977	1.3069	1.3160	1.3254	1.3342
10	1.2483	1.2563	1.2645	1.2727	1.2816	1.2901	1.2986	1.3068	1.3146
20	1.2368	1.2443	1.2519	1.2595	1.2672	1.2753	1.2833	1.2906	1.2976
30	1.2268	1.2341	1.2412	1.2482	1.2554	1.2625	1.2704	1.2771	1.2836
40	1.2179	1.2246	1.2314	1.2381	1.2448	1.2516	1.2584	1.2651	1.2712
50	1.2096	1.2159	1.2225	1.2289	1.2352	1.2416	1.2481	1.2539	1.2602
60	1.2021	1.2080	1.2143	1.2204	1.2264	1.2327	1.2389	1.2442	1.2499
70	1.1950	1.2007	1.2066	1.2126	1.2183	1.2243	1.2303	1.2354	1.2407
80	1.1885	1.1938	1.1997	1.2053	1.2108	1.2166	1.2223	1.2272	1.2324
90	1.1823	1.1874	1.1930	1.1985	1.2038	1.2094	1.2149	1.2194	1.2245
100	1.1764	1.1813	1.1867	1.1922	1.1972	1.2028	1.2080	1.2124	1.2172
110	1.1709	1.1756	1.1808	1.1861	1.1911	1.1965	1.2016	1.2056	1.2104
120	1.1656	1.1701	1.1752	1.1804	1.1852	1.1904	1.1954	1.1993	1.2039
130	1.1605	1.1649	1.1698	1.1748	1.1795	1.1846	1.1896	1.1933	1.1977
140	1.1556	1.1599	1.1646	1.1696	1.1742	1.1792	1.1840	1.1876	1.1919
150	1.1511	1.1552	1.1597	1.1647	1.1691	1.1739	1.1787	1.1822	1.1862
160	1.1465	1.1505	1.1551	1.1599	1.1642	1.1689	1.1736	1.1769	1.1809
170	1.1422	1.1461	1.1505	1.1552	1.1595	1.1641	1.1687	1.1719	1.1758
180	1.1381	1.1418	1.1462	1.1508	1.1550	1.1596	1.1641	1.1672	1.1709
190	1.1342	1.1379	1.1420	1.1465	1.1507	1.1550	1.1594	1.1627	1.1662
200	1.1303	1.1340	1.1379	1.1424	1.1464	1.1508	1.1551	1.1582	1.1616

continue **E-B/88w**

P/MPa	*T*/K			
	514.84	525.16	535.15	545.25
	V_{spez}/cm^{3}g^{-1}			
0.1	1.3439	1.3538	1.3642	1.3743
10	1.3232	1.3320	1.3411	1.3498
20	1.3054	1.3132	1.3214	1.3291
30	1.2906	1.2979	1.3054	1.3126
40	1.2778	1.2845	1.2915	1.2981
50	1.2664	1.2727	1.2791	1.2855
60	1.2562	1.2621	1.2682	1.2742
70	1.2463	1.2524	1.2583	1.2639
80	1.2378	1.2430	1.2490	1.2546
90	1.2297	1.2348	1.2402	1.2460
100	1.2223	1.2271	1.2323	1.2375
110	1.2151	1.2200	1.2250	1.2299
120	1.2085	1.2132	1.2179	1.2229
130	1.2023	1.2068	1.2113	1.2161
140	1.1963	1.2006	1.2052	1.2097
150	1.1906	1.1948	1.1991	1.2037
160	1.1852	1.1892	1.1935	1.1979
170	1.1800	1.1840	1.1880	1.1924
180	1.1750	1.1789	1.1828	1.1872
190	1.1702	1.1740	1.1778	1.1822
200	1.1657	1.1694	1.1731	1.1773

Copolymers from ethylene and methacrylic acid

Average chemical composition of the copolymers, acronyms, range of data, and references:

Copolymer	Acronym	Range of data *T*/K	*P*/MPa	Ref.
Ethylene/methacrylic acid copolymer				
4.0 wt% methacrylic acid	E-MAA/04w	418-524	0.1-200	95ZOL
9.0 wt% methacrylic acid	E-MAA/09w	410-524	0.1-200	95ZOL
11.5 wt% methacrylic acid	E-MAA/11w	409-524	0.1-200	95ZOL

continue

continue

Copolymer	Acronym	Range of data T/K	P/MPa	Ref.
12.0 wt% methacrylic acid	E-MAA/12w	410-524	0.1-200	95ZOL
15.0 wt% methacrylic acid	E-MAA/15w	401-526	0.1-200	95ZOL
20.0 wt% methacrylic acid	E-MAA/20w	402-524	0.1-200	95ZOL

Characterization:

Copolymer (B)	M_n/ g/mol	M_w/ g/mol	M_η/ g/mol	Further information
E-MAA/04w				DuPont, Nucrel, MI = 11
E-MAA/09w				DuPont, Nucrel, MI = 10
E-MAA/11w				DuPont, Nucrel, MI = 1.5
E-MAA/12w				DuPont, Nucrel, MI = 14
E-MAA/15w				DuPont, Nucrel, MI = 25
E-MAA/20w				DuPont, Nucrel, MI = 60

Experimental *PVT* data:

copolymer (B): **E-MAA/04w** **95ZOL**

P/MPa	T/K 418.05	426.05	434.35	442.45	450.55	458.55	466.65	474.45	482.65
	V_{spez}/cm^{3}g^{-1}								
0.1	1.2600	1.2677	1.2754	1.2827	1.2904	1.2982	1.3060	1.3136	1.3217
20	1.2392	1.2459	1.2525	1.2590	1.2655	1.2722	1.2788	1.2854	1.2924
40	1.2218	1.2283	1.2338	1.2397	1.2456	1.2518	1.2574	1.2634	1.2695
60	1.2069	1.2126	1.2179	1.2233	1.2288	1.2341	1.2396	1.2451	1.2508
80	1.1943	1.1992	1.2040	1.2090	1.2140	1.2190	1.2241	1.2292	1.2343
100	1.1822	1.1871	1.1916	1.1964	1.2010	1.2058	1.2104	1.2155	1.2199
120	1.1717	1.1762	1.1804	1.1849	1.1895	1.1939	1.1982	1.2028	1.2072
140	1.1619	1.1661	1.1701	1.1744	1.1789	1.1828	1.1870	1.1914	1.1955
160	1.1527	1.1570	1.1608	1.1648	1.1688	1.1729	1.1769	1.1810	1.1849
180	1.1444	1.1482	1.1519	1.1561	1.1597	1.1635	1.1673	1.1716	1.1751
200	1.1364	1.1403	1.1439	1.1474	1.1512	1.1549	1.1585	1.1624	1.1660

continue **E-MAA/04w**

P/MPa	*T*/K				
	490.95	498.15	507.75	516.05	523.95
	V_{spez}/cm^{3}g^{-1}				
0.1	1.3301	1.3380	1.3475	1.3563	1.3648
20	1.2993	1.3058	1.3136	1.3205	1.3278
40	1.2755	1.2815	1.2881	1.2940	1.3002
60	1.2563	1.2617	1.2671	1.2727	1.2784
80	1.2391	1.2449	1.2494	1.2545	1.2596
100	1.2248	1.2297	1.2340	1.2385	1.2435
120	1.2116	1.2164	1.2204	1.2246	1.2292
140	1.1995	1.2042	1.2081	1.2120	1.2164
160	1.1887	1.1932	1.1967	1.2003	1.2046
180	1.1788	1.1831	1.1863	1.1899	1.1939
200	1.1695	1.1736	1.1767	1.1801	1.1838

copolymer (B): **E-MAA/09w** **95ZOL**

P/MPa	*T*/K								
	410.55	418.85	426.95	435.25	443.05	451.35	459.45	467.35	475.45
	V_{spez}/cm^{3}g^{-1}								
0.1	1.2303	1.2373	1.2446	1.2518	1.2588	1.2667	1.2748	1.2818	1.2894
20	1.2108	1.2172	1.2234	1.2297	1.2361	1.2427	1.2493	1.2555	1.2622
40	1.1945	1.2003	1.2060	1.2118	1.2175	1.2234	1.2291	1.2349	1.2408
60	1.1805	1.1858	1.1911	1.1963	1.2016	1.2071	1.2122	1.2175	1.2229
80	1.1683	1.1730	1.1779	1.1828	1.1877	1.1925	1.1976	1.2024	1.2076
100	1.1569	1.1616	1.1662	1.1707	1.1754	1.1799	1.1846	1.1891	1.1938
120	1.1468	1.1511	1.1555	1.1598	1.1642	1.1684	1.1729	1.1773	1.1817
140	1.1375	1.1415	1.1458	1.1499	1.1540	1.1580	1.1622	1.1665	1.1707
160	1.1290	1.1328	1.1369	1.1407	1.1446	1.1485	1.1524	1.1566	1.1604
180	1.1208	1.1245	1.1284	1.1322	1.1360	1.1396	1.1434	1.1472	1.1513
200	1.1135	1.1170	1.1206	1.1241	1.1279	1.1314	1.1351	1.1388	1.1426

continue **E-MAA/09w**

P/MPa	T/K					
	484.05	491.55	500.35	508.35	516.15	524.45
	V_{spez}/cm^{3}g^{-1}					
0.1	1.2974	1.3050	1.3137	1.3221	1.3294	1.3389
20	1.2690	1.2755	1.2827	1.2890	1.2959	1.3031
40	1.2466	1.2528	1.2586	1.2644	1.2708	1.2767
60	1.2284	1.2337	1.2390	1.2441	1.2501	1.2556
80	1.2123	1.2175	1.2220	1.2269	1.2323	1.2371
100	1.1982	1.2031	1.2074	1.2119	1.2170	1.2213
120	1.1858	1.1905	1.1943	1.1987	1.2031	1.2075
140	1.1745	1.1788	1.1826	1.1866	1.1909	1.1949
160	1.1643	1.1683	1.1718	1.1756	1.1797	1.1835
180	1.1549	1.1588	1.1620	1.1656	1.1694	1.1731
200	1.1459	1.1496	1.1528	1.1564	1.1601	1.1636

copolymer (B): E-MAA/11w 95ZOL

P/MPa	T/K								
	409.25	417.95	425.75	433.85	441.95	449.95	458.35	465.95	474.85
	V_{spez}/cm^{3}g^{-1}								
0.1	1.2149	1.2220	1.2291	1.2362	1.2434	1.2506	1.2581	1.2657	1.2737
20	1.1967	1.2030	1.2093	1.2158	1.2222	1.2282	1.2348	1.2410	1.2479
40	1.1812	1.1870	1.1927	1.1983	1.2042	1.2101	1.2157	1.2214	1.2274
60	1.1679	1.1730	1.1786	1.1837	1.1890	1.1942	1.1996	1.2048	1.2102
80	1.1559	1.1610	1.1658	1.1706	1.1756	1.1806	1.1854	1.1905	1.1953
100	1.1452	1.1498	1.1545	1.1590	1.1637	1.1684	1.1729	1.1776	1.1823
120	1.1354	1.1398	1.1442	1.1485	1.1530	1.1573	1.1617	1.1661	1.1703
140	1.1265	1.1305	1.1347	1.1389	1.1430	1.1472	1.1514	1.1555	1.1595
160	1.1181	1.1219	1.1262	1.1300	1.1341	1.1380	1.1419	1.1460	1.1497
180	1.1104	1.1141	1.1180	1.1218	1.1256	1.1295	1.1332	1.1372	1.1407
200	1.1033	1.1068	1.1107	1.1145	1.1180	1.1217	1.1252	1.1289	1.1324

continue **E-MAA/11w**

P/MPa	*T*/K					
	483.15	491.45	499.95	507.95	516.75	524.65
	V_{spez}/cm^{3}g^{-1}					
0.1	1.2812	1.2892	1.2974	1.3059	1.3140	1.3226
20	1.2546	1.2612	1.2683	1.2748	1.2818	1.2888
40	1.2335	1.2392	1.2453	1.2513	1.2573	1.2635
60	1.2157	1.2209	1.2266	1.2317	1.2373	1.2430
80	1.2004	1.2052	1.2101	1.2152	1.2202	1.2254
100	1.1868	1.1914	1.1963	1.2006	1.2053	1.2101
120	1.1748	1.1790	1.1832	1.1877	1.1921	1.1966
140	1.1638	1.1678	1.1718	1.1760	1.1802	1.1844
160	1.1538	1.1575	1.1614	1.1653	1.1692	1.1733
180	1.1445	1.1481	1.1518	1.1558	1.1593	1.1632
200	1.1363	1.1396	1.1431	1.1468	1.1502	1.1542

copolymer (B): E-MAA/12w 95ZOL

P/MPa	*T*/K								
	409.75	418.45	426.05	434.55	442.75	450.55	458.45	466.45	474.65
	V_{spez}/cm^{3}g^{-1}								
0.1	1.2161	1.2234	1.2298	1.2369	1.2442	1.2513	1.2589	1.2664	1.2746
20	1.1973	1.2035	1.2096	1.2159	1.2222	1.2284	1.2350	1.2412	1.2478
40	1.1817	1.1874	1.1928	1.1984	1.2040	1.2098	1.2154	1.2212	1.2270
60	1.1679	1.1731	1.1783	1.1835	1.1885	1.1937	1.1990	1.2042	1.2095
80	1.1560	1.1606	1.1655	1.1701	1.1750	1.1800	1.1847	1.1896	1.1946
100	1.1450	1.1495	1.1540	1.1584	1.1629	1.1675	1.1720	1.1767	1.1812
120	1.1351	1.1391	1.1436	1.1476	1.1519	1.1563	1.1606	1.1650	1.1692
140	1.1260	1.1298	1.1340	1.1380	1.1419	1.1463	1.1502	1.1543	1.1584
160	1.1175	1.1211	1.1252	1.1289	1.1327	1.1368	1.1408	1.1446	1.1484
180	1.1098	1.1131	1.1170	1.1205	1.1242	1.1281	1.1318	1.1355	1.1395
200	1.1024	1.1055	1.1095	1.1127	1.1162	1.1201	1.1237	1.1272	1.1307

continue **E-MAA/12w**

P/MPa	T/K					
	482.95	491.05	498.85	507.35	515.75	524.15
	V_{spez}/cm^{3}g^{-1}					
0.1	1.2823	1.2900	1.2975	1.3060	1.3142	1.3240
20	1.2546	1.2613	1.2677	1.2746	1.2814	1.2889
40	1.2330	1.2389	1.2447	1.2507	1.2567	1.2633
60	1.2149	1.2202	1.2255	1.2309	1.2363	1.2421
80	1.1994	1.2044	1.2093	1.2143	1.2191	1.2245
100	1.1858	1.1904	1.1951	1.1996	1.2041	1.2091
120	1.1735	1.1779	1.1824	1.1870	1.1909	1.1954
140	1.1625	1.1666	1.1707	1.1748	1.1788	1.1833
160	1.1523	1.1567	1.1603	1.1642	1.1678	1.1720
180	1.1430	1.1467	1.1505	1.1542	1.1577	1.1617
200	1.1345	1.1381	1.1416	1.1453	1.1488	1.1522

copolymer (B): **E-MAA/15w** **95ZOL**

P/MPa	T/K								
	401.55	409.75	418.15	426.35	434.25	442.25	450.05	458.95	467.35
	V_{spez}/cm^{3}g^{-1}								
0.1	1.1917	1.1980	1.2049	1.2119	1.2193	1.2263	1.2333	1.2414	1.2484
20	1.1742	1.1801	1.1861	1.1925	1.1988	1.2052	1.2113	1.2179	1.2242
40	1.1600	1.1652	1.1706	1.1762	1.1820	1.1879	1.1935	1.1991	1.2048
60	1.1469	1.1522	1.1571	1.1622	1.1676	1.1730	1.1782	1.1834	1.1884
80	1.1356	1.1402	1.1451	1.1498	1.1551	1.1599	1.1649	1.1696	1.1743
100	1.1253	1.1296	1.1344	1.1387	1.1436	1.1483	1.1531	1.1574	1.1616
120	1.1158	1.1200	1.1243	1.1287	1.1334	1.1378	1.1421	1.1462	1.1502
140	1.1072	1.1111	1.1152	1.1193	1.1237	1.1282	1.1322	1.1360	1.1398
160	1.0991	1.1029	1.1068	1.1108	1.1150	1.1190	1.1231	1.1266	1.1303
180	1.0917	1.0953	1.0990	1.1029	1.1071	1.1108	1.1147	1.1181	1.1218
200	1.0848	1.0844	1.0919	1.0955	1.0996	1.1031	1.1072	1.1101	1.1135

continue **E-MAA/15w**

P/MPa	T/K						
	475.65	484.05	492.15	500.85	509.25	517.25	525.75
	V_{spez}/cm^{3}g^{-1}						
0.1	1.2563	1.2642	1.2715	1.2800	1.2886	1.2969	1.3048
20	1.2309	1.2376	1.2439	1.2508	1.2575	1.2644	1.2713
40	1.2109	1.2166	1.2224	1.2285	1.2343	1.2402	1.2463
60	1.1939	1.1995	1.2046	1.2099	1.2152	1.2207	1.2261
80	1.1792	1.1842	1.1891	1.1940	1.1989	1.2039	1.2089
100	1.1663	1.1710	1.1754	1.1799	1.1846	1.1892	1.1938
120	1.1548	1.1590	1.1634	1.1676	1.1718	1.1762	1.1804
140	1.1441	1.1484	1.1522	1.1562	1.1603	1.1644	1.1684
160	1.1345	1.1383	1.1422	1.1460	1.1498	1.1538	1.1578
180	1.1255	1.1292	1.1329	1.1365	1.1402	1.1440	1.1477
200	1.1172	1.1208	1.1245	1.1279	1.1314	1.1350	1.1385

copolymer (B): E-MAA/20w 95ZOL

P/MPa	T/K								
	401.95	410.15	417.95	426.25	434.25	442.25	450.55	458.35	466.55
	V_{spez}/cm^{3}g^{-1}								
0.1	1.1685	1.1742	1.1809	1.1879	1.1943	1.2015	1.2082	1.2150	1.2224
20	1.1501	1.1556	1.1616	1.1674	1.1733	1.1792	1.1853	1.1913	1.1978
40	1.1357	1.1409	1.1465	1.1514	1.1577	1.1621	1.1675	1.1730	1.1787
60	1.1233	1.1279	1.1330	1.1378	1.1426	1.1477	1.1526	1.1579	1.1630
80	1.1122	1.1166	1.1212	1.1257	1.1304	1.1349	1.1395	1.1443	1.1492
100	1.1023	1.1063	1.1108	1.1149	1.1193	1.1237	1.1279	1.1325	1.1369
120	1.0931	1.0970	1.1011	1.1052	1.1093	1.1133	1.1174	1.1218	1.1261
140	1.0847	1.0884	1.0924	1.0961	1.1001	1.1040	1.1081	1.1119	1.1158
160	1.0768	1.0804	1.0841	1.0877	1.0915	1.0952	1.0991	1.1029	1.1067
180	1.0695	1.0729	1.0765	1.0800	1.0836	1.0871	1.0906	1.0944	1.0980
200	1.0627	1.0658	1.0693	1.0726	1.0762	1.0794	1.0830	1.0866	1.0899

continue **E-MAA/20w**

P/MPa	T/K 474.85	482.95	490.95	499.45	507.45	515.55	523.95
	V_{spez}/cm^{3}g^{-1}						
0.1	1.2302	1.2375	1.2454	1.2531	1.2616	1.2684	1.2770
20	1.2040	1.2104	1.2170	1.2233	1.2295	1.2360	1.2429
40	1.1843	1.1901	1.1955	1.2014	1.2068	1.2126	1.2186
60	1.1680	1.1732	1.1782	1.1834	1.1885	1.1938	1.1989
80	1.1539	1.1586	1.1633	1.1679	1.1725	1.1776	1.1824
100	1.1413	1.1458	1.1501	1.1545	1.1590	1.1632	1.1678
120	1.1301	1.1343	1.1382	1.1426	1.1464	1.1506	1.1551
140	1.1197	1.1236	1.1276	1.1314	1.1354	1.1393	1.1436
160	1.1102	1.1143	1.1177	1.1215	1.1252	1.1289	1.1327
180	1.1015	1.1054	1.1087	1.1122	1.1156	1.1193	1.1231
200	1.0934	1.0970	1.1003	1.1037	1.1069	1.1104	1.1138

Copolymers from ethylene and 1-octene

Average chemical composition of the copolymers, acronyms, range of data, and references:

Copolymer	Acronym	Range of data T/K	P/MPa	Ref.
Ethylene/1-octene copolymer				
2.0 wt% 1-octene	E-O/02w	423-543	0.1-200	98SUH, 99MAI
7.5 wt% 1-octene	E-O/08w	423-543	0.1-200	98SUH, 99MAI
11.4 wt% 1-octene	E-O/11w	423-543	0.1-200	98SUH, 99MAI
12.0 wt% 1-octene	E-O/12w	423-543	0.1-200	98SUH, 99MAI
14.0 wt% 1-octene	E-O/14w	423-543	0.1-200	98SUH, 99MAI
25.0 wt% 1-octene	E-O/25w	423-543	0.1-200	98SUH, 99MAI
39.4 wt% 1-octene	E-O/39w	423-543	0.1-200	98SUH, 99MAI
55.0 wt% 1-octene	E-O/55w	423-543	0.1-200	98SUH, 99MAI
64.0 wt% 1-octene	E-O/64w	423-543	0.1-200	98SUH, 99MAI

Characterization:

Copolymer (B)	M_n/ g/mol	M_w/ g/mol	M_η/ g/mol	Further information
E-O/02w	35200	73900		Dow Chemical, ρ_B = 0.935 g/cm^3
E-O/08w	36400	87400		Dow Chemical, ρ_B = 0.915 g/cm^3
E-O/11w	230000	598000		synthesized in the laboratory, ρ_B = 0.8975 g/cm^3
E-O/12w	34100	81800		Dow Chemical, ρ_B = 0.905 g/cm^3
E-O/14w	30200	90600		Dow Chemical, ρ_B = 0.895 g/cm^3
E-O/25w	41400	86900		Dow Chemical, ρ_B = 0.875 g/cm^3
E-O/39w	123400	271500		synthesized in the laboratory, ρ_B = 0.8626 g/cm^3
E-O/55w	105100	220700		synthesized in the laboratory, ρ_B = 0.8485 g/cm^3
E-O/64w	82400	197800		synthesized in the laboratory, ρ_B = 0.8503 g/cm^3

Experimental *PVT* data:

copolymer (B): E-O/02w 98SUH, 99MAI

P/MPa	*T*/K 423.97	433.40	443.51	453.42	463.64	473.72	483.63	493.11	502.84
	V_{spez}/cm^{3}g^{-1}								
0.1	1.2703	1.2793	1.2885	1.2982	1.3085	1.3183	1.3286	1.3381	1.3486
10	1.2590	1.2673	1.2759	1.2848	1.2943	1.3031	1.3124	1.3214	1.3308
20	1.2486	1.2564	1.2645	1.2727	1.2816	1.2896	1.2981	1.3068	1.3152
30	1.2394	1.2467	1.2543	1.2620	1.2704	1.2778	1.2857	1.2937	1.3019
40	1.2303	1.2377	1.2448	1.2521	1.2602	1.2672	1.2747	1.2824	1.2898
50	1.2224	1.2290	1.2361	1.2431	1.2508	1.2576	1.2645	1.2719	1.2789
60	1.2151	1.2214	1.2279	1.2350	1.2422	1.2487	1.2553	1.2624	1.2691
70	1.2082	1.2143	1.2205	1.2273	1.2343	1.2405	1.2467	1.2536	1.2601
80	1.2017	1.2076	1.2135	1.2198	1.2270	1.2329	1.2390	1.2455	1.2517
90	1.1957	1.2012	1.2070	1.2131	1.2197	1.2258	1.2316	1.2379	1.2438
100	1.1898	1.1952	1.2008	1.2067	1.2131	1.2187	1.2247	1.2307	1.2365
110	1.1842	1.1895	1.1949	1.2006	1.2068	1.2122	1.2181	1.2239	1.2296
120	1.1790	1.1841	1.1894	1.1949	1.2009	1.2061	1.2114	1.2175	1.2230

continue

continue

P/MPa	T/K 423.97	433.40	443.51	453.42	463.64	473.72	483.63	493.11	502.84
	V_{spez}/cm^{3}g^{-1}								
130		1.1789	1.1840	1.1894	1.1951	1.2003	1.2056	1.2109	1.2167
140		1.1738	1.1788	1.1841	1.1897	1.1947	1.1999	1.2050	1.2102
150		1.1689	1.1738	1.1789	1.1844	1.1894	1.1944	1.1994	1.2045
160		1.1642	1.1691	1.1741	1.1793	1.1842	1.1891	1.1940	1.1989
170			1.1645	1.1694	1.1745	1.1792	1.1841	1.1888	1.1938
180			1.1600	1.1649	1.1699	1.1745	1.1793	1.1840	1.1887
190			1.1558	1.1606	1.1654	1.1700	1.1747	1.1793	1.1838
200			1.1518	1.1565	1.1612	1.1656	1.1703	1.1748	1.1791

continue **E-O/02w**

P/MPa	T/K 513.17	522.94	533.06	542.96
	V_{spez}/cm^{3}g^{-1}			
0.1	1.3597	1.3703	1.3816	1.3926
10	1.3407	1.3503	1.3601	1.3698
20	1.3242	1.3330	1.3417	1.3505
30	1.3102	1.3181	1.3263	1.3344
40	1.2976	1.3051	1.3127	1.3201
50	1.2864	1.2934	1.3005	1.3076
60	1.2762	1.2830	1.2896	1.2964
70	1.2669	1.2733	1.2798	1.2861
80	1.2582	1.2644	1.2704	1.2765
90	1.2502	1.2561	1.2618	1.2677
100	1.2426	1.2483	1.2538	1.2594
110	1.2354	1.2410	1.2463	1.2518
120	1.2286	1.2340	1.2392	1.2444
130	1.2222	1.2274	1.2324	1.2375
140	1.2160	1.2210	1.2260	1.2309
150	1.2100	1.2151	1.2197	1.2246
160	1.2040	1.2093	1.2139	1.2186
170	1.1987	1.2034	1.2082	1.2128
180	1.1935	1.1981	1.2025	1.2074
190	1.1884	1.1931	1.1973	1.2018
200	1.1837	1.1882	1.1925	1.1968

copolymer (B): E-O/08w 98SUH, 99MAI

P/MPa	*T*/K 423.39	433.78	443.83	453.70	464.04	474.46	484.38	494.28	504.08
	V_{spez}/cm^{3}g^{-1}								
0.1	1.2736	1.2829	1.2925	1.3017	1.3114	1.3218	1.3317	1.3421	1.3523
10	1.2615	1.2704	1.2792	1.2878	1.2966	1.3059	1.3151	1.3243	1.3335
20	1.2503	1.2590	1.2671	1.2751	1.2833	1.2917	1.3004	1.3086	1.3169
30	1.2408	1.2490	1.2568	1.2642	1.2719	1.2797	1.2878	1.2955	1.3033
40	1.2321	1.2394	1.2473	1.2543	1.2616	1.2689	1.2767	1.2839	1.2912
50	1.2242	1.2311	1.2381	1.2453	1.2522	1.2592	1.2666	1.2734	1.2804
60	1.2168	1.2233	1.2302	1.2366	1.2436	1.2503	1.2574	1.2639	1.2705
70	1.2099	1.2161	1.2227	1.2289	1.2353	1.2421	1.2487	1.2551	1.2615
80	1.2034	1.2094	1.2157	1.2217	1.2278	1.2341	1.2408	1.2470	1.2530
90	1.1974	1.2030	1.2092	1.2149	1.2208	1.2269	1.2334	1.2393	1.2452
100	1.1915	1.1970	1.2029	1.2085	1.2142	1.2201	1.2259	1.2322	1.2379
110	1.1860	1.1912	1.1970	1.2024	1.2079	1.2137	1.2193	1.2249	1.2308
120	1.1808	1.1857	1.1913	1.1966	1.2021	1.2076	1.2130	1.2184	1.2239
130	1.1756	1.1804	1.1858	1.1910	1.1963	1.2018	1.2070	1.2123	1.2177
140	1.1708	1.1754	1.1807	1.1857	1.1908	1.1962	1.2011	1.2064	1.2117
150	1.1660	1.1704	1.1757	1.1805	1.1855	1.1909	1.1957	1.2009	1.2058
160	1.1614	1.1658	1.1708	1.1756	1.1805	1.1857	1.1904	1.1955	1.2004
170	1.1571	1.1613	1.1661	1.1708	1.1756	1.1807	1.1853	1.1902	1.1950
180	1.1528	1.1570	1.1617	1.1663	1.1710	1.1761	1.1805	1.1854	1.1900
190	1.1487	1.1528	1.1575	1.1620	1.1665	1.1715	1.1758	1.1806	1.1851
200	1.1448	1.1489	1.1533	1.1577	1.1623	1.1671	1.1713	1.1761	1.1806

continue **E-O/08w**

P/MPa	*T*/K 514.22	524.46	534.71	543.97
	V_{spez}/cm^{3}g^{-1}			
0.1	1.3628	1.3738	1.3853	1.3957
10	1.3429	1.3526	1.3627	1.3718
20	1.3254	1.3342	1.3433	1.3513
30	1.3111	1.3192	1.3275	1.3350
40	1.2985	1.3060	1.3137	1.3207
50	1.2871	1.2943	1.3016	1.3080
60	1.2769	1.2837	1.2905	1.2966
70	1.2674	1.2740	1.2805	1.2863
80	1.2588	1.2651	1.2713	1.2769

continue

continue

P/MPa	*T*/K 514.22	524.46	534.71	543.97
	V_{spez}/cm^{3}g^{-1}			
90	1.2506	1.2569	1.2627	1.2680
100	1.2430	1.2492	1.2546	1.2599
110	1.2358	1.2418	1.2470	1.2520
120	1.2290	1.2348	1.2397	1.2448
130	1.2222	1.2281	1.2330	1.2378
140	1.2160	1.2219	1.2264	1.2312
150	1.2102	1.2153	1.2203	1.2250
160	1.2046	1.2096	1.2139	1.2190
170	1.1991	1.2042	1.2081	1.2127
180	1.1940	1.1988	1.2027	1.2073
190	1.1890	1.1937	1.1976	1.2021
200	1.1844	1.1890	1.1927	1.1970

copolymer (B): E-O/11w 98SUH, 99MAI

P/MPa	*T*/K 422.14	431.98	441.99	451.98	462.44	472.53	482.35	492.16	502.32
	V_{spez}/cm^{3}g^{-1}								
0.1	1.2819	1.2925	1.3017	1.3110	1.3210	1.3308	1.3408	1.3506	1.3611
10	1.2702	1.2797	1.2884	1.2971	1.3063	1.3153	1.3243	1.3332	1.3424
20	1.2594	1.2679	1.2763	1.2846	1.2929	1.3013	1.3096	1.3178	1.3260
30	1.2499	1.2579	1.2655	1.2738	1.2817	1.2894	1.2971	1.3047	1.3122
40	1.2411	1.2487	1.2560	1.2638	1.2712	1.2788	1.2858	1.2929	1.3000
50	1.2330	1.2404	1.2473	1.2544	1.2618	1.2688	1.2756	1.2824	1.2890
60	1.2256	1.2327	1.2393	1.2459	1.2527	1.2597	1.2664	1.2728	1.2790
70	1.2186	1.2255	1.2319	1.2383	1.2447	1.2511	1.2578	1.2640	1.2700
80	1.2121	1.2188	1.2250	1.2310	1.2373	1.2434	1.2494	1.2559	1.2617
90	1.2059	1.2124	1.2183	1.2244	1.2301	1.2361	1.2420	1.2477	1.2538
100	1.2001	1.2064	1.2121	1.2179	1.2235	1.2292	1.2350	1.2406	1.2460
110	1.1945	1.2006	1.2062	1.2118	1.2174	1.2227	1.2283	1.2337	1.2391
120	1.1893	1.1950	1.2005	1.2060	1.2113	1.2166	1.2220	1.2271	1.2325
130	1.1842	1.1898	1.1950	1.2005	1.2057	1.2107	1.2160	1.2210	1.2261
140	1.1792	1.1847	1.1899	1.1951	1.2002	1.2051	1.2102	1.2150	1.2202
150	1.1745	1.1798	1.1849	1.1900	1.1948	1.1999	1.2047	1.2095	1.2144
160	1.1701	1.1751	1.1800	1.1852	1.1898	1.1947	1.1995	1.2041	1.2090
170	1.1657	1.1707	1.1754	1.1805	1.1849	1.1898	1.1943	1.1990	1.2038
180	1.1615	1.1663	1.1710	1.1759	1.1803	1.1850	1.1895	1.1940	1.1988
190	1.1574	1.1623	1.1667	1.1716	1.1758	1.1805	1.1850	1.1893	1.1939
200	1.1536	1.1584	1.1626	1.1673	1.1715	1.1763	1.1805	1.1849	1.1894

continue **E-O/11w**

P/MPa	*T*/K 512.35	522.40	532.43	542.16
	V_{spez}/cm^{3}g^{-1}			
0.1	1.3715	1.3826	1.3935	1.4045
10	1.3518	1.3621	1.3717	1.3810
20	1.3344	1.3443	1.3530	1.3610
30	1.3202	1.3293	1.3373	1.3444
40	1.3076	1.3158	1.3234	1.3303
50	1.2962	1.3039	1.3109	1.3176
60	1.2858	1.2932	1.2999	1.3061
70	1.2765	1.2836	1.2898	1.2958
80	1.2678	1.2745	1.2806	1.2862
90	1.2595	1.2662	1.2720	1.2774
100	1.2519	1.2583	1.2638	1.2691
110	1.2448	1.2509	1.2563	1.2614
120	1.2376	1.2439	1.2491	1.2541
130	1.2313	1.2369	1.2423	1.2472
140	1.2250	1.2305	1.2354	1.2407
150	1.2193	1.2244	1.2293	1.2339
160	1.2137	1.2186	1.2234	1.2280
170	1.2083	1.2131	1.2177	1.2224
180	1.2032	1.2078	1.2122	1.2169
190	1.1983	1.2028	1.2071	1.2117
200	1.1936	1.1981	1.2022	1.2068

copolymer (B): **E-O/12w** **98SUH, 99MAI**

P/MPa	*T*/K 424.40	433.97	444.27	454.30	464.81	475.00	484.88	494.98	505.43
	V_{spez}/cm^{3}g^{-1}								
0.1	1.2798	1.2887	1.2983	1.3081	1.3184	1.3283	1.3387	1.3487	1.3594
10	1.2674	1.2758	1.2848	1.2939	1.3032	1.3122	1.3217	1.3307	1.3402
20	1.2560	1.2642	1.2726	1.2810	1.2895	1.2978	1.3066	1.3148	1.3232
30	1.2464	1.2538	1.2620	1.2700	1.2779	1.2857	1.2939	1.3014	1.3095
40	1.2377	1.2444	1.2520	1.2599	1.2675	1.2748	1.2824	1.2896	1.2971
50	1.2295	1.2360	1.2431	1.2503	1.2577	1.2649	1.2721	1.2790	1.2859
60	1.2221	1.2281	1.2350	1.2419	1.2486	1.2559	1.2627	1.2692	1.2760
70	1.2150	1.2207	1.2274	1.2340	1.2404	1.2474	1.2540	1.2604	1.2668
80	1.2084	1.2140	1.2203	1.2267	1.2329	1.2392	1.2460	1.2520	1.2582

continue

continue

P/MPa	*T*/K								
	424.40	433.97	444.27	454.30	464.81	475.00	484.88	494.98	505.43
	V_{spez}/cm^{3}g^{-1}								
90	1.2022	1.2075	1.2137	1.2199	1.2258	1.2318	1.2381	1.2443	1.2502
100	1.1963	1.2014	1.2073	1.2133	1.2192	1.2249	1.2309	1.2367	1.2427
110	1.1905	1.1956	1.2013	1.2072	1.2129	1.2184	1.2243	1.2297	1.2352
120	1.1852	1.1900	1.1955	1.2012	1.2068	1.2122	1.2178	1.2232	1.2284
130	1.1800	1.1847	1.1900	1.1956	1.2010	1.2063	1.2118	1.2171	1.2221
140	1.1750	1.1796	1.1847	1.1902	1.1956	1.2006	1.2060	1.2111	1.2160
150	1.1701	1.1746	1.1796	1.1851	1.1902	1.1951	1.2005	1.2054	1.2101
160	1.1656	1.1699	1.1747	1.1800	1.1851	1.1899	1.1951	1.2000	1.2044
170	1.1610	1.1653	1.1700	1.1752	1.1802	1.1849	1.1899	1.1948	1.1990
180	1.1568	1.1609	1.1655	1.1706	1.1755	1.1801	1.1850	1.1897	1.1939
190	1.1525	1.1566	1.1612	1.1662	1.1710	1.1755	1.1802	1.1849	1.1889
200	1.1486	1.1526	1.1571	1.1620	1.1668	1.1710	1.1758	1.1803	1.1842

continue **E-O/12w**

P/MPa	*T*/K			
	515.58	525.44	535.59	545.01
	V_{spez}/cm^{3}g^{-1}			
0.1	1.3705	1.3814	1.3928	1.4032
10	1.3501	1.3597	1.3696	1.3788
20	1.3324	1.3411	1.3498	1.3580
30	1.3176	1.3257	1.3337	1.3414
40	1.3047	1.3123	1.3196	1.3269
50	1.2933	1.3004	1.3074	1.3141
60	1.2828	1.2896	1.2962	1.3025
70	1.2733	1.2797	1.2861	1.2920
80	1.2644	1.2707	1.2768	1.2824
90	1.2562	1.2622	1.2680	1.2734
100	1.2484	1.2543	1.2599	1.2651
110	1.2411	1.2468	1.2522	1.2574
120	1.2337	1.2397	1.2450	1.2497
130	1.2271	1.2329	1.2380	1.2428
140	1.2210	1.2261	1.2314	1.2361
150	1.2149	1.2200	1.2248	1.2297
160	1.2092	1.2141	1.2187	1.2233
170	1.2038	1.2084	1.2131	1.2174
180	1.1986	1.2031	1.2076	1.2118
190	1.1935	1.1980	1.2024	1.2065
200	1.1888	1.1931	1.1974	1.2013

copolymer (B): E-O/14w 98SUH, 99MAI

P/MPa	*T*/K 423.68	433.84	443.84	453.96	463.69	474.13	483.96	493.93	503.56
	V_{spez}/cm^{3}g^{-1}								
0.1	1.2820	1.2913	1.3007	1.3104	1.3201	1.3303	1.3401	1.3502	1.3608
10	1.2696	1.2784	1.2872	1.2961	1.3050	1.3142	1.3232	1.3323	1.3419
20	1.2582	1.2668	1.2750	1.2833	1.2915	1.2999	1.3083	1.3165	1.3253
30	1.2485	1.2563	1.2644	1.2721	1.2798	1.2877	1.2954	1.3031	1.3113
40	1.2397	1.2469	1.2543	1.2620	1.2691	1.2767	1.2840	1.2911	1.2988
50	1.2315	1.2383	1.2454	1.2523	1.2596	1.2666	1.2736	1.2804	1.2877
60	1.2240	1.2305	1.2372	1.2438	1.2504	1.2575	1.2642	1.2706	1.2776
70	1.2169	1.2230	1.2296	1.2360	1.2423	1.2488	1.2555	1.2617	1.2682
80	1.2103	1.2161	1.2225	1.2287	1.2347	1.2411	1.2472	1.2534	1.2596
90	1.2041	1.2097	1.2158	1.2218	1.2276	1.2337	1.2396	1.2453	1.2516
100	1.1981	1.2035	1.2094	1.2152	1.2208	1.2268	1.2326	1.2380	1.2442
110	1.1925	1.1976	1.2034	1.2091	1.2144	1.2203	1.2258	1.2310	1.2367
120	1.1871	1.1919	1.1976	1.2031	1.2085	1.2140	1.2194	1.2245	1.2300
130	1.1819	1.1866	1.1921	1.1975	1.2027	1.2080	1.2133	1.2184	1.2235
140	1.1769	1.1814	1.1868	1.1920	1.1971	1.2024	1.2074	1.2123	1.2174
150	1.1720	1.1764	1.1816	1.1869	1.1918	1.1968	1.2019	1.2067	1.2116
160	1.1674	1.1718	1.1768	1.1819	1.1866	1.1916	1.1965	1.2012	1.2060
170	1.1629	1.1672	1.1720	1.1771	1.1817	1.1867	1.1914	1.1959	1.2007
180	1.1587	1.1627	1.1675	1.1725	1.1771	1.1818	1.1865	1.1909	1.1955
190	1.1545	1.1585	1.1632	1.1681	1.1725	1.1771	1.1817	1.1862	1.1906
200	1.1506	1.1545	1.1590	1.1639	1.1682	1.1727	1.1772	1.1816	1.1859

continue **E-O/14w**

P/MPa	*T*/K 513.94	524.07	533.83	544.30
	V_{spez}/cm^{3}g^{-1}			
0.1	1.3721	1.3831	1.3938	1.4051
10	1.3519	1.3617	1.3711	1.3808
20	1.3343	1.3433	1.3517	1.3602
30	1.3196	1.3279	1.3356	1.3433
40	1.3066	1.3144	1.3216	1.3288
50	1.2950	1.3024	1.3092	1.3157
60	1.2845	1.2915	1.2979	1.3041
70	1.2750	1.2815	1.2877	1.2936
80	1.2660	1.2725	1.2783	1.2839

continue

continue

P/MPa	513.94	524.07	533.83	544.30
	T/K			
	V_{spez}/cm^{3}g^{-1}			
90	1.2577	1.2639	1.2695	1.2750
100	1.2500	1.2560	1.2614	1.2665
110	1.2427	1.2484	1.2536	1.2586
120	1.2354	1.2413	1.2463	1.2512
130	1.2288	1.2342	1.2394	1.2441
140	1.2224	1.2277	1.2325	1.2375
150	1.2165	1.2216	1.2262	1.2307
160	1.2107	1.2157	1.2202	1.2245
170	1.2052	1.2100	1.2144	1.2187
180	1.2000	1.2047	1.2090	1.2132
190	1.1950	1.1996	1.2037	1.2078
200	1.1902	1.1947	1.1987	1.2026

copolymer (B): E-O/25w 98SUH, 99MAI

P/MPa	423.10	433.01	443.28	453.42	463.80	473.36	483.63	493.64	503.31
	T/K								
	V_{spez}/cm^{3}g^{-1}								
0.1	1.2825	1.2930	1.3026	1.3122	1.3227	1.3324	1.3431	1.3533	1.3639
10	1.2704	1.2798	1.2887	1.2977	1.3072	1.3163	1.3256	1.3349	1.3444
20	1.2592	1.2677	1.2763	1.2845	1.2933	1.3019	1.3101	1.3188	1.3275
30	1.2494	1.2574	1.2650	1.2733	1.2814	1.2895	1.2971	1.3050	1.3133
40	1.2403	1.2479	1.2550	1.2625	1.2706	1.2781	1.2853	1.2926	1.3004
50	1.2320	1.2392	1.2459	1.2531	1.2603	1.2679	1.2745	1.2815	1.2888
60	1.2242	1.2311	1.2375	1.2443	1.2514	1.2582	1.2648	1.2714	1.2783
70	1.2169	1.2236	1.2296	1.2363	1.2430	1.2496	1.2556	1.2623	1.2687
80	1.2101	1.2166	1.2224	1.2286	1.2352	1.2416	1.2474	1.2533	1.2600
90	1.2038	1.2099	1.2155	1.2216	1.2279	1.2341	1.2397	1.2454	1.2515
100	1.1977	1.2038	1.2091	1.2150	1.2209	1.2270	1.2324	1.2380	1.2439
110	1.1919	1.1978	1.2030	1.2086	1.2145	1.2203	1.2256	1.2310	1.2366
120	1.1863	1.1922	1.1971	1.2025	1.2083	1.2138	1.2190	1.2243	1.2298
130	1.1811	1.1866	1.1915	1.1968	1.2023	1.2078	1.2127	1.2179	1.2233
140	1.1759	1.1814	1.1861	1.1913	1.1966	1.2019	1.2069	1.2118	1.2172
150	1.1710	1.1765	1.1809	1.1860	1.1912	1.1964	1.2012	1.2061	1.2112
160	1.1662	1.1716	1.1760	1.1808	1.1859	1.1911	1.1958	1.2005	1.2055
170	1.1617	1.1670	1.1712	1.1759	1.1809	1.1860	1.1907	1.1951	1.2001
180	1.1575	1.1624	1.1666	1.1713	1.1761	1.1811	1.1857	1.1902	1.1949
190	1.1533	1.1583	1.1623	1.1669	1.1715	1.1764	1.1809	1.1853	1.1900
200	1.1492	1.1541	1.1581	1.1624	1.1672	1.1719	1.1764	1.1806	1.1853

continue **E-O/25w**

P/MPa	*T*/K 514.02	524.03	533.91	543.88
	V_{spez}/cm^{3}g^{-1}			
0.1	1.3754	1.3865	1.3977	1.4094
10	1.3546	1.3644	1.3745	1.3846
20	1.3368	1.3455	1.3546	1.3637
30	1.3216	1.3296	1.3384	1.3465
40	1.3082	1.3156	1.3239	1.3314
50	1.2963	1.3033	1.3108	1.3181
60	1.2854	1.2924	1.2995	1.3063
70	1.2755	1.2821	1.2890	1.2953
80	1.2665	1.2729	1.2791	1.2852
90	1.2580	1.2641	1.2704	1.2761
100	1.2496	1.2558	1.2620	1.2675
110	1.2422	1.2481	1.2542	1.2596
120	1.2352	1.2406	1.2467	1.2520
130	1.2285	1.2337	1.2392	1.2448
140	1.2221	1.2273	1.2328	1.2377
150	1.2161	1.2211	1.2263	1.2311
160	1.2102	1.2152	1.2202	1.2249
170	1.2045	1.2095	1.2145	1.2189
180	1.1992	1.2041	1.2091	1.2134
190	1.1941	1.1989	1.2038	1.2080
200	1.1894	1.1940	1.1987	1.2028

copolymer (B): **E-O/39w** **98SUH, 99MAI**

P/MPa	*T*/K 423.60	433.65	443.42	453.53	464.23	474.19	483.79	493.75	503.81
	V_{spez}/cm^{3}g^{-1}								
0.1	1.2785	1.2879	1.2973	1.3070	1.3177	1.3273	1.3366	1.3458	1.3558
10	1.2664	1.2752	1.2838	1.2926	1.3023	1.3113	1.3197	1.3279	1.3368
20	1.2554	1.2636	1.2715	1.2796	1.2883	1.2970	1.3046	1.3121	1.3201
30	1.2458	1.2535	1.2611	1.2686	1.2767	1.2846	1.2920	1.2989	1.3061
40	1.2369	1.2442	1.2513	1.2585	1.2662	1.2735	1.2801	1.2870	1.2936
50	1.2288	1.2356	1.2426	1.2492	1.2566	1.2635	1.2697	1.2758	1.2826
60	1.2212	1.2278	1.2344	1.2407	1.2478	1.2544	1.2602	1.2661	1.2722
70	1.2142	1.2206	1.2269	1.2330	1.2396	1.2461	1.2516	1.2570	1.2628
80	1.2077	1.2136	1.2198	1.2256	1.2320	1.2382	1.2433	1.2487	1.2542

continue

continue

P/MPa	*T*/K								
	423.60	433.65	443.42	453.53	464.23	474.19	483.79	493.75	503.81
	V_{spez}/cm^{3}g^{-1}								
90	1.2015	1.2072	1.2132	1.2188	1.2249	1.2309	1.2358	1.2409	1.2463
100	1.1956	1.2011	1.2068	1.2124	1.2183	1.2240	1.2288	1.2338	1.2389
110	1.1899	1.1953	1.2008	1.2063	1.2119	1.2175	1.2220	1.2269	1.2318
120	1.1845	1.1898	1.1952	1.2004	1.2058	1.2113	1.2156	1.2202	1.2250
130	1.1794	1.1844	1.1898	1.1949	1.2001	1.2055	1.2096	1.2141	1.2187
140	1.1744	1.1793	1.1845	1.1895	1.1945	1.1997	1.2038	1.2082	1.2127
150	1.1698	1.1744	1.1794	1.1844	1.1893	1.1944	1.1983	1.2026	1.2069
160	1.1651	1.1697	1.1747	1.1794	1.1843	1.1891	1.1930	1.1972	1.2015
170	1.1607	1.1651	1.1700	1.1746	1.1794	1.1843	1.1879	1.1921	1.1962
180	1.1565	1.1608	1.1655	1.1701	1.1747	1.1795	1.1831	1.1871	1.1911
190	1.1524	1.1568	1.1613	1.1657	1.1702	1.1750	1.1784	1.1825	1.1863
200	1.1485	1.1528	1.1572	1.1617	1.1659	1.1706	1.1741	1.1780	1.1818

continue **E-O/39w**

P/MPa	*T*/K			
	514.05	523.82	533.95	543.80
	V_{spez}/cm^{3}g^{-1}			
0.1	1.3664	1.3770	1.3878	1.3993
10	1.3463	1.3558	1.3653	1.3754
20	1.3287	1.3374	1.3460	1.3551
30	1.3142	1.3222	1.3303	1.3385
40	1.3012	1.3086	1.3162	1.3236
50	1.2896	1.2965	1.3034	1.3106
60	1.2791	1.2859	1.2921	1.2991
70	1.2692	1.2760	1.2820	1.2888
80	1.2604	1.2669	1.2727	1.2790
90	1.2522	1.2581	1.2641	1.2699
100	1.2444	1.2501	1.2557	1.2617
110	1.2372	1.2428	1.2481	1.2535
120	1.2305	1.2358	1.2408	1.2461
130	1.2241	1.2292	1.2339	1.2390
140	1.2178	1.2227	1.2276	1.2324
150	1.2119	1.2167	1.2214	1.2262
160	1.2063	1.2110	1.2155	1.2200
170	1.2008	1.2057	1.2099	1.2142
180	1.1957	1.2005	1.2046	1.2088
190	1.1909	1.1954	1.1995	1.2037
200	1.1862	1.1906	1.1947	1.1988

copolymer (B): E-O/55w 98SUH, 99MAI

P/MPa	T/K 423.15	433.05	443.06	453.05	463.51	473.26	483.30	493.10	503.26
	V_{spez}/cm^{3}g^{-1}								
0.1	1.2860	1.2957	1.3054	1.3151	1.3262	1.3357	1.3443	1.3543	1.3650
10	1.2738	1.2827	1.2918	1.3006	1.3105	1.3190	1.3268	1.3359	1.3456
20	1.2626	1.2709	1.2793	1.2874	1.2964	1.3040	1.3113	1.3198	1.3287
30	1.2527	1.2606	1.2686	1.2762	1.2844	1.2914	1.2978	1.3061	1.3145
40	1.2436	1.2510	1.2586	1.2658	1.2735	1.2803	1.2862	1.2934	1.3016
50	1.2352	1.2424	1.2496	1.2562	1.2637	1.2699	1.2756	1.2822	1.2898
60	1.2275	1.2343	1.2413	1.2476	1.2547	1.2605	1.2659	1.2724	1.2796
70	1.2202	1.2270	1.2334	1.2397	1.2464	1.2519	1.2571	1.2633	1.2701
80	1.2134	1.2199	1.2263	1.2322	1.2387	1.2440	1.2490	1.2549	1.2613
90	1.2071	1.2132	1.2194	1.2252	1.2315	1.2365	1.2413	1.2470	1.2532
100	1.2010	1.2071	1.2130	1.2186	1.2247	1.2295	1.2341	1.2396	1.2455
110	1.1953	1.2012	1.2069	1.2123	1.2183	1.2229	1.2274	1.2327	1.2385
120	1.1898	1.1956	1.2011	1.2063	1.2121	1.2166	1.2209	1.2261	1.2315
130	1.1845	1.1901	1.1955	1.2007	1.2064	1.2106	1.2149	1.2198	1.2251
140	1.1795	1.1849	1.1901	1.1952	1.2007	1.2047	1.2090	1.2137	1.2189
150	1.1747	1.1799	1.1851	1.1899	1.1953	1.1993	1.2034	1.2080	1.2131
160	1.1700	1.1754	1.1801	1.1848	1.1902	1.1940	1.1981	1.2024	1.2075
170	1.1656	1.1706	1.1754	1.1801	1.1852	1.1888	1.1928	1.1973	1.2020
180	1.1612	1.1662	1.1708	1.1755	1.1805	1.1841	1.1880	1.1922	1.1969
190	1.1572	1.1621	1.1665	1.1710	1.1758	1.1795	1.1832	1.1875	1.1921
200	1.1532	1.1581	1.1623	1.1668	1.1716	1.1750	1.1787	1.1829	1.1873

continue **E-O/55w**

P/MPa	T/K 513.39	523.14	532.76	542.82
	V_{spez}/cm^{3}g^{-1}			
0.1	1.3760	1.3866	1.3970	1.4090
10	1.3556	1.3646	1.3741	1.3845
20	1.3380	1.3457	1.3547	1.3638
30	1.3229	1.3300	1.3383	1.3468
40	1.3095	1.3161	1.3237	1.3317
50	1.2973	1.3037	1.3110	1.3183
60	1.2864	1.2929	1.2997	1.3064
70	1.2765	1.2825	1.2895	1.2959
80	1.2675	1.2731	1.2796	1.2859

continue

continue

P/MPa	*T*/K 513.39	523.14	532.76	542.82
	V_{spez}/cm^{3}g^{-1}			
90	1.2591	1.2646	1.2709	1.2765
100	1.2512	1.2566	1.2627	1.2682
110	1.2438	1.2490	1.2549	1.2601
120	1.2369	1.2418	1.2475	1.2527
130	1.2301	1.2351	1.2407	1.2456
140	1.2238	1.2286	1.2341	1.2388
150	1.2180	1.2226	1.2278	1.2324
160	1.2123	1.2167	1.2217	1.2262
170	1.2067	1.2112	1.2160	1.2204
180	1.2015	1.2059	1.2106	1.2148
190	1.1965	1.2007	1.2053	1.2095
200	1.1917	1.1959	1.2005	1.2045

copolymer (B): E-O/64w 98SUH, 99MAI

P/MPa	*T*/K 423.50	433.48	443.33	453.37	463.49	474.04	483.94	493.91	503.70
	V_{spez}/cm^{3}g^{-1}								
0.1	1.2741	1.2839	1.2936	1.3034	1.3145	1.3255	1.3355	1.3464	1.3572
10	1.2728	1.2825	1.2921	1.3019	1.3128	1.3237	1.3336	1.3443	1.3550
20	1.2716	1.2811	1.2906	1.3003	1.3111	1.3219	1.3317	1.3423	1.3528
30	1.2703	1.2798	1.2892	1.2988	1.3094	1.3201	1.3298	1.3403	1.3507
40	1.2690	1.2784	1.2878	1.2973	1.3078	1.3183	1.3280	1.3383	1.3486
50	1.2678	1.2771	1.2864	1.2958	1.3061	1.3165	1.3261	1.3364	1.3466
60	1.2665	1.2758	1.2850	1.2944	1.3045	1.3148	1.3244	1.3345	1.3446
70	1.2653	1.2745	1.2836	1.2929	1.3030	1.3131	1.3226	1.3326	1.3426
80	1.2641	1.2732	1.2822	1.2915	1.3014	1.3115	1.3209	1.3308	1.3406
90	1.2629	1.2719	1.2809	1.2901	1.2998	1.3098	1.3191	1.3290	1.3387
100	1.2617	1.2707	1.2796	1.2887	1.2983	1.3082	1.3175	1.3272	1.3368
110	1.2605	1.2693	1.2782	1.2873	1.2968	1.3064	1.3158	1.3253	1.3349
120	1.2593	1.2681	1.2769	1.2858	1.2953	1.3048	1.3140	1.3235	1.3329
130	1.2581	1.2669	1.2756	1.2845	1.2938	1.3033	1.3123	1.3218	1.3310
140	1.2570	1.2657	1.2742	1.2830	1.2920	1.3016	1.3108	1.3199	1.3292
150	1.2558	1.2645	1.2729	1.2817	1.2906	1.3002	1.3091	1.3182	1.3274
160	1.2547	1.2633	1.2717	1.2804	1.2891	1.2987	1.3076	1.3167	1.3257
170	1.2536	1.2621	1.2705	1.2791	1.2877	1.2972	1.3060	1.3150	1.3240
180	1.2525	1.2609	1.2692	1.2780	1.2864	1.2958	1.3045	1.3135	1.3224
190	1.2515	1.2598	1.2681	1.2767	1.2851	1.2944	1.3031	1.3119	1.3207
200	1.2504	1.2586	1.2669	1.2755	1.2838	1.2930	1.3016	1.3104	1.3191

continue **E-O/64w**

P/MPa	*T*/K 513.61	523.74	533.61	543.85
	V_{spez}/cm^{3}g^{-1}			
0.1	1.3687	1.3806	1.3925	1.4044
10	1.3663	1.3781	1.3898	1.4014
20	1.3640	1.3756	1.3871	1.3986
30	1.3617	1.3732	1.3846	1.3958
40	1.3595	1.3708	1.3820	1.3931
50	1.3573	1.3684	1.3795	1.3904
60	1.3551	1.3661	1.3771	1.3878
70	1.3530	1.3639	1.3747	1.3852
80	1.3509	1.3617	1.3724	1.3827
90	1.3489	1.3595	1.3701	1.3802
100	1.3469	1.3574	1.3678	1.3778
110	1.3447	1.3551	1.3653	1.3753
120	1.3427	1.3529	1.3630	1.3729
130	1.3407	1.3508	1.3609	1.3704
140	1.3388	1.3488	1.3587	1.3682
150	1.3369	1.3467	1.3566	1.3660
160	1.3351	1.3448	1.3546	1.3637
170	1.3333	1.3429	1.3526	1.3617
180	1.3315	1.3411	1.3507	1.3595
190	1.3298	1.3393	1.3489	1.3575
200	1.3282	1.3376	1.3470	1.3557

Copolymers from ethylene and propylene

Average chemical composition of the copolymers, acronyms, range of data, and references:

Copolymer	Acronym	Range of data *T*/K	*P*/MPa	Ref.
Ethylene/propylene copolymer				
10.8 wt% propylene	E-P/11w	422-542	0.1-200	99MAI
17.4 wt% propylene	E-P/17w	422-542	0.1-200	99MAI

continue

continue

Copolymer	Acronym	Range of data T/K	P/MPa	Ref.
23.0 wt% propylene	E-P/23w	398-570	0.1-200	95ZOL
31.8 wt% propylene	E-P/32w	422-542	0.1-200	99MAI
50.0 wt% propylene	E-P/50w	413-523	0.1-63	93ROD
57.0 wt% propylene	E-P/57w	377-581	0.1-200	95ZOL
60.8 wt% propylene	E-P/61w	422-542	0.1-200	99MAI
76.0 wt% propylene	E-P/76w(1)	388-575	0.1-200	95ZOL
76.2 wt% propylene	E-P/76w(2)	422-542	0.1-200	99MAI
83.5 wt% propylene	E-P/84w(1)	422-542	0.1-200	99MAI
84.0 wt% propylene	E-P/84w(2)	388-570	0.1-200	95ZOL
96.4 wt% propylene	E-P/96w	422-533	0.1-200	99MAI
unknown propylene content	E-P/?	413-523	0.1-62	73RAO

Characterization:

Copolymer (B)	M_n/ g/mol	M_w/ g/mol	M_η/ g/mol	Further information
E-P/11w	143600	273000		synthesized in the laboratory, $\rho_B = 0.9007$ g/cm^3
E-P/17w	152000	258000		synthesized in the laboratory, $\rho_B = 0.9008$ g/cm^3
E-P/23w	53300	160000		random, most probable distribution
E-P/32w	104500	178000		synthesized in the laboratory, $\rho_B = 0.8871$ g/cm^3
E-P/57w	51500	134000		random, most probable distribution
E-P/61w	91900	138000		synthesized in the laboratory, $\rho_B = 0.8582$ g/cm^3
E-P/76w(1)	40200	84400		random, most probable distribution
E-P/76w(2)	77800	117000		synthesized in the laboratory, $\rho_B = 0.8600$ g/cm^3
E-P/84w(1)	49600	94200		synthesized in the laboratory, $\rho_B = 0.8783$ g/cm^3
E-P/84w(2)	35700	75000		random, most probable distribution
E-P/96w	96900	145400		synthesized in the laboratory, $\rho_B = 0.8889$ g/cm^3
E-P/?				no data

Experimental *PVT* data:

copolymer (B): E-P/11w **99MAI**

P/MPa	*T*/K 421.89	431.91	441.92	452.05	462.47	472.46	482.53	492.22	501.54
	V_{spez}/cm^{3}g^{-1}								
0.1	1.2776	1.2871	1.2975	1.3068	1.3164	1.3266	1.3362	1.3462	1.3559
10	1.2662	1.2750	1.2844	1.2929	1.3020	1.3113	1.3200	1.3291	1.3380
20	1.2557	1.2638	1.2723	1.2806	1.2889	1.2976	1.3056	1.3138	1.3220
30	1.2465	1.2542	1.2622	1.2695	1.2779	1.2858	1.2934	1.3012	1.3089
40	1.2381	1.2452	1.2530	1.2599	1.2674	1.2755	1.2826	1.2899	1.2971
50	1.2303	1.2370	1.2444	1.2511	1.2582	1.2653	1.2727	1.2798	1.2866
60	1.2230	1.2295	1.2364	1.2429	1.2498	1.2566	1.2636	1.2704	1.2769
70	1.2164	1.2224	1.2291	1.2353	1.2420	1.2486	1.2549	1.2617	1.2680
80	1.2101	1.2158	1.2222	1.2283	1.2347	1.2410	1.2471	1.2532	1.2597
90	1.2040	1.2097	1.2157	1.2216	1.2277	1.2339	1.2398	1.2459	1.2516
100	1.1983	1.2037	1.2097	1.2153	1.2213	1.2272	1.2329	1.2387	1.2444
110	1.1929	1.1981	1.2038	1.2094	1.2152	1.2208	1.2264	1.2322	1.2375
120	1.1877	1.1928	1.1983	1.2037	1.2093	1.2149	1.2202	1.2258	1.2310
130	1.1826	1.1876	1.1930	1.1983	1.2037	1.2092	1.2144	1.2198	1.2247
140	1.1779	1.1826	1.1878	1.1930	1.1985	1.2036	1.2087	1.2139	1.2190
150	1.1732	1.1779	1.1829	1.1882	1.1932	1.1985	1.2034	1.2085	1.2132
160	1.1688	1.1733	1.1783	1.1834	1.1884	1.1935	1.1983	1.2032	1.2078
170	1.1645	1.1688	1.1737	1.1787	1.1836	1.1885	1.1934	1.1982	1.2026
180	1.1602	1.1646	1.1695	1.1742	1.1792	1.1841	1.1885	1.1932	1.1978
190	1.1563	1.1605	1.1652	1.1700	1.1747	1.1796	1.1841	1.1886	1.1929
200	1.1524	1.1566	1.1612	1.1659	1.1705	1.1753	1.1798	1.1842	1.1884

continue **E-P/11w**

P/MPa	*T*/K 512.33	521.91	531.78	541.21
	V_{spez}/cm^{3}g^{-1}			
0.1	1.3669	1.3778	1.3882	1.3999
10	1.3475	1.3573	1.3662	1.3764
20	1.3305	1.3394	1.3471	1.3561
30	1.3166	1.3248	1.3318	1.3402
40	1.3040	1.3118	1.3185	1.3263
50	1.2930	1.3004	1.3067	1.3140
60	1.2832	1.2901	1.2959	1.3029
70	1.2739	1.2807	1.2861	1.2928

continue

continue

P/MPa	*T*/K 512.33	521.91	531.78	541.21
	V_{spez}/cm^{3}g^{-1}			
80	1.2654	1.2719	1.2773	1.2836
90	1.2576	1.2638	1.2689	1.2751
100	1.2496	1.2562	1.2611	1.2670
110	1.2427	1.2485	1.2537	1.2594
120	1.2360	1.2417	1.2467	1.2523
130	1.2296	1.2352	1.2397	1.2456
140	1.2236	1.2291	1.2333	1.2387
150	1.2179	1.2232	1.2275	1.2326
160	1.2124	1.2176	1.2216	1.2267
170	1.2070	1.2121	1.2162	1.2211
180	1.2020	1.2070	1.2109	1.2158
190	1.1972	1.2021	1.2058	1.2106
200	1.1925	1.1973	1.2011	1.2057

copolymer (B): E-P/17w 99MAI

P/MPa	*T*/K 422.29	432.39	442.28	452.35	462.56	472.70	482.67	492.24	501.91
	V_{spez}/cm^{3}g^{-1}								
0.1	1.2665	1.2756	1.2849	1.2951	1.3050	1.3143	1.3237	1.3336	1.3435
10	1.2551	1.2634	1.2720	1.2811	1.2901	1.2990	1.3076	1.3166	1.3255
20	1.2447	1.2524	1.2603	1.2682	1.2770	1.2854	1.2931	1.3015	1.3097
30	1.2353	1.2426	1.2502	1.2576	1.2654	1.2736	1.2811	1.2888	1.2964
40	1.2269	1.2337	1.2408	1.2478	1.2553	1.2625	1.2702	1.2773	1.2846
50	1.2191	1.2254	1.2323	1.2391	1.2459	1.2530	1.2598	1.2672	1.2740
60	1.2118	1.2178	1.2243	1.2308	1.2376	1.2442	1.2506	1.2573	1.2642
70	1.2050	1.2107	1.2171	1.2232	1.2298	1.2361	1.2423	1.2486	1.2549
80	1.1987	1.2040	1.2102	1.2161	1.2224	1.2285	1.2346	1.2406	1.2466
90	1.1926	1.1977	1.2038	1.2095	1.2156	1.2214	1.2273	1.2332	1.2389
100	1.1869	1.1919	1.1977	1.2031	1.2090	1.2148	1.2204	1.2262	1.2316
110	1.1814	1.1861	1.1920	1.1972	1.2028	1.2084	1.2139	1.2195	1.2248
120	1.1762	1.1809	1.1865	1.1915	1.1971	1.2024	1.2076	1.2131	1.2183
130	1.1712	1.1756	1.1812	1.1860	1.1914	1.1967	1.2018	1.2072	1.2122
140	1.1665	1.1706	1.1760	1.1809	1.1860	1.1912	1.1962	1.2014	1.2063
150	1.1618	1.1660	1.1713	1.1759	1.1809	1.1860	1.1908	1.1958	1.2006
160	1.1574	1.1613	1.1666	1.1712	1.1761	1.1809	1.1856	1.1906	1.1952
170	1.1531	1.1569	1.1621	1.1666	1.1713	1.1761	1.1807	1.1856	1.1900
180	1.1491	1.1526	1.1578	1.1622	1.1666	1.1714	1.1759	1.1807	1.1851
190	1.1450	1.1486	1.1535	1.1579	1.1623	1.1669	1.1714	1.1760	1.1802
200	1.1413	1.1447	1.1496	1.1539	1.1581	1.1627	1.1669	1.1714	1.1757

continue **E-P/17w**

P/MPa	512.38	522.39	532.94	542.69
	T/K; V_{spez}/cm^{3}g^{-1}			
0.1	1.3540	1.3643	1.3757	1.3869
10	1.3348	1.3440	1.3539	1.3637
20	1.3180	1.3262	1.3351	1.3440
30	1.3041	1.3117	1.3199	1.3279
40	1.2916	1.2989	1.3063	1.3139
50	1.2805	1.2874	1.2943	1.3015
60	1.2705	1.2769	1.2836	1.2902
70	1.2614	1.2675	1.2737	1.2801
80	1.2522	1.2586	1.2647	1.2709
90	1.2443	1.2500	1.2562	1.2622
100	1.2369	1.2422	1.2485	1.2541
110	1.2298	1.2352	1.2405	1.2465
120	1.2231	1.2284	1.2336	1.2389
130	1.2168	1.2219	1.2270	1.2320
140	1.2109	1.2157	1.2206	1.2257
150	1.2050	1.2098	1.2147	1.2196
160	1.1995	1.2042	1.2088	1.2137
170	1.1942	1.1988	1.2033	1.2081
180	1.1894	1.1937	1.1981	1.2028
190	1.1844	1.1886	1.1930	1.1975
200	1.1797	1.1838	1.1883	1.1926

copolymer (B): E-P/23w **95ZOL**

P/MPa	398.45	408.65	418.85	428.65	438.95	449.25	458.95	469.05	479.15
	T/K; V_{spez}/cm^{3}g^{-1}								
0.1	1.2632	1.2730	1.2825	1.2914	1.3010	1.3109	1.3195	1.3296	1.3392
20	1.2433	1.2524	1.2607	1.2688	1.2768	1.2852	1.2932	1.3017	1.3097
40	1.2266	1.2350	1.2427	1.2500	1.2571	1.2649	1.2716	1.2791	1.2863
60	1.2121	1.2199	1.2271	1.2338	1.2407	1.2473	1.2537	1.2605	1.2670
80	1.1993	1.2065	1.2136	1.2200	1.2262	1.2322	1.2383	1.2445	1.2505
100	1.1878	1.1948	1.2014	1.2075	1.2133	1.2191	1.2246	1.2304	1.2360
120	1.1773	1.1839	1.1901	1.1962	1.2014	1.2071	1.2173	1.2179	1.2230
140	1.1677	1.1740	1.1799	1.1861	1.1908	1.1963	1.2012	1.2065	1.2113
160	1.1588	1.1650	1.1708	1.1762	1.1812	1.1862	1.1912	1.1963	1.2007
180	1.1505	1.1567	1.1620	1.1671	1.1720	1.1770	1.1818	1.1864	1.1909
200	1.1427	1.1488	1.1537	1.1587	1.1637	1.1685	1.1730	1.1777	1.1817

continue **E-P/23w**

P/MPa	T/K								
	489.55	499.35	509.45	519.95	529.75	539.75	550.35	559.65	570.45
	V_{spez}/cm^{3}g^{-1}								
0.1	1.3496	1.3599	1.3698	1.3808	1.3913	1.4023	1.4137	1.4242	1.4632
20	1.3180	1.3265	1.3350	1.3435	1.3524	1.3608	1.3699	1.3786	1.3878
40	1.2937	1.3010	1.3085	1.3160	1.3232	1.3307	1.3384	1.3459	1.3538
60	1.2738	1.2805	1.2869	1.2933	1.3004	1.3067	1.3135	1.3206	1.3271
80	1.2566	1.2629	1.2687	1.2746	1.2807	1.2867	1.2928	1.2990	1.3054
100	1.2415	1.2475	1.2529	1.2583	1.2640	1.2696	1.2752	1.2808	1.2865
120	1.2283	1.2337	1.2389	1.2442	1.2493	1.2545	1.2597	1.2650	1.2701
140	1.2163	1.2214	1.2264	1.2309	1.2361	1.2409	1.2457	1.2508	1.2557
160	1.2055	1.2102	1.2148	1.2192	1.2242	1.2286	1.2332	1.2381	1.2426
180	1.1955	1.2000	1.2044	1.2085	1.2131	1.2175	1.2218	1.2264	1.2307
200	1.1863	1.1904	1.1948	1.1985	1.2032	1.2072	1.2112	1.2158	1.2198

copolymer (B): E-P/32w 99MAI

P/MPa	T/K								
	422.35	432.50	442.35	452.05	462.27	472.37	482.18	492.51	502.28
	V_{spez}/cm^{3}g^{-1}								
0.1	1.2740	1.2836	1.2938	1.3028	1.3122	1.3220	1.3317	1.3422	1.3523
10	1.2624	1.2711	1.2804	1.2886	1.2975	1.3066	1.3155	1.3249	1.3339
20	1.2517	1.2598	1.2681	1.2761	1.2843	1.2929	1.3010	1.3095	1.3178
30	1.2423	1.2498	1.2578	1.2648	1.2731	1.2810	1.2886	1.2966	1.3042
40	1.2337	1.2408	1.2484	1.2550	1.2624	1.2702	1.2776	1.2851	1.2922
50	1.2259	1.2324	1.2399	1.2460	1.2531	1.2601	1.2675	1.2746	1.2814
60	1.2186	1.2247	1.2319	1.2377	1.2445	1.2512	1.2578	1.2651	1.2716
70	1.2118	1.2177	1.2245	1.2301	1.2366	1.2431	1.2494	1.2558	1.2626
80	1.2053	1.2110	1.2176	1.2229	1.2292	1.2354	1.2415	1.2477	1.2536
90	1.1993	1.2047	1.2110	1.2163	1.2222	1.2282	1.2342	1.2400	1.2458
100	1.1935	1.1987	1.2048	1.2100	1.2157	1.2214	1.2272	1.2329	1.2385
110	1.1881	1.1930	1.1990	1.2040	1.2095	1.2151	1.2206	1.2262	1.2315
120	1.1828	1.1876	1.1932	1.1984	1.2036	1.2090	1.2144	1.2197	1.2250
130	1.1777	1.1825	1.1880	1.1930	1.1980	1.2033	1.2086	1.2136	1.2187
140	1.1729	1.1775	1.1828	1.1877	1.1926	1.1978	1.2028	1.2079	1.2128
150	1.1683	1.1726	1.1777	1.1827	1.1875	1.1925	1.1976	1.2023	1.2071
160	1.1639	1.1680	1.1730	1.1778	1.1826	1.1876	1.1923	1.1972	1.2016
170	1.1595	1.1635	1.1684	1.1733	1.1779	1.1828	1.1873	1.1921	1.1966
180	1.1553	1.1593	1.1640	1.1687	1.1733	1.1781	1.1826	1.1872	1.1916
190	1.1513	1.1553	1.1597	1.1645	1.1689	1.1737	1.1780	1.1826	1.1868
200	1.1475	1.1513	1.1557	1.1604	1.1647	1.1694	1.1736	1.1783	1.1822

continue **E-P/32w**

P/MPa	*T*/K 512.33	522.19	532.19	542.01
	V_{spez}/cm^{3}g^{-1}			
0.1	1.3631	1.3732	1.3843	1.3959
10	1.3434	1.3523	1.3619	1.3720
20	1.3262	1.3340	1.3426	1.3515
30	1.3119	1.3193	1.3271	1.3353
40	1.2993	1.3062	1.3135	1.3211
50	1.2880	1.2947	1.3015	1.3085
60	1.2779	1.2843	1.2906	1.2973
70	1.2685	1.2747	1.2808	1.2871
80	1.2599	1.2659	1.2717	1.2777
90	1.2513	1.2576	1.2633	1.2691
100	1.2438	1.2500	1.2553	1.2609
110	1.2367	1.2421	1.2479	1.2533
120	1.2300	1.2353	1.2403	1.2461
130	1.2237	1.2288	1.2336	1.2386
140	1.2175	1.2226	1.2273	1.2322
150	1.2118	1.2167	1.2213	1.2259
160	1.2064	1.2109	1.2155	1.2201
170	1.2010	1.2055	1.2100	1.2143
180	1.1960	1.2003	1.2047	1.2090
190	1.1912	1.1954	1.1997	1.2039
200	1.1866	1.1907	1.1949	1.1989

copolymer (B): E-P/57w 95ZOL

P/MPa	*T*/K 377.65	385.55	394.05	401.95	410.05	418.45	426.45	434.85	442.45
	V_{spez}/cm^{3}g^{-1}								
0.1	1.2411	1.2479	1.2558	1.2629	1.2703	1.2782	1.2856	1.2933	1.3006
20	1.2222	1.2284	1.2353	1.2420	1.2481	1.2548	1.2612	1.2678	1.2724
40	1.2067	1.2124	1.2185	1.2244	1.2301	1.2361	1.2421	1.2478	1.2535
60	1.1931	1.1984	1.2040	1.2095	1.2147	1.2201	1.2257	1.2307	1.2362
80	1.1810	1.1860	1.1911	1.1962	1.2012	1.2061	1.2111	1.2161	1.2211
100	1.1702	1.1749	1.1797	1.1844	1.1892	1.1937	1.1984	1.2031	1.2079
120	1.1604	1.1647	1.1693	1.1739	1.1780	1.1824	1.1869	1.1912	1.1958
140	1.1511	1.1555	1.1596	1.1639	1.1681	1.1721	1.1766	1.1806	1.1850
160	1.1427	1.1469	1.1510	1.1549	1.1587	1.1625	1.1666	1.1707	1.1748
180	1.1347	1.1386	1.1427	1.1463	1.1502	1.1537	1.1575	1.1616	1.1655
200	1.1274	1.1311	1.1348	1.1385	1.1422	1.1454	1.1491	1.1528	1.1569

continue **E-P/57w**

P/MPa	T/K								
	451.05	459.35	467.15	475.35	483.65	491.65	499.85	508.25	516.65
	V_{spez}/cm^{3}g^{-1}								
0.1	1.3088	1.3169	1.3250	1.3331	1.3414	1.3497	1.3580	1.3678	1.3756
20	1.2810	1.2878	1.2945	1.3014	1.3084	1.3153	1.3220	1.3292	1.3362
40	1.2595	1.2654	1.2713	1.2775	1.2835	1.2899	1.2956	1.3016	1.3078
60	1.2417	1.2473	1.2523	1.2578	1.2632	1.2687	1.2742	1.2795	1.2852
80	1.2262	1.2311	1.2362	1.2415	1.2461	1.2510	1.2561	1.2609	1.2659
100	1.2128	1.2171	1.2218	1.2265	1.2310	1.2360	1.2407	1.2448	1.2494
120	1.2003	1.2045	1.2087	1.2133	1.2176	1.2220	1.2264	1.2308	1.2349
140	1.1891	1.1933	1.1972	1.2015	1.2056	1.2097	1.2138	1.2178	1.2221
160	1.1790	1.1826	1.1868	1.9088	1.9945	1.9986	1.2026	1.2060	1.2099
180	1.1695	1.1729	1.1768	1.1806	1.1843	1.1880	1.1920	1.1953	1.1988
200	1.1607	1.1639	1.1676	1.1714	1.1748	1.1787	1.1821	1.1853	1.1888

continue **E-P/57w**

P/MPa	T/K							
	524.85	533.25	541.15	549.05	557.05	565.25	573.35	581.55
	V_{spez}/cm^{3}g^{-1}							
0.1	1.3850	1.3938	1.4029	1.4118	1.4209	1.4307	1.4405	1.4512
20	1.3434	1.3506	1.3575	1.3647	1.3716	1.3790	1.3865	1.3939
40	1.3137	1.3201	1.3259	1.3318	1.3380	1.3441	1.3503	1.3566
60	1.2904	1.2956	1.3010	1.3063	1.3118	1.3171	1.3227	1.3283
80	1.2709	1.2757	1.2808	1.2854	1.2902	1.2951	1.3002	1.3051
100	1.2539	1.2587	1.2632	1.2674	1.2719	1.2764	1.2811	1.2855
120	1.2391	1.2432	1.2476	1.2519	1.2557	1.2599	1.2642	1.2684
140	1.2258	1.2300	1.2336	1.2376	1.2414	1.2453	1.2493	1.2532
160	1.2137	1.2174	1.2210	1.2248	1.2287	1.2322	1.2358	1.2397
180	1.2027	1.2064	1.2098	1.2132	1.2165	1.2201	1.2237	1.2272
200	1.1922	1.1958	1.1990	1.2024	1.2057	1.2092	1.2124	1.2158

copolymer (B): **E-P/61w** **99MAI**

P/MPa	*T*/K 422.35	432.82	442.90	452.71	462.75	473.22	483.07	492.81	502.36
	V_{spez}/cm^{3}g^{-1}								
0.1	1.2670	1.2766	1.2859	1.2954	1.3046	1.3159	1.3255	1.3352	1.3450
10	1.2550	1.2638	1.2724	1.2810	1.2895	1.2993	1.3081	1.3170	1.3259
20	1.2441	1.2521	1.2601	1.2680	1.2760	1.2844	1.2929	1.3010	1.3093
30	1.2344	1.2419	1.2495	1.2567	1.2643	1.2722	1.2795	1.2875	1.2953
40	1.2257	1.2325	1.2397	1.2466	1.2537	1.2612	1.2682	1.2751	1.2829
50	1.2176	1.2240	1.2309	1.2375	1.2441	1.2511	1.2578	1.2645	1.2713
60	1.2100	1.2161	1.2226	1.2289	1.2354	1.2420	1.2484	1.2547	1.2613
70	1.2030	1.2087	1.2150	1.2211	1.2274	1.2337	1.2398	1.2458	1.2521
80	1.1965	1.2019	1.2078	1.2137	1.2198	1.2259	1.2317	1.2376	1.2435
90	1.1903	1.1954	1.2011	1.2069	1.2128	1.2186	1.2243	1.2298	1.2357
100	1.1843	1.1893	1.1948	1.2004	1.2061	1.2119	1.2172	1.2227	1.2282
110	1.1787	1.1836	1.1889	1.1942	1.1999	1.2052	1.2106	1.2159	1.2213
120	1.1734	1.1781	1.1831	1.1885	1.1939	1.1991	1.2042	1.2094	1.2147
130	1.1683	1.1727	1.1776	1.1829	1.1882	1.1933	1.1982	1.2032	1.2084
140	1.1635	1.1677	1.1726	1.1775	1.1827	1.1877	1.1925	1.1975	1.2025
150	1.1588	1.1629	1.1675	1.1725	1.1775	1.1823	1.1870	1.1918	1.1967
160	1.1543	1.1582	1.1627	1.1675	1.1725	1.1772	1.1818	1.1866	1.1913
170	1.1501	1.1537	1.1581	1.1628	1.1677	1.1722	1.1768	1.1814	1.1860
180	1.1458	1.1493	1.1537	1.1583	1.1631	1.1675	1.1719	1.1765	1.1810
190	1.1418	1.1452	1.1495	1.1539	1.1586	1.1630	1.1672	1.1718	1.1763
200	1.1380	1.1411	1.1455	1.1499	1.1543	1.1586	1.1629	1.1672	1.1717

continue **E-P/61w**

P/MPa	*T*/K 512.82	522.52	532.78	542.86
	V_{spez}/cm^{3}g^{-1}			
0.1	1.3560	1.3663	1.3773	1.3890
10	1.3356	1.3448	1.3543	1.3643
20	1.3179	1.3263	1.3347	1.3435
30	1.3032	1.3110	1.3186	1.3267
40	1.2903	1.2974	1.3046	1.3122
50	1.2787	1.2855	1.2922	1.2993
60	1.2679	1.2747	1.2811	1.2878
70	1.2583	1.2645	1.2709	1.2773
80	1.2494	1.2555	1.2613	1.2677

continue

continue

P/MPa	*T*/K			
	512.82	522.52	532.78	542.86
	V_{spez}/cm^{3}g^{-1}			
90	1.2414	1.2471	1.2527	1.2584
100	1.2337	1.2392	1.2446	1.2501
110	1.2265	1.2319	1.2369	1.2424
120	1.2197	1.2249	1.2299	1.2349
130	1.2133	1.2183	1.2231	1.2280
140	1.2072	1.2121	1.2167	1.2215
150	1.2013	1.2061	1.2106	1.2153
160	1.1958	1.2005	1.2047	1.2092
170	1.1905	1.1950	1.1992	1.2035
180	1.1852	1.1898	1.1938	1.1982
190	1.1804	1.1847	1.1886	1.1930
200	1.1757	1.1800	1.1838	1.1879

copolymer (B): E-P/76w(1) 95ZOL

P/MPa	*T*/K								
	387.65	397.85	408.05	418.45	428.75	438.75	448.65	458.95	469.35
	V_{spez}/cm^{3}g^{-1}								
0.1	1.2496	1.2588	1.2678	1.2772	1.2868	1.2962	1.3058	1.3158	1.3260
20	1.2300	1.2380	1.2459	1.2540	1.2622	1.2702	1.2785	1.2868	1.2956
40	1.2136	1.2207	1.2278	1.2351	1.2422	1.2496	1.2569	1.2643	1.2717
60	1.1993	1.2059	1.2126	1.2192	1.2255	1.2322	1.2389	1.2455	1.2522
80	1.1686	1.1928	1.1989	1.2050	1.2110	1.2175	1.2234	1.2298	1.2357
100	1.1755	1.1812	1.1868	1.1924	1.1980	1.2040	1.2097	1.2155	1.2212
120	1.1655	1.1707	1.1758	1.1812	1.1864	1.1922	1.1975	1.2028	1.2082
140	1.1557	1.1607	1.1659	1.1708	1.1760	1.1811	1.1865	1.1915	1.1965
160	1.1470	1.1519	1.1566	1.1614	1.1661	1.1711	1.1760	1.1810	1.1858
180	1.1389	1.1433	1.1481	1.1524	1.1571	1.1619	1.1666	1.1712	1.1758
200	1.1311	1.1355	1.1401	1.1442	1.1486	1.1534	1.1579	1.1623	1.1668

continue **E-P/76w(1)**

P/MPa	*T*/K								
	479.65	489.55	499.75	509.75	519.85	530.15	550.35	560.25	569.75
	V_{spez}/cm^{3}g^{-1}								
0.1	1.3363	1.3467	1.3570	1.3672	1.3780	1.3899	1.4115	1.4230	1.4342
20	1.3037	1.3121	1.3207	1.3292	1.3380	1.3470	1.3640	1.3731	1.3815
40	1.2791	1.2863	1.2938	1.3012	1.3087	1.3164	1.3307	1.3381	1.3458
60	1.2588	1.2654	1.2721	1.2788	1.2855	1.2920	1.3050	1.3114	1.3181
80	1.2417	1.2477	1.2539	1.2602	1.2660	1.2724	1.2837	1.2895	1.2957
100	1.2267	1.2323	1.2381	1.2439	1.2494	1.2548	1.2656	1.2711	1.2767
120	1.2132	1.2186	1.2240	1.2296	1.2346	1.2399	1.2498	1.2546	1.2600
140	1.2013	1.2064	1.2115	1.2165	1.2214	1.2265	1.2359	1.2401	1.2452
160	1.1904	1.1952	1.1999	1.2048	1.2094	1.2141	1.2227	1.2270	1.2321
180	1.1800	1.1851	1.1895	1.1941	1.1985	1.2028	1.2111	1.2152	1.2198
200	1.1706	1.1752	1.1797	1.1841	1.1882	1.1926	1.2006	1.2042	1.2086

copolymer (B): **E-P/76w(2)** **99MAI**

P/MPa	*T*/K								
	422.46	432.42	442.40	452.55	462.61	472.81	482.66	492.45	502.25
	V_{spez}/cm^{3}g^{-1}								
0.1	1.2641	1.2736	1.2827	1.2923	1.3015	1.3115	1.3224	1.3322	1.3417
10	1.2519	1.2604	1.2689	1.2776	1.2861	1.2950	1.3044	1.3133	1.3221
20	1.2407	1.2484	1.2564	1.2644	1.2723	1.2803	1.2884	1.2971	1.3050
30	1.2308	1.2381	1.2457	1.2531	1.2604	1.2681	1.2756	1.2830	1.2905
40	1.2219	1.2286	1.2356	1.2429	1.2497	1.2570	1.2640	1.2711	1.2780
50	1.2136	1.2200	1.2267	1.2336	1.2401	1.2471	1.2537	1.2605	1.2669
60	1.2062	1.2120	1.2186	1.2251	1.2313	1.2379	1.2442	1.2507	1.2569
70	1.1992	1.2047	1.2109	1.2172	1.2233	1.2295	1.2356	1.2417	1.2476
80	1.1925	1.1978	1.2039	1.2099	1.2158	1.2219	1.2276	1.2335	1.2392
90	1.1863	1.1915	1.1973	1.2029	1.2087	1.2145	1.2203	1.2258	1.2313
100	1.1805	1.1854	1.1910	1.1965	1.2022	1.2077	1.2132	1.2186	1.2239
110	1.1749	1.1796	1.1850	1.1903	1.1959	1.2014	1.2066	1.2119	1.2170
120	1.1695	1.1742	1.1795	1.1846	1.1899	1.1953	1.2004	1.2054	1.2104
130	1.1644	1.1690	1.1741	1.1791	1.1841	1.1895	1.1944	1.1995	1.2043
140	1.1595	1.1639	1.1690	1.1737	1.1790	1.1840	1.1888	1.1937	1.1984
150	1.1549	1.1591	1.1642	1.1686	1.1736	1.1786	1.1834	1.1882	1.1928
160	1.1506	1.1545	1.1595	1.1638	1.1688	1.1735	1.1782	1.1829	1.1873
170	1.1463	1.1501	1.1548	1.1593	1.1640	1.1688	1.1732	1.1778	1.1821
180	1.1420	1.1458	1.1504	1.1547	1.1595	1.1640	1.1685	1.1730	1.1773
190	1.1381	1.1416	1.1464	1.1506	1.1552	1.1595	1.1639	1.1683	1.1725
200	1.1342	1.1376	1.1423	1.1464	1.1509	1.1553	1.1596	1.1639	1.1680

continue **E-P/76w(2)**

P/MPa	*T*/K			
	512.63	522.53	532.52	542.51
	V_{spez}/cm^{3}g^{-1}			
0.1	1.3523	1.3631	1.3735	1.3849
10	1.3315	1.3409	1.3499	1.3597
20	1.3134	1.3217	1.3297	1.3385
30	1.2986	1.3062	1.3134	1.3216
40	1.2851	1.2928	1.2995	1.3069
50	1.2735	1.2802	1.2871	1.2940
60	1.2631	1.2695	1.2756	1.2825
70	1.2537	1.2597	1.2655	1.2721
80	1.2450	1.2508	1.2564	1.2621
90	1.2368	1.2425	1.2478	1.2532
100	1.2293	1.2348	1.2399	1.2452
110	1.2222	1.2274	1.2326	1.2374
120	1.2156	1.2205	1.2255	1.2302
130	1.2092	1.2140	1.2189	1.2235
140	1.2030	1.2078	1.2125	1.2171
150	1.1974	1.2019	1.2065	1.2109
160	1.1918	1.1963	1.2006	1.2052
170	1.1864	1.1909	1.1951	1.1996
180	1.1814	1.1856	1.1899	1.1941
190	1.1767	1.1808	1.1849	1.1889
200	1.1720	1.1760	1.1800	1.1841

copolymer (B): E-P/84w(1) 99MAI

P/MPa	*T*/K								
	422.88	433.00	443.00	453.12	463.34	473.59	483.52	493.17	503.21
	V_{spez}/cm^{3}g^{-1}								
0.1	1.2720	1.2816	1.2909	1.3007	1.3113	1.3209	1.3304	1.3409	1.3509
10	1.2593	1.2681	1.2766	1.2855	1.2950	1.3037	1.3124	1.3217	1.3307
20	1.2479	1.2558	1.2636	1.2719	1.2803	1.2885	1.2966	1.3049	1.3133
30	1.2377	1.2452	1.2525	1.2602	1.2681	1.2753	1.2833	1.2911	1.2987
40	1.2285	1.2355	1.2424	1.2497	1.2571	1.2639	1.2710	1.2785	1.2860
50	1.2200	1.2267	1.2332	1.2401	1.2472	1.2536	1.2603	1.2670	1.2744
60	1.2121	1.2185	1.2247	1.2313	1.2382	1.2444	1.2507	1.2571	1.2636
70	1.2050	1.2109	1.2170	1.2233	1.2298	1.2357	1.2418	1.2480	1.2542
80	1.1982	1.2039	1.2097	1.2158	1.2221	1.2278	1.2337	1.2396	1.2455

continue

continue

P/MPa	*T*/K								
	422.88	433.00	443.00	453.12	463.34	473.59	483.52	493.17	503.21
	V_{spez}/cm^{3}g^{-1}								
90	1.1918	1.1972	1.2029	1.2088	1.2150	1.2204	1.2260	1.2318	1.2375
100	1.1858	1.1910	1.1965	1.2022	1.2081	1.2134	1.2189	1.2245	1.2299
110	1.1801	1.1850	1.1906	1.1959	1.2017	1.2068	1.2121	1.2175	1.2228
120	1.1747	1.1793	1.1847	1.1900	1.1956	1.2005	1.2058	1.2111	1.2161
130	1.1695	1.1740	1.1792	1.1843	1.1898	1.1947	1.1996	1.2048	1.2097
140	1.1645	1.1688	1.1739	1.1790	1.1842	1.1890	1.1939	1.1991	1.2037
150	1.1598	1.1639	1.1689	1.1738	1.1790	1.1837	1.1885	1.1934	1.1978
160	1.1552	1.1591	1.1641	1.1689	1.1739	1.1785	1.1832	1.1881	1.1923
170	1.1508	1.1546	1.1595	1.1642	1.1690	1.1736	1.1782	1.1830	1.1871
180	1.1467	1.1503	1.1549	1.1597	1.1645	1.1687	1.1734	1.1781	1.1820
190	1.1426	1.1461	1.1507	1.1553	1.1601	1.1642	1.1688	1.1734	1.1773
200	1.1387	1.1420	1.1467	1.1511	1.1559	1.1598	1.1644	1.1687	1.1727

continue **E-P/84w(1)**

P/MPa	*T*/K			
	513.39	523.53	533.36	543.24
	V_{spez}/cm^{3}g^{-1}			
0.1	1.3618	1.3721	1.3830	1.3939
10	1.3404	1.3494	1.3588	1.3683
20	1.3218	1.3299	1.3384	1.3468
30	1.3067	1.3140	1.3218	1.3293
40	1.2934	1.3002	1.3074	1.3144
50	1.2816	1.2879	1.2947	1.3013
60	1.2708	1.2771	1.2833	1.2897
70	1.2607	1.2670	1.2731	1.2790
80	1.2517	1.2574	1.2636	1.2693
90	1.2435	1.2489	1.2543	1.2603
100	1.2357	1.2410	1.2462	1.2520
110	1.2285	1.2335	1.2385	1.2437
120	1.2216	1.2265	1.2314	1.2364
130	1.2150	1.2198	1.2246	1.2294
140	1.2089	1.2136	1.2181	1.2229
150	1.2031	1.2076	1.2119	1.2167
160	1.1974	1.2019	1.2062	1.2106
170	1.1920	1.1964	1.2006	1.2051
180	1.1868	1.1912	1.1952	1.1996
190	1.1818	1.1862	1.1902	1.1944
200	1.1771	1.1814	1.1853	1.1894

copolymer (B): E-P/84w(2) 95ZOL

P/MPa	T/K								
	387.65	398.15	408.55	418.65	429.45	440.05	450.75	461.15	471.65
	V_{spez}/cm^{3}g^{-1}								
0.1	1.2480	1.2572	1.2666	1.2757	1.2857	1.2961	1.3063	1.3162	1.3267
20	1.2286	1.2365	1.2448	1.2528	1.2613	1.2704	1.2785	1.2875	1.2961
40	1.2123	1.2193	1.2264	1.2338	1.2414	1.2493	1.2567	1.2644	1.2718
60	1.1980	1.2042	1.2108	1.2179	1.2459	1.2316	1.2383	1.2451	1.2521
80	1.1849	1.1911	1.1972	1.2035	1.2099	1.2165	1.2226	1.2291	1.2353
100	1.1738	1.1795	1.1852	1.1911	1.1971	1.2032	1.2089	1.2148	1.2207
120	1.1634	1.1688	1.1743	1.1799	1.1855	1.1912	1.1965	1.2021	1.2078
140	1.1539	1.1591	1.1643	1.1695	1.1748	1.1803	1.1853	1.1905	1.1958
160	1.1452	1.1501	1.1550	1.1600	1.1654	1.1704	1.1750	1.1801	1.1851
180	1.1371	1.1418	1.1464	1.1515	1.1562	1.1610	1.1655	1.1705	1.1752
200	1.1295	1.1341	1.1387	1.1433	1.1480	1.1524	1.1569	1.1613	1.1659

continue **E-P/84w(2)**

P/MPa	T/K								
	481.45	492.35	502.45	513.25	523.85	534.45	554.55	565.05	575.15
	V_{spez}/cm^{3}g^{-1}								
0.1	1.3365	1.3474	1.3578	1.3700	1.3813	1.3917	1.4149	1.4266	1.4388
20	1.3041	1.3129	1.3216	1.3305	1.3396	1.3486	1.3663	1.3754	1.3847
40	1.2790	1.2868	1.2944	1.3020	1.3097	1.3173	1.3325	1.3399	1.3480
60	1.2586	1.2654	1.2722	1.2790	1.2862	1.2926	1.3060	1.3123	1.3195
80	1.2412	1.2475	1.2538	1.2600	1.2659	1.2723	1.2847	1.2903	1.2964
100	1.2262	1.2321	1.2380	1.2436	1.2491	1.2550	1.2660	1.2715	1.2769
120	1.2129	1.2182	1.2238	1.2290	1.2341	1.2397	1.2499	1.2546	1.2601
140	1.2006	1.2057	1.2112	1.2159	1.2207	1.2258	1.2354	1.2400	1.2449
160	1.1896	1.1945	1.1994	1.2040	1.2088	1.2134	1.2229	1.2268	1.2315
180	1.1795	1.1841	1.1889	1.1934	1.1976	1.2021	1.2108	1.2147	1.2193
200	1.1704	1.1748	1.1792	1.1833	1.1874	1.1918	1.2001	1.2038	1.2081

copolymer (B): **E-P/96w** **99MAI**

P/MPa	*T*/K								
	422.53	432.68	442.81	452.84	463.23	472.84	482.90	492.87	502.57
	V_{spez}/cm^{3}g^{-1}								
0.1	1.2759	1.2855	1.2946	1.3042	1.3154	1.3243	1.3341	1.3441	1.3543
10	1.2619	1.2706	1.2790	1.2877	1.2975	1.3056	1.3147	1.3235	1.3325
20	1.2493	1.2569	1.2647	1.2728	1.2812	1.2893	1.2975	1.3055	1.3135
30	1.2387	1.2461	1.2534	1.2607	1.2688	1.2758	1.2838	1.2912	1.2985
40	1.2292	1.2361	1.2430	1.2500	1.2576	1.2641	1.2714	1.2785	1.2854
50	1.2207	1.2270	1.2339	1.2404	1.2476	1.2538	1.2607	1.2669	1.2741
60	1.2129	1.2189	1.2253	1.2317	1.2383	1.2443	1.2509	1.2568	1.2631
70	1.2056	1.2113	1.2176	1.2237	1.2299	1.2358	1.2421	1.2478	1.2538
80	1.1989	1.2043	1.2104	1.2162	1.2222	1.2279	1.2339	1.2393	1.2452
90	1.1927	1.1977	1.2036	1.2091	1.2149	1.2206	1.2263	1.2317	1.2373
100	1.1866	1.1915	1.1972	1.2027	1.2083	1.2137	1.2193	1.2244	1.2297
110	1.1808	1.1856	1.1911	1.1965	1.2017	1.2072	1.2125	1.2176	1.2227
120	1.1754	1.1800	1.1852	1.1907	1.1957	1.2009	1.2063	1.2111	1.2162
130	1.1702	1.1749	1.1799	1.1852	1.1899	1.1952	1.2003	1.2050	1.2099
140	1.1653	1.1696	1.1747	1.1799	1.1845	1.1896	1.1945	1.1991	1.2040
150	1.1606	1.1647	1.1697	1.1747	1.1791	1.1843	1.1891	1.1936	1.1982
160	1.1561	1.1601	1.1649	1.1698	1.1742	1.1792	1.1838	1.1883	1.1928
170	1.1517	1.1555	1.1601	1.1651	1.1692	1.1743	1.1787	1.1832	1.1877
180	1.1475	1.1513	1.1557	1.1607	1.1647	1.1696	1.1739	1.1783	1.1826
190	1.1435	1.1472	1.1515	1.1564	1.1601	1.1651	1.1693	1.1736	1.1779
200	1.1395	1.1431	1.1474	1.1522	1.1559	1.1609	1.1650	1.1691	1.1733

continue **E-P/96w**

P/MPa	*T*/K		
	512.93	522.96	532.67
	V_{spez}/cm^{3}g^{-1}		
0.1	1.3651	1.3756	1.3864
10	1.3419	1.3510	1.3604
20	1.3219	1.3300	1.3384
30	1.3064	1.3137	1.3213
40	1.2927	1.2996	1.3067
50	1.2808	1.2873	1.2940
60	1.2701	1.2762	1.2827
70	1.2599	1.2662	1.2722
80	1.2510	1.2564	1.2628

continue

continue

P/MPa	T/K 512.93	522.96	532.67
	V_{spez}/cm^{3}g^{-1}		
90	1.2429	1.2481	1.2536
100	1.2352	1.2402	1.2457
110	1.2280	1.2328	1.2380
120	1.2213	1.2259	1.2309
130	1.2148	1.2193	1.2241
140	1.2088	1.2133	1.2178
150	1.2031	1.2073	1.2119
160	1.1975	1.2017	1.2061
170	1.1922	1.1962	1.2006
180	1.1870	1.1910	1.1952
190	1.1823	1.1860	1.1903
200	1.1775	1.1813	1.1854

copolymer (B): E-P/? 73RAO

P/atm	T/K 413.15	448.15	483.15	523.15
	V_{spez}/cm^{3}g^{-1}			
1	1.2758	1.2992	1.3281	1.3663
79	1.2722	1.2936	1.3197	1.3512
159	1.2655	1.2855	1.3099	1.3389
232	1.2594	1.2775	1.3008	1.3279
316	1.2506	1.2677	1.2902	1.3153
474	1.2353	1.2531	1.2737	1.2953
618	1.2258	1.2433	1.2632	1.2847

Tait equation parameter functions:

Copolymer $V(P/\text{MPa}, T/\text{K}) = V(0, T/\text{K})\{1 - C^*\ln[1 + (P/\text{MPa})/B(T/\text{K})]\}$
with $C = 0.0894$ and $\theta = T/\text{K} - 273.15$

	$V(0,T/\text{K})$/cm^{3}g^{-1}	B(T/K)/MPa
E-P/50w	$1.2291 + 5.799\ 10^{-5}\theta + 1.964\ 10^{-6}\theta^2$	$487.0\ \exp(-8.103\ 10^{-3}\theta)$

Copolymers from ethylene and styrene

Average chemical composition of the copolymers, acronyms, range of data, and references:

Copolymer	Acronym	Range of data T/K	P/MPa	Ref.
Ethylene/styrene copolymer				
1.7 mol% styrene	E-S/02x	403-473	0.1-170	96XUY
3.5 mol% styrene	E-S/04x	403-473	0.1-200	96XUY
6.0 mol% styrene	E-S/06x	403-473	0.1-200	96XUY
13.8 mol% styrene	E-S/14x	403-473	0.1-200	96XUY
21.5 mol% styrene	E-S/22x	353-473	0.1-200	96XUY
29.4 mol% styrene	E-S/29x	343-473	0.1-200	96XUY

Characterization:

Copolymer (B)	M_n/ g/mol	M_w/ g/mol	M_η/ g/mol	Further information
E-S/02x	175000	455000		synthesized in the laboratory
E-S/04x	77000	177000		synthesized in the laboratory
E-S/06x	168000	437000		synthesized in the laboratory
E-S/14x	39000	97500		synthesized in the laboratory
E-S/22x	53000	111300		synthesized in the laboratory
E-S/29x	19000	47500		synthesized in the laboratory

Experimental *PVT* data:

copolymer (B): **E-S/02x** **96XUY**

P/MPa	T/K							
	403.15	413.15	423.15	433.15	443.15	453.15	463.15	473.15
	V_{spez}/cm^{3}g^{-1}							
0.1	1.2103	1.2188	1.2272	1.2363	1.2451	1.2546	1.2653	1.2752
10	1.2012	1.2092	1.2171	1.2256	1.2336	1.2423	1.2514	1.2604
20		1.2001	1.2075	1.2153	1.2228	1.2307	1.2383	1.2471
30		1.1919	1.1989	1.2065	1.2133	1.2207	1.2280	1.2353
40		1.1844	1.1912	1.1983	1.2048	1.2118	1.2188	1.2251
50		1.1776	1.1840	1.1908	1.1970	1.2036	1.2102	1.2160
60		1.1712	1.1772	1.1837	1.1898	1.1961	1.2023	1.2076
70			1.1708	1.1771	1.1830	1.1887	1.1949	1.1999
80			1.1649	1.1709	1.1766	1.1821	1.1879	1.1927
90			1.1590	1.1648	1.1705	1.1756	1.1813	1.1859
100			1.1534	1.1591	1.1646	1.1695	1.1751	1.1794
110				1.1535	1.1591	1.1638	1.1691	1.1734
120				1.1482	1.1536	1.1583	1.1635	1.1675
130				1.1433	1.1486	1.1531	1.1582	1.1619
140				1.1385	1.1436	1.1480	1.1528	1.1566
150					1.1390	1.1432	1.1479	1.1516
160					1.1348	1.1387	1.1432	1.1467
170					1.1305	1.1343	1.1388	1.1422

copolymer (B): **E-S/04x** **96XUY**

P/MPa	T/K							
	403.15	413.15	423.15	433.15	443.15	453.15	463.15	473.15
	V_{spez}/cm^{3}g^{-1}							
0.1	1.1956	1.2041	1.2122	1.2216	1.2298	1.2394	1.2489	1.2628
10	1.1864	1.1941	1.2021	1.2106	1.2183	1.2268	1.2355	1.2473
20	1.1775	1.1847	1.1923	1.2001	1.2076	1.2151	1.2232	1.2335
30	1.1697	1.1765	1.1837	1.1911	1.1981	1.2052	1.2125	1.2210
40	1.1625	1.1690	1.1760	1.1831	1.1896	1.1961	1.2033	1.2107
50	1.1560	1.1620	1.1688	1.1755	1.1817	1.1880	1.1946	1.2011
60	1.1498	1.1557	1.1619	1.1685	1.1744	1.1804	1.1866	1.1925
70	1.1437	1.1495	1.1556	1.1620	1.1676	1.1732	1.1795	1.1844
80		1.1437	1.1494	1.1557	1.1609	1.1666	1.1723	1.1770

continue

continue

P/MPa	*T*/K 403.15	413.15	423.15	433.15	443.15	453.15	463.15	473.15
	V_{spez}/cm^{3}g^{-1}							
90		1.1382	1.1436	1.1498	1.1547	1.1600	1.1658	1.1700
100		1.1329	1.1383	1.1443	1.1488	1.1540	1.1594	1.1633
110		1.1278	1.1328	1.1389	1.1431	1.1483	1.1534	1.1570
120		1.1229	1.1277	1.1336	1.1377	1.1427	1.1477	1.1512
130			1.1228	1.1286	1.1326	1.1375	1.1423	1.1457
140			1.1180	1.1239	1.1278	1.1325	1.1372	1.1403
150			1.1136	1.1192	1.1229	1.1277	1.1323	1.1351
160				1.1151	1.1184	1.1231	1.1276	1.1303
170				1.1108	1.1141	1.1186	1.1230	1.1256
180				1.1070	1.1102	1.1146	1.1187	1.1211
190					1.1063	1.1105	1.1146	1.1170
200					1.1027	1.1068	1.1108	1.1131

copolymer (B): E-S/06x 96XUY

P/MPa	*T*/K 403.15	413.15	423.15	433.15	443.15	453.15	463.15	473.15
	V_{spez}/cm^{3}g^{-1}							
0.1	1.1696	1.1771	1.1849	1.1929	1.2008	1.2088	1.2173	1.2272
10	1.1611	1.1682	1.1756	1.1829	1.1901	1.1974	1.2052	1.2139
20	1.1529	1.1596	1.1664	1.1734	1.1801	1.1866	1.1938	1.2021
30	1.1456	1.1522	1.1587	1.1654	1.1715	1.1776	1.1847	1.1918
40	1.1393	1.1457	1.1517	1.1581	1.1638	1.1696	1.1761	1.1827
50	1.1334	1.1393	1.1453	1.1515	1.1568	1.1623	1.1683	1.1743
60	1.1276	1.1335	1.1393	1.1452	1.1503	1.1556	1.1613	1.1667
70	1.1224	1.1281	1.1334	1.1391	1.1443	1.1492	1.1545	1.1596
80	1.1173	1.1229	1.1281	1.1334	1.1382	1.1431	1.1484	1.1530
90	1.1124	1.1178	1.1227	1.1281	1.1326	1.1373	1.1424	1.1466
100	1.1076	1.1131	1.1176	1.1230	1.1274	1.1318	1.1367	1.1410
110		1.1082	1.1129	1.1179	1.1221	1.1266	1.1315	1.1353
120		1.1039	1.1082	1.1132	1.1174	1.1214	1.1263	1.1300
130		1.0997	1.1037	1.1087	1.1128	1.1166	1.1214	1.1249
140		1.0955	1.0996	1.1045	1.1083	1.1120	1.1166	1.1201
150		1.0915	1.0955	1.1003	1.1040	1.1076	1.1122	1.1155
160			1.0914	1.0963	1.0998	1.1034	1.1078	1.1111
170			1.0878	1.0923	1.0959	1.0994	1.1036	1.1069
180			1.0841	1.0888	1.0921	1.0955	1.0997	1.1028
190			1.0808	1.0853	1.0888	1.0920	1.0960	1.0992
200			1.0778	1.0821	1.0852	1.0887	1.0926	1.0956

copolymer (B): **E-S/14x** **96XUY**

P/MPa	T/K 403.15	413.15	423.15	433.15	443.15	453.15	463.15	473.15
	V_{spez}/cm^{3}g^{-1}							
0.1	1.1220	1.1294	1.1368	1.1449	1.1526	1.1606	1.1690	1.1787
10	1.1139	1.1208	1.1279	1.1353	1.1427	1.1501	1.1576	1.1661
20	1.1062	1.1126	1.1192	1.1261	1.1331	1.1401	1.1470	1.1543
30	1.0992	1.1053	1.1117	1.1183	1.1251	1.1316	1.1380	1.1449
40	1.0931	1.0991	1.1051	1.1112	1.1178	1.1239	1.1301	1.1364
50	1.0874	1.0931	1.0988	1.1047	1.1111	1.1169	1.1227	1.1287
60	1.0820	1.0875	1.0931	1.0986	1.1047	1.1103	1.1159	1.1215
70	1.0769	1.0821	1.0874	1.0931	1.0988	1.1041	1.1096	1.1148
80	1.0720	1.0771	1.0822	1.0876	1.0932	1.0983	1.1035	1.1086
90	1.0673	1.0723	1.0773	1.0824	1.0878	1.0927	1.0978	1.1026
100	1.0628	1.0676	1.0725	1.0773	1.0826	1.0873	1.0923	1.0969
110	1.0584	1.0631	1.0679	1.0725	1.0777	1.0823	1.0872	1.0916
120	1.0543	1.0589	1.0635	1.0680	1.0730	1.0774	1.0823	1.0865
130	1.0503	1.0548	1.0593	1.0636	1.0685	1.0728	1.0775	1.0816
140	1.0464	1.0507	1.0551	1.0595	1.0642	1.0684	1.0730	1.0769
150	1.0428	1.0471	1.0513	1.0554	1.0601	1.0641	1.0686	1.0725
160	1.0391	1.0433	1.0475	1.0516	1.0560	1.0601	1.0644	1.0681
170	1.0356	1.0398	1.0439	1.0479	1.0523	1.0561	1.0604	1.0641
180	1.0324	1.0364	1.0405	1.0444	1.0486	1.0525	1.0566	1.0602
190	1.0293	1.0333	1.0372	1.0410	1.0451	1.0489	1.0530	1.0565
200	1.0262	1.0302	1.0341	1.0379	1.0418	1.0456	1.0496	1.0531

copolymer (B): **E-S/22x** **96XUY**

P/MPa	T/K 353.15	363.15	373.15	383.15	393.15	403.15	413.15	423.15	433.15
	V_{spez}/cm^{3}g^{-1}								
0.1	1.0512	1.0573	1.0636	1.0700	1.0766	1.0832	1.0903	1.0973	1.1051
10	1.0458	1.0516	1.0576	1.0637	1.0699	1.0761	1.0828	1.0573	1.0626
20	1.0404	1.0459	1.0515	1.0574	1.0633	1.0692	1.0756	1.0895	1.0966
30	1.0353	1.0406	1.0461	1.0517	1.0574	1.0630	1.0690	1.0817	1.0883
40	1.0306	1.0358	1.0409	1.0464	1.0519	1.0572	1.0629	1.0749	1.0812
50	1.0261	1.0312	1.0362	1.0415	1.0466	1.0517	1.0574	1.0685	1.0746
60	1.0219	1.0267	1.0316	1.0367	1.0418	1.0467	1.0521	1.0627	1.0684
70	1.0178	1.0226	1.0271	1.0322	1.0371	1.0418	1.0471	1.0520	1.0571
80	1.0139	1.0184	1.0230	1.0279	1.0326	1.0372	1.0423	1.0471	1.0521

continue

continue

P/MPa	*T*/K								
	353.15	363.15	373.15	383.15	393.15	403.15	413.15	423.15	433.15
	V_{spez}/cm^3g^{-1}								
90	1.0102	1.0145	1.0190	1.0238	1.0283	1.0329	1.0377	1.0424	1.0471
100	1.0066	1.0109	1.0151	1.0198	1.0241	1.0287	1.0332	1.0378	1.0423
110	1.0031	1.0072	1.0114	1.0159	1.0200	1.0245	1.0290	1.0335	1.0380
120	0.9997	1.0036	1.0077	1.0122	1.0162	1.0205	1.0249	1.0293	1.0335
130	0.9964	1.0003	1.0042	1.0086	1.0125	1.0167	1.0209	1.0254	1.0293
140	0.9932	0.9970	1.0009	1.0051	1.0089	1.0130	1.0171	1.0214	1.0253
150	0.9901	0.9939	0.9976	1.0018	1.0055	1.0096	1.0136	1.0177	1.0215
160	0.9870	0.9909	0.9944	0.9987	1.0023	1.0063	1.0101	1.0141	1.0178
170	0.9841	0.9879	0.9914	0.9956	0.9991	1.0030	1.0066	1.0107	1.0144
180	0.9812	0.9851	0.9885	0.9927	0.9961	0.9999	1.0036	1.0074	1.0110
190	0.9784	0.9824	0.9857	0.9898	0.9932	0.9969	1.0005	1.0043	1.0078
200	0.9759	0.9798	0.9831	0.9871	0.9904	0.9940	0.9976	1.0014	1.0047

continue **E-S/22x**

P/MPa	*T*/K			
	443.15	453.15	463.15	473.15
	V_{spez}/cm^3g^{-1}			
0.1	1.1126	1.1201	1.1274	1.1348
10	1.1037	1.1106	1.1175	1.1243
20	1.0951	1.1015	1.1082	1.1144
30	1.0876	1.0936	1.0999	1.1058
40	1.0808	1.0864	1.0923	1.0981
50	1.0744	1.0797	1.0854	1.0909
60	1.0684	1.0736	1.0790	1.0843
70	1.0627	1.0677	1.0730	1.0782
80	1.0574	1.0622	1.0672	1.0721
90	1.0522	1.0569	1.0618	1.0666
100	1.0473	1.0518	1.0566	1.0613
110	1.0427	1.0470	1.0517	1.0563
120	1.0382	1.0425	1.0470	1.0514
130	1.0338	1.0380	1.0426	1.0468
140	1.0298	1.0339	1.0383	1.0424
150	1.0258	1.0298	1.0341	1.0381
160	1.0221	1.0259	1.0302	1.0340
170	1.0184	1.0221	1.0262	1.0301
180	1.0149	1.0185	1.0226	1.0264
190	1.0117	1.0151	1.0191	1.0229
200	1.0086	1.0119	1.0159	1.0196

copolymer (B): **E-S/29x** **96XUY**

P/MPa	343.15	353.15	363.15	373.15	383.15	393.15	403.15	413.15	423.15
	T/K; V_{spez}/cm^{3}g^{-1}								
0.1	1.0319	1.0378	1.0437	1.0499	1.0559	1.0624	1.0691	1.0760	1.0829
10	1.0268	1.0324	1.0382	1.0440	1.0498	1.0559	1.0623	1.0687	1.0753
20	1.0218	1.0270	1.0327	1.0381	1.0435	1.0493	1.0556	1.0616	1.0677
30	1.0172	1.0221	1.0277	1.0328	1.0382	1.0436	1.0497	1.0554	1.0614
40	1.0128	1.0176	1.0229	1.0277	1.0332	1.0382	1.0441	1.0496	1.0551
50	1.0087	1.0132	1.0184	1.0231	1.0283	1.0333	1.0387	1.0442	1.0497
60	1.0047	1.0092	1.0142	1.0187	1.0238	1.0284	1.0338	1.0391	1.0443
70	1.0007	1.0052	1.0099	1.0145	1.0193	1.0238	1.0290	1.0341	1.0391
80	0.9969	1.0015	1.0060	1.0103	1.0151	1.0193	1.0245	1.0294	1.0343
90	0.9934	0.9977	1.0022	1.0063	1.0110	1.0151	1.0201	1.0250	1.0297
100	0.9898	0.9940	0.9984	1.0024	1.0068	1.0110	1.0157	1.0206	1.0251
110	0.9864	0.9904	0.9950	0.9987	1.0032	1.0069	1.0117	1.0164	1.0208
120	0.9830	0.9871	0.9915	0.9950	0.9993	1.0031	1.0076	1.0123	1.0168
130	0.9798	0.9837	0.9881	0.9916	0.9958	0.9995	1.0039	1.0084	1.0127
140	0.9767	0.9806	0.9847	0.9881	0.9923	0.9958	1.0002	1.0047	1.0090
150	0.9737	0.9774	0.9816	0.9850	0.9889	0.9923	0.9965	1.0011	1.0051
160	0.9708	0.9745	0.9785	0.9818	0.9858	0.9889	0.9933	0.9977	1.0015
170	0.9680	0.9717	0.9757	0.9787	0.9827	0.9858	0.9899	0.9944	0.9981
180	0.9654	0.9689	0.9730	0.9759	0.9797	0.9829	0.9869	0.9911	0.9949
190	0.9628	0.9663	0.9702	0.9731	0.9769	0.9799	0.9839	0.9881	0.9918
200	0.9604	0.9638	0.9676	0.9705	0.9742	0.9772	0.9812	0.9853	0.9888

continue **E-S/29x**

P/MPa	433.15	443.15	453.15	463.15	473.15
	T/K; V_{spez}/cm^{3}g^{-1}				
0.1	1.0899	1.0972	1.1041	1.1126	1.1225
10	1.0819	1.0885	1.0952	1.1026	1.1104
20	1.0740	1.0802	1.0866	1.0930	1.0995
30	1.0672	1.0729	1.0792	1.0851	1.0898
40	1.0609	1.0665	1.0722	1.0779	1.0811
50	1.0551	1.0604	1.0658	1.0714	1.0732
60	1.0497	1.0547	1.0599	1.0652	1.0663
70	1.0443	1.0492	1.0542	1.0593	1.0598
80	1.0392	1.0441	1.0487	1.0537	1.0540

continue

continue

P/MPa	T/K				
	433.15	443.15	453.15	463.15	473.15
	V_{spez}/cm^{3}g^{-1}				
90	1.0345	1.0392	1.0435	1.0484	1.0485
100	1.0298	1.0345	1.0385	1.0431	1.0434
110	1.0253	1.0299	1.0338	1.0383	1.0385
120	1.0210	1.0254	1.0293	1.0329	1.0366
130	1.0168	1.0212	1.0250	1.0285	1.0320
140	1.0128	1.0171	1.0208	1.0248	1.0251
150	1.0092	1.0131	1.0169	1.0205	1.0241
160	1.0053	1.0095	1.0130	1.0160	1.0191
170	1.0019	1.0058	1.0094	1.0126	1.0158
180	0.9986	1.0025	1.0058	1.0086	1.0113
190	0.9954	0.9992	1.0025	1.0055	1.0084
200	0.9923	0.9962	0.9994	1.0021	1.0049

Tait equation parameter functions:

Copolymer	$V(P/\text{MPa}, T/\text{K}) = V(0, T/\text{K})\{1 - C^*\ln[1 + (P/\text{MPa})/B(T/\text{K})]\}$ with C = 0.0894 and $\theta = T/\text{K} - 273.15$	
	$V(0,T/\text{K})$/cm^{3}g^{-1}	B(T/K)/MPa
E-S/02x	$1.17640 - 6.3389\ 10^{-4}T + 1.7815\ 10^{-6}T^2$	$278.02\ \exp(-6.449\ 10^{-3}\theta)$
E-S/04x	$1.36913 - 1.5928\ 10^{-3}T + 2.8859\ 10^{-6}T^2$	$281.50\ \exp(-6.694\ 10^{-3}\theta)$
E-S/06x	$1.00930 + 4.2123\ 10^{-5}T + 8.8038\ 10^{-7}T^2$	$258.51\ \exp(-5.502\ 10^{-3}\theta)$
E-S/14x	$0.98570 - 5.9468\ 10^{-5}T + 9.8460\ 10^{-7}T^2$	$240.72\ \exp(-5.059\ 10^{-3}\theta)$
E-S/22x	$0.93220 + 6.4243\ 10^{-5}T + 7.7144\ 10^{-7}T^2$	$241.50\ \exp(-4.732\ 10^{-3}\theta)$
E-S/22x	$0.96947 - 1.8184\ 10^{-4}T + 1.0626\ 10^{-6}T^2$	$250.23\ \exp(-4.948\ 10^{-3}\theta)$

Comments: Tait equation parameters - this work

Copolymers from ethylene and vinyl acetate

Average chemical composition of the copolymers, acronyms, range of data, and references:

Copolymer	Acronym	Range of data T/K	P/MPa	Ref.
Ethylene/vinyl acetate copolymer				
9 wt% vinyl acetate	E-VA/09w	393-473	10-80	90BU1
14 wt% vinyl acetate	E-VA/14w	393-535	0.1-200	95ZOL
18 wt% vinyl acetate	E-VA/18w(1)	385-517	0.1-177	86ZOL
18 wt% vinyl acetate	E-VA/18w(2)	393-523	0.1-200	95ZOL
20 wt% vinyl acetate	E-VA/20w	353-422	5-40	90BU1
25 wt% vinyl acetate	E-VA/25w(1)	367-515	0.1-177	86ZOL
25 wt% vinyl acetate	E-VA/25w(2)	394-532	0.1-200	95ZOL
28 wt% vinyl acetate	E-VA/28w(B)	373-473	10-80	90BU1
28 wt% vinyl acetate	E-VA/28w(B)	373-473	10-80	90BU2
28 wt% vinyl acetate	E-VA/28w(Z1)	367-517	0.1-177	86ZOL
28 wt% vinyl acetate	E-VA/28w(Z2)	377-519	0.1-200	95ZOL
33 wt% vinyl acetate	E-VA/33w	385-526	0.1-200	95ZOL
40 wt% vinyl acetate	E-VA/40w(1)	348-517	0.1-177	86ZOL
40 wt% vinyl acetate	E-VA/40w(2)	368-497	0.1-200	95ZOL
65 wt% vinyl acetate	E-VA/65w	343-502	0.1-200	95ZOL
87.1 wt% vinyl acetate	E-VA/87w	303-443	0.1	90CIM

Characterization:

Copolymer (B)	M_n/ g/mol	M_w/ g/mol	M_η/ g/mol	Further information
E-VA/09w		6500		EVA-1-wax, BASF, Ger.
E-VA/14w				Sci. Polym. Products
E-VA/18w(1)				Sci. Polym. Products, catalog #244
E-VA/18w(2)				Sci. Polym. Products
E-VA/20w				C-Elvax 210, DSM, Netherlands
E-VA/25w(1)				Sci. Polym. Products, catalog #245
E-VA/25w(2)				Sci. Polym. Products
E-VA/28w(B)		6000		Dodiflow 2846-1, Hoechst, Ger.
E-VA/28w(Z1)				Sci. Polym. Products, catalog #316
E-VA/28w(Z2)				Sci. Polym. Products
E-VA/33w				Sci. Polym. Products

continue

continue

Copolymer (B)	M_n/ g/mol	M_w/ g/mol	M_η/ g/mol	Further information
E-VA/40w(1)				Sci. Polym. Products, catalog #326
E-VA/40w(2)				Sci. Polym. Products
E-VA/65w				no data
E-VA/87w	67000	200000		Kuraray Co. (Japan)

Experimental *PVT* data:

copolymer (B): **E-VA/09w** **90BU1**

P/MPa	*T*/K 393.15	403.15	413.15	423.15	433.15	443.15	453.15	463.15	473.15
	V_{spez}/cm^{3}g^{-1}								
10	1.2201	1.2283	1.2374	1.2474	1.2592	1.2691	1.2785	1.2896	1.2984
20	1.2100	1.2175	1.2255	1.2354	1.2462	1.2560	1.2654	1.2751	1.2824
30	1.2014	1.2080	1.2155	1.2250	1.2354	1.2447	1.2534	1.2627	1.2688
40	1.1935	1.2002	1.2061	1.2151	1.2257	1.2347	1.2426	1.2518	1.2565
50	1.1853	1.1923	1.1979	1.2069	1.2165	1.2253	1.2341	1.2409	1.2467
60	1.1782	1.1849	1.1905	1.1995	1.2078	1.2164	1.2248	1.2310	1.2368
70	1.1722	1.1784	1.1844	1.1925	1.2007	1.2094	1.2166	1.2225	1.2289
80	1.1667	1.1729	1.1781	1.1864	1.1943	1.2028	1.2105	1.2155	1.2218

copolymer (B): **E-VA/20w** **90BU1**

P/MPa	*T*/K 353.9	363.5	373.8	383.3	393.0	402.7	412.3	421.9
	V_{spez}/cm^{3}g^{-1}							
5	1.1273	1.1338	1.1408	1.1472	1.1538	1.1604	1.1669	1.1734
10	1.1229	1.1292	1.1359	1.1422	1.1485	1.1549	1.1612	1.1675
15	1.1184	1.1245	1.1311	1.1371	1.1433	1.1494	1.1555	1.1616
20	1.1140	1.1199	1.1262	1.1320	1.1380	1.1439	1.1498	1.1557
25	1.1096	1.1153	1.1214	1.1270	1.1327	1.1385	1.1441	1.1498
30	1.1051	1.1106	1.1165	1.1229	1.1275	1.1330	1.1385	1.1439
35	1.1007	1.1059	1.1116	–	1.1222	1.1275	1.1328	1.1381
40	1.0963	1.1013	1.1068	1.1118	1.1169	1.1220	1.1271	1.1322

copolymer (B): **E-VA/14w** **95ZOL**

P/MPa	*T*/K								
	393.75	401.75	410.05	418.65	426.35	435.15	442.95	451.95	459.15
	V_{spez}/cm^{3}g^{-1}								
0.1	1.2055	1.2128	1.2199	1.2278	1.2351	1.2436	1.2507	1.2589	1.2664
20	1.1877	1.1939	1.2003	1.2071	1.2138	1.2208	1.2270	1.2339	1.2403
40	1.1724	1.1779	1.1837	1.1899	1.1959	1.2021	1.2079	1.2140	1.2199
60	1.1586	1.1640	1.1693	1.1749	1.1807	1.1861	1.1915	1.1968	1.2026
80	1.1466	1.1516	1.1565	1.1617	1.1671	1.1721	1.1773	1.1821	1.1875
100	1.1360	1.1407	1.1452	1.1501	1.1552	1.1599	1.1647	1.1694	1.1744
120	1.1261	1.1305	1.1350	1.1395	1.1446	1.1488	1.1533	1.1575	1.1625
140	1.1169	1.1214	1.1257	1.1300	1.1344	1.1384	1.1430	1.1467	1.1516
160	1.1086	1.1126	1.1166	1.1209	1.1252	1.1290	1.1332	1.1368	1.1416
180	1.1007	1.1045	1.1084	1.1124	1.1167	1.1203	1.1242	1.1277	1.1323
200	1.0931	1.0972	1.1008	1.1046	1.1089	1.1121	1.1159	1.1193	1.1238

continue **E-VA/14w**

P/MPa	*T*/K								
	468.65	477.05	484.95	492.65	501.85	510.25	518.75	527.15	535.25
	V_{spez}/cm^{3}g^{-1}								
0.1	1.2749	1.2833	1.2913	1.2999	1.3089	1.3175	1.3272	1.3361	1.3461
20	1.2476	1.2548	1.2615	1.2682	1.2761	1.2831	1.2906	1.2984	1.3067
40	1.2260	1.2323	1.2386	1.2444	1.2507	1.2570	1.2636	1.2703	1.2775
60	1.2078	1.2135	1.2192	1.2247	1.2302	1.2362	1.2417	1.2477	1.2546
80	1.1924	1.1975	1.2029	1.2079	1.2129	1.2182	1.2235	1.2291	1.2349
100	1.1786	1.1835	1.1886	1.1935	1.1979	1.2027	1.2077	1.2127	1.2184
120	1.1665	1.1709	1.1756	1.1803	1.1845	1.1890	1.1936	1.1983	1.2036
140	1.1552	1.1596	1.1638	1.1683	1.1723	1.1764	1.1808	1.1853	1.1902
160	1.1449	1.1490	1.1533	1.1575	1.1611	1.1652	1.1693	1.1735	1.1782
180	1.1356	1.1394	1.1435	1.1475	1.1511	1.1548	1.1589	1.1628	1.1673
200	1.1266	1.1306	1.1345	1.1385	1.1415	1.1453	1.1490	1.1528	1.1575

copolymer (B): E-VA/18w(2) 95ZOL

P/MPa	T/K								
	393.75	402.05	409.45	417.35	426.15	434.15	441.65	449.95	457.65
	V_{spez}/cm^{3}g^{-1}								
0.1	1.2023	1.2091	1.2163	1.2232	1.2313	1.2394	1.2459	1.2538	1.2618
20	1.1844	1.1906	1.1968	1.2033	1.2104	1.2174	1.2233	1.2300	1.2368
40	1.1691	1.1748	1.1806	1.1868	1.1928	1.1993	1.2046	1.2106	1.2166
60	1.1556	1.1611	1.1664	1.1720	1.1776	1.1835	1.1885	1.1942	1.1994
80	1.1436	1.1484	1.1537	1.1589	1.1641	1.1693	1.1744	1.1793	1.1844
100	1.1331	1.1376	1.1427	1.1478	1.1525	1.1575	1.1620	1.1667	1.1714
120	1.1233	1.1276	1.1325	1.1371	1.1417	1.1464	1.1510	1.1551	1.1596
140	1.1142	1.1183	1.1232	1.1275	1.1317	1.1361	1.1403	1.1446	1.1488
160	1.1058	1.1098	1.1142	1.1187	1.1226	1.1269	1.1308	1.1347	1.1389
180	1.0980	1.1018	1.1061	1.1103	1.1142	1.1181	1.1220	1.1257	1.1296
200	1.0905	1.0943	1.0985	1.1025	1.1062	1.1099	1.1137	1.1173	1.1210

continue **E-VA/18w(2)**

P/MPa	T/K							
	466.45	474.15	483.85	491.25	499.85	507.95	515.45	523.85
	V_{spez}/cm^{3}g^{-1}							
0.1	1.2699	1.2782	1.2882	1.2943	1.3035	1.3120	1.3196	1.3288
20	1.2437	1.2503	1.2579	1.2641	1.2715	1.2786	1.2851	1.2926
40	1.2225	1.2285	1.2349	1.2406	1.2470	1.2531	1.2587	1.2651
60	1.2047	1.2101	1.2158	1.2211	1.2267	1.2325	1.2372	1.2431
80	1.1893	1.1944	1.1993	1.2043	1.2094	1.2145	1.2192	1.2243
100	1.1758	1.1808	1.1853	1.1900	1.1946	1.1993	1.2036	1.2087
120	1.1638	1.1684	1.1724	1.1771	1.1814	1.1857	1.1897	1.1942
140	1.1526	1.1570	1.1609	1.1651	1.1691	1.1732	1.1771	1.1815
160	1.1425	1.1467	1.1504	1.1543	1.1581	1.1623	1.1657	1.1696
180	1.1331	1.1372	1.1404	1.1446	1.1480	1.1521	1.1552	1.1590
200	1.1246	1.1283	1.1313	1.1351	1.1386	1.1422	1.1458	1.1494

copolymer (B): E-VA/18w(1) 86ZOL

P/kg cm^{-2}	T/K 385.55	393.85	403.15	412.55	421.45	430.75	439.15	447.85	457.65
	V_{spez}/cm^{3}g^{-1}								
1	1.2063	1.2142	1.2204	1.2286	1.2372	1.2440	1.2519	1.2578	1.2674
100	1.1977	1.2045	1.2114	1.2189	1.2263	1.2329	1.2398	1.2464	1.2548
200	1.1898	1.1953	1.2024	1.2094	1.2160	1.2224	1.2293	1.2354	1.2429
300	1.1818	1.1870	1.1939	1.2006	1.2071	1.2131	1.2194	1.2253	1.2328
400	1.1746	1.1797	1.1863	1.1926	1.1985	1.2046	1.2105	1.2163	1.2231
500	1.1676	1.1727	1.1786	1.1848	1.1907	1.1965	1.2023	1.2075	1.2142
600	1.1611	1.1662	1.1719	1.1778	1.1833	1.1889	1.1944	1.1996	1.2058
700	1.1554	1.1599	1.1655	1.1716	1.1767	1.1820	1.1872	1.1924	1.1984
800	1.1494	1.1540	1.1592	1.1653	1.1702	1.1751	1.1803	1.1848	1.1913
900	1.1441	1.1485	1.1536	1.1592	1.1638	1.1689	1.1740	1.1783	1.1845
1000	1.1388	1.1432	1.1481	1.1535	1.1580	1.1630	1.1679	1.1722	1.1783
1100	1.1340	1.1380	1.1431	1.1482	1.1526	1.1573	1.1618	1.1666	1.1721
1200	1.1292	1.1332	1.1379	1.1432	1.1474	1.1520	1.1562	1.1609	1.1665
1300	1.1246	1.1285	1.1332	1.1381	1.1425	1.1469	1.1512	1.1557	1.1611
1400	1.1202	1.1242	1.1285	1.1336	1.1375	1.1419	1.1464	1.1505	1.1557
1500	1.1161	1.1196	1.1242	1.1290	1.1329	1.1372	1.1416	1.1457	1.1506
1600	1.1122	1.1155	1.1201	1.1243	1.1287	1.1326	1.1369	1.1410	1.1456
1700	1.1082	1.1119	1.1159	1.1207	1.1243	1.1286	1.1325	1.1362	1.1414
1800	1.1042	1.1078	1.1120	1.1165	1.1203	1.1240	1.1285	1.1322	1.1370

continue **E-VA/18w(1)**

P/kg cm^{-2}	T/K 466.25	475.35	483.35	492.45	500.75	508.85	517.15
	V_{spez}/cm^{3}g^{-1}						
1	1.2748	1.2819	1.2893	1.2977	1.3083	1.3132	1.3208
100	1.2617	1.2684	1.2761	1.2831	1.2905	1.2972	1.3048
200	1.2496	1.2558	1.2624	1.2698	1.2759	1.2824	1.2900
300	1.2383	1.2447	1.2510	1.2580	1.2634	1.2694	1.2762
400	1.2290	1.2345	1.2405	1.2471	1.2519	1.2581	1.2646
500	1.2198	1.2250	1.2302	1.2371	1.2420	1.2474	1.2533
600	1.2115	1.2163	1.2218	1.2275	1.2326	1.2378	1.2436
700	1.2035	1.2082	1.2136	1.2191	1.2238	1.2287	1.2345
800	1.1962	1.2004	1.2057	1.2109	1.2158	1.2203	1.2260

continue

continue

P/kg cm^{-2}	T/K 466.25	475.35	483.35	492.45	500.75	508.85	517.15
	V_{spez}/cm^{3}g^{-1}						
900	1.1892	1.1931	1.1983	1.2033	1.2076	1.2126	1.2182
1000	1.1828	1.1866	1.1918	1.1966	1.2005	1.2053	1.2106
1100	1.1767	1.1800	1.1853	1.1899	1.1938	1.1984	1.2036
1200	1.1705	1.1743	1.1790	1.1839	1.1876	1.1918	1.1969
1300	1.1651	1.1682	1.1732	1.1776	1.1816	1.1859	1.1907
1400	1.1597	1.1636	1.1678	1.1719	1.1759	1.1801	1.1845
1500	1.1542	1.1576	1.1628	1.1663	1.1703	1.1746	1.1791
1600	1.1493	1.1531	1.1573	1.1609	1.1648	1.1689	1.1741
1700	1.1445	1.1482	1.1525	1.1567	1.1599	1.1638	1.1682
1800	1.1402	1.1437	1.1480	1.1517	1.1551	1.1589	1.1630

copolymer (B): E-VA/25w(2) 95ZOL

P/MPa	T/K 393.85	401.35	409.25	417.75	425.85	432.75	441.65	449.95	457.15
	V_{spez}/cm^{3}g^{-1}								
0.1	1.1796	1.1866	1.1938	1.2014	1.2089	1.2158	1.2238	1.2314	1.2385
20	1.1622	1.1687	1.1752	1.1816	1.1883	1.1947	1.2015	1.2080	1.2140
40	1.1474	1.1532	1.1590	1.1648	1.1711	1.1764	1.1827	1.1885	1.1939
60	1.1341	1.1395	1.1449	1.1503	1.1557	1.1609	1.1666	1.1719	1.1768
80	1.1220	1.1274	1.1324	1.1373	1.1423	1.1472	1.1524	1.1574	1.1619
100	1.1119	1.1170	1.1216	1.1262	1.1309	1.1356	1.1404	1.1451	1.1493
120	1.1023	1.1070	1.1114	1.1158	1.1203	1.1247	1.1293	1.1338	1.1377
140	1.0935	1.0980	1.1022	1.1064	1.1108	1.1150	1.1191	1.1233	1.1270
160	1.0855	1.0898	1.0937	1.0976	1.1019	1.1058	1.1099	1.1137	1.1174
180	1.0778	1.0819	1.0858	1.0895	1.0934	1.0976	1.1014	1.1052	1.1085
200	1.0707	1.0747	1.0785	1.0821	1.0857	1.0896	1.0933	1.0970	1.1002

continue **E-VA/25w(2)**

P/MPa	*T*/K								
	465.75	474.15	482.55	490.75	499.05	507.25	515.65	523.45	531.75
	V_{spez}/cm^{3}g^{-1}								
0.1	1.2461	1.2547	1.2628	1.2713	1.2802	1.2899	1.2985	1.3075	1.3188
20	1.2206	1.2279	1.2345	1.2415	1.2489	1.2563	1.2638	1.2715	1.2805
40	1.1996	1.2059	1.2120	1.2181	1.2250	1.2309	1.2377	1.2446	1.2524
60	1.1824	1.1877	1.1931	1.1987	1.2046	1.2102	1.2162	1.2222	1.2293
80	1.1667	1.1718	1.1770	1.1821	1.1876	1.1926	1.1980	1.2036	1.2102
100	1.1540	1.1589	1.1634	1.1681	1.1730	1.1779	1.1829	1.1881	1.1941
120	1.1419	1.1464	1.1508	1.1553	1.1598	1.1645	1.1692	1.1741	1.1797
140	1.1310	1.1352	1.1393	1.1436	1.1479	1.1522	1.1567	1.1613	1.1668
160	1.1211	1.1251	1.1291	1.1332	1.1372	1.1413	1.1455	1.1498	1.1551
180	1.1121	1.1162	1.1197	1.1235	1.1274	1.1315	1.1353	1.1395	1.1445
200	1.1037	1.1073	1.1110	1.1148	1.1185	1.1221	1.1259	1.1301	1.1347

copolymer (B): **E-VA/25w(1)** **86ZOL**

P/kg cm^{-2}	*T*/K								
	367.45	376.35	385.45	393.85	403.05	412.85	421.45	430.55	439.65
	V_{spez}/cm^{3}g^{-1}								
1	1.1648	1.1720	1.1785	1.1872	1.1950	1.2029	1.2097	1.2186	1.2254
100	1.1564	1.1637	1.1701	1.1775	1.1847	1.1926	1.1990	1.2062	1.2136
200	1.1486	1.1555	1.1624	1.1687	1.1754	1.1829	1.1891	1.1956	1.2021
300	1.1414	1.1480	1.1544	1.1608	1.1669	1.1734	1.1801	1.1864	1.1923
400	1.1347	1.1411	1.1475	1.1530	1.1591	1.1657	1.1717	1.1773	1.1834
500	1.1284	1.1344	1.1405	1.1460	1.1519	1.1578	1.1639	1.1695	1.1749
600	1.1224	1.1280	1.1341	1.1396	1.1451	1.1506	1.1564	1.1624	1.1671
700	1.1169	1.1227	1.1280	1.1333	1.1388	1.1442	1.1495	1.1550	1.1598
800	1.1115	1.1170	1.1224	1.1275	1.1325	1.1378	1.1436	1.1484	1.1529
900	1.1063	1.1116	1.1169	1.1217	1.1267	1.1320	1.1369	1.1419	1.1465
1000	1.1015	1.1068	1.1120	1.1165	1.1217	1.1266	1.1315	1.1359	1.1403
1100	1.0970	1.1017	1.1069	1.1113	1.1163	1.1214	1.1261	1.1304	1.1344
1200	1.0924	1.0975	1.1020	1.1067	1.1113	1.1164	1.1208	1.1251	1.1284
1300	1.0884	1.0930	1.0977	1.1023	1.1067	1.1115	1.1157	1.1198	1.1234
1400	1.0841	1.0885	1.0932	1.0974	1.1023	1.1070	1.1110	1.1151	1.1184
1500	1.0799	1.0847	1.0896	1.0933	1.0977	1.1023	1.1063	1.1104	1.1138
1600	1.0763	1.0807	1.0852	1.0889	1.0935	1.0982	1.1017	1.1057	1.1092
1700	1.0724	1.0771	1.0812	1.0851	1.0897	1.0941	1.0977	1.1012	1.1050
1800	1.0689	1.0732	1.0775	1.0814	1.0856	1.0897	1.0936	1.0971	1.1008

continue **E-VA/25w(1)**

P/kg cm^{-2}	448.15	457.85	466.45	473.05	482.25	490.65	499.05	507.05	515.55
	T/K, V_{spez}/cm^{3}g^{-1}								
1	1.2325	1.2408	1.2497	1.2568	1.2646	1.2702	1.2773	1.2859	1.2926
100	1.2204	1.2282	1.2356	1.2408	1.2482	1.2547	1.2624	1.2692	1.2754
200	1.2090	1.2163	1.2230	1.2275	1.2345	1.2405	1.2474	1.2543	1.2604
300	1.1988	1.2058	1.2121	1.2161	1.2218	1.2286	1.2348	1.2414	1.2471
400	1.1895	1.1961	1.2019	1.2055	1.2117	1.2175	1.2237	1.2296	1.2350
500	1.1807	1.1870	1.1928	1.1964	1.2022	1.2075	1.2129	1.2189	1.2240
600	1.1727	1.1787	1.1843	1.1873	1.1933	1.1982	1.2036	1.2092	1.2140
700	1.1653	1.1713	1.1764	1.1797	1.1847	1.1898	1.1949	1.2004	1.2048
800	1.1581	1.1638	1.1689	1.1723	1.1769	1.1817	1.1867	1.1918	1.1966
900	1.1514	1.1569	1.1618	1.1651	1.1696	1.1746	1.1791	1.1840	1.1886
1000	1.1451	1.1504	1.1552	1.1583	1.1632	1.1676	1.1722	1.1766	1.1811
1100	1.1392	1.1444	1.1489	1.1523	1.1565	1.1609	1.1653	1.1700	1.1740
1200	1.1335	1.1386	1.1430	1.1462	1.1505	1.1545	1.1591	1.1633	1.1673
1300	1.1283	1.1332	1.1375	1.1405	1.1444	1.1489	1.1531	1.1572	1.1612
1400	1.1229	1.1278	1.1322	1.1349	1.1390	1.1434	1.1472	1.1514	1.1552
1500	1.1181	1.1228	1.1269	1.1294	1.1339	1.1378	1.1417	1.1456	1.1496
1600	1.1135	1.1181	1.1220	1.1245	1.1286	1.1326	1.1364	1.1406	1.1438
1700	1.1088	1.1134	1.1174	1.1199	1.1237	1.1274	1.1315	1.1354	1.1389
1800	1.1046	1.1089	1.1127	1.1156	1.1191	1.1232	1.1265	1.1301	1.1340

copolymer (B): **E-VA/28w(Z2)** **95ZOL**

P/MPa	376.85	385.05	393.65	401.85	410.05	418.55	427.05	434.85	443.65
	T/K, V_{spez}/cm^{3}g^{-1}								
0.1	1.1509	1.1576	1.1647	1.1712	1.1786	1.1895	1.1931	1.2010	1.2089
20	1.1352	1.1413	1.1476	1.1535	1.1599	1.1664	1.1727	1.1796	1.1862
40	1.1215	1.1271	1.1328	1.1382	1.1440	1.1499	1.1555	1.1619	1.1675
60	1.1094	1.1146	1.1197	1.1248	1.1303	1.1355	1.1408	1.1463	1.1516
80	1.0982	1.1031	1.1080	1.1127	1.1177	1.1227	1.1276	1.1328	1.1377
100	1.0888	1.0933	1.0978	1.1024	1.1069	1.1116	1.1163	1.1213	1.1258
120	1.0798	1.0841	1.0884	1.0926	1.0971	1.1015	1.1058	1.1106	1.1147
140	1.0715	1.0757	1.0797	1.0838	1.0878	1.0922	1.0962	1.1007	1.1046
160	1.0637	1.0676	1.0715	1.0755	1.0795	1.0834	1.0874	1.0917	1.0954
180	1.0564	1.0602	1.0641	1.0681	1.0717	1.0756	1.0792	1.0834	1.0868
200	1.0499	1.0534	1.0569	1.0607	1.0644	1.0678	1.0717	1.0757	1.0791

continue **E-VA/28w(Z2)**

P/MPa	*T*/K 451.85	459.95	468.25	476.35	485.25	493.45	502.05	510.45	518.95
	V_{spez}/cm^{3}g^{-1}								
0.1	1.2162	1.2244	1.2315	1.2392	1.2486	1.2562	1.2644	1.2733	1.2827
20	1.1926	1.1989	1.2056	1.2123	1.2193	1.2264	1.2331	1.2406	1.2478
40	1.1732	1.1791	1.1849	1.1909	1.1970	1.2029	1.2092	1.2154	1.2221
60	1.1569	1.1621	1.1675	1.1729	1.1784	1.1838	1.1895	1.1953	1.2008
80	1.1427	1.1475	1.1524	1.1575	1.1624	1.1674	1.1725	1.1776	1.1831
100	1.1303	1.1350	1.1395	1.1444	1.1489	1.1535	1.1583	1.1632	1.1680
120	1.1190	1.1233	1.1279	1.1322	1.1365	1.1409	1.1451	1.1499	1.1544
140	1.1087	1.1128	1.1169	1.1211	1.1251	1.1294	1.1336	1.1376	1.1421
160	1.0992	1.1032	1.1071	1.1112	1.1151	1.1190	1.1227	1.1269	1.1312
180	1.0905	1.0945	1.0982	1.1018	1.1056	1.1095	1.1130	1.1169	1.1209
200	1.0824	1.0860	1.0897	1.0933	1.0969	1.1006	1.1039	1.1078	1.1116

copolymer (B): **E-VA/28w(Z1)** **86ZOL**

P/kg cm^{-2}	*T*/K 367.25	376.75	385.65	393.95	403.25	412.75	421.65	430.45	439.35
	V_{spez}/cm^{3}g^{-1}								
1	1.1600	1.1661	1.1744	1.1818	1.1901	1.1973	1.2054	1.2123	1.2201
100	1.1516	1.1580	1.1658	1.1728	1.1797	1.1872	1.1942	1.2013	1.2086
200	1.1440	1.1501	1.1572	1.1642	1.1704	1.1774	1.1845	1.1910	1.1977
300	1.1370	1.1432	1.1498	1.1563	1.1623	1.1689	1.1758	1.1815	1.1883
400	1.1305	1.1363	1.1429	1.1490	1.1547	1.1612	1.1673	1.1733	1.1795
500	1.1245	1.1300	1.1363	1.1421	1.1478	1.1538	1.1598	1.1654	1.1714
600	1.1186	1.1242	1.1300	1.1358	1.1411	1.1469	1.1527	1.1582	1.1638
700	1.1133	1.1180	1.1241	1.1295	1.1349	1.1406	1.1458	1.1515	1.1571
800	1.1077	1.1131	1.1185	1.1241	1.1289	1.1342	1.1394	1.1446	1.1502
900	1.1026	1.1077	1.1133	1.1186	1.1232	1.1287	1.1336	1.1387	1.1442
1000	1.0980	1.1029	1.1083	1.1134	1.1179	1.1233	1.1278	1.1329	1.1382
1100	1.0935	1.0980	1.1036	1.1083	1.1130	1.1179	1.1225	1.1271	1.1325
1200	1.0890	1.0937	1.0988	1.1039	1.1081	1.1131	1.1175	1.1221	1.1272
1300	1.0849	1.0895	1.0946	1.0995	1.1035	1.1080	1.1127	1.1173	1.1222
1400	1.0808	1.0854	1.0904	1.0948	1.0993	1.1038	1.1081	1.1121	1.1173
1500	1.0769	1.0815	1.0863	1.0906	1.0947	1.0994	1.1035	1.1078	1.1126
1600	1.0733	1.0775	1.0823	1.0870	1.0907	1.0952	1.0993	1.1036	1.1082
1700	1.0697	1.0740	1.0786	1.0825	1.0866	1.0911	1.0951	1.0991	1.1036
1800	1.0663	1.0701	1.0752	1.0793	1.0832	1.0875	1.0910	1.0951	1.0999

continue **E-VA/28w(Z1)**

P/kg cm^{-2}	T/K								
	447.55	457.25	465.85	474.75	483.35	492.15	500.75	508.55	517.05
	V_{spez}/cm^{3}g^{-1}								
1	1.2279	1.2365	1.2426	1.2500	1.2585	1.2663	1.2743	1.2809	1.2900
100	1.2157	1.2237	1.2302	1.2371	1.2442	1.2520	1.2595	1.2671	1.2739
200	1.2047	1.2120	1.2180	1.2240	1.2317	1.2386	1.2447	1.2521	1.2584
300	1.1946	1.2018	1.2072	1.2133	1.2198	1.2263	1.2328	1.2395	1.2454
400	1.1857	1.1924	1.1979	1.2034	1.2098	1.2158	1.2218	1.2283	1.2337
500	1.1774	1.1837	1.1886	1.1940	1.2003	1.2056	1.2114	1.2168	1.2226
600	1.1695	1.1757	1.1802	1.1854	1.1915	1.1968	1.2020	1.2079	1.2132
700	1.1623	1.1683	1.1729	1.1774	1.1837	1.1886	1.1936	1.1993	1.2041
800	1.1551	1.1611	1.1655	1.1701	1.1757	1.1807	1.1856	1.1907	1.1953
900	1.1491	1.1547	1.1585	1.1631	1.1688	1.1736	1.1785	1.1832	1.1873
1000	1.1432	1.1485	1.1521	1.1564	1.1622	1.1667	1.1712	1.1757	1.1799
1100	1.1374	1.1426	1.1462	1.1505	1.1559	1.1602	1.1646	1.1692	1.1733
1200	1.1317	1.1368	1.1406	1.1447	1.1499	1.1540	1.1583	1.1629	1.1671
1300	1.1269	1.1317	1.1346	1.1392	1.1440	1.1486	1.1526	1.1571	1.1612
1400	1.1217	1.1265	1.1301	1.1339	1.1390	1.1426	1.1472	1.1511	1.1553
1500	1.1171	1.1217	1.1245	1.1290	1.1336	1.1378	1.1421	1.1455	1.1497
1600	1.1128	1.1174	1.1201	1.1237	1.1284	1.1324	1.1371	1.1403	1.1443
1700	1.1083	1.1129	1.1154	1.1193	1.1239	1.1277	1.1320	1.1356	1.1397
1800	1.1042	1.1086	1.1110	1.1148	1.1194	1.1231	1.1270	1.1306	1.1351

copolymer (B): E-VA/28w(B) 90BU1

P/MPa	T/K								
	373.15	383.15	393.15	403.15	413.15	423.15	433.15	443.15	453.15
	V_{spez}/cm^{3}g^{-1}								
10	1.1378	1.1443	1.1563	1.1636	1.1749	1.1822	1.1884	1.1978	1.2057
20	1.1288	1.1356	1.1462	1.1529	1.1631	1.1696	1.1755	1.1848	1.1922
30	1.1208	1.1277	1.1372	1.1439	1.1529	1.1589	1.1644	1.1736	1.1805
40	1.1131	1.1200	1.1288	1.1355	1.1433	1.1488	1.1542	1.1635	1.1698
50	1.1061	1.1128	1.1210	1.1277	1.1348	1.1400	1.1462	1.1547	1.1612
60	1.0997	1.1069	1.1139	1.1203	1.1275	1.1321	1.1383	1.1472	1.1526
70	1.0940	1.1008	1.1079	1.1145	1.1205	1.1249	1.1311	1.1399	1.1454
80	1.0888	1.0956	1.1022	1.1086	1.1148	1.1189	1.1244	1.1338	1.1395

copolymer (B): **E-VA/33w** **95ZOL**

P/MPa	T/K 385.25	393.75	402.05	409.95	418.45	427.15	435.15	443.55	451.95
	V_{spez}/cm^{3}g^{-1}								
0.1	1.1403	1.1473	1.1544	1.1610	1.1686	1.1765	1.1836	1.1909	1.1983
20	1.1238	1.1301	1.1362	1.1421	1.1487	1.1554	1.1617	1.1680	1.1746
40	1.1100	1.1155	1.1211	1.1265	1.1324	1.1383	1.1441	1.1498	1.1555
60	1.0975	1.1026	1.1080	1.1130	1.1185	1.1237	1.1290	1.1342	1.1394
80	1.0863	1.0911	1.0959	1.1008	1.1057	1.1106	1.1156	1.1206	1.1251
100	1.0766	1.0811	1.0857	1.0902	1.0949	1.0995	1.1042	1.1088	1.1131
120	1.0676	1.0718	1.0760	1.0805	1.0848	1.0892	1.0935	1.0979	1.1019
140	1.0591	1.0632	1.0672	1.0718	1.0756	1.0796	1.0839	1.0878	1.0916
160	1.0513	1.0555	1.0590	1.0630	1.0669	1.0710	1.0748	1.0787	1.0826
180	1.0441	1.0478	1.0514	1.0554	1.0591	1.0628	1.0665	1.0703	1.0737
200	1.0371	1.0409	1.0443	1.0481	1.0515	1.0551	1.0586	1.0623	1.0656

continue **E-VA/33w**

P/MPa	T/K 459.65	468.05	476.75	484.85	492.85	501.65	510.35	518.35	526.55
	V_{spez}/cm^{3}g^{-1}								
0.1	1.2057	1.2136	1.2216	1.2296	1.2377	1.2466	1.2551	1.2646	1.2738
20	1.1810	1.1874	1.1942	1.2010	1.2080	1.2149	1.2222	1.2294	1.2373
40	1.1611	1.1670	1.1732	1.1789	1.1849	1.1910	1.1975	1.2039	1.2106
60	1.1445	1.1500	1.1550	1.1607	1.1661	1.1715	1.1776	1.1829	1.1892
80	1.1301	1.1349	1.1399	1.1450	1.1499	1.1551	1.1604	1.1654	1.1712
100	1.1177	1.1221	1.1266	1.1315	1.1361	1.1408	1.1456	1.1505	1.1559
120	1.1063	1.1105	1.1150	1.1194	1.1238	1.1279	1.1327	1.1369	1.1422
140	1.0958	1.0999	1.1038	1.1082	1.1124	1.1163	1.1207	1.1249	1.1298
160	1.0863	1.0902	1.0939	1.0981	1.1019	1.1058	1.1099	1.1138	1.1186
180	1.0776	1.0811	1.0849	1.0887	1.0925	1.0961	1.1000	1.1038	1.1082
200	1.0694	1.0728	1.0762	1.0800	1.0836	1.0871	1.0909	1.0946	1.0988

copolymer (B): E-VA/40w(2) 95ZOL

P/MPa	T/K								
	367.75	375.95	383.05	391.55	399.45	408.35	415.95	424.35	431.95
	V_{spez}/cm^{3}g^{-1}								
0.1	1.1010	1.1072	1.1135	1.1205	1.1275	1.1349	1.1415	1.1491	1.1555
20	1.0872	1.0927	1.0986	1.1048	1.1108	1.1170	1.1233	1.1299	1.1359
40	1.0746	1.0798	1.0854	1.0910	1.0967	1.1019	1.1078	1.1139	1.1191
60	1.0634	1.0683	1.0736	1.0788	1.0839	1.0887	1.0943	1.0998	1.1047
80	1.0532	1.0578	1.0630	1.0678	1.0726	1.0771	1.0822	1.0871	1.0921
100	1.0444	1.0488	1.0536	1.0583	1.0629	1.0670	1.0718	1.0764	1.0810
120	1.0361	1.0402	1.0450	1.0493	1.0536	1.0575	1.0621	1.0665	1.0709
140	1.0284	1.0324	1.0368	1.0409	1.0451	1.0486	1.0534	1.0574	1.0616
160	1.0213	1.0250	1.0296	1.0334	1.0373	1.0406	1.0450	1.0488	1.0532
180	1.0148	1.0183	1.0226	1.0263	1.0301	1.0332	1.0375	1.0412	1.0451
200	1.0084	1.0118	1.0161	1.0198	1.0233	1.0264	1.0305	1.0338	1.0378

continue **E-VA/40w(2)**

P/MPa	T/K							
	441.05	448.75	455.65	464.85	473.75	481.65	489.15	496.85
	V_{spez}/cm^{3}g^{-1}							
0.1	1.1635	1.1708	1.1772	1.1855	1.1938	1.2012	1.2087	1.2164
20	1.1424	1.1485	1.1545	1.1613	1.1682	1.1747	1.1811	1.1878
40	1.1248	1.1304	1.1360	1.1419	1.1479	1.1536	1.1595	1.1655
60	1.1097	1.1151	1.1203	1.1255	1.1307	1.1360	1.1415	1.1469
80	1.0966	1.1016	1.1067	1.1112	1.1161	1.1209	1.1262	1.1313
100	1.0851	1.0899	1.0948	1.0989	1.1032	1.1080	1.1128	1.1177
120	1.0748	1.0792	1.0838	1.0876	1.0918	1.0962	1.1007	1.1052
140	1.0651	1.0694	1.0739	1.0774	1.0815	1.0857	1.0899	1.0941
160	1.0564	1.0607	1.0649	1.0681	1.0719	1.0757	1.0800	1.0841
180	1.0484	1.0524	1.0565	1.0596	1.0632	1.0670	1.0709	1.0750
200	1.0407	1.0447	1.0487	1.0517	1.0550	1.0587	1.0625	1.0663

copolymer (B): E-VA/40w(1) 86ZOL

P/kg cm^{-2}	T/K 348.25	357.05	366.75	376.05	385.15	393.55	402.65	412.15	421.05
	V_{spez}/cm^{3}g^{-1}								
1	1.0901	1.0964	1.1041	1.1112	1.1174	1.1267	1.1339	1.1407	1.1474
100	1.0837	1.0901	1.0973	1.1040	1.1107	1.1180	1.1240	1.1315	1.1377
200	1.0773	1.0835	1.0902	1.0965	1.1030	1.1101	1.1156	1.1229	1.1285
300	1.0713	1.0774	1.0837	1.0901	1.0962	1.1027	1.1082	1.1148	1.1202
400	1.0658	1.0717	1.0778	1.0838	1.0898	1.0959	1.1009	1.1076	1.1127
500	1.0603	1.0658	1.0718	1.0778	1.0831	1.0895	1.0947	1.1006	1.1056
600	1.0553	1.0608	1.0661	1.0720	1.0777	1.0833	1.0881	1.0941	1.0985
700	1.0505	1.0556	1.0612	1.0669	1.0724	1.0777	1.0826	1.0883	1.0924
800	1.0460	1.0510	1.0562	1.0617	1.0670	1.0722	1.0771	1.0823	1.0864
900	1.0415	1.0466	1.0513	1.0568	1.0623	1.0668	1.0719	1.0769	1.0809
1000	1.0373	1.0419	1.0470	1.0521	1.0575	1.0622	1.0668	1.0714	1.0755
1100	1.0335	1.0380	1.0426	1.0475	1.0530	1.0575	1.0620	1.0664	1.0704
1200	1.0296	1.0339	1.0386	1.0434	1.0486	1.0532	1.0575	1.0620	1.0657
1300	1.0259	1.0300	1.0345	1.0394	1.0443	1.0489	1.0532	1.0576	1.0609
1400	1.0223	1.0264	1.0308	1.0357	1.0405	1.0449	1.0489	1.0532	1.0567
1500	1.0187	1.0230	1.0273	1.0317	1.0366	1.0408	1.0449	1.0490	1.0522
1600	1.0154	1.0197	1.0234	1.0282	1.0329	1.0375	1.0410	1.0450	1.0484
1700	1.0123	1.0164	1.0206	1.0249	1.0295	1.0333	1.0375	1.0414	1.0446
1800	1.0092	1.0133	1.0172	1.0216	1.0262	1.0301	1.0338	1.0373	1.0408

continue **E-VA/40w(1)**

P/kg cm^{-2}	T/K 430.25	439.45	447.85	457.45	466.15	474.95	484.05	500.55	516.75
	V_{spez}/cm^{3}g^{-1}								
1	1.1544	1.1598	1.1694	1.1769	1.1863	1.1917	1.1987	1.2138	1.2288
100	1.1446	1.1508	1.1576	1.1654	1.1724	1.1789	1.1857	1.1998	1.2138
200	1.1348	1.1410	1.1472	1.1549	1.1610	1.1671	1.1734	1.1867	1.1999
300	1.1264	1.1324	1.1374	1.1451	1.1509	1.1566	1.1629	1.1752	1.1876
400	1.1186	1.1245	1.1297	1.1364	1.1413	1.1472	1.1531	1.1648	1.1766
500	1.1111	1.1165	1.1221	1.1281	1.1332	1.1385	1.1440	1.1550	1.1663
600	1.1045	1.1096	1.1148	1.1204	1.1252	1.1302	1.1357	1.1465	1.1568
700	1.0978	1.1029	1.1081	1.1136	1.1179	1.1231	1.1280	1.1382	1.1489
800	1.0917	1.0966	1.1016	1.1071	1.1108	1.1159	1.1208	1.1304	1.1407

continue

continue

P/kg cm^{-2}	T/K								
	430.25	439.45	447.85	457.45	466.15	474.95	484.05	500.55	516.75
	V_{spez}/cm^{3}g^{-1}								
900	1.0861	1.0905	1.0952	1.1008	1.1046	1.1094	1.1141	1.1229	1.1330
1000	1.0805	1.0851	1.0898	1.0950	1.0984	1.1032	1.1078	1.1170	1.1263
1100	1.0755	1.0796	1.0847	1.0891	1.0928	1.0971	1.1019	1.1106	1.1196
1200	1.0707	1.0748	1.0792	1.0839	1.0871	1.0914	1.0960	1.1045	1.1132
1300	1.0661	1.0699	1.0741	1.0790	1.0821	1.0861	1.0906	1.0989	1.1076
1400	1.0618	1.0654	1.0695	1.0742	1.0770	1.0811	1.0856	1.0940	1.1023
1500	1.0574	1.0608	1.0651	1.0693	1.0724	1.0761	1.0806	1.0885	1.0967
1600	1.0528	1.0568	1.0609	1.0648	1.0678	1.0715	1.0757	1.0838	1.0908
1700	1.0488	1.0526	1.0564	1.0612	1.0636	1.0668	1.0713	1.0793	1.0864
1800	1.0455	1.0486	1.0525	1.0570	1.0591	1.0624	1.0671	1.0746	1.0823

copolymer (B): E-VA/65w 95ZOL

P/MPa	T/K								
	343.65	351.65	360.35	368.55	376.75	385.15	393.55	401.85	409.35
	V_{spez}/cm^{3}g^{-1}								
0.1	1.0072	1.0132	1.0195	1.0256	1.0314	1.0378	1.0437	1.0503	1.0563
20	0.9962	1.0018	1.0076	1.0131	1.0184	1.0240	1.0295	1.0353	1.0408
40	0.9860	0.9914	0.9967	1.0018	1.0067	1.0117	1.0171	1.0222	1.0276
60	0.9769	0.9819	0.9870	0.9917	0.9962	1.0009	1.0058	1.0108	1.0154
80	0.9683	0.9733	0.9780	0.9826	0.9867	0.9911	0.9958	1.0004	1.0049
100	0.9609	0.9656	0.9701	0.9744	0.9784	0.9826	0.9870	0.9912	0.9956
120	0.9538	0.9584	0.9627	0.9669	0.9706	0.9746	0.9788	0.9829	0.9870
140	0.9471	0.9515	0.9556	0.9598	0.9643	0.9672	0.9712	0.9752	0.9790
160	0.9409	0.9453	0.9492	0.9533	0.9567	0.9603	0.9641	0.9678	0.9717
180	0.9352	0.9393	0.9432	0.9471	0.9505	0.9541	0.9577	0.9615	0.9649
200	0.9297	0.9338	0.9375	0.9414	0.9447	0.9480	0.9516	0.9550	0.9585

continue **E-VA/65w**

P/MPa	*T*/K 418.45	426.65	434.75	442.85	451.25	459.65	476.35	493.25	501.75
	V_{spez}/cm^{3}g^{-1}								
0.1	1.0631	1.0702	1.0763	1.0834	1.0908	1.0982	1.1144	1.1336	1.1455
20	1.0468	1.0529	1.0587	1.0647	1.0711	1.0777	1.0916	1.1082	1.1180
40	1.0327	1.0383	1.0436	1.0491	1.0548	1.0604	1.0729	1.0878	1.0969
60	1.0205	1.0256	1.0304	1.0355	1.0406	1.0459	1.0574	1.0711	1.0790
80	1.0096	1.0143	1.0190	1.0236	1.0284	1.0333	1.0439	1.0564	1.0640
100	0.9999	1.0046	1.0088	1.0133	1.0176	1.0222	1.0321	1.0440	1.0510
120	0.9911	0.9953	0.9994	1.0036	1.0079	1.0123	1.0216	1.0329	1.0394
140	0.9829	0.9869	0.9909	0.9949	0.9988	1.0029	1.0120	1.0225	1.0289
160	0.9754	0.9793	0.9830	0.9868	0.9906	0.9945	1.0030	1.0133	1.0193
180	0.9685	0.9721	0.9759	0.9793	0.9830	0.9867	0.9949	1.0049	1.0107
200	0.9619	0.9654	0.9689	0.9726	0.9759	0.9796	0.9875	0.9969	1.0027

copolymer (B): **E-VA/87w** **90CIM**

P/MPa	*T*/K 303.15	313.15	323.15	333.15	343.15	353.15	363.15	373.15	383.15
	V_{spez}/cm^{3}g^{-1}								
0.1	0.879	0.885	0.891	0.897	0.903	0.909	0.916	0.922	0.929

continue **E-VA/87w**

P/MPa	*T*/K 393.15	403.15	413.15	423.15	433.15	443.15
	V_{spez}/cm^{3}g^{-1}					
0.1	0.936	0.943	0.950	0.958	0.965	0.973

Tait equation parameter functions:

Copolymer	$V(P/\text{MPa}, T/\text{K}) = V(0, T/\text{K})\{1 - C^*\ln[1 + (P/\text{MPa})/B(T/\text{K})]\}$	
	$V(0,T/\text{K})/\text{cm}^3\text{g}^{-1}$	$B(T/\text{K})/\text{MPa}$
	C = 0.0894 and $\theta = T/\text{K} - 273.15$	
E-VA/18w(1)	$1.02391 \exp(2.173\ 10^{-5}T^{1.5})$	$188.2 \exp(-4.537\ 10^{-3}\theta)$
E-VA/25w(1)	$1.00416 \exp(2.244\ 10^{-5}T^{1.5})$	$184.4 \exp(-4.734\ 10^{-3}\theta)$
E-VA/28w(Z1)	$1.00832 \exp(2.241\ 10^{-5}T^{1.5})$	$183.5 \exp(-4.457\ 10^{-3}\theta)$
E-VA/40w(1)	$1.06332 \exp(2.288\ 10^{-5}T^{1.5})$	$205.1 \exp(-4.989\ 10^{-3}\theta)$
	C = 0.065504	
E-VA/09w	$0.96060 + 0.35674\ 10^{-3}T + 8.4287\ 10^{-7}T^2$	$667.89 \exp(-5.9742\ 10^{-3}T)$
	C = 0.064626	
E-VA/28w(B)	$0.75181 + 1.20228\ 10^{-3}T - 3.6156\ 10^{-7}T^2$	$386.92 \exp(-4.9032\ 10^{-3}T)$

Copolymers from ethylene and vinyl alcohol

Average chemical composition of the copolymers, acronyms, range of data, and references:

Copolymer	Acronym	Range of data T/K	P/MPa	Ref.
Ethylene/vinyl alcohol copolymer				
56 wt% vinyl alcohol	E-VAL/56w	468-535	0.1-200	95ZOL
62 wt% vinyl alcohol	E-VAL/62w	466-532	0.1-200	95ZOL
70 wt% vinyl alcohol	E-VAL/70w	469-536	0.1-200	95ZOL

Characterization:

Copolymer (B)	M_n/ g/mol	M_w/ g/mol	M_η/ g/mol	Further information
E-VAL/56w				no data
E-VAL/62w				no data
E-VAL/70w				no data

Experimental *PVT* data:

copolymer (B): E-VAL/56w **95ZOL**

P/MPa	*T*/K								
	468.35	476.85	485.35	493.45	501.95	510.25	518.75	527.25	535.55
	V_{spez}/cm^{3}g^{-1}								
0.1	0.9808	0.9860	0.9911	0.9960	1.0017	1.0067	1.0122	1.0178	1.0241
20	0.9711	0.9759	0.9806	0.9853	0.9902	0.9949	1.0000	1.0051	1.0107
40	0.9629	0.9673	0.9718	0.9762	0.9806	0.9852	0.9897	0.9944	0.9997
60	0.9553	0.9595	0.9636	0.9678	0.9721	0.9764	0.9807	0.9850	0.9898
80	0.9483	0.9523	0.9562	0.9602	0.9643	0.9684	0.9723	0.9765	0.9811
100	0.9421	0.9458	0.9496	0.9534	0.9573	0.9611	0.9648	0.9689	0.9731
120	0.9360	0.9396	0.9432	0.9468	0.9507	0.9542	0.9578	0.9615	0.9659
140	0.9302	0.9338	0.9373	0.9409	0.9443	0.9478	0.9513	0.9549	0.9589
160	0.9249	0.9283	0.9316	0.9351	0.9385	0.9419	0.9452	0.9486	0.9525
180	0.9199	0.9232	0.9265	0.9297	0.9330	0.9363	0.9394	0.9428	0.9465
200	0.9151	0.9182	0.9214	0.9246	0.9278	0.9309	0.9340	0.9374	0.9409

copolymer (B): E-VAL/62w **95ZOL**

P/MPa	*T*/K								
	466.45	475.05	483.35	491.55	499.05	507.75	516.05	524.15	531.75
	V_{spez}/cm^{3}g^{-1}								
0.1	0.9527	0.9576	0.9621	0.9671	0.9713	0.9767	0.9818	0.9889	0.9982

continue

continue **E-VAL/62w**

P/MPa	*T*/K								
	466.45	475.05	483.35	491.55	499.05	507.75	516.05	524.15	531.75
	V_{spez}/cm^3g^{-1}								
20	0.9445	0.9489	0.9531	0.9575	0.9617	0.9663	0.9712	0.9779	0.9865
40	0.9372	0.9413	0.9453	0.9493	0.9535	0.9575	0.9621	0.9686	0.9768
60	0.9306	0.9344	0.9382	0.9420	0.9460	0.9497	0.9541	0.9603	0.9680
80	0.9243	0.9280	0.9315	0.9351	0.9390	0.9426	0.9467	0.9528	0.9600
100	0.9186	0.9221	0.9256	0.9289	0.9326	0.9359	0.9400	0.9458	0.9528
120	0.9132	0.9164	0.9197	0.9230	0.9266	0.9298	0.9336	0.9394	0.9462
140	0.9081	0.9114	0.9144	0.9176	0.9210	0.9240	0.9279	0.9334	0.9397
160	0.9032	0.9063	0.9093	0.9125	0.9158	0.9186	0.9224	0.9277	0.9338
180	0.8986	0.9016	0.9046	0.9075	0.9108	0.9136	0.9171	0.9224	0.9283
200	0.8943	0.8973	0.9000	0.9029	0.9060	0.9087	0.9122	0.9174	0.9231

copolymer (B): **E-VAL/70w** **95ZOL**

P/MPa	*T*/K								
	469.25	477.75	486.15	494.35	502.65	511.35	519.55	527.85	536.15
	V_{spez}/cm^3g^{-1}								
0.1	0.9386	0.9431	0.9479	0.9514	0.9561	0.9610	0.9658	0.9711	0.9765
20	0.9299	0.9342	0.9385	0.9419	0.9462	0.9506	0.9551	0.9597	0.9648
40	0.9228	0.9269	0.9308	0.9340	0.9381	0.9422	0.9464	0.9508	0.9555
60	0.9163	0.9201	0.9238	0.9268	0.9307	0.9346	0.9386	0.9429	0.9472
80	0.9102	0.9139	0.9174	0.9203	0.9240	0.9276	0.9314	0.9354	0.9397
100	0.9047	0.9082	0.9115	0.9142	0.9178	0.9214	0.9251	0.9288	0.9328
120	0.8994	0.9028	0.9059	0.9086	0.9120	0.9154	0.9188	0.9224	0.9263
140	0.8944	0.8976	0.9007	0.9032	0.9065	0.9097	0.9130	0.9165	0.9202
160	0.8896	0.8928	0.8957	0.8981	0.9013	0.9043	0.9076	0.9110	0.9146
180	0.8851	0.8882	0.8909	0.8933	0.8964	0.8994	0.9025	0.9058	0.9092
200	0.8807	0.8838	0.8864	0.8887	0.8918	0.8946	0.8976	0.9008	0.9042

Copolymers from perfluoroethylene oxide and perfluoromethylene oxide

Average chemical composition of the copolymers, acronyms, range of data, and references:

Copolymer	Acronym	Range of data T/K	P/MPa	Ref.
Perfluoro(oxyethylene/oxymethylene) copolymer				
43 mol% perfluorooxymethylene	PF(OE-OM)/43x	253-333	0.1	95DAN
44 mol% perfluorooxymethylene	PF(OE-OM)/44x	253-333	0.1	95DAN
47 mol% perfluorooxymethylene	PF(OE-OM)/47x	253-333	0.1	95DAN
48 mol% perfluorooxymethylene	PF(OE-OM)/48x	253-333	0.1	95DAN
49 mol% perfluorooxymethylene	PF(OE-OM)/49x(1)	253-333	0.1	95DAN
49 mol% perfluorooxymethylene	PF(OE-OM)/49x(2)	253-333	0.1	95DAN
49 mol% perfluorooxymethylene	PF(OE-OM)/49x(3)	253-333	0.1	95DAN
51 mol% perfluorooxymethylene	PF(OE-OM)/51x(1)	253-333	0.1	95DAN
51 mol% perfluorooxymethylene	PF(OE-OM)/51x(2)	253-333	0.1	95DAN
52 mol% perfluorooxymethylene	PF(OE-OM)/52x	253-333	0.1	95DAN

Characterization:

Copolymer (B)	M_n/ g/mol	M_w/ g/mol	M_η/ g/mol	Further information
PF(OE-OM)/43x	465	500		endgroup $-CH_2$-OH on each side of the chain, i.e., = 12.4 wt% in this sample
PF(OE-OM)/44x	850	1045		endgroup $-CH_2$-OH on each side of the chain, i.e., = 7.3 wt% in this sample
PF(OE-OM)/47x	400	450		endgroup $-CH_2$-OH on each side of the chain, i.e., = 15.5 wt% in this sample
PF(OE-OM)/48x	750	855		endgroup $-CH_2$-OH on each side of the chain, i.e., = 8.3 wt% in this sample
PF(OE-OM)/49x(1)	2550	3290		endgroup $-CH_2$-OH on each side of the chain, i.e., = 2.4 wt% in this sample
PF(OE-OM)/49x(2)	3700	4480		endgroup $-CH_2$-OH on each side of the chain, i.e., = 1.7 wt% in this sample
PF(OE-OM)/49x(3)	5300	6520		endgroup $-CH_2$-OH on each side of the chain, i.e., = 1.2 wt% in this sample

continue

continue

Copolymer (B)	M_n/ g/mol	M_w/ g/mol	M_η/ g/mol	Further information
PF(OE-OM)/51x(1)	9400	10900		endgroup $-CH_2$-OH on each side of the chain, i.e., = 0.7 wt% in this sample
PF(OE-OM)/51x(2)	14900	17900		endgroup $-CH_2$-OH on each side of the chain, i.e., = 0.4 wt% in this sample
PF(OE-OM)/52x	1100	1230		endgroup $-CH_2$-OH on each side of the chain, i.e., = 5.6 wt% in this sample

Experimental *PVT* data:

copolymer (B): PF(OE-OM)/43x 95DAN

P/MPa = 0.1 = constant

T/K	253.15	263.15	273.15	283.15	293.15	303.15	313.15	323.15
V_{spez}/cm^{3}g^{-1}	0.5585	0.5636	0.5687	0.5738	0.5789	0.5843	0.5895	0.5949
T/K	333.15							
V_{spez}/cm^{3}g^{-1}	0.6003							

copolymer (B): PF(OE-OM)/44x 95DAN

P/MPa = 0.1 = constant

T/K	253.15	263.15	273.15	283.15	293.15	303.15	313.15	323.15
V_{spez}/cm^{3}g^{-1}	0.5457	0.5512	0.5566	0.5621	0.5674	0.5729	0.5785	0.5840
T/K	333.15							
V_{spez}/cm^{3}g^{-1}	0.5896							

copolymer (B): PF(OE-OM)/47x 95DAN

P/MPa = 0.1 = constant

T/K	253.15	263.15	273.15	283.15	293.15	303.15	313.15	323.15
V_{spez}/cm^{3}g^{-1}	0.5625	0.5674	0.5723	0.5773	0.5823	0.5874	0.5924	0.5973
T/K	333.15							
V_{spez}/cm^{3}g^{-1}	0.6027							

copolymer (B): **PF(OE-OM)/48x** **95DAN**

P/MPa = 0.1 = constant

T/K	253.15	263.15	273.15	283.15	293.15	303.15	313.15	323.15
V_{spez}/cm^{3}g^{-1}	0.5476	0.5529	0.5583	0.5638	0.5691	0.5748	0.5803	0.5858
T/K	333.15							
V_{spez}/cm^{3}g^{-1}	0.5914							

copolymer (B): **PF(OE-OM)/49x(1)** **95DAN**

P/MPa = 0.1 = constant

T/K	253.15	263.15	273.15	283.15	293.15	303.15	313.15	323.15
V_{spez}/cm^{3}g^{-1}	0.5311	0.5367	0.5423	0.5479	0.5534	0.5592	0.5646	0.5705
T/K	333.15							
V_{spez}/cm^{3}g^{-1}	0.5764							

copolymer (B): **PF(OE-OM)/49x(2)** **95DAN**

P/MPa = 0.1 = constant

T/K	253.15	263.15	273.15	283.15	293.15	303.15	313.15	323.15
V_{spez}/cm^{3}g^{-1}	0.5280	0.5336	0.5393	0.5450	0.5506	0.5566	0.5624	0.5681
T/K	333.15							
V_{spez}/cm^{3}g^{-1}	0.5738							

copolymer (B): **PF(OE-OM)/49x(3)** **95DAN**

P/MPa = 0.1 = constant

T/K	253.15	263.15	273.15	283.15	293.15	303.15	313.15	323.15
V_{spez}/cm^{3}g^{-1}	0.5255	0.5312	0.5369	0.5427	0.5483	0.5543	0.5599	0.5656
T/K	333.15							
V_{spez}/cm^{3}g^{-1}	0.5711							

copolymer (B): **PF(OE-OM)/51x(1)** **95DAN**

P/MPa = 0.1 = constant

T/K	253.15	263.15	273.15	283.15	293.15	303.15	313.15	323.15
V_{spez}/cm^{3}g^{-1}	0.5246	0.5302	0.5359	0.5416	0.5471	0.5529	0.5588	0.5644
T/K	333.15							
V_{spez}/cm^{3}g^{-1}	0.5701							

copolymer (B): PF(OE-OM)/51x(2) 95DAN

P/MPa = 0.1 = constant

T/K	253.15	263.15	273.15	283.15	293.15	303.15	313.15	323.15
V_{spez}/cm^{3}g^{-1}	0.5223	0.5280	0.5337	0.5395	0.5452	0.5512	0.5570	0.5626
T/K	333.15							
V_{spez}/cm^{3}g^{-1}	0.5682							

copolymer (B): PF(OE-OM)/52x 95DAN

P/MPa = 0.1 = constant

T/K	253.15	263.15	273.15	283.15	293.15	303.15	313.15	323.15
V_{spez}/cm^{3}g^{-1}	0.5405	0.5461	0.5518	0.5575	0.5631	0.5692	0.5747	0.5806
T/K	333.15							
V_{spez}/cm^{3}g^{-1}	0.5862							

Copolymers from styrene and methyl methacrylate

Average chemical composition of the copolymers, acronyms, range of data, and references:

Copolymer	Acronym	Range of data T/K	P/MPa	Ref.
Styrene/methyl methacrylate copolymer				
20.5 wt% methyl methacrylate	S-MMA/21w	383-543	0.1-200	92KIM
58.5 wt% methyl methacrylate	S-MMA/59w	383-543	0.1-200	92KIM

Characterization:

Copolymer (B)	M_n/ g/mol	M_w/ g/mol	M_η/ g/mol	Further information
S-MMA/21w	110000	268000		Richardson Polymer Noan 81
S-MMA/59w				Richardson Polymer RPC 100

Experimental *PVT* data:

Tait equation parameter functions:

Copolymer	$V(P/\text{MPa}, T/\text{K}) = V(0, T/\text{K})\{1 - C^*\ln[1 + (P/\text{MPa})/B(T/\text{K})]\}$ with C = 0.0894 and $\theta = T/\text{K} - 273.15$	
	$V(0,T/\text{K})/\text{cm}^3\text{g}^{-1}$	$B(T/\text{K})/\text{MPa}$
S-MMA/21w	$0.9063 + 3.5702\ 10^{-4}\theta + 6.5323\ 10^{-7}\theta^2$	$232.0\ \exp(-4.1430\ 10^{-3}\theta)$
S-MMA/59w	$0.8610 + 3.3500\ 10^{-4}\theta + 6.9801\ 10^{-7}\theta^2$	$261.0\ \exp(-4.6112\ 10^{-3}\theta)$

Copolymers from N-vinylcarbazole and 4-alkylstyrenes

Average chemical composition of the copolymers, acronyms, range of data, and references:

Copolymer	Acronym	Range of data *T*/K	*P*/MPa	Ref.
N-Vinylcarbazole/4-ethylstyrene copolymer				
50 mol% 4-ethylstyrene	VCZ-ES/50x	393-443	30-100	94PRI
N-Vinylcarbazole/4-hexylstyrene copolymer				
20 mol% 4-hexylstyrene	VCZ-HS/20x	473-523	30-100	94PRI
33 mol% 4-hexylstyrene	VCZ-HS/33x	463-523	30-100	94PRI
40 mol% 4-hexylstyrene	VCZ-HS/40x	423-493	30-100	94PRI
50 mol% 4-hexylstyrene	VCZ-HS/50x	373-443	30-100	94PRI
60 mol% 4-hexylstyrene	VCZ-HS/60x	383-453	30-100	94PRI
67 mol% 4-hexylstyrene	VCZ-HS/67x	333-423	30-100	94PRI
80 mol% 4-hexylstyrene	VCZ-HS/80x	313-423	30-100	94PRI
N-Vinylcarbazole/4-octylstyrene copolymer				
50 mol% 4-octylstyrene	VCZ-OS/50x	403-453	30-100	94PRI
N-Vinylcarbazole/4-pentylstyrene copolymer				
50 mol% 4-pentylstyrene	VCZ-PS/50x	383-443	30-100	94PRI

Characterization:

Copolymer (B)	M_n/ g/mol	M_w/ g/mol	M_η/ g/mol	Further information
VCZ-ES/50x				synthesized in the laboratory
VCZ-HS/20x				synthesized in the laboratory
VCZ-HS/33x				synthesized in the laboratory
VCZ-HS/40x				synthesized in the laboratory
VCZ-HS/50x				synthesized in the laboratory
VCZ-HS/60x				synthesized in the laboratory
VCZ-HS/67x				synthesized in the laboratory
VCZ-HS/80x				synthesized in the laboratory
VCZ-OS/50x				synthesized in the laboratory
VCZ-PS/50x				synthesized in the laboratory

Experimental *PVT* data:

Tait equation parameter functions:

Copolymer	$V(P/\text{MPa}, T/\text{K}) = V(0, T/\text{K})\{1 - \text{C}*\ln[1 + (P/\text{MPa})/\text{B}(T/\text{K})]\}$ with C = 0.0894 and $\theta = T/\text{K} - 273.15$	
	$V(0,T/\text{K})/\text{cm}^3\text{g}^{-1}$	$\text{B}(T/\text{K})/\text{MPa}$
VCZ-ES/50x	$0.6676 + 6.63\ 10^{-4}T$	$5281.7\ \exp(-9.264\ 10^{-3}\theta)$
VCZ-HS/20x	$0.6416 + 5.42\ 10^{-4}T$	$489.8\ \exp(-3.193\ 10^{-3}\theta)$
VCZ-HS/33x	$0.7710 + 4.86\ 10^{-4}T$	$460.4\ \exp(-3.453\ 10^{-3}\theta)$
VCZ-HS/40x	$0.7805 + 4.92\ 10^{-4}T$	$155.0\ \exp(-1.605\ 10^{-3}\theta)$
VCZ-HS/50x	$0.7827 + 5.05\ 10^{-4}T$	$136.0\ \exp(-1.083\ 10^{-3}\theta)$
VCZ-HS/60x	$0.8213 + 6.23\ 10^{-4}T$	$229.1\ \exp(-2.133\ 10^{-3}\theta)$
VCZ-HS/67x	$0.8028 + 6.50\ 10^{-4}T$	$581.7\ \exp(-4.553\ 10^{-3}\theta)$
VCZ-HS/80x	$0.7753 + 6.17\ 10^{-4}T$	$247.6\ \exp(-2.604\ 10^{-3}\theta)$
VCZ-OS/50x	$0.7081 + 7.40\ 10^{-4}T$	$666.5\ \exp(-4.503\ 10^{-3}\theta)$
VCZ-PS/50x	$0.7814 + 4.36\ 10^{-4}T$	$880.1\ \exp(-4.393\ 10^{-3}\theta)$

6.2. References

73RAO Rao, K. and Griskey, R.G., An equation of state for molten polymers, *J. Appl. Polym. Sci.*, 17, 3293, 1973.

77REN Renuncio, J.A.R. and Prausnitz, J.M., Volumetric properties of block and random copolymers of butadiene and styrene at pressures to 1 kilobar, *J. Appl. Polym. Sci.*, 21, 2867, 1977.

86ZOL Zoller, P., Jain, R. K., and Simha, R., Equation of state of copolymer melts: The poly(vinyl acetate)-polyethylene pair, *J. Polym. Sci., Part B: Polym. Phys.*, 24, 687, 1986. (experimental data by private communication with P. Zoller)

90BU1 Busch, B., Das spezifische Volumen geschmolzener Polymere und Copolymere in Abhängigkeit von Druck und Temperatur, *PhD-Thesis*, Univ. Osnabrück, 1990.

90BU2 Busch, B., Lechner, M.D., and Kleintjens, L.A., The specific volume of polymer melts as a function of temperature and pressure, *Thermochim. Acta*, 160, 131, 1990.

90CIM Cimmino, S., Martuscelli, E., Saviano, M., and Silvestre, C., A comparison of theoretical predictions and experimental evidences of miscibility for poly(ethylene oxide)-based blends, *Makromol. Chem., Macromol. Symp.*, 38, 61,1990.

92KIM Kim, C.K. and Paul, D.R., Interaction parameters for blends containing polycarbonates. 2. Tetramethyl bisphenol-A polycarbonate - styrene copolymers, *Polymer*, 33, 2089, 1992. (experimental data by private communication with D. R. Paul)

92WIN Wind, R.W., Untersuchungen zum Phasenverhalten von Mischungen aus Ethylen, Acrylsäure und Ethylen-Acrylsäure-Copolymeren unter hohem Druck, *PhD-Thesis*, TH Darmstadt, 1992.

93ROD Rodgers, P.A., Pressure-volume-temperature relationships for polymeric liquids: A review of equations of state and their characteristic parameters for 56 polymers, *J. Appl. Polym. Sci.*, 48, 1061, 1993.

94PRI Privalko, V.P., Arbuzova, A.P., Korskanov, V.V., and Zagdanskaya, N.E., Thermodynamics of random copolymers of N-vinylcarbazole and 4-alkylstyrenes, *Polym. Intern.*, 35, 161, 1994.

94WAN Wang, Y.Z., Hsieh, K.H., Chen, L.W., and Tseng, H.C., Effect of compatibility on specific volume of molten polyblends, *J. Appl. Polym. Sci.*, 53, 1191, 1994.

95DAN Danusso, F., Levi, M., Gianotti, G., and Turri, S., Thermovolumetric properties versus glass transition temperature of diolic perfluoro-poly(oxyethylene-ran-oxymethylene) oligomers, *Macromol. Chem. Phys.*, 196, 2855, 1995.

95ZOL Zoller, P. and Walsh, D.J., *Standard Pressure-Volume-Temperature Data for Polymers*, Technomic Publishing, Lancaster, 1995.

96XUY Xu, Y., Charakterisierung von Styrol-Ethylen-Copolymeren und deren Mischungen mit Poly(2,6-dimethyl-1,4-phenylen ether), *Diploma paper*, Albert-Ludwigs-Univ., Freiburg i. Br., 1996. (experimental data by personal communication with Y. Thomann)

98SUH Suhm, J., Maier, R.-D., Kressler, J., and Mühlhaupt, R., Microstructure and pressure-volume-temperature properties of ethene/1-octene random copolymers, *Acta Polym.*, 49, 80, 1998. (experimental data by personal communication with R.-D. Maier)

99MAI Maier, R.-D., Thermodynamik und Kristallisation von Metallocen-Polyolefinen, *PhD-Thesis*, Albert-Ludwigs-Univ., Freiburg i. Br., 1999. (experimental data by personal communication with R.-D. Maier)

7. SECOND VIRIAL COEFFICIENTS (A_2) OF COPOLYMER SOLUTIONS

7.1. Experimental A_2 data

Average chemical composition of the copolymers, acronyms and references:

Copolymer (B)	Acronym	Ref.
Acrylonitrile/α-methylstyrene copolymer		
46.0 mol% acrylonitrile	AN-αMS/46x	79GL2
Acrylonitrile/vinylidene chloride copolymer		
59.0 wt% acrylonitrile	AN-VdC/59w	86KAM
Butadiene/styrene copolymer		
14.6 mol% styrene	B-S/15x(1)	79STA
15.5 mol% styrene	B-S/15x(2)	79STA
23.9 mol% styrene	B-S/24x	79STA
25.0 mol% styrene	B-S/25x	79STA
25.0 mol% styrene (9 mol% 1,4-*cis*, 51 mol% 1,4-*trans*)	B-S/25x(9/51)	61COO
25.0 mol% styrene (9 mol% 1,4-*cis*, 54 mol% 1,4-*trans*)	B-S/25x(9/54)	61COO
29.0 mol% styrene (7.8 mol% 1,4-*cis*, 50.4 mol% 1,4-*trans*)	B-S/29x(8/50)	61COO
30.0 mol% styrene (12.6 mol% 1,4-*cis*, 44.8 mol% 1,4-*trans*)	B-S/30x(13/45)	61COO
38.7 mol% styrene	B-S/39x(1)	79STA
39.4 mol% styrene	B-S/39x(2)	79STA
62.4 mol% styrene	B-S/62x	79STA
63.6 mol% styrene	B-S/64x	79STA
Butyl acrylate/methyl methacrylate copolymer		
29.5 wt% methyl methacrylate	BA-MMA/30w	70WU1
56.0 wt% methyl methacrylate	BA-MMA/56w	70WU1
79.5 wt% methyl methacrylate	BA-MMA/80w	70WU1

Copolymer (B)	Acronym	Ref.
Butyl methacrylate/methyl methacrylate copolymer		
50.0 wt% methyl methacrylate	BMA-MMA/50w	70WU2
Butylene/ethylethylene copolymer		
1:1, alternating	B-EE/50x	84MAY
Ethyl acrylate/methyl methacrylate copolymer		
50.0 wt% methyl methacrylate	EA-MMA/50w	70WU2
Ethylene/propylene copolymer		
1:1, alternating	E-P/50x	84MAY
Ethylene/tetrafluoroethylene copolymer		
1:1, alternating	E-TFE/50x	89CHU
Ethylene oxide/propylene oxide copolymer		
72.4 mol% ethylene oxide	EO-PO/72x	91LOU
79.5 mol% ethylene oxide	EO-PO/80x	91LOU
86.6 mol% ethylene oxide	EO-PO/87x	91LOU
9,9-bis(4-Hydroxyphenyl)fluorene/terephthaloyl chloride copolymer		
1:1, alternating	HPF-TPC/50x	77TA1
N-Isopropylacrylamide/acrylamide copolymer		
15 mol% acrylamide	NIPAM-AM/15x	94MUM
30 mol% acrylamide	NIPAM-AM/30x	94MUM
45 mol% acrylamide	NIPAM-AM/45x	94MUM
60 mol% acrylamide	NIPAM-AM/60x	94MUM
Isopropylethylene/1-methylbutylene copolymer		
1:1, alternating	IPE-MB/50x	84MAY
Isopropylethylene/1-methyl-1-ethylethylene copolymer		
1:1, alternating	IPE-MEE/50x	84MAY

Copolymer (B)	Acronym	Ref.
Methyl acrylate/methyl methacrylate copolymer		
50.0 wt% methyl methacrylate	MA-MMA/50w	70WU2
Methyl methacrylate/acrylonitrile copolymer		
23.6 mol% acrylonitrile	MMA-AN/24x	82MA2
50.0 mol% acrylonitrile	MMA-AN/50x	82MA2
74.0 mol% acrylonitrile	MMA-AN/74x	82MA2
1-Octadecene/maleic anhydride copolymer		
1:1, alternating	ODC-MAH/50x	85MAT
Oxyethylene/oxymethylene copolymer		
<5 wt% oxyethylene	OE-OM/05w	65WAG
Pyromellitic acid/4-amino-1,4-phenyleneoxy-1,4-phenyleneamine copolymer		
1:1, alternating	P-AA1/50x	77BIR
Pyromellitic acid/4-amino-1,4-phenyleneoxy-1,4-phenylenoxy-1,4-phenyleneamine copolymer		
1:1, alternating	P-AA2/50x	77BIR
Styrene/acrylonitrile copolymer		
25.0 wt% acrylonitrile	S-AN/25w	98SCH
38.5 mol% acrylonitrile	S-AN/38x	77GL1
38.5 mol% acrylonitrile	S-AN/38x	77GL2
38.5 mol% acrylonitrile	S-AN/38x	79GL1
40.0 mol% acrylonitrile	S-AN/40x	89EGO
41.0 mol% acrylonitrile	S-AN/41x	89EGO
51.0 mol% acrylonitrile	S-AN/51x	82MA1
Styrene/p-methoxystyrene copolymer		
26.0 mol% p-methoxystyrene	S-pMOS/26x(1)	72PIZ
26.4 mol% p-methoxystyrene	S-pMOS/26x(2)	72PIZ
53.0 mol% p-methoxystyrene	S-pMOS/53x	72PIZ
53.8 mol% p-methoxystyrene	S-pMOS/54x	72PIZ
75.6 mol% p-methoxystyrene	S-pMOS/76x(1)	72PIZ
75.6 mol% p-methoxystyrene	S-pMOS/76x(2)	72PIZ

Copolymer (B)	Acronym	Ref.
Styrene/methyl methacrylate copolymer		
29.8 mol% methyl methacrylate	S-MMA/30x	66KOT
43.8 mol% methyl methacrylate	S-MMA/44x	66KOT
46.4 wt% methyl methacrylate	S-MMA/46w	55STO
70.7 mol% methyl methacrylate	S-MMA/71x	66KOT
Tetrafluoroethylene/trifluoronitrosomethane copolymer		
1:1, alternating	TFE-TFNM/50x	61MOR
Vinyl acetate/vinyl chloride copolymer		
11.4 mol% vinyl acetate	VA-VC/11x	85MIN
25.0 mol% vinyl acetate	VA-VC/25x	85MIN
40.0 mol% vinyl acetate	VA-VC/40x	85MIN
66.0 mol% vinyl acetate	VA-VC/66x	85MIN
Vinyl acetate/1-vinyl-2-pyrrolidinone copolymer		
65.0 mol% 1-vinyl-2-pyrrolidinone	VA-NVP/65x	89EGO
71.0 mol% 1-vinyl-2-pyrrolidinone	VA-NVP/71x	89EGO
81.0 mol% 1-vinyl-2-pyrrolidinone	VA-NVP/81x	89EGO
86.0 mol% 1-vinyl-2-pyrrolidinone	VA-NVP/86x	89EGO
Vinyl chloride/butyl acrylate copolymer		
16.6 mol% butyl acrylate	VC-BA/17x	85MIN
40.0 mol% butyl acrylate	VC-BA/40x	85MIN
Vinyl chloride/butyl ether copolymer		
9.2 mol% butyl ether	VC-BE/09x	85MIN
Vinyl chloride/methyl acrylate copolymer		
25.5 mol% methyl acrylate	VC-MA/25x	85MIN
41.6 mol% methyl acrylate	VC-MA/42x	85MIN
64.0 mol% methyl acrylate	VC-MA/64x	85MIN
Vinyl chloride/vinylidene chloride copolymer		
15.0 mol% vinylidene chloride	VC-VdC/15x	85MIN
42.0 mol% vinylidene chloride	VC-VdC/42x	85MIN

Copolymer-solvent systems, measuring temperatures and A_2-values:

Copolymer (B)	M_n/ g/mol	M_w/ g/mol	Solvent (A)	T/ K	A_2 10^4/ cm^3mol/g^2
AN-αMS/46x		45000	tetrahydrofuran	298.15	7.19
AN-αMS/46x		54000	tetrahydrofuran	298.15	7.14
AN-αMS/46x		65000	tetrahydrofuran	298.15	7.28
AN-αMS/46x		72000	tetrahydrofuran	298.15	9.42
AN-αMS/46x		75000	tetrahydrofuran	298.15	9.06
AN-αMS/46x		79000	tetrahydrofuran	298.15	7.99
AN-αMS/46x		93000	tetrahydrofuran	298.15	8.35
AN-αMS/46x		104000	tetrahydrofuran	298.15	8.44
AN-αMS/46x		105000	tetrahydrofuran	298.15	6.48
AN-αMS/46x		137000	tetrahydrofuran	298.15	7.83
AN-αMS/46x		160000	tetrahydrofuran	298.15	8.26
AN-αMS/46x		170000	tetrahydrofuran	298.15	7.64
AN-αMS/46x		208000	tetrahydrofuran	298.15	8.88
AN-αMS/46x		218000	tetrahydrofuran	298.15	7.55
AN-αMS/46x		226000	tetrahydrofuran	298.15	6.40
AN-αMS/46x		268000	tetrahydrofuran	298.15	7.64
AN-αMS/46x		313000	tetrahydrofuran	298.15	6.93
AN-αMS/46x		332000	tetrahydrofuran	298.15	7.19
AN-αMS/46x		417000	tetrahydrofuran	298.15	6.66
AN-VdC/59w		204000	γ-butyrolactone	298.15	0.119
AN-VdC/59w		42000	N,N-dimethylacetamide	298.15	0.132
AN-VdC/59w		62000	N,N-dimethylacetamide	298.15	0.121
AN-VdC/59w		90000	N,N-dimethylacetamide	298.15	0.120
AN-VdC/59w		140000	N,N-dimethylacetamide	298.15	0.120
AN-VdC/59w		204000	N,N-dimethylacetamide	298.15	0.115
AN-VdC/59w		270000	N,N-dimethylacetamide	298.15	0.110
AN-VdC/59w		348000	N,N-dimethylacetamide	298.15	0.095
AN-VdC/59w		500000	N,N-dimethylacetamide	298.15	0.090
AN-VdC/59w		204000	N,N-dimethylformamide	298.15	0.137
B-S/15x(1)	30300	54800	tetrahydrofuran	296.15	14.4
B-S/15x(2)	131000	143000	tetrahydrofuran	296.15	11.2
B-S/24x	122000	132000	tetrahydrofuran	296.15	10.9
B-S/25x	50400	52900	tetrahydrofuran	296.15	13.9
B-S/25x	153000	200000	tetrahydrofuran	296.15	9.2

Copolymer (B)	M_n/ g/mol	M_w/ g/mol	Solvent (A)	T/ K	A_2 10^4/ cm^3mol/g^2
B-S/25x(9/51)	112000	800000	benzene	301.75	14.2
B-S/25x(9/51)	112000	800000	cyclohexane	301.75	0.46
B-S/25x(9/54)	84100	740000	benzene	301.75	12.0
B-S/25x(9/54)	84100	740000	cyclohexane	301.75	1.39
B-S/29x(8/50)	116000	1700000	benzene	301.75	12.8
B-S/29x(8/50)	116000	1700000	cyclohexane	301.75	1.62
B-S/30x(13/45)	58400	2670000	benzene	301.75	14.9
B-S/30x(13/45)	58400	2670000	cyclohexane	301.75	0.14
B-S/39x(2)	49700	52200	tetrahydrofuran	296.15	12.9
B-S/39x(1)	118000	131000	tetrahydrofuran	296.15	8.8
B-S/62x	137000	148000	tetrahydrofuran	296.15	7.8
B-S/64x	47200	50500	tetrahydrofuran	296.15	10.5
BA-MMA/30x		27000	2-propanone	293.15	9.5
BA-MMA/30x		46000	2-propanone	293.15	7.8
BA-MMA/30x		150000	2-propanone	293.15	5.0
BA-MMA/30x		240000	2-propanone	293.15	4.5
BA-MMA/30x		320000	2-propanone	293.15	4.3
BA-MMA/30x		500000	2-propanone	293.15	3.9
BA-MMA/30x		800000	2-propanone	293.15	3.6
BA-MMA/56x		24000	2-propanone	293.15	11.0
BA-MMA/56x		49000	2-propanone	293.15	8.5
BA-MMA/56x		91000	2-propanone	293.15	6.4
BA-MMA/56x		160000	2-propanone	293.15	5.3
BA-MMA/56x		300000	2-propanone	293.15	4.7
BA-MMA/56x		530000	2-propanone	293.15	4.25
BA-MMA/56x		710000	2-propanone	293.15	4.0
BA-MMA/80x		22000	2-propanone	293.15	8.0
BA-MMA/80x		43000	2-propanone	293.15	6.0
BA-MMA/80x		87000	2-propanone	293.15	5.2
BA-MMA/80x		125000	2-propanone	293.15	4.6
BA-MMA/80x		280000	2-propanone	293.15	3.5
BA-MMA/80x		530000	2-propanone	293.15	3.2
BA-MMA/80x		740000	2-propanone	293.15	2.6
BMA-MMA/50x		70000	2-propanone	293.15	3.2
BMA-MMA/50x		160000	2-propanone	293.15	2.8

Copolymer (B)	M_n/ g/mol	M_w/ g/mol	Solvent (A)	T/ K	A_2 10^4/ cm^3mol/g^2
BMA-MMA/50x		210000	2-propanone	293.15	2.5
BMA-MMA/50x		330000	2-propanone	293.15	2.3
BMA-MMA/50x		430000	2-propanone	293.15	2.2
B-EE/50x	19500	19600	toluene	310.15	9.5
B-EE/50x	54500	56700	toluene	310.15	8.4
B-EE/50x	95500	103000	toluene	310.15	7.8
B-EE/50x	129000	137000	toluene	310.15	7.55
B-EE/50x	209000	228000	toluene	310.15	7.1
EA-MMA/50w		95000	2-propanone	293.15	5.0
EA-MMA/50w		140000	2-propanone	293.15	4.3
EA-MMA/50w		220000	2-propanone	293.15	3.8
EA-MMA/50w		240000	2-propanone	293.15	3.8
EA-MMA/50w		500000	2-propanone	293.15	3.5
EA-MMA/50w		660000	2-propanone	293.15	3.5
E-P/50x	8200	8400	toluene	310.15	22.3
E-P/50x	14500	14900	toluene	310.15	18.4
E-P/50x	22500	24500	toluene	310.15	15.8
E-P/50x	34600	36300	toluene	310.15	13.7
E-P/50x	41600	46400	toluene	310.15	12.85
E-P/50x	91400	96000	toluene	310.15	9.8
E-P/50x	267000	301000	toluene	310.15	6.8
E-P/50x	293000	316000	toluene	310.15	6.6
E-P/50x	545000	600000	toluene	310.15	5.35
E-TFE/50x		540000	diisobutyl adipate	513.15	1.97
E-TFE/50x		1160000	diisobutyl adipate	513.15	1.02
E-TFE/50x	1605000	3210000	diisobutyl adipate	513.15	0.511
EO-PO/72x		36000	water	298.15	6.0
EO-PO/80x		30800	water	298.15	10.5
EO-PO/80x		32500	water	298.15	12.5
EO-PO/87x		30100	water	298.15	21.5
HPF-TPC/50x		90000	trichloromethane	298.15	5.0
HPF-TPC/50x		90000	1,1,2,2-tetrachloroethane	298.15	7.8

Copolymer (B)	M_n/ g/mol	M_w/ g/mol	Solvent (A)	T/ K	A_2 10^4/ cm^3mol/g^2
NIPAM-AM/15x		3100000	water	298.15	3.01
NIPAM-AM/30x		4500000	water	298.15	3.31
NIPAM-AM/45x		3900000	water	298.15	3.46
NIPAM-AM/60x		2200000	water	298.15	3.62
IPE-MB/50x	11000	11300	toluene	310.15	13.2
IPE-MB/50x	17400	17600	toluene	310.15	11.5
IPE-MB/50x	37100	37800	toluene	310.15	9.2
IPE-MB/50x	97500	101000	toluene	310.15	6.9
IPE-MB/50x	115000	119000	toluene	310.15	6.55
IPE-MB/50x	216000	227000	toluene	310.15	5.4
IPE-MB/50x	275000	286000	toluene	310.15	5.0
IPE-MB/50x	393000	425000	toluene	310.15	4.5
IPE-MEE/50x	14600	15000	toluene	310.15	5.9
IPE-MEE/50x	27000	29200	toluene	310.15	5.6
IPE-MEE/50x	68300	72900	toluene	310.15	5.1
IPE-MEE/50x	90300	107000	toluene	310.15	5.0
IPE-MEE/50x	127000	133000	toluene	310.15	4.9
IPE-MEE/50x	157000	170000	toluene	310.15	4.8
IPE-MEE/50x	258000	278000	toluene	310.15	4.6
MA-MMA/50w		80000	2-propanone	293.15	5.5
MA-MMA/50w		115000	2-propanone	293.15	4.8
MA-MMA/50w		135000	2-propanone	293.15	4.5
MA-MMA/50w		155000	2-propanone	293.15	4.2
MA-MMA/50w		200000	2-propanone	293.15	4.0
MA-MMA/50w		240000	2-propanone	293.15	3.9
MA-MMA/50w		420000	2-propanone	293.15	3.4
MA-MMA/50w		525000	2-propanone	293.15	3.5
MMA-AN/24x		400000	acetonitrile	303.15	1.522
MMA-AN/24x		440000	acetonitrile	303.15	1.52
MMA-AN/24x		580000	acetonitrile	303.15	1.46
MMA-AN/24x		750000	acetonitrile	303.15	1.415
MMA-AN/24x		1016000	acetonitrile	303.15	1.304
MMA-AN/24x		1240000	acetonitrile	303.15	1.24
MMA-AN/24x		1560000	acetonitrile	303.15	0.991

Copolymer (B)	M_n/ g/mol	M_w/ g/mol	Solvent (A)	T/ K	A_2 10^4/ cm^3mol/g^2
MMA-AN/50x		270000	acetonitrile	303.15	0.525
MMA-AN/50x		460000	acetonitrile	303.15	0.51
MMA-AN/50x		676000	acetonitrile	303.15	0.478
MMA-AN/50x		758000	acetonitrile	303.15	0.45
MMA-AN/50x		1150000	acetonitrile	303.15	0.422
MMA-AN/50x		1200000	acetonitrile	303.15	0.40
MMA-AN/50x		1830000	acetonitrile	303.15	0.315
MMA-AN/74x		600000	acetonitrile	303.15	0.292
MMA-AN/74x		832000	acetonitrile	303.15	0.255
MMA-AN/74x		871000	acetonitrile	303.15	0.248
MMA-AN/74x		1000000	acetonitrile	303.15	0.234
MMA-AN/74x		1047000	acetonitrile	303.15	0.225
MMA-AN/74x		1175000	acetonitrile	303.15	0.2075
MMA-AN/74x		1352000	acetonitrile	303.15	0.185
MMA-AN/74x		1530000	acetonitrile	303.15	0.16
MMA-AN/74x		1700000	acetonitrile	303.15	0.14
ODC-MAH/50x		28000	ethyl acetate	298.15	3.0
ODC-MAH/50x		37000	ethyl acetate	298.15	2.9
ODC-MAH/50x		51000	ethyl acetate	298.15	2.8
ODC-MAH/50x		68000	ethyl acetate	298.15	2.5
ODC-MAH/50x		99000	ethyl acetate	298.15	2.1
OE-OM/05w	34000	62000	1H,1H,5H-octafluoro-1-pentanol	383.15	22.0
OE-OM/05w	40000	72000	1H,1H,5H-octafluoro-1-pentanol	383.15	19.0
OE-OM/05w	44000	96000	1H,1H,5H-octafluoro-1-pentanol	383.15	16.0
OE-OM/05w	56000	129000	1H,1H,5H-octafluoro-1-pentanol	383.15	17.5
P-AA1/50x		90000	N,N-dimethylacetamide	298.15	11.0
P-AA1/50x		130000	N,N-dimethylacetamide	298.15	10.0
P-AA1/50x		150000	N,N-dimethylacetamide	298.15	10.5
P-AA1/50x		350000	N,N-dimethylacetamide	298.15	9.5

Copolymer (B)	M_n/ g/mol	M_w/ g/mol	Solvent (A)	T/ K	A_2 10^4/ cm^3mol/g^2
P-AA2/50x		20000	N,N-dimethylacetamide	298.15	20.0
P-AA2/50x		50000	N,N-dimethylacetamide	298.15	17.0
P-AA2/50x		80000	N,N-dimethylacetamide	298.15	15.0
P-AA2/50x		90000	N,N-dimethylacetamide	298.15	20.0
S-AN/25w	90000	147000	tetrahydrofuran	298.15	11.5
S-AN/38x		96000	2-butanone	298.15	4.72
S-AN/38x		197000	2-butanone	298.15	7.70
S-AN/38x		197000	2-butanone	298.15	7.72
S-AN/38x		238000	2-butanone	298.15	4.15
S-AN/38x		264000	2-butanone	298.15	5.45
S-AN/38x		265000	2-butanone	298.15	5.43
S-AN/38x		323000	2-butanone	298.15	4.04
S-AN/38x		543000	2-butanone	298.15	3.67
S-AN/38x		601000	2-butanone	298.15	5.25
S-AN/38x		602000	2-butanone	298.15	5.23
S-AN/38x		647000	2-butanone	298.15	4.62
S-AN/38x		648000	2-butanone	298.15	4.62
S-AN/38x		756000	2-butanone	298.15	4.80
S-AN/38x		761000	2-butanone	298.15	4.77
S-AN/38x		832000	2-butanone	298.15	5.03
S-AN/38x		836000	2-butanone	298.15	5.01
S-AN/38x		1130000	2-butanone	298.15	4.89
S-AN/38x		1390000	2-butanone	298.15	3.59
S-AN/38x		144000	tetrahydrofuran	298.15	8.47
S-AN/38x		145000	tetrahydrofuran	298.15	8.29
S-AN/38x		150000	tetrahydrofuran	298.15	8.88
S-AN/38x		230000	tetrahydrofuran	298.15	7.40
S-AN/38x		286000	tetrahydrofuran	298.15	6.11
S-AN/38x		605000	tetrahydrofuran	298.15	7.81
S-AN/38x		645000	tetrahydrofuran	298.15	7.38
S-AN/38x		825000	tetrahydrofuran	298.15	7.55
S-AN/38x		1072000	tetrahydrofuran	298.15	7.74
S-AN/38x		1123000	tetrahydrofuran	298.15	6.83
S-AN/38x		1873000	tetrahydrofuran	298.15	4.30
S-AN/40x	77000		50 mol% acrylonitrile + 50 mol% styrene	298.15	12.0

Copolymer (B)	M_n/ g/mol	M_w/ g/mol	Solvent (A)	T/ K	A_2 10^4/ cm^3mol/g^2
S-AN/41x	382000		50 mol% acrylonitrile + 50 mol% styrene	298.15	6.7
S-AN/41x	842000		50 mol% acrylonitrile + 50 mol% styrene	298.15	5.5
S-AN/41x	1076000		50 mol% acrylonitrile + 50 mol% styrene	298.15	3.6
S-AN/51x		269000	ethyl acetate	303.15	4.275
S-AN/51x		347000	ethyl acetate	303.15	2.36
S-AN/51x		457000	ethyl acetate	303.15	1.89
S-AN/51x		794000	ethyl acetate	303.15	1.23
S-AN/51x		912000	ethyl acetate	303.15	1.204
S-AN/51x		1365000	ethyl acetate	303.15	1.138
S-AN/51x		2240000	ethyl acetate	303.15	1.108
S-pMOS/26x(1)		61000	toluene	298.15	5.34
S-pMOS/26x(2)	71000		toluene	298.15	5.0
S-pMOS/26x(2)	86000		toluene	298.15	5.6
S-pMOS/26x(1)		93000	toluene	298.15	5.62
S-pMOS/26x(2)	105000		toluene	298.15	4.6
S-pMOS/26x(1)		131000	toluene	298.15	5.54
S-pMOS/26x(2)	137000		toluene	298.15	4.3
S-pMOS/26x(2)	171000		toluene	298.15	4.3
S-pMOS/26x(1)		200000	toluene	298.15	4.99
S-pMOS/26x(1)		245000	toluene	298.15	4.64
S-pMOS/26x(1)		285000	toluene	298.15	4.69
S-pMOS/26x(2)	308000		toluene	298.15	3.8
S-pMOS/26x(1)		348000	toluene	298.15	4.67
S-pMOS/26x(2)	350000		toluene	298.15	3.0
S-pMOS/26x(1)		420000	toluene	298.15	4.49
S-pMOS/26x(1)		665000	toluene	298.15	4.29
S-pMOS/53x		66000	toluene	298.15	5.46
S-pMOS/53x		104000	toluene	298.15	5.33
S-pMOS/53x		168000	toluene	298.15	4.68
S-pMOS/53x		269000	toluene	298.15	4.26
S-pMOS/53x		292000	toluene	298.15	4.22
S-pMOS/53x		446000	toluene	298.15	3.97
S-pMOS/53x		564000	toluene	298.15	3.98
S-pMOS/53x		642000	toluene	298.15	3.86
S-pMOS/53x		920000	toluene	298.15	3.82
S-pMOS/53x		1346000	toluene	298.15	3.81
S-pMOS/53x		1783000	toluene	298.15	3.41

Copolymer (B)	M_n/ g/mol	M_w/ g/mol	Solvent (A)	T/ K	A_2 10^4/ cm^{3}mol/g^2
S-pMOS/54x	35500		toluene	298.15	4.1
S-pMOS/54x	50500		toluene	298.15	5.2
S-pMOS/54x	60000		toluene	298.15	4.2
S-pMOS/54x	72000		toluene	298.15	3.6
S-pMOS/54x	94000		toluene	298.15	4.0
S-pMOS/54x	100000		toluene	298.15	3.6
S-pMOS/54x	121000		toluene	298.15	3.5
S-pMOS/54x	154000		toluene	298.15	3.2
S-pMOS/54x	234000		toluene	298.15	4.2
S-pMOS/54x	400000		toluene	298.15	3.8
S-pMOS/54x	660000		toluene	298.15	3.9
S-pMOS/54x	701000		toluene	298.15	3.2
S-pMOS/76x(2)	44000		toluene	298.15	5.9
S-pMOS/76x(2)	61000		toluene	298.15	5.4
S-pMOS/76x(1)		79000	toluene	298.15	4.54
S-pMOS/76x(1)		118000	toluene	298.15	4.50
S-pMOS/76x(2)	119000		toluene	298.15	4.1
S-pMOS/76x(2)	169000		toluene	298.15	3.9
S-pMOS/76x(1)		184000	toluene	298.15	3.92
S-pMOS/76x(2)	187000		toluene	298.15	4.2
S-pMOS/76x(2)	232000		toluene	298.15	3.2
S-pMOS/76x(1)		279000	toluene	298.15	3.72
S-pMOS/76x(2)	294000		toluene	298.15	3.0
S-pMOS/76x(2)	421000		toluene	298.15	3.5
S-pMOS/76x(1)		431000	toluene	298.15	3.55
S-pMOS/76x(1)		730000	toluene	298.15	3.30
S-pMOS/76x(1)		1149000	toluene	298.15	3.05
S-pMOS/76x(1)		1717000	toluene	298.15	3.00
S-MMA/30x	342000		cyclohexanol	339.15	0.0
S-MMA/30x	342000		2-ethoxyethanol	345.95	0.0
S-MMA/44x	350000		cyclohexanol	334.45	0.0
S-MMA/44x	350000		2-ethoxyethanol	331.55	0.0
S-MMA/46w		1330000	2-butanone	298.15	1.8
S-MMA/46w		1330000	1,4-dioxane	298.15	3.3
S-MMA/46w		1330000	nitroethane	298.15	1.5
S-MMA/46w		1330000	tetrachloromethane	298.15	2.3
S-MMA/46w		1900000	2-butanone	298.15	1.5
S-MMA/46w		1900000	1,4-dioxane	298.15	3.6
S-MMA/46w		1900000	nitroethane	298.15	1.2
S-MMA/46w		1900000	tetrachloromethane	298.15	2.4

Copolymer (B)	M_n/ g/mol	M_w/ g/mol	Solvent (A)	T/ K	A_2 10^4/ cm^3mol/g^2
S-MMA/71x	354000		cyclohexanol	341.15	0.0
S-MMA/71x	354000		2-ethoxyethanol	313.15	0.0
TFE-TFNM/50x		283000	1,1,2-trichloro-1,2,2-trifluoroethane	308.15	0.745
TFE-TFNM/50x		437000	1,1,2-trichloro-1,2,2-trifluoroethane	308.15	0.0855
TFE-TFNM/50x		861000	1,1,2-trichloro-1,2,2-trifluoroethane	308.15	0.100
TFE-TFNM/50x		1080000	1,1,2-trichloro-1,2,2-trifluoroethane	308.15	0.235
TFE-TFNM/50x		1270000	1,1,2-trichloro-1,2,2-trifluoroethane	308.15	0.214
TFE-TFNM/50x		1390000	1,1,2-trichloro-1,2,2-trifluoroethane	308.15	0.238
TFE-TFNM/50x		1970000	1,1,2-trichloro-1,2,2-trifluoroethane	308.15	0.173
VA-NVP65x	61000		30 mol% vinylacetate/70 mol% 1-vinyl-2-pyrrolidinone	298.15	−11.1
VA-NVP65x	126000		30 mol% vinylacetate/70 mol% 1-vinyl-2-pyrrolidinone	298.15	− 4.3
VA-NVP65x	148000		30 mol% vinylacetate/70 mol% 1-vinyl-2-pyrrolidinone	298.15	− 0.6
VA-NVP65x	631000		30 mol% vinylacetate/70 mol% 1-vinyl-2-pyrrolidinone	298.15	+ 1.1
VA-VC/11x		29500	tetrahydrofuran	298.15	0.119
VA-VC/11x		42500	tetrahydrofuran	298.15	0.118
VA-VC/11x		96700	tetrahydrofuran	298.15	0.089
VA-VC/25x		21000	tetrahydrofuran	298.15	0.148
VA-VC/25x		49900	tetrahydrofuran	298.15	0.082
VA-VC/25x		185000	tetrahydrofuran	298.15	0.050
VA-VC/40x		43000	tetrahydrofuran	298.15	0.118
VA-VC/40x		79000	tetrahydrofuran	298.15	0.092
VA-VC/40x		179000	tetrahydrofuran	298.15	0.064

Copolymer (B)	M_n/ g/mol	M_w/ g/mol	Solvent (A)	T/ K	A_2 10^4/ cm^3mol/g^2
VA-VC/66x		100000	tetrahydrofuran	298.15	0.028
VA-VC/66x		248000	tetrahydrofuran	298.15	0.025
VA-VC/66x		708000	tetrahydrofuran	298.15	0.012
VC-BA/17x		173000	tetrahydrofuran	298.15	0.052
VC-BA/17x		242000	tetrahydrofuran	298.15	0.055
VC-BA/17x		302000	tetrahydrofuran	298.15	0.065
VC-BA/40x		205000	tetrahydrofuran	298.15	0.024
VC-BA/40x		632000	tetrahydrofuran	298.15	0.018
VC-BE/09x		30000	tetrahydrofuran	298.15	0.115
VC-BE/09x		53000	tetrahydrofuran	298.15	0.083
VC-BE/09x		98000	tetrahydrofuran	298.15	0.077
VC-MA/25x		221000	tetrahydrofuran	298.15	0.057
VC-MA/25x		476000	tetrahydrofuran	298.15	0.052
VC-MA/42x		121000	tetrahydrofuran	298.15	0.080
VC-MA/42x		320000	tetrahydrofuran	298.15	0.047
VC-MA/42x		891000	tetrahydrofuran	298.15	0.032
VC-MA/64x		379000	tetrahydrofuran	298.15	0.014
VC-MA/64x		1060000	tetrahydrofuran	298.15	0.022
VC-MA/64x		1340000	tetrahydrofuran	298.15	0.033
VC-MA/64x		2510000	tetrahydrofuran	298.15	0.022
VC-VdC/15x		50000	tetrahydrofuran	298.15	0.091
VC-VdC/15x		121000	tetrahydrofuran	298.15	0.070
VC-VdC/15x		223000	tetrahydrofuran	298.15	0.045
VC-VdC/42x		79000	tetrahydrofuran	298.15	0.026
VC-VdC/42x		133000	tetrahydrofuran	298.15	0.027
VC-VdC/42x		187000	tetrahydrofuran	298.15	0.040

7.2. References

55STO Stockmayer, W.H., Moore, L.D., Fixman, M., and Epstein, B.N., Copolymers in dilute solution. I. Preliminary results for styrene-methyl methacrylate, *J. Polym. Sci.*, 16, 517, 1955.

61COO Cooper, W., Vaughan, G., Eaves, D.E., and Madden, R.W., Molecular weight distribution and branching in butadiene polymers and copolymers, *J. Polym. Sci.*, 50, 159, 1961.

61MOR Morneau, G.A., Roth, P.I., and Shultz, A.R., Trifluoronitrosomethane/tetrafluoroethylene elastomers dilute solution properties and molecular weight, *J. Polym. Sci.*, 55, 609, 1961.

65WAG Wagner, H.L. and Wissbrun, K.F, Molecular weight and rheology of acetal copolymers, *Makromol. Chem.*, 81, 14, 1965.

66KOT Kotaka, T., Ohnuma, H., and Murakami, S., The theta-condition for random and block copolymers of styrene and methyl methacrylate, Y, *J. Phys. Chem.*, 70, 4099, 1966.

70WU1 Wunderlich, W., Lösungseigenschaften von statistischen Copolymeren aus Methylmethacrylat und n-Butylacrylat, *Angew. Makromol. Chem.*, 11, 73, 1970.

70WU2 Wunderlich, W., Flexibilität und thermodynamische Eigenschaften von Polyacrylaten, Polymethacrylaten und statistischen Acrylat/Methacrylat-Copolymeren, *Angew. Makromol. Chem.*, 11, 189, 1970.

72PIZ Pizzoli, M. and Ceccorulli, G., Solution properties of styrene-p-methoxystyrene random copolymers II, *Eur. Polym. J.*, 8, 769, 1972.

77BIR Birshtein, T.M., Zubkov, V.A., Milevskaya, I.S., Eskin, V.E., and Baranovskaya, I.A., Flexibility of aromatic polyimides and polyamidoacids, *Eur. Polym. J.*, 13, 375, 1977.

77GL1 Gloeckner, G. and Mauksch, D., Streulichtmessungen an azeotropen Styrol-Acrylnitril-Copolymeren in Butanon, *Z. Phys. Chem., Leipzig*, 258, 1142, 1977.

77GL2 Gloeckner, G., Über das Verhalten von Styrol-Acrylnitril-Copolymeren in Lösung III, *Faserforsch. Textiltechn.*, 28, 111, 1977.

77TA1 Tager, A.A., Kolmakova, L.K., Anufriev, V.A., Bessonov, Yu.S., Zhigunova, O.A., Vinogradova, S.V., Salazkin, S.N., and Tsilipotkina, M.V., Thermodynamics of the dissolution of cardo polyarylates in chloroform and tetrachloroethane (Russ.), *Vysokomol. Soedin., Ser. A*, 19, 2367, 1977.

79GL1 Gloeckner, G., Francuskiewicz, F., and Reichardt, H.-U., On the behaviour of styrene-acrylonitrile copolymers in solution IV. Light scattering investigation of azeotropic copolymers in tetrahydrofuran, *Acta Polym.*, 30, 551, 1979.

79GL2 Gloeckner, G., Francuskiewicz, F., and Reichardt, H.-U., On the behaviour of styrene-acrylonitrile copolymers in solution V. Light scattering investigation of α-methylstyrene copolymers with about 46 mol% acrylonitrile in tetrahydrofuran as a solvent, *Acta Polym.*, 30, 628, 1979.

79STA Stacy, C.J. and Kraus, G., Second virial coefficients of homopolymers and copolymers of butadiene and styrene in tetrahydrofuran, *J. Polym. Sci., Polym. Phys. Ed.*, 17, 2007, 1979.

82MA1 Mangalam, P.V. and Kalpagam, V., Styrene-acrylonitrile random copolymer in ethyl acetate, *J. Polym. Sci., Polym. Phys. Ed.*, 20, 773, 1982.

82MA2 Mangalam, P.V. and Kalpagam, V., Behaviour of methylmethacrylate-acrylonitrile random copolymers in dilute solution, *Polymer*, 23, 991, 1982.

84MAY Mays, J., Hadjichristidis, N., and Fetters, L.J., Characteristic ratios of model polydienes and polyolefins, *Macromolecules*, 17, 2723, 1984.

85MAT Matsuo, K., Stockmayer, W.H., and Bangerter, F., Conformational properties of poly(1-octadecene/maleic anhydride) in solution, *Macromolecules*, 18, 1346, 1985.

85MIN	Minsker, K.S., Panchesnikova, R.B., Monakov, Yu.B., and Zaikov, G.E., Thermodynamic and hydrodynamic properties of chlorine-containing carbochain polymers in solutions, *Eur. Polym. J.*, 21, 981, 1985.
86KAM	Kamide, K., Miyazaki, Y., and Yamazaki, H., Dilute solution properties of acrylonitrile/ vinylidene chloride copolymer, *Polym. J.*, 18, 645, 1986.
89EGO	Egorochkin, G.A., Semchikov, Yu.D., Smirnova, L.A., Knyazeva, T.E., Tokhonova, Z.A., Karyakin, N.V., and Sveshnikova, T.G., Thermodynamic analysis of the copolymerization of styrene with acrylonitrile and N-vinylpyrrolidone with vinyl acetate (Russ.), *Vysokomol. Soedin., Ser. B*, 31, 46, 1989.
89CHU	Chu, B., Wu, C., and Buek, W., Light scattering characterization of an alternating copolymer of ethylene and tetrafluoroethylene. 3. Temperature dependence of polymer size, *Macromolecules*, 22, 371, 1989.
91LOU	Louai, A., Sarazin, D., Pollet, G., Francois, J., and Moreaux, F., Properties of ethylene oxide-propylene oxide statistical copolymers in aqueous solution, *Polymer*, 32, 703, 1991.
94MUM	Mumick, P.S., and McCormick, C.L., Water soluble copolymers. 54: N-isopropyl-acrylamide-co-acrylamide copolymers in drag reduction: synthesis, characterization, and dilute solution behavior, *Polym. Eng. Sci.*, 34, 1419, 1994.
98SCH	Schneider, A., Homopolymer- und Copolymerlösungen im Vergleich: Wechselwirkungs-parameter und Grenzflächenspannung, *PhD-Thesis*, Univ. Mainz, 1998.

8. APPENDICES

8.1. List of copolymer acronyms

Acronym	Copolymer	Page(s)
AN-B	Acrylonitrile/butadiene copolymer	22, 24, 29-32, 88-91, 269 270, 307-308, 346-348
AN-IP	Acrylonitrile/isoprene copolymer	271
AN-MA	Acrylonitrile/methyl acrylate copolymer	308-309
AN-MPCHEMA	Acrylonitrile/2-(3-methyl-3-phenylcyclobutyl)-2-hydroxyethyl methacrylate copolymer	91-93, 309-310
AN-αMS	Acrylonitrile/α-methylstyrene copolymer	310-311, 431, 435
AN-VC	Acrylonitrile/vinyl chloride copolymer	272-273
AN-VdC	Acrylonitrile/vinylidene chloride copolymer	95, 273-274, 431, 435
B-EE	Butylene/ethylethylene copolymer	432, 436-437
BA-MMA	Butyl acrylate/methyl methacrylate copolymer	431, 436
BAIPC-TPC	Bisphenol A-isophthaloyl chloride/terephthaloyl chloride copolymer	274-275
BMA-IBMA	Butyl methacrylate/isobutyl methacrylate copolymer	281
BMA-MMA	Butyl methacrylate/methyl methacrylate copolymer	282, 431, 436-437
BO-EO	Poly(butylene oxide)-b-poly(ethylene oxide) diblock copolymer	187
pBS-pMS	p-Bromostyrene/p-methylstyrene copolymer	35-36
CBL-PHA	ε-Carbobenzoxy-L-lysine/L-phenylalanine copolymer	283-284
CP-MMA	Chloroprene/methyl methacrylate copolymer	314
CP-MMA-MA	Chloroprene/methyl methacrylate/methacrylic acid terpolymer	314-315
DEA-VFe	N,N-Diethylacrylamide/vinylferrocene copolymer	148
DEM-VA	Diethylmaleate/vinyl acetate copolymer	315-316
DMA-BA	N,N-Dimethylacrylamide/butyl acrylate copolymer	175, 178
DMA-BOEA	N,N-Dimethylacrylamide/2-butoxyethyl acrylate copolymer	175, 178
DMA-EA	N,N-Dimethylacrylamide/ethyl acrylate copolymer	175, 178
DMA-EOEA	N,N-Dimethylacrylamide/2-ethoxyethyl acrylate copolymer	175, 178
DMA-MA	N,N-Dimethylacrylamide/methyl acrylate copolymer	176, 178
DMA-MOEA	N,N-Dimethylacrylamide/2-methoxyethyl acrylate copolymer	175, 178
DMA-PA	N,N-Dimethylacrylamide/propyl acrylate copolymer	175, 178
DMS-BAC	Dimethylsiloxane/bisphenol-A carbonate copolymer	106-108
DMS-b-S	Poly(dimethylsiloxane)-b-polystyrene diblock copolymer	163
DMS-MPS	Dimethylsiloxane/methylphenylsiloxane copolymer	47-50, 149-151, 176, 179
E-AA	Ethylene/acrylic acid copolymer	188-191, 229-233, 350-355
E-B	Ethylene/1-butene copolymer	356-363

Acronym	Copolymer	Page(s)
E-BA	Ethylene/butyl acrylate copolymer	192-194
E-CO	Ethylene/carbon monoxide copolymer	108-109, 316-317
E-H	Ethylene/1-hexene copolymer	194-196, 233-240
E-MA	Ethylene/methyl acrylate copolymer	196-198
E-MAA	Ethylene/methacrylic acid copolymer	363-370
E-O	Ethylene/1-octene copolymer	198-199, 370-383
E-P	Ethylene/propylene copolymer	22, 109-110, 176, 179-181, 200-204, 241-243, 284-292, 317-318, 383-398, 432, 437
E-P-D	Ethylene/propylene/diene terpolymer	109-110, 317-318
E-P-IP	Ethylene/propylene/isoprene terpolymer	200, 205-206
E-S	Ethylene/styrene copolymer	399-405
E-TFE	Ethylene/tetrafluoroethylene copolymer	432, 437
E-VA	Ethylene/vinyl acetate copolymer	50-62, 111-117, 153, 176, 181, 206-221, 243-263, 292-293, 319-321, 406-421
E-VAL	Ethylene/vinyl alcohol copolymer	176, 181, 321-322, 421-423
EA-MMA	Ethyl acrylate/methyl methacrylate copolymer	432, 437
EMA-MA	Ethyl methacrylate/methacrylic acid copolymer	117-118
EO-b-MMA	Poly(ethylene oxide)-b-poly(methyl methacrylate) diblock copolymer	64-65
EO-b-tBMA	Poly(ethylene oxide)-b-poly(*tert*-butyl methacrylate) diblock copolymer	62-63
EO-MMA-EO	Poly(ethylene oxide)-b-poly(methyl methacrylate)-b-poly(ethylene oxide) triblock copolymer	64-65
EO-PO	Ethylene oxide/propylene oxide copolymer	66-67, 164-170, 176, 181, 432, 437
EO-b-PO	Poly(ethylene oxide)-b-poly(propylene oxide) diblock copolymer	66-67, 221-223, 293-294
EO-PO-EO	Poly(ethylene oxide)-b-poly(propylene oxide)-b-poly(ethylene oxide) triblock copolymer	66-68, 154, 222-224, 293-295
EO-S-EO	Poly(ethylene oxide)-b-polystyrene-b-poly(ethylene oxide) triblock copolymer	69-70
ET-pHBA	Ethylene terephthalate/p-hydroxybenzoic acid copolymer	170-171
HBA-HPA	3-Hydroxybutanoic acid/3-hydroxypentanoic acid copolymer	70-72
HPF-TPC	9,9-bis(4-Hydroxyphenyl)fluorene/terephthaloyl chloride copolymer	432, 437
IBMA-AA	Isobutyl methacrylate/acrylic acid copolymer	118-119
IBMA-MMA	Isobutyl methacrylate/methyl methacrylate copolymer	295-296
IPE-MB	1-Methylbutylene/isopropylethylene copolymer	432
IPE-MEE	Isopropylethylene/1-methyl-1-ethylethylene copolymer	432, 438
MA-MMA	Methyl acrylate/methyl methacrylate copolymer	432, 438
MAH-DEG	Maleic anhydride/diethylene glycol copolymer	177, 181
MMA-AN	Methyl methacrylate/acrylonitrile copolymer	433, 438-439
NCPAM-VFe	N-Cyclopropylacrylamide/vinylferrocene copolymer	147

Acronym	Copolymer	Page(s)
NEAM-VFe	N-Ethylacrylamide/vinylferrocene copolymer	151-152
NIPAM-AM	N-Isopropylacrylamide/acrylamide copolymer	177, 181, 432, 438
NIPAM-IA	N-Isopropylacrylamide/itaconic acid copolymer	155
NIPAM-VFe	N-Isopropylacrylamide/vinylferrocene copolymer	156
ODC-MAH	1-Octadecene/maleic anhydride copolymer	433, 439
OE-OM	Oxyethylene/oxymethylene copolymer	433, 439
P-AA1	Pyromellitic acid/4-amino-1,4-phenyleneoxy-1,4-phenyleneamine copolymer	433, 439
P-AA2	Pyromellitic acid/4-amino-1,4-phenyleneoxy-1,4-phenylenoxy-1,4-phenyleneamine copolymer	433, 440
PF(OE-OM)	Perfluoro(oxyethylene/oxymethylene) copolymer	424-427
PHTH-TC	Phenolphthalein/terephthaloyl chloride copolymer	336
PO-EO-PO	Poly(propylene oxide)-b-poly(ethylene oxide)-b-poly(propylene oxide) triblock copolymer	66, 68, 222-225
S-αMS	Styrene/α-methylstyrene copolymer	157-159, 177, 182
S-AA	Styrene/acrylic acid copolymer	119-120
S-AM	Styrene/acrylamide copolymer	173
S-AN	Acrylonitrile/styrene copolymer	32-35, 93-94, 145-146, 177, 181-182, 341-346, 433, 439-441
S-BR	Butadiene/styrene copolymer	37-44, 96-104, 275-280, 311-313, 348-349
S-B-S	Polystyrene-b-polybutadiene-b-polystyrene triblock copolymer	37-38, 46, 96-97, 105, 173, 311-313, 348-349
S-B(h)-S	Polystyrene-b-polybutadiene(hydrogenated)-b-polystyrene triblock copolymer	37-38, 47
S-?-B	Butadiene/styrene copolymer (block copolymer without specification)	37-38, 44-46, 348-349
S-BMA	Styrene/butyl methacrylate copolymer	72-73, 120-121, 296-298, 322-323
gS-CDA	Styrene/cellulose diacetate graft copolymer	177, 182
S-DCM	Styrene/docosyl maleate copolymer	73-74
S-DDM	Styrene/dodecyl maleate copolymer	75-76
S-DMAM	Styrene/N,N-dimethyl aminoethyl methacrylate copolymer	122-123
S-DVB	Styrene/divinylbenzene copolymer	323-324
S-EA	Styrene/ethyl acrylate copolymer	325-326
S-b-EO	Polystyrene-b-poly(ethylene oxide) diblock copolymer	69-70
S-I-S	Polystyrene-b-polyisoprene-b-polystyrene triblock copolymer	76-77
S-IBMA	Styrene/isobutyl methacrylate copolymer	121-122
S-MAA	Styrene/methacrylic acid copolymer	173
S-MAH	Styrene/maleic anhydride copolymer	78
S-MAH-MA	Styrene/maleic anhydride/methacrylic acid terpolymer	327
S-MEMA	Styrene/2-methoxyethyl methacrylate copolymer	174
S-b-MMA	Polystyrene-b-poly(methyl methacrylate) diblock copolymer	79, 81-82

Acronym	Copolymer	Page(s)
S-MMA	Styrene/methyl methacrylate copolymer	79-81, 177, 182, 328, 427-428, 434, 442-443
S-pMOS	Styrene/p-methoxystyrene copolymer	433, 441-442
S-MPCHEMA	Styrene/2-(3-methyl-3-phenylcyclobutyl)-2-hydroxyethyl methacrylate copolymer	124-125, 329
S-NMA	Styrene/nonyl methacrylate copolymer	125-126, 330
S-NVP	Styrene/1-vinyl-2-pyrrolidinone copolymer	174
S-PM	Styrene/pentyl maleate copolymer	83-84
S-4VP	Styrene/4-vinylpyridine copolymer	174
TFE-HFP	Tetrafluoroethylene/hexafluoropropylene copolymer	228, 265
TFE-PFD	Tetrafluoroethylene/2,2-bis(trifluoromethyl)-4,5-difluoro-1,3-dioxole copolymer	126-127, 228, 331
TFE-TFNM	Tetrafluoroethylene/trifluoronitrosomethane copolymer	177, 182, 434, 443
VA-MAA	Vinyl acetate/methacrylic acid copolymer	174
VA-NVP	Vinyl acetate/1-vinyl-2-pyrrolidinone copolymer	162, 434, 443
VA-VAL	Vinyl acetate/vinyl alcohol copolymer	87-88, 127-134, 298-304,
VA-VC	Vinyl acetate/vinyl chloride copolymer	84-86, 134-139, 305, 434, 443-444
VA-VC-GMA	Vinyl acetate/vinyl chloride/glycidyl methacrylate terpolymer	134-135, 139
VA-VC-HPA	Vinyl acetate/vinyl chloride/hydroxypropyl acrylate terpolymer	134-135, 140
VA-VCPL	Vinyl acetate/vinyl caprolactam copolymer	162
VAA-VIBA	N-Vinylacetamide/N-vinylisobutyramide copolymer	162
VAL-VBU	Vinyl alcohol/vinyl butyrate copolymer	162, 177, 182
VAMN-VCPL	Vinyl amine/vinyl caprolactam copolymer	177, 182
VC-BA	Vinyl chloride/butyl acrylate copolymer	434, 444
VC-BE	Vinyl chloride/butyl ether copolymer	434, 444
VC-MA	Vinyl chloride/methyl acrylate copolymer	434, 444
VC-VdC	Vinyl chloride/vinylidene chloride copolymer	140-141, 306, 332, 434, 444
VCPL-VMA	Vinyl caprolactam/N-vinyl-N-methylacetamide copolymer	162
VCZ-ES	N-Vinylcarbazole/4-ethylstyrene copolymer	428-429
VCZ-HS	N-Vinylcarbazole/4-hexylstyrene copolymer	428-429
VCZ-OS	N-Vinylcarbazole/4-octylstyrene copolymer	428-429
VCZ-PS	N-Vinylcarbazole/4-pentylstyrene copolymer	428-429
VdF-HFP	Vinylidene fluoride/hexafluoropropylene copolymer	228
VIBA-VVA	N-Vinylisobutyramide/N-vinylvaleramide copolymer	162
VTZ-MVTZ	5-Vinyltetrazole/2-methyl-5-vinyltetrazole copolymer	336

8.2. List of systems and properties in order of the copolymers

Copolymer (B)	Solvent (A)	Property	Page(s)
Acrylamide/N,N-dimethylaminoethyl methacrylate copolymer			
	water	LLE	160
Acrylamide/N-isopropylacrylamide copolymer *see* N-Isopropylacrylamide/acrylamide copolymer			
Acrylamide/styrene copolymer *see* Styrene/acrylamide copolymer			
Acrylic acid/ethylene copolymer *see* Ethylene/acrylic acid copolymer			
Acrylic acid/isobutyl methacrylate copolymer *see* Isobutyl methacrylate/acrylic acid copolymer			
Acrylic acid/N-isopropylacrylamide copolymer			
	water	LLE	160
Acrylic acid/methyl acrylate copolymer			
	water	LLE	160
	water	Enthalpy	333
Acrylic acid/methyl methacrylate copolymer			
	water	Enthalpy	333
Acrylic acid/2-methyl-5-vinylpyridine copolymer			
	water	LLE	160
Acrylic acid/nonyl acrylate copolymer			
	bisphenol-A-diglycidyl ether	LLE	160

Copolymer (B)	Solvent (A)	Property	Page(s)
Acrylic acid/styrene copolymer *see* Styrene/acrylic acid copolymer			
Acrylonitrile/butadiene copolymer			
	acetonitrile	VLE	29-31
	benzene	Henry	89-91
	benzene	Enthalpy	270, 307
	benzene	Enthalpy	308
	1,3-butadiene	Henry	91
	n-butane	Henry	91
	butylcyclohexane	Henry	90
	1-chlorobutane	Henry	90
	cyclohexane	VLE	30-31
	cyclohexane	Henry	90-91
	cyclohexane	Enthalpy	308
	n-decane	Henry	89-91
	2,2-dimethylbutane	Henry	91
	1,4-dioxane	Henry	89-90
	ethyl acetate	Henry	89-90
	ethylbenzene	Henry	91
	n-heptane	Henry	89-90
	n-heptane	Enthalpy	307-308
	n-hexane	VLE	30-31
	n-hexane	Henry	89-90
	n-hexane	Enthalpy	308
	hydrogen	Enthalpy	333
	mesitylene	Henry	91
	2-methylpropane	Henry	91
	nitrogen	Enthalpy	333
	n-nonane	Henry	89-90
	n-nonane	Enthalpy	307
	n-octane	VLE	30-31
	n-octane	Henry	89-91
	n-octane	Enthalpy	307
	1-octene	Henry	90
	n-pentane	VLE	30-32
	2-pentanone	Henry	90
	2-propanone	Henry	89-90
	tetrachloromethane	Henry	90
	tetrahydrofuran	Henry	89-90
	toluene	Henry	89-91
	toluene	Enthalpy	307-308
	trichloromethane	VLE	30
	trichloromethane	Henry	89-90
	without	PVT	346-348

Copolymer (B)	Solvent (A)	Property	Page(s)
Acrylonitrile/isoprene copolymer			
	carbon dioxide	Enthalpy	333
	N,N-dimethylformamide	Enthalpy	271
	helium	Enthalpy	333
	hydrogen	Enthalpy	333
	nitrogen	Enthalpy	333
	oxygen	Enthalpy	333
Acrylonitrile/methyl acrylate copolymer			
	1-butanol	Enthalpy	309
	ethanol	Enthalpy	309
	1-hexanol	Enthalpy	309
	1-pentanol	Enthalpy	309
	1-propanol	Enthalpy	309
Acrylonitrile/2-methylfuran copolymer			
	N,N-dimethylformamide	Enthalpy	333
Acrylonitrile/methyl methacrylate copolymer *see* Methyl methacrylate/acrylonitrile copolymer			
Acrylonitrile/2-(3-methyl-3-phenylcyclobutyl)-2-hydroxyethyl methacrylate copolymer			
	benzene	Henry	92
	benzene	Enthalpy	310
	2-butanone	Henry	92
	2-butanone	Enthalpy	310
	n-decane	Henry	92
	n-decane	Enthalpy	310
	n-dodecane	Henry	92
	n-dodecane	Enthalpy	310
	ethanol	Henry	92
	ethanol	Enthalpy	310
	ethyl acetate	Henry	93
	ethyl acetate	Enthalpy	310
	methanol	Henry	93
	methanol	Enthalpy	310
	methyl acetate	Henry	92
	methyl acetate	Enthalpy	310
	n-nonane	Henry	92
	n-nonane	Enthalpy	310
	n-octane	Henry	92
	n-octane	Enthalpy	310

Copolymer (B)	Solvent (A)	Property	Page(s)
Acrylonitrile/2-(3-methyl-3-phenylcyclobutyl)-2-hydroxyethyl methacrylate copolymer			
	2-propanone	Henry	92
	2-propanone	Enthalpy	310
	toluene	Henry	92
	toluene	Enthalpy	310
	n-undecane	Henry	93
	n-undecane	Enthalpy	310
	o-xylene	Henry	93
	o-xylene	Enthalpy	310
Acrylonitrile/α-methylstyrene copolymer			
	benzene	Enthalpy	311
	1-butanol	Enthalpy	311
	2-butanone	Enthalpy	311
	butyl acetate	Enthalpy	311
	butylbenzene	Enthalpy	311
	chlorobenzene	Enthalpy	311
	cyclohexanol	Enthalpy	311
	dichloromethane	Enthalpy	311
	1,4-dioxane	Enthalpy	311
	n-dodecane	Enthalpy	311
	n-tetradecane	Enthalpy	311
	tetrahydrofuran	Enthalpy	311
	tetrahydrofuran	A_2	435
	toluene	Enthalpy	311
	trichloromethane	Enthalpy	311
Acrylonitrile/styrene copolymer			
	acrylonitrile	Enthalpy	333
	acrylonitrile	A_2	440-441
	benzene	VLE	33
	benzene	Henry	94
	2-butanone	LLE	161
	2-butanone	A_2	439-440
	cyclohexane	LLE	161
	1,2-dichloroethane	VLE	35
	N,N-dimethylformamide	Enthalpy	333
	dimethyl phthalate	LLE	173
	ethyl acetate	Henry	94
	ethyl acetate	LLE	161, 181
	ethyl acetate	LLE	182
	ethyl acetate	A_2	440-441
	propylbenzene	VLE	33
	propylbenzene	Henry	94
	styrene	Enthalpy	333
	styrene	A_2	440-441

Copolymer (B)	Solvent (A)	Property	Page(s)
Acrylonitrile/styrene copolymer			
	tetrahydrofuran	A_2	439-440
	toluene	VLE	32-34
	toluene	Henry	94
	toluene	LLE	145-146
	toluene	LLE	162, 181
	trichloromethane	Henry	94
	trichloromethane	Enthalpy	336
	water	Enthalpy	333
	o-xylene	VLE	34
	o-xylene	Henry	94
	m-xylene	VLE	34
	m-xylene	Henry	94
	p-xylene	VLE	34
	p-xylene	Henry	94
	without	PVT	341-346
Acrylonitrile/styrenesulfonate copolymer			
	N,N-dimethylformamide	Enthalpy	334
	water	Enthalpy	334
Acrylonitrile/vinyl chloride copolymer			
	N,N-dimethylformamide	Enthalpy	272-273
	N,N-dimethylformamide	Enthalpy	334
	water	Enthalpy	334
Acrylonitrile/vinylidene chloride copolymer			
	γ-butyrolactone	A_2	435
	1,1-dichloroethene	Henry	95
	N,N-dimethylacetamide	A_2	435
	N,N-dimethylformamide	A_2	435
	2-propanone	Enthalpy	274
N-Acryloylpyrrolidine/vinylferrocene copolymer			
	water	LLE	160
Arylene sulfoxide/butadiene polyblock copolymer			
	trichloromethane	LLE	172
	1,1,2,2-tetrachloroethane	LLE	172
Bisphenol-A carbonate/dimethylsiloxane copolymer *see* Dimethylsiloxane/bisphenol-A carbonate copolymer			

Copolymer (B)	Solvent (A)	Property	Page(s)
Bisphenol A-isophthaloyl chloride/terephthaloyl chloride copolymer			
	N,N-dimethylacetamide	Enthalpy	275
	1,1,2,2-tetrachloroethane	Enthalpy	275
p-Bromostyrene/p-methylstyrene copolymer			
	toluene	VLE	36
Butadiene/acrylonitrile copolymer *see* Acrylonitrile/butadiene copolymer			
Butadiene/methyl methacrylate copolymer			
	argon	Enthalpy	334
	hydrogen	Enthalpy	334
	nitrogen	Enthalpy	334
Butadiene/styrene copolymer			
	argon	Enthalpy	334
	benzene	VLE	40
	benzene	Henry	97-105
	benzene	Enthalpy	276-280
	benzene	Enthalpy	312, 334
	benzene	A_2	435-436
	1,3-butadiene	Henry	101-104
	n-butane	Henry	101-104
	2-butanone	Henry	105
	cyclohexane	VLE	38-42
	cyclohexane	Henry	97-105
	cyclohexane	Enthalpy	312-313
	cyclohexane	A_2	435-436
	n-decane	Henry	98-104
	2,2-dimethylbutane	Henry	101, 104
	ethylbenzene	VLE	41-42
	ethylbenzene	Henry	101-105
	ethylbenzene	Enthalpy	277-280
	n-heptane	Henry	98-105
	n-heptane	Enthalpy	312-313
	n-hexane	VLE	39, 42
	n-hexane	Henry	97-105
	n-hexane	Enthalpy	312-313
	hydrogen	Enthalpy	334
	mesitylene	VLE	41
	mesitylene	Henry	101-104
	2-methylpropane	Henry	101, 104
	nitrogen	Enthalpy	334

Copolymer (B)	Solvent (A)	Property	Page(s)
Butadiene/styrene copolymer			
	n-nonane	VLE	43
	n-nonane	Henry	98-102
	n-octane	Henry	98-104
	n-pentane	VLE	39, 43
	2-propanone	VLE	38-40, 43
	styrene	Henry	103
	styrene	LLE	173
	tetrahydrofuran	LLE	173
	tetrahydrofuran	A_2	435-436
	toluene	VLE	38-41
	toluene	Henry	97-105
	toluene	LLE	173
	toluene	Enthalpy	312-313
	trichloromethane	VLE	39-40, 44
	trichloromethane	Henry	97, 105
	2,2,4-trimethylpentane	Enthalpy	277-280
	p-xylene	VLE	41
	p-xylene	Henry	103-105
	without	PVT	348-349
Butadiene/styrene block copolymer (unspecified)			
	ethylbenzene	VLE	44-45
	n-nonane	VLE	44-46
	without	PVT	348-349
1-Butene/ethylene copolymer *see* Ethylene/1-butene copolymer			
2-Butoxyethyl acrylate/N,N-dimethylacrylamide copolymer *see* N,N-Dimethylacrylamide/2-butoxyethyl acrylate copolymer			
Butyl acrylate/N,N-dimethylacrylamide copolymer *see* N,N-Dimethylacrylamide/butyl acrylate copolymer			
Butyl acrylate/ethylene copolymer *see* Ethylene/butyl acrylate copolymer			

Copolymer (B)	Solvent (A)	Property	Page(s)
Butyl acrylate/methyl methacrylate copolymer			
	2-propanone	A_2	436
Butyl acrylate/vinyl chloride copolymer *see* Vinyl chloride/butyl acrylate copolymer			
Butylene/ethylethylene copolymer			
	toluene	A_2	436-437
1,2-Butylene oxide/ethylene oxide copolymer *see* Ethylene oxide/1,2-butylene oxide copolymer			
Butyl ether/vinyl chloride copolymer *see* Vinyl chloride/butyl ether copolymer			
Butyl methacrylate/isobutyl methacrylate copolymer			
	cyclohexanone	Enthalpy	281
Butyl methacrylate/N-isopropylacrylamide copolymer *see* N-Isopropylacrylamide/butyl methacrylate copolymer			
Butyl methacrylate/methyl methacrylate copolymer			
	cyclohexanone	Enthalpy	282
	2-propanone	A_2	436-437
Butyl methacrylate/styrene copolymer *see* Styrene/butyl methacrylate copolymer			
tert-Butylstyrene/dimethylsiloxane triblock copolymer			
	2-butanone	LLE	160
	1-nitropropane	LLE	160

Copolymer (B)	Solvent (A)	Property	Page(s)
Carbon monoxide/ethylene copolymer *see* Ethylene/carbon monoxide copolymer			
ε-Carbobenzoxy-L-lysine/L-phenylalanine copolymer			
	1,2-dichloroethane	Enthalpy	283-284
Chloroprene/methyl methacrylate copolymer			
	dichloromethane	Enthalpy	314
	tetrachloromethane	Enthalpy	314
	trichloromethane	Enthalpy	314
Chloroprene/methyl methacrylate/methacrylic acid terpolymer			
	dichloromethane	Enthalpy	315
	tetrachloromethane	Enthalpy	315
	trichloromethane	Enthalpy	315
Cyclopentene/maleic acid copolymer			
	water	Enthalpy	334
N-Cyclopropylacrylamide/vinylferrocene copolymer			
	water	LLE	147
N,N-Diethylacrylamide/vinylferrocene copolymer			
	water	LLE	148
Diethylene glycol/maleic anhydride copolymer *see* Maleic anhydride/diethylene glycol copolymer			
Diethylmaleate/vinyl acetate copolymer			
	n-heptane	Enthalpy	316
	n-hexane	Enthalpy	316
	n-nonane	Enthalpy	316
	n-octane	Enthalpy	316
N,N-Dimethylacrylamide/2-butoxyethyl acrylate copolymer			
	water	LLE	178

Copolymer (B)	Solvent (A)	Property	Page(s)
N,N-Dimethylacrylamide/butyl acrylate copolymer	water	LLE	178
N,N-Dimethylacrylamide/2-ethoxyethyl acrylate copolymer	water	LLE	178
N,N-Dimethylacrylamide/ethyl acrylate copolymer	water	LLE	178
N,N-Dimethylacrylamide/2-methoxyethyl acrylate copolymer	water	LLE	160, 178
N,N-Dimethylacrylamide/methyl acrylate copolymer	water	LLE	178
N,N-Dimethylacrylamide/N-phenylacrylamide copolymer	water	LLE	160
N,N-Dimethylacrylamide/propyl acrylate copolymer	water	LLE	178
N,N-Dimethylaminoethyl methacrylate/acrylamide copolymer *see* Acrylamide/N,N-dimethylaminoethyl methacrylate copolymer			
N,N-Dimethylaminoethyl methacrylate/styrene copolymer *see* Styrene/N,N-dimethylaminoethyl methacrylate copolymer			
Dimethylsiloxane/bisphenol-A carbonate copolymer	n-decane	Henry	106-108

Copolymer (B)	Solvent (A)	Property	Page(s)
Dimethylsiloxane/bisphenol-A carbonate copolymer			
	chlorobenzene	Henry	106-108
	o-dichlorobenzene	Henry	106-108
	toluene	Henry	106-108
Dimethylsiloxane/methylphenylsiloxane copolymer			
	anisole	VLE	48-49
	anisole	LLE	149-150
	anisole	LLE	160, 179
	2-propanone	VLE	49-50
	2-propanone	LLE	150-151
	2-propanone	LLE	160, 179
Dimethylsiloxane/1,1,3,3-tetramethyldisiloxanyl-ethylene diblock copolymer			
	ethoxybenzene	LLE	160
Docosyl maleate/styrene copolymer *see* Styrene/docosyl maleate copolymer			
Dodecyl maleate/styrene copolymer *see* Styrene/dodecyl maleate copolymer			
2-Ethoxyethyl acrylate/N,N-dimethylacrylamide copolymer *see* N,N-Dimethylacrylamide/2-ethoxyethyl acrylate copolymer			
N-Ethylacrylamide/vinylferrocene copolymer			
	water	LLE	152
Ethyl acrylate/N,N-dimethylacrylamide copolymer *see* N,N-Dimethylacrylamide/ethyl acrylate copolymer			

Copolymer (B)	Solvent (A)	Property	Page(s)
Ethyl acrylate/ethylene copolymer *see* Ethylene/ethyl acrylate copolymer			
Ethyl acrylate/methyl methacrylate copolymer	2-propanone	A_2	437
Ethyl acrylate/styrene copolymer *see* Styrene/ethyl acrylate copolymer			
Ethyl acrylate/tetraethylene glycol dimethacrylate copolymer	argon	Enthalpy	334
	carbon dioxide	Enthalpy	334
	hydrogen	Enthalpy	334
	krypton	Enthalpy	334
	methane	Enthalpy	334
	neon	Enthalpy	334
	nitrogen	Enthalpy	334
	oxygen	Enthalpy	334
Ethyl acrylate/4-vinylpyridine copolymer	tetrahydrofuran	LLE	172
Ethylene/acrylic acid copolymer	acrylic acid	HPPE	229-233
	n-butane	HPPE	225, 264
	1-butene	HPPE	225
	dimethyl ether	HPPE	225, 264
	ethanol	HPPE	264
	ethene	HPPE	188-191
	ethene	HPPE	225, 229
	ethene	HPPE	230-233
	propane	HPPE	225
	propene	HPPE	225
	without	PVT	350-355
Ethylene/1-butene copolymer	n-heptane	HPPE	226
	propane	HPPE	226
	without	PVT	355-363

Copolymer (B)	Solvent (A)	Property	Page(s)
Ethylene/butyl acrylate copolymer			
	ethene	HPPE	192-194
	ethene	HPPE	226
Ethylene/carbon monoxide copolymer			
	n-octane	Henry	109
	n-octane	Enthalpy	317
Ethylene/ethyl acrylate copolymer			
	1-butene	HPPE	264
Ethylene/2-ethylhexyl acrylate copolymer			
	ethene	HPPE	264
	2-ethylhexyl acrylate	HPPE	264
Ethylene/1-hexene copolymer			
	n-butane	HPPE	233-237
	n-butane	HPPE	264
	carbon dioxide	HPPE	264
	ethane	HPPE	234
	ethene	HPPE	195, 233
	ethene	HPPE	234-240
	ethene	HPPE	264
	helium	HPPE	234-239
	n-heptane	HPPE	226
	1-hexene	HPPE	235-237
	1-hexene	HPPE	264
	methane	HPPE	235-240
	2-methylpropane	HPPE	226, 264
	nitrogen	HPPE	264
	propane	HPPE	195-196
	propane	HPPE	226, 236
Ethylene/maleic acid copolymer			
	water	Enthalpy	334
Ethylene/methacrylic acid copolymer			
	n-butane	HPPE	226, 264
	1-butene	HPPE	226
	dimethyl ether	HPPE	226, 264
	ethanol	HPPE	264
	ethene	HPPE	226
	propane	HPPE	226
	without	PVT	363-370

Copolymer (B)	Solvent (A)	Property	Page(s)
Ethylene/methyl acrylate copolymer			
	n-butane	HPPE	226, 264
	1-butanol	HPPE	264
	1-butene	HPPE	226, 264
	carbon dioxide	HPPE	226
	chlorodifluoromethane	HPPE	226, 264
	dimethyl ether	HPPE	226
	ethane	HPPE	226
	ethanol	HPPE	264-265
	ethene	HPPE	197-198
	ethene	HPPE	226-227
	n-hexane	HPPE	227, 265
	1-hexene	HPPE	265
	methanol	HPPE	227, 265
	methyl acrylate	HPPE	264
	propane	HPPE	227, 264
	propane	HPPE	265
	1-propanol	HPPE	265
	2-propanone	HPPE	264-265
	propene	HPPE	227
Ethylene/4-methyl-1-pentene copolymer			
	n-heptane	HPPE	227
Ethylene/1-octene copolymer			
	cyclohexane	HPPE	227
	ethene	HPPE	199
	n-heptane	HPPE	227
	n-hexane	HPPE	227
	2-methylpentane	HPPE	227
	propane	HPPE	227
	propene	HPPE	227
	without	PVT	370-383
Ethylene/propylene copolymer			
	benzene	Henry	110
	benzene	Enthalpy	318, 335
	1-butene	HPPE	201-202
	1-butene	HPPE	227, 241
	1-butene	HPPE	265
	tert-butylbenzene	Enthalpy	318
	carbon dioxide	HPPE	265
	chlorobenzene	Enthalpy	335
	cyclohexane	Henry	110
	cyclohexane	LLE	179-180
	cyclohexane	Enthalpy	285, 318

Copolymer (B)	Solvent (A)	Property	Page(s)
Ethylene/propylene copolymer			
	cyclohexane	Enthalpy	335
	cyclooctane	Enthalpy	285-286
	cyclopentane	LLE	179-180
	cyclopentane	Enthalpy	286
	cis-decahydronaphthalene	Enthalpy	286
	trans-decahydronaphthalene	Enthalpy	286-287
	n-decane	Enthalpy	318
	dichloromethane	Henry	110
	dichloromethane	Enthalpy	335
	3,3-diethylpentane	Enthalpy	287
	2,2-dimethylbutane	LLE	179-180
	2,3-dimethylbutane	LLE	179-180
	2,4-dimethylhexane	LLE	180
	2,5-dimethylhexane	LLE	180
	3,4-dimethylhexane	LLE	179-180
	2,2-dimethylpentane	LLE	179-180
	2,2-dimethylpentane	Enthalpy	287
	2,3-dimethylpentane	LLE	179-180
	2,3-dimethylpentane	Enthalpy	287-288
	2,4-dimethylpentane	LLE	179-180
	2,4-dimethylpentane	Enthalpy	288
	3,3-dimethylpentane	Enthalpy	288
	n-dodecane	Enthalpy	288-289
	ethene	HPPE	202, 227
	ethene	HPPE	241-243
	ethene	HPPE	265
	ethylbenzene	Enthalpy	318
	3-ethylpentane	LLE	179-180
	3-ethylpentane	Enthalpy	289
	n-heptane	Henry	110
	n-heptane	LLE	179-180
	n-heptane	HPPE	227
	n-heptane	Enthalpy	335
	2,2,4,4,6,8,8-heptamethylnonane	Enthalpy	289
	n-hexadecane	Enthalpy	290
	n-hexane	Henry	110
	n-hexane	LLE	160, 179
	n-hexane	LLE	180
	n-hexane	HPPE	265
	n-hexane	Enthalpy	318, 335
	1-hexene	HPPE	202-203
	1-hexene	HPPE	241-243
	1-hexene	HPPE	265
	methane	HPPE	265
	2-methylbutane	LLE	160, 179
	2-methylbutane	LLE	180
	2-methylbutane	HPPE	265

Copolymer (B)	Solvent (A)	Property	Page(s)
Ethylene/propylene copolymer			
	methylcyclohexane	LLE	179-180
	methylcyclopentane	LLE	160, 179
	methylcyclopentane	LLE	180
	2-methylhexane	LLE	179-180
	3-methylhexane	LLE	181
	3-methylhexane	Enthalpy	290
	2-methylpentane	LLE	179
	3-methylpentane	LLE	161
	n-nonane	LLE	179-181
	n-octane	Henry	110
	n-octane	LLE	179-181
	n-octane	Enthalpy	290-291
	n-octane	Enthalpy	318
	2,2,4,6,6-pentamethylheptane	Enthalpy	291
	n-pentane	Henry	110
	n-pentane	LLE	161, 179
	n-pentane	LLE	181
	n-pentane	Enthalpy	335
	propane	HPPE	227
	propene	HPPE	203-204
	propene	HPPE	227, 265
	tetrachloromethane	Henry	110
	tetrachloromethane	Enthalpy	335
	2,2,4,4-tetramethylpentane	LLE	179-181
	2,2,4,4-tetramethylpentane	Enthalpy	291-292
	toluene	Henry	110
	toluene	A_2	437
	trichloromethane	Henry	110
	1,1,2-trichlorotrifluoroethane	Enthalpy	335
	2,2,3-trimethylbutane	LLE	179-181
	2,3,4-trimethylhexane	LLE	179
	2,2,4-trimethylpentane	LLE	179-181
	2,2,4-trimethylpentane	Enthalpy	292, 318
	without	PVT	383-398
Ethylene/propylene/diene terpolymer			
	benzene	Henry	110
	benzene	Enthalpy	318
	n-heptane	Henry	110
	n-heptane	Enthalpy	318
	n-hexane	Henry	110
	n-hexane	Enthalpy	318
	n-nonane	Henry	110
	n-nonane	Enthalpy	318
	n-octane	Henry	110
	n-octane	Enthalpy	318

Copolymer (B)	Solvent (A)	Property	Page(s)
Ethylene/propylene/diene terpolymer			
	toluene	Henry	110
	toluene	Enthalpy	318
Ethylene/propylene/isoprene terpolymer			
	dimethyl ether	HPPE	205
	ethane	HPPE	205
	ethene	HPPE	205
	propane	HPPE	206
	propene	HPPE	206
Ethylene/styrene copolymer			
	without	PVT	399-405
Ethylene/tetrafluoroethylene copolymer			
	diisobutyl adipate	A_2	437
Ethylene/vinyl acetate copolymer			
	acetaldehyde	Henry	114
	acetaldehyde	Enthalpy	320
	acetic acid	Henry	114
	acetic acid	Enthalpy	320
	acetonitrile	Henry	114
	acetonitrile	Enthalpy	320
	acrylonitrile	Henry	114
	acrylonitrile	Enthalpy	320
	benzene	VLE	51-53
	benzene	Henry	113-117
	benzene	Enthalpy	319-320
	1-bromobutane	Henry	114
	1-bromobutane	Enthalpy	320
	2-bromobutane	Henry	114
	2-bromobutane	Enthalpy	320
	n-butane	HPPE	265
	1-butanol	VLE	53
	1-butanol	Henry	114-116
	1-butanol	Enthalpy	319-320
	2-butanol	Henry	114
	2-butanol	Enthalpy	320
	2-butanone	Henry	112-116
	2-butanone	Enthalpy	320
	butyl acetate	VLE	53-54
	butyl acetate	Enthalpy	320
	carbon dioxide	Henry	112-115
	carbon dioxide	HPPE	251, 255-256

Copolymer (B)	Solvent (A)	Property	Page(s)
Ethylene/vinyl acetate copolymer			
	carbon dioxide	HPPE	265
	chlorobenzene	Henry	114-116
	chlorobenzene	Enthalpy	319
	1-chlorobutane	Henry	115-116
	1-chlorobutane	Enthalpy	319
	chlorocyclohexane	Enthalpy	319
	cyclohexane	VLE	54
	cyclohexane	Henry	113-117
	cyclohexane	Enthalpy	319-320
	cyclohexanol	Enthalpy	319
	cyclohexanone	Enthalpy	319
	n-decane	Henry	115-116
	n-decane	Enthalpy	320
	n-dodecane	Enthalpy	335
	dichloroethane	Henry	114
	dichloroethane	Enthalpy	320
	dichloromethane	Henry	114-116
	dichloromethane	Enthalpy	321
	diethyl ether	Henry	114-116
	diethyl ether	Enthalpy	321
	1,4-dioxane	Henry	113-117
	1,4-dioxane	Enthalpy	321
	diphenyl ether	LLE	161, 181
	dipropyl ether	Enthalpy	321
	ethane	Henry	112-115
	ethane	HPPE	251-252
	ethanol	Henry	114
	ethanol	Enthalpy	320-321
	ethene	Henry	112-115
	ethene	HPPE	208-221, 227
	ethene	HPPE	244-263, 265
	ethyl acetate	VLE	54-55
	ethyl acetate	Henry	113-117
	ethyl acetate	Enthalpy	320-321
	formic acid	Henry	114
	formic acid	Enthalpy	321
	furan	Henry	114
	furan	Enthalpy	321
	helium	HPPE	252, 257-258
	n-heptane	Henry	114
	n-hexane	Henry	114-116
	n-hexane	Enthalpy	319-321
	methanol	Henry	114-116
	methanol	Enthalpy	320
	methyl acetate	VLE	55-56
	methyl acetate	LLE	153, 181
	methyl chloride	Henry	112-115

Copolymer (B)	Solvent (A)	Property	Page(s)
Ethylene/vinyl acetate copolymer			
	methylcyclohexane	Henry	114
	methylcyclohexane	Enthalpy	320
	nitroethane	Henry	114
	nitroethane	Enthalpy	320
	nitrogen	HPPE	252-253, 258
	nitrogen	HPPE	259, 265
	nitromethane	Henry	114
	nitromethane	Enthalpy	320
	1-nitropropane	Henry	114
	1-nitropropane	Enthalpy	320
	2-nitropropane	Henry	114
	2-nitropropane	Enthalpy	320
	n-octadecane	Enthalpy	335
	n-octane	Henry	114-117
	n-octane	Enthalpy	320
	1-octene	Henry	114-116
	1-octene	Enthalpy	320
	n-pentane	Henry	114-116
	n-pentane	Enthalpy	320
	2-pentanone	Henry	115-116
	2-pentanone	Enthalpy	319-320
	3-pentanone	Henry	114
	phenol	Enthalpy	319
	propane	HPPE	253
	2-propanol	Henry	112-115
	2-propanol	Enthalpy	320
	2-propanone	Henry	112-116
	2-propanone	Enthalpy	320
	propionitrile	Henry	114
	propionitrile	Enthalpy	320
	propyl acetate	VLE	56
	propyl acetate	Henry	114
	propyl acetate	Enthalpy	321
	sulfur dioxide	Henry	112-115
	tetrachloromethane	Henry	113-117
	tetrachloromethane	Enthalpy	319-321
	tetrahydrofuran	Henry	113-117
	tetrahydrofuran	Enthalpy	293, 321
	toluene	VLE	57-59
	toluene	Henry	113-117
	toluene	Enthalpy	320
	trichloromethane	VLE	59
	trichloromethane	Henry	114-116
	trichloromethane	Enthalpy	319-321
	2,2,2-trifluoroethanol	Henry	114
	2,2,2-trifluoroethanol	Enthalpy	321
	vinyl acetate	Henry	112-115

Copolymer (B)	Solvent (A)	Property	Page(s)
Ethylene/vinyl acetate copolymer			
	vinyl acetate	HPPE	215, 244-263
	vinyl acetate	HPPE	265
	water	Henry	114
	water	Enthalpy	321
	o-xylene	VLE	59-60
	m-xylene	Henry	114
	m-xylene	Enthalpy	321
	p-xylene	VLE	60-62
	p-xylene	Enthalpy	320
	without	PVT	406-421
Ethylene/vinyl alcohol copolymer			
	n-butane	HPPE	265
	1-butanol	Enthalpy	322
	2-butanol	Enthalpy	322
	dimethylsulfoxide	LLE	172
	ethanol	Enthalpy	322
	2-methyl-1-propanol	Enthalpy	322
	2-methyl-2-propanol	Enthalpy	322
	methanol	Enthalpy	322
	1-propanol	Enthalpy	322
	1-propanol	LLE	172
	2-propanol	LLE	161, 172
	2-propanol	Enthalpy	322
	water	LLE	161, 172, 181
	without	PVT	421-423
Ethylene glycol/N-isopropylacrylamide copolymer			
	water	LLE	161
Ethylene oxide/1,2-butylene oxide copolymer			
	water	LLE	161
Ethylene oxide/dimethylsiloxane triblock copolymer			
	water	LLE	161
Ethylene oxide/propylene oxide copolymer			
	acetic acid	LLE	161
	butanoic acid	LLE	161
	propionic acid	LLE	161
	water	LLE	161, 164
	water	LLE	165-172

Copolymer (B)	Solvent (A)	Property	Page(s)
Ethylene oxide/propylene oxide copolymer			
	water	LLE	181
	water	A_2	437
Ethylene terephthalate/p-hydroxybenzoic acid copolymer			
	trichloromethane	LLE	171, 172
Ethylethylene/butylene copolymer *see* Butylene/ethylethylene copolymer			
2-Ethylhexyl acrylate/ethylene copolymer *see* Ethylene/2-ethylhexyl acrylate copolymer			
Ethyl methacrylate/methacrylic acid copolymer			
	benzene	Henry	118
	n-decane	Henry	118
4-Ethylstyrene/N-vinylcarbazole copolymer *see* N-Vinylcarbazole/4-ethylstyrene copolymer			
Hexafluoropropylene/tetrafluoroethylene copolymer *see* Tetrafluoroethylene/hexafluoropropylene copolymer			
Hexafluoropropylene/vinylidene fluoride copolymer *see* Vinylidene fluoride/hexafluoropropylene copolymer			
1-Hexene/ethylene copolymer *see* Ethylene/1-hexene copolymer			
4-Hexylstyrene/N-vinylcarbazole copolymer *see* N-Vinylcarbazole/4-hexylstyrene copolymer			

Copolymer (B)	Solvent (A)	Property	Page(s)
p-Hydroxybenzoic acid/ethylene terephthalate copolymer *see* Ethylene terephthalate/p-hydroxybenzoic acid copolymer			
3-Hydroxybutanoic acid/3-hydroxypentanoic acid copolymer			
	chlorobenzene	VLE	71-72
2-Hydroxyethyl methacrylate/methyl methacrylate copolymer *see* Methyl methacrylate/2-hydroxyethyl methacrylate copolymer			
9,9-bis(4-Hydroxyphenyl)fluorene/terephthaloyl chloride copolymer			
	1,1,2,2-tetrachloroethane	Enthalpy	335
	1,1,2,2-tetrachloroethane	A_2	437
	trichloromethane	Enthalpy	335
	trichloromethane	A_2	437
Isobutylene/maleic acid copolymer			
	water	Enthalpy	335
Isobutyl methacrylate/acrylic acid copolymer			
	benzene	Henry	119
	n-decane	Henry	119
Isobutyl methacrylate/butyl methacrylate copolymer *see* Butyl methacrylate/isobutyl methacrylate copolymer			
Isobutyl methacrylate/methyl methacrylate copolymer			
	cyclohexanone	Enthalpy	296
Isobutyl methacrylate/styrene copolymer *see* Styrene/isobutyl methacrylate copolymer			

Copolymer (B)	Solvent (A)	Property	Page(s)
Isoprene/acrylonitrile copolymer *see* Acrylonitrile/isoprene copolymer			
N-Isopropylacrylamide/acrylamide copolymer			
	water	LLE	181
	water	A_2	438
N-Isopropylacrylamide/acrylic acid copolymer *see* Acrylic acid/N-isopropylacrylamide copolymer			
N-Isopropylacrylamide/butyl methacrylate copolymer			
	water	LLE	161
N-Isopropylacrylamide/ethylene glycol copolymer *see* Ethylene glycol/N-isopropylacrylamide copolymer			
N-Isopropylacrylamide/itaconic acid copolymer			
	water	LLE	155
N-Isopropylacrylamide/N-octadecylacrylamide copolymer			
	water	Enthalpy	335
N-Isopropylacrylamide/vinylferrocene copolymer			
	water	LLE	156
Isopropylethylene/1-methylbutylene copolymer			
	toluene	A_2	438
Isopropylethylene/1-methyl-1-ethylethylene copolymer			
	toluene	A_2	438
Itaconic acid/N-isopropylacrylamide copolymer *see* N-Isopropylacrylamide/itaconic acid copolymer			

Copolymer (B)	Solvent (A)	Property	Page(s)
Maleic acid/cyclopentene copolymer *see* Cyclopentene/maleic acid copolymer			
Maleic acid/ethylene copolymer *see* Ethylene/maleic acid copolymer			
Maleic acid/isobutylene copolymer *see* Isobutylene/maleic acid copolymer			
Maleic acid/4-methyl-1-pentene copolymer *see* 4-Methyl-1-pentene/maleic acid copolymer			
Maleic acid/propylene copolymer *see* Propylene/maleic acid copolymer			
Maleic acid/trimethyl-1-pentene copolymer *see* Trimethyl-1-pentene/maleic acid copolymer			
Maleic anhydride/diethylene glycol copolymer			
	styrene	LLE	161, 181
Maleic anhydride/1-octadecene copolymer *see* 1-Octadecene/maleic anhydride copolymer			
Maleic anhydride/styrene copolymer *see* Styrene/maleic anhydride copolymer			
Methacrylic acid/ethylene copolymer *see* Ethylene/methacrylic acid copolymer			
Methacrylic acid/ethyl methacrylate copolymer *see* Ethyl methacrylate/methacrylic acid copolymer			

Copolymer (B)	Solvent (A)	Property	Page(s)
Methacrylic acid/styrene copolymer *see* Styrene/methacrylic acid copolymer			
Methacrylic acid/vinyl acetate copolymer *see* Vinyl acetate/methacrylic acid copolymer			
N-Methacryloyl-L-alanine/N-phenyl-methacrylamide copolymer			
	water	Enthalpy	336
2-Methoxyethyl acrylate/N,N-dimethyl-acrylamide copolymer *see* N,N-Dimethylacrylamide/2-methoxyethyl acrylate copolymer			
2-Methoxyethyl methacrylate/styrene copolymer *see* Styrene/2-methoxyethyl methacrylate copolymer			
p-Methoxystyrene/styrene copolymer *see* Styrene/p-methoxystyrene copolymer			
Methyl acrylate/acrylic acid copolymer *see* Acrylic acid/methyl acrylate copolymer			
Methyl acrylate/acrylonitrile copolymer *see* Acrylonitrile/methyl acrylate copolymer			
Methyl acrylate/N,N-dimethylacrylamide copolymer *see* N,N-Dimethylacrylamide/methyl acrylate copolymer			
Methyl acrylate/ethylene copolymer *see* Ethylene/methyl acrylate copolymer			

Copolymer (B)	Solvent (A)	Property	Page(s)
Methyl acrylate/methyl methacrylate copolymer	2-propanone	A_2	438
Methyl acrylate/vinyl chloride copolymer *see* Vinyl chloride/methyl acrylate copolymer			
1-Methylbutylene/isopropylethylene copolymer *see* Isopropylethylene/1-methylbutylene copolymer			
1-Methyl-1-ethylethylene/isopropylethylene copolymer *see* Isopropylethylene/1-methyl-1-ethylethylene copolymer			
Methyl methacrylate/acrylic acid copolymer *see* Acrylic acid/methyl methacrylate copolymer			
Methyl methacrylate/acrylonitrile copolymer	acetonitrile	A_2	438-439
Methyl methacrylate/butadiene copolymer *see* Butadiene/methyl methacrylate copolymer			
Methyl methacrylate/butyl acrylate copolymer *see* Butyl acrylate/methyl methacrylate copolymer			
Methyl methacrylate/butyl methacrylate copolymer *see* Butyl methacrylate/methyl methacrylate copolymer			
Methyl methacrylate/chloroprene copolymer *see* Chloroprene/methyl methacrylate copolymer			

Copolymer (B)	Solvent (A)	Property	Page(s)
Methyl methacrylate/ethyl acrylate copolymer *see* Ethyl acrylate/methyl methacrylate copolymer			
Methyl methacrylate/2-hydroxyethyl methacrylate copolymer			
	carbon dioxide	HPPE	227
Methyl methacrylate/isobutyl methacrylate copolymer *see* Isobutyl methacrylate/methyl methacrylate copolymer			
Methyl methacrylate/methyl acrylate copolymer *see* Methyl acrylate/methyl methacrylate copolymer			
Methyl methacrylate/styrene copolymer *see* Styrene/methyl methacrylate copolymer			
Methyl methacrylate/4-vinylpyridine copolymer			
	2-butanone	LLE	173
	1,4-dioxane	LLE	172
	trichloromethane	LLE	173
4-Methyl-1-pentene/ethylene copolymer *see* Ethylene/4-methyl-1-pentene copolymer			
4-Methyl-1-pentene/maleic acid copolymer			
	water	Enthalpy	336
Methylphenylsiloxane/dimethylsiloxane copolymer *see* Dimethylsiloxane/methylphenylsiloxane copolymer			
α-Methylstyrene/acrylonitrile copolymer *see* Acrylonitrile/α-methylstyrene copolymer			

Copolymer (B)	Solvent (A)	Property	Page(s)
α-Methylstyrene/styrene copolymer *see* Styrene/α-methylstyrene copolymer			
p-Methylstyrene/p-bromostyrene copolymer *see* p-Bromostyrene/p-methylstyrene copolymer			
2-Methyl-5-vinylpyridine/acrylic acid copolymer *see* Acrylic acid/2-methyl-5-vinylpyridine copolymer			
Nonyl acrylate/acrylic acid copolymer *see* Acrylic acid/nonyl acrylate copolymer			
Nonyl methacrylate/styrene copolymer *see* Styrene/nonyl methacrylate copolymer			
1-Octadecene/maleic anhydride copolymer	ethyl acetate	A_2	439
N-Octadecylacrylamide/N-isopropylacrylamide copolymer *see* N-Isopropylacrylamide/N-octadecylacrylamide copolymer			
1-Octene/ethylene copolymer *see* Ethylene/1-octene copolymer			
4-Octylstyrene/N-vinylcarbazole copolymer *see* N-Vinylcarbazole/4-octylstyrene copolymer			
Oxyethylene/oxymethylene copolymer	1H,1H,5H-octafluoro-1-pentanol	A_2	439
Pentyl maleate/styrene copolymer *see* Styrene/pentyl maleate copolymer			

Copolymer (B)	Solvent (A)	Property	Page(s)
4-Pentylstyrene/N-vinylcarbazole copolymer *see* N-Vinylcarbazole/4-pentylstyrene copolymer			
Perfluoro(oxyethylene/oxymethylene) copolymer			
	without	PVT	424-427
Phenolphthalein/terephthaloyl chloride copolymer			
	1,1,2,2-tetrachloroethane	Enthalpy	336
	trichloromethane	Enthalpy	336
N-Phenylacrylamide/N,N-dimethylacrylamide copolymer *see* N,N-Dimethylacrylamide/N-phenylacrylamide copolymer			
Poly(butylene oxide)-b-poly(ethylene oxide) diblock copolymer			
	carbon dioxide	HPPE	187
Poly(dimethylsiloxane)-b-polystyrene diblock copolymer			
	cyclohexane	LLE	163, 173
Poly(ethylene oxide)-b-poly(*tert*-butyl methacrylate) diblock copolymer			
	toluene	VLE	63
Poly(ethylene oxide)-b-poly(methyl methacrylate) diblock copolymer			
	toluene	VLE	64
Poly(ethylene oxide)-b-poly(methyl methacrylate)-b-poly(ethylene oxide) triblock copolymer			
	toluene	VLE	65
Poly(ethylene oxide)-b-poly(propylene oxide) diblock copolymer			
	carbon dioxide	HPPE	223
	ethylbenzene	VLE	67

Copolymer (B)	Solvent (A)	Property	Page(s)
Poly(ethylene oxide)-b-poly(propylene oxide) diblock copolymer			
	formamide	LLE	161
	tetrachloromethane	VLE	67
	tetrachloromethane	Enthalpy	294
	water	LLE	161
Poly(ethylene oxide)-b-poly(propylene oxide)-b-poly(ethylene oxide) triblock copolymer			
	1-butanol	LLE	172
	carbon dioxide	HPPE	223-224
	carbon dioxide	HPPE	228
	tetrachloromethane	VLE	67
Poly(ethylene oxide)-b-poly(propylene oxide)-b-poly(ethylene oxide) triblock copolymer			
	tetrachloromethane	Enthalpy	294-295
	toluene	VLE	67-68
	water	LLE	154, 172
Poly(ethylene oxide)-b-polystyrene-b-poly (ethylene oxide) triblock copolymer			
	toluene	VLE	70
Poly(ethylene oxide)-b-poly(tetrahydrofuran)-b-poly(ethylene oxide) triblock copolymer			
	1-butanol	LLE	172
	water	LLE	172
Poly(2-ethyl-2-oxazoline)-b-poly(ε-caprolactone) diblock copolymer			
	water	LLE	161
Poly(propylene oxide)-b-poly(ethylene oxide)-b-poly(propylene oxide) triblock copolymer			
	carbon dioxide	HPPE	224-225
	toluene	VLE	68
Polystyrene-b-polybutadiene-b-polystyrene triblock copolymer			
	benzene	Enthalpy	313
	2-butanone	Enthalpy	313
	cyclohexane	VLE	46
	cyclohexane	Enthalpy	313

Copolymer (B)	Solvent (A)	Property	Page(s)
Polystyrene-b-polybutadiene-b-polystyrene triblock copolymer			
	ethylbenzene	Enthalpy	313
	n-heptane	Enthalpy	313
	n-hexane	Enthalpy	313
	4-methyl-2-pentanone	LLE	173
	toluene	Enthalpy	313
	trichloromethane	Enthalpy	313
	p-xylene	Enthalpy	313
	without	PVT	348-349
Polystyrene-b-polybutadiene (hydrogenated)-b-polystyrene triblock copolymer			
	cyclohexane	VLE	47
Polystyrene-b-poly(butyl methacrylate) diblock copolymer			
	2-propanol	LLE	162
Polystyrene-b-poly(ethylene oxide) diblock copolymer			
	toluene	VLE	69-70
Polystyrene-b-polyisoprene diblock copolymer			
	cyclohexane	Enthalpy	335
	4-methyl-2-pentanone	Enthalpy	335
	toluene	Enthalpy	335
Polystyrene-b-polyisoprene diblock copolymer (hydrogenated)			
	N,N-dimethylformamide	LLE	173
	methylcyclohexane	LLE	173
Polystyrene-b-polyisoprene-b-polystyrene triblock copolymer			
	cyclohexane	VLE	77
Polystyrene-b-poly(methyl methacrylate) diblock copolymer			
	benzene	VLE	81
	ethylbenzene	VLE	81-82
	mesitylene	VLE	82
	toluene	VLE	82
	p-xylene	VLE	82

Copolymer (B)	Solvent (A)	Property	Page(s)
Propyl acrylate/N,N-dimethylacrylamide copolymer *see* N,N-Dimethylacrylamide/propyl acrylate copolymer			
Propylene/ethylene copolymer *see* Ethylene/propylene copolymer			
Propylene/maleic acid copolymer	water	Enthalpy	336
Propylene oxide/ethylene oxide copolymer *see* Ethylene oxide/propylene oxide copolymer			
Pyromellitic acid/4-amino-1,4-phenyleneoxy-1,4-phenyleneamine copolymer	N,N-dimethylacetamide	A_2	439
Pyromellitic acid/4-amino-1,4-phenyleneoxy-1-4-phenylenoxy-1,4-phenyleneamine copolymer	N,N-dimethylacetamide	A_2	440
Styrene/acrylamide copolymer	tetrahydrofuran	LLE	173
Styrene/acrylic acid copolymer	benzene	Henry	120
	n-decane	Henry	120
	tetrahydrofuran	LLE	173
Styrene/acrylonitrile copolymer *see* Acrylonitrile/styrene copolymer			
Styrene/butyl methacrylate copolymer	benzene	Henry	121
	benzene	Enthalpy	323
	1-butanol	Enthalpy	323
	2-butanone	LLE	162
	2-butanone	Enthalpy	297-298

Copolymer (B)	Solvent (A)	Property	Page(s)
Styrene/butyl methacrylate copolymer			
	butyl acetate	Enthalpy	323
	butylbenzene	Enthalpy	323
	tert-butylbenzene	Henry	121
	tert-butylbenzene	Enthalpy	323
	butylcyclohexane	Enthalpy	323
	chlorobenzene	Enthalpy	323
	1-chlorobutane	Henry	121
	1-chlorobutane	Enthalpy	323
	cyclohexane	Henry	121
	cyclohexane	Enthalpy	323
	cyclohexanol	Enthalpy	323
	n-decane	Henry	121
	n-decane	Enthalpy	323
	dichloromethane	Henry	121
	dichloromethane	Enthalpy	323
	n-dodecane	Enthalpy	323
	ethylbenzene	Enthalpy	323
	methylcyclohexane	Enthalpy	323
	n-octane	Henry	121
	n-octane	Enthalpy	323
	2-pentanone	Enthalpy	323
	2-propanol	LLE	162
	2-propanone	VLE	73
	tetrachloromethane	Henry	121
	tetrachloromethane	Enthalpy	323
	trichloromethane	VLE	73
	trichloromethane	Henry	121
	trichloromethane	Enthalpy	323
	3,4,5-trimethylheptane	Henry	121
	3,4,5-trimethylheptane	Enthalpy	323
	2,2,4-trimethylpentane	Henry	121
	2,2,4-trimethylpentane	Enthalpy	323
Styrene/cellulose diacetate graft copolymer			
	2-propanone	LLE	162
	N,N-dimethylformamide	LLE	162, 182
	tetrahydrofuran	LLE	162, 182
Styrene/N,N-dimethyl aminoethyl methacrylate copolymer			
	benzene	Henry	123
	n-decane	Henry	123
Styrene/divinylbenzene copolymer			
	1-butanol	Enthalpy	324

Copolymer (B)	Solvent (A)	Property	Page(s)
Styrene/divinylbenzene copolymer			
	ethanol	Enthalpy	324
	methanol	Enthalpy	324
	1-pentanol	Enthalpy	324
	1-propanol	Enthalpy	324
Styrene/docosyl maleate copolymer			
	cyclohexane	VLE	74
	methanol	VLE	74
	2-propanone	VLE	74
Styrene/dodecyl maleate copolymer			
	cyclohexane	VLE	75
	methanol	VLE	75
	2-propanone	VLE	76
Styrene/ethyl acrylate copolymer			
	n-decane	Enthalpy	325-326
	n-hexane	Enthalpy	325-326
	n-nonane	Enthalpy	325-326
Styrene/ethylene copolymer *see* Ethylene/styrene copolymer			
Styrene/isobutyl methacrylate copolymer			
	benzene	Henry	122
	tert-butylbenzene	Henry	122
	1-chlorobutane	Henry	122
	cyclohexane	Henry	122
	n-decane	Henry	122
	dichloromethane	Henry	122
	n-octane	Henry	122
	tetrachloromethane	Henry	122
	trichloromethane	Henry	122
	3,4,5-trimethylheptane	Henry	122
	2,2,4-trimethylpentane	Henry	122
Styrene/maleic anhydride copolymer			
	methanol	VLE	78
	2-propanone	VLE	78

Copolymer (B)	Solvent (A)	Property	Page(s)
Styrene/maleic anhydride/methacrylic acid terpolymer			
	1-butanol	Enthalpy	327
	n-decane	Enthalpy	327
	dibutyl ether	Enthalpy	327
	1,4-dioxane	Enthalpy	327
	n-heptane	Enthalpy	327
	n-nonane	Enthalpy	327
	n-octane	Enthalpy	327
Styrene/methacrylic acid copolymer			
	benzene	LLE	173
	tetrahydrofuran	LLE	173
	trichloromethane	LLE	173
Styrene/2-methoxyethyl methacrylate copolymer			
	dimethyl sulfoxide	LLE	174
	1,1,2,2-tetrachloroethene	LLE	174
Styrene/p-methoxystyrene copolymer			
	toluene	A_2	441-442
Styrene/methyl methacrylate copolymer			
	benzene	VLE	79
	2-butanone	A_2	442
	cyclohexanol	LLE	182
	cyclohexanol	A_2	442-443
	decahydronaphthalene	LLE	174
	1,4-dioxane	A_2	442
	2-ethoxyethanol	A_2	442-443
	ethylbenzene	VLE	80
	n-heptane	Enthalpy	328
	n-hexane	Enthalpy	328
	mesitylene	VLE	80
	methyl acetate	VLE	81
	nitroethane	LLE	174
	nitroethane	A_2	442
	n-nonane	Enthalpy	328
	n-octane	Enthalpy	328
	2-propanone	VLE	81
	tetrachloromethane	A_2	442
	toluene	VLE	80
	trichloromethane	VLE	81
	p-xylene	VLE	81
	without	PVT	427-428

Copolymer (B)	Solvent (A)	Property	Page(s)
Styrene/2-(3-methyl-3-phenylcyclobutyl)-2-hydroxyethyl methacrylate copolymer			
	benzene	Henry	124
	benzene	Enthalpy	329
	2-butanone	Henry	125
	2-butanone	Enthalpy	329
	n-decane	Henry	125
	n-decane	Enthalpy	329
	n-dodecane	Henry	125
	n-dodecane	Enthalpy	329
	ethanol	Henry	125
	ethanol	Enthalpy	329
	ethyl acetate	Henry	125
	ethyl acetate	Enthalpy	329
	methanol	Henry	125
	methanol	Enthalpy	329
	methyl acetate	Henry	124
	methyl acetate	Enthalpy	329
	n-nonane	Henry	125
	n-nonane	Enthalpy	329
	n-octane	Henry	125
	n-octane	Enthalpy	329
	2-propanone	Henry	125
	2-propanone	Enthalpy	329
	toluene	Henry	125
	toluene	Enthalpy	329
	n-undecane	Henry	125
	n-undecane	Enthalpy	329
	o-xylene	Henry	125
	o-xylene	Enthalpy	329
Styrene/α-methylstyrene copolymer			
	butyl acetate	LLE	157, 182
	cyclohexane	LLE	158, 182
	cyclopentane	LLE	158, 182
	trans-decahydronaphthalene	LLE	158, 182
	hexyl acetate	LLE	159, 182
	pentyl acetate	LLE	159, 182
Styrene/nonyl methacrylate copolymer			
	1-butanol	Henry	126
	1-butanol	Enthalpy	330
	dibutyl ether	Henry	126
	dibutyl ether	Enthalpy	330
	1,4-dioxane	Henry	126
	1,4-dioxane	Enthalpy	330
	n-heptane	Henry	126

Copolymer (B)	Solvent (A)	Property	Page(s)
Styrene/nonyl methacrylate copolymer			
	n-heptane	Enthalpy	330
	n-nonane	Enthalpy	330
	n-octane	Henry	126
	n-octane	Enthalpy	330
Styrene/pentyl maleate copolymer			
	cyclohexane	VLE	83
	methanol	VLE	83
	2-propanone	VLE	84
Styrene/4-vinylpyridine copolymer			
	2-butanone	LLE	174
	1,4-dioxane	LLE	174
	tetrahydrofuran	LLE	174
	trichloromethane	LLE	174
Styrene/1-vinyl-2-pyrrolidinone copolymer			
	tetrahydrofuran	LLE	174
Tetrafluoroethylene/ethylene copolymer *see* Ethylene/tetrafluoroethylene copolymer			
Tetrafluoroethylene/hexafluoropropylene copolymer			
	carbon dioxide	HPPE	228, 265
	chlorodifluoromethane	HPPE	228
	hexafluoroethane	HPPE	228
	hexafluoropropene	HPPE	228
	octafluoropropane	HPPE	228
	sulfur hexafluoride	HPPE	228, 265
	tetrafluoromethane	HPPE	228
	trifluoromethane	HPPE	228, 265
Tetrafluoroethylene/2,2-bis(trifluoromethyl)-4,5-difluoro-1,3-dioxole copolymer			
	benzene	Henry	127
	benzene	Enthalpy	331
	n-butane	Enthalpy	331
	carbon dioxide	HPPE	228
	n-heptane	Henry	127
	n-heptane	Enthalpy	331
	hexafluorobenzene	Enthalpy	331
	n-hexane	Henry	127

Copolymer (B)	Solvent (A)	Property	Page(s)
Tetrafluoroethylene/2,2-bis(trifluoromethyl)-4,5-difluoro-1,3-dioxole copolymer			
	n-hexane	Enthalpy	331
	2-methylpropane	Enthalpy	331
	octafluorotoluene	Enthalpy	331
	n-pentane	Enthalpy	331
Tetrafluoroethylene/trifluoronitrosomethane copolymer			
	1,1,2-trichlorotrifluoroethane	LLE	182
	1,1,2-trichlorotrifluoroethane	A_2	443
Trimethyl-1-pentene/maleic acid copolymer			
	water	Enthalpy	336
N-Vinylacetamide/N-vinylisobutyramide copolymer			
	water	LLE	162
Vinyl acetate/diethylmaleate copolymer *see* Diethylmaleate/vinyl acetate copolymer			
Vinyl acetate/ethylene copolymer *see* Ethylene/vinyl acetate copolymer			
Vinyl acetate/methacrylic acid copolymer			
	tetrahydrofuran	LLE	174
Vinyl acetate/vinyl alcohol copolymer			
	benzene	Henry	128-133
	1-butanol	Henry	128-133
	2-butanol	Henry	128-133
	tert-butanol	Henry	128-133
	cyclohexanol	Henry	128-133
	n-decane	Henry	128-133
	1-decanol	Henry	129-133
	n-dodecane	Henry	129-133
	ethanol	Enthalpy	303-304
	ethyl acetate	Enthalpy	303-304
	ethylbenzene	Henry	129-133
	1-heptanol	Henry	129-133
	1-hexanol	Henry	128-133

Copolymer (B)	Solvent (A)	Property	Page(s)
Vinyl acetate/vinyl alcohol copolymer			
	n-nonane	Henry	128-133
	1-octanol	Henry	128-133
	1-pentanol	Henry	128-133
	2-pentanol	Henry	128-133
	1-propanol	Henry	128-133
	2-propanol	Henry	128-133
	2-propanone	Enthalpy	303-304
	n-tetradecane	Henry	128-133
	tetrahydrofuran	LLE	174
	toluene	Henry	128-133
	n-undecane	Henry	128-133
	water	VLE	87-88
	water	LLE	162, 174
	water	Enthalpy	299-303
Vinyl acetate/vinyl caprolactam copolymer			
	water	LLE	162
Vinyl acetate/vinyl chloride copolymer			
	acetaldehyde	Henry	136-138
	acetic acid	Henry	136-138
	acetonitrile	Henry	136-138
	acrylonitrile	Henry	137
	benzene	VLE	85
	benzene	Henry	135-139
	1-butanol	VLE	85
	2-butanone	Henry	136-139
	chlorobenzene	VLE	85
	chlorobenzene	Henry	137
	cyclohexane	Henry	135-139
	cyclohexanone	Enthalpy	305
	n-decane	Henry	137
	1,2-dichloroethane	Henry	135-139
	1,4-dioxane	Henry	135-139
	ethylbenzene	VLE	85-86
	ethylbenzene	Henry	137-138
	fluorobenzene	Henry	138
	n-heptane	Henry	135-139
	methanol	VLE	86
	methanol	Henry	136-138
	nitroethane	Henry	135-139
	n-nonane	Henry	137
	n-octane	VLE	86
	2-propanol	Henry	135-139
	2-propanone	Henry	136-139
	propyl acetate	Henry	138

Copolymer (B)	Solvent (A)	Property	Page(s)
Vinyl acetate/vinyl chloride copolymer			
	styrene	Henry	137
	tetrachloromethane	Henry	137
	tetrahydrofuran	Henry	136-139
	tetrahydrofuran	A_2	443
	toluene	Henry	135-139
	trichloromethane	Henry	137-138
	vinyl acetate	Henry	138
	vinyl chloride	Henry	138
	p-xylene	VLE	86
	p-xylene	Henry	138
Vinyl acetate/vinyl chloride/glycidyl methacrylate terpolymer			
	2-propanone	Henry	139
	propyl acetate	Henry	139
	vinyl acetate	Henry	139
	vinyl chloride	Henry	139
Vinyl acetate/vinyl chloride/hydroxypropyl acrylate terpolymer			
	2-propanone	Henry	140
	propyl acetate	Henry	140
	vinyl chloride	Henry	140
Vinyl acetate/1-vinyl-2-pyrrolidinone copolymer			
	water	LLE	162
	vinyl acetate	Enthalpy	336
	vinyl acetate	A_2	443
	1-vinyl-2-pyrrolidinone	Enthalpy	336
	1-vinyl-2-pyrrolidinone	A_2	443
Vinyl alcohol/ethylene copolymer *see* Ethylene/vinyl alcohol copolymer			
Vinyl alcohol/vinyl acetate copolymer *see* Vinyl acetate/vinyl alcohol copolymer			
Vinyl alcohol/vinyl butyrate copolymer			
	water	LLE	162, 182

Copolymer (B)	Solvent (A)	Property	Page(s)
Vinyl amine/vinyl caprolactam copolymer			
	water	LLE	162, 182
	water	Enthalpy	336
Vinyl caprolactam/vinyl acetate copolymer *see* Vinyl acetate/vinyl caprolactam copolymer			
Vinyl caprolactam/N-vinyl-N-methylacetamide copolymer			
	water	LLE	162
N-Vinylcarbazole/4-ethylstyrene copolymer			
	without	PVT	428-429
N-Vinylcarbazole/4-hexylstyrene copolymer			
	without	PVT	428-429
N-Vinylcarbazole/4-octylstyrene copolymer			
	without	PVT	428-429
N-Vinylcarbazole/4-pentylstyrene copolymer			
	without	PVT	428-429
Vinyl chloride/acrylonitrile copolymer *see* Acrylonitrile/vinyl chloride copolymer			
Vinyl chloride/butyl acrylate copolymer			
	tetrahydrofuran	A_2	444
Vinyl chloride/butyl ether copolymer			
	tetrahydrofuran	A_2	444
Vinyl chloride/methyl acrylate copolymer			
	tetrahydrofuran	A_2	444
Vinyl chloride/vinyl acetate copolymer *see* Vinyl acetate/vinyl chloride copolymer			

Copolymer (B)	Solvent (A)	Property	Page(s)
Vinyl chloride/vinylidene chloride copolymer			
	benzene	Henry	141
	cyclohexane	Henry	141
	n-decane	Henry	141
	1,1-dichloroethene	Henry	141
	n-heptane	Henry	141
	n-nonane	Henry	141
	n-octane	Henry	141
	tetrachloromethane	Henry	141
	tetrahydrofuran	Henry	141
	tetrahydrofuran	A_2	444
	toluene	Henry	141
	trichloromethane	Henry	141
	trichloromethane	Enthalpy	306
	water	Enthalpy	332
Vinylferrocene/N-acryloylpyrrolidine copolymer *see* N-Acryloylpyrrolidine/vinylferrocene copolymer			
Vinylferrocene/N-cyclopropylacrylamide copolymer *see* N-Cyclopropylacrylamide/vinylferrocene copolymer			
Vinylferrocene/N,N-diethylacrylamide copolymer *see* N,N-Diethylacrylamide/vinylferrocene copolymer			
Vinylferrocene/N-ethylacrylamide copolymer *see* N-Ethylacrylamide/vinylferrocene copolymer			
Vinylferrocene/N-isopropylacrylamide copolymer *see* N-Isopropylacrylamide/vinylferrocene copolymer			
Vinylidene chloride/acrylonitrile copolymer *see* Acrylonitrile/vinylidene chloride copolymer			
Vinylidene chloride/vinyl chloride copolymer *see* Vinyl chloride/vinylidene chloride copolymer			

Copolymer (B)	Solvent (A)	Property	Page(s)
Vinylidene fluoride/hexafluoropropylene copolymer			
	carbon dioxide	HPPE	228
	1-chloro-1,1-difluoroethane	HPPE	228
	chlorodifluoromethane	HPPE	228
	chlorotrifluoromethane	HPPE	228
	1,1-difluoroethane	HPPE	228
	difluoromethane	HPPE	228
	hexafluoropropene	HPPE	228
	pentafluoroethane	HPPE	228
	1,1,2,2-tetrafluoroethane	HPPE	228
	trifluoromethane	HPPE	228
N-Vinylisobutyramide/N-vinylvaleramide copolymer			
	water	LLE	162
4-Vinylpyridine/ethyl acrylate copolymer *see* Ethyl acrylate/4-vinylpyridine copolymer			
4-Vinylpyridine/methyl methacrylate copolymer *see* Methyl methacrylate/4-vinylpyridine copolymer			
4-Vinylpyridine/styrene copolymer *see* Styrene/4-vinylpyridine copolymer			
1-Vinyl-2-pyrrolidinone/styrene copolymer *see* Styrene/1-vinyl-2-pyrrolidinone copolymer			
1-Vinyl-2-pyrrolidinone/vinyl acetate copolymer *see* Vinyl acetate/1-vinyl-2-pyrrolidinone copolymer			
5-Vinyltetrazole/2-methyl-5-vinyltetrazole copolymer			
	N,N-dimethylformamide	Enthalpy	336
	water	Enthalpy	336

8.3. List of solvents in alphabetical order

Name	Formula	CAS-RN	Page(s)
acetaldehyde	C_2H_4O	75-07-0	114, 136, 138, 320
acetic acid	$C_2H_4O_2$	64-19-7	114, 136, 138, 161, 320
acetonitrile	C_2H_3N	75-05-8	24, 29-31, 114, 136, 138, 320, 438-439
acrylic acid	$C_3H_4O_2$	79-10-7	229-232
acrylonitrile	C_3H_3N	107-13-1	114, 137, 320, 333, 440-441
anisole	C_7H_8O	100-66-3	48-49, 149-150, 160, 179
argon	Ar	7440-37-1	334
benzene	C_6H_6	71-43-2	33, 40, 51-53, 79, 81, 85, 89-94, 97-105, 110, 113-124, 127-141, 173, 270, 276-280, 307-313, 318-320, 323, 329, 331, 333-335, 435-436
bisphenol-A-diglycidyl ether	$C_{21}H_{24}O_4$	1675-54-3	160
1-bromobutane	C_4H_9Br	109-65-9	114, 320
2-bromobutane	C_4H_9Br	78-76-2	114, 320
1,3-butadiene	C_4H_6	106-99-0	91, 101-104
n-butane	C_4H_{10}	106-97-8	91, 101-104, 225-226, 233, 236-237, 264-265, 331
butanoic acid	$C_4H_8O_2$	107-92-6	161
1-butanol	$C_4H_{10}O$	71-36-3	53, 85, 114-116, 126-133, 172, 264, 309, 311, 319-324, 327, 330
2-butanol	$C_4H_{10}O$	78-92-2	114, 128-133, 320, 322
tert-butanol	$C_4H_{10}O$	75-65-0	128-133
2-butanone	C_4H_8O	78-93-3	92, 105, 112-116, 125, 136-139, 160-162, 173-174, 297-298, 310-313, 320, 329, 333, 439-442
1-butene	C_4H_8	106-98-9	201-202, 225-227, 241, 264-265
butyl acetate	$C_6H_{12}O_2$	123-86-4	53-54, 157, 182, 311, 320, 323
butylbenzene	$C_{10}H_{14}$	104-51-8	311, 323
tert-butyl-benzene	$C_{10}H_{14}$	98-06-6	121-122, 318, 323
butylcyclohexane	$C_{10}H_{20}$	1678-93-9	90, 323
γ-butyrolactone	$C_4H_6O_2$	96-48-0	435
carbon dioxide	CO_2	124-38-9	112-115, 187, 223-228, 251, 255-256, 264-265, 333-334
chlorobenzene	C_6H_5Cl	108-90-7	71-72, 85, 106-108, 114-116, 137, 311, 319-320, 323, 335
1-chlorobutane	C_4H_9Cl	109-69-3	90, 115-116, 121-122, 319, 323
chlorocyclohexane	$C_6H_{11}Cl$	542-18-7	319

Name	Formula	CAS-RN	Page(s)
1-chloro-1,1-difluoroethane	$C_2H_3ClF_2$	75-68-3	228
chlorodifluoromethane	$CHClF_2$	75-45-6	226, 228, 264
chlorotrifluoromethane	$CClF_3$	75-72-9	228
cyclohexane	C_6H_{12}	110-82-7	30-31, 38-42, 46-47, 54, 74-77, 83, 90-91, 97-105, 110, 113-114, 117, 121-122, 135-141, 158, 161, 163, 173, 179-182, 227, 285, 308, 312-313, 318-320, 323, 333-335, 435-436
cyclohexanol	$C_6H_{12}O$	108-93-0	128-133, 182, 311, 319, 323, 442-443
cyclohexanone	$C_6H_{10}O$	108-94-1	281-282, 296, 305, 319
cyclooctane	C_8H_{16}	292-64-8	285-286
cyclopentane	C_5H_{10}	287-92-3	158, 179-182, 286
decahydronaphthalene	$C_{10}H_{18}$	91-17-8	174
cis-decahydronahpthalene	$C_{10}H_{18}$	493-01-6	286
trans-decahydronahpthalene	$C_{10}H_{18}$	493-02-7	158, 286-287
n-decane	$C_{10}H_{22}$	124-18-5	89-92, 98-108, 115-125, 128-133, 137, 141, 310, 318, 320, 323-330
1-decanol	$C_{10}H_{22}O$	112-30-1	129-133
dibutyl ether	$C_8H_{18}O$	142-96-1	126, 327, 330
1,2-dichloroethane	$C_2H_4Cl_2$	107-06-2	35, 114, 135-139, 283-284
dichloromethane	CH_2Cl_2	75-09-2	110, 114, 116, 121-122, 311, 314-315, 321-323, 335
diethyl ether	$C_4H_{10}O$	60-29-7	114, 116, 321
3,3-diethylpentane	C_9H_{20}	4032-86-4	287
1,1-difluoroethane	$C_2H_4F_2$	75-37-6	228
difluoromethane	CH_2F_2	75-10-5	228
diisobutyl adipate	$C_{14}H_{26}O_4$	141-04-8	437
N,N-dimethylacetamide	C_4H_9NO	127-19-5	275, 435, 439-440
2,2-dimethylbutane	C_6H_{14}	75-83-2	91, 101, 104, 179-180
2,3-dimethylbutane	C_6H_{14}	79-29-8	179-180
dimethyl ether	C_2H_6O	115-10-6	205, 225-226, 264
N,N-dimethylformamide	C_3H_7NO	68-12-2	162, 173, 271-273, 435
2,4-dimethylhexane	C_8H_{18}	589-43-5	180
2,5-dimethylhexane	C_8H_{18}	592-13-2	180
3,4-dimethylhexane	C_8H_{18}	583-48-2	179-180
2,2-dimethylpentane	C_7H_{16}	590-35-2	179-180, 287
2,3-dimethylpentane	C_7H_{16}	565-59-3	179-180, 287-288
2,4-dimethylpentane	C_7H_{16}	108-08-7	179-180, 288
3,3-dimethylpentane	C_7H_{16}	562-49-2	288
dimethyl phthalate	$C_{10}H_{10}O_4$	131-11-3	173
dimethyl sulfoxide	C_2H_6OS	67-68-5	174
1,4-dioxane	$C_4H_8O_2$	123-91-1	89-90, 113-114, 117, 126, 135-139, 174, 311, 321, 327, 330, 442

Name	Formula	CAS-RN	Page(s)
diphenyl ether	$C_{12}H_{10}O$	101-84-8	161, 181
dipropyl ether	$C_6H_{14}O$	111-43-3	114, 321
n-dodecane	$C_{12}H_{26}$	112-40-3	92, 125, 129-133, 288-289, 310-311, 323, 329, 335
ethane	C_2H_6	74-84-0	112-115, 205, 226, 234, 251
ethanol	C_2H_6O	64-17-5	92, 114, 125, 264-265, 303-304, 309-310, 320-324, 329
ethene	C_2H_4	74-85-1	112-115, 188-195, 197-199, 202, 205, 208-221, 225-227, 229-265
ethoxybenzene	$C_8H_{10}O$	103-73-1	160
2-ethoxyethanol	$C_4H_{10}O_2$	110-80-5	442-443
ethyl acetate	$C_4H_8O_2$	141-78-6	54-55, 89-90, 93-94, 113-117, 125, 161, 181-182, 303-304, 310, 320-321, 329, 439-441
ethylbenzene	C_8H_{10}	100-41-4	41-45, 67, 80-82, 85-86, 91, 101-105, 129-133, 137-138, 277-280, 313, 318, 323
2-ethylhexyl acrylate	$C_{11}H_{20}O_2$	103-11-7	264
3-ethylpentane	C_7H_{16}	617-78-7	179-180, 289
fluorobenzene	C_6H_5F	462-06-6	138
formamide	CH_3NO	75-12-7	161
formic acid	CH_2O_2	64-18-6	114, 321
furan	C_4H_4O	110-00-9	114, 321
helium	He	7440-59-7	234, 238, 252, 257, 333
2,2,4,4,6,8,8-heptamethylnonane	$C_{16}H_{34}$	4390-04-9	289
n-heptane	C_7H_{16}	142-82-5	89-90, 98-105, 110, 114, 126-127, 135-141, 179-180, 226-227, 307-308, 312-313, 316, 318, 321, 327-331, 335
1-heptanol	$C_7H_{16}O$	111-70-6	129-133
n-hexadecane	$C_{16}H_{34}$	544-76-3	129-134, 290
hexafluorobenzene	C_6F_6	392-56-3	331
hexafluoroethane	C_2F_6	76-16-4	228
hexafluoropropene	C_3F_6	116-15-4	228
n-hexane	C_6H_{14}	110-54-3	30-31, 39, 42, 89-91, 97-105, 110, 114-116, 127, 160, 179-180, 227, 265, 308, 312-313, 316, 318-320, 325-331, 335
1-hexanol	$C_6H_{14}O$	111-27-3	128-134, 309
1-hexene	C_6H_{12}	592-41-6	202-203, 235-242, 264-265
hexyl acetate	$C_8H_{16}O_2$	142-92-7	159, 182
hydrogen	H_2	1333-74-0	333-334
krypton	Kr	7439-90-9	334
mesitylene	C_9H_{12}	108-67-8	41, 80, 82, 91, 101-104
methane	CH_4	74-82-8	235, 239-240, 265, 334

Name	Formula	CAS-RN	Page(s)
methanol	CH_4O	67-56-1	74-75, 78, 83, 86, 93, 114, 116, 125, 136, 138, 227, 265, 310, 320-324, 329
methyl acetate	$C_3H_6O_2$	79-20-9	55-56, 81, 92, 124, 153, 181, 310, 329
methyl acrylate	$C_4H_6O_2$	96-33-3	264
2-methylbutane	C_5H_{12}	78-78-4	160, 179-180, 265
methyl chloride	CH_3Cl	74-87-3	112-115
methylcyclohexane	C_7H_{14}	108-87-2	114, 173, 179-180, 320, 323
methylcyclopentane	C_6H_{12}	96-37-7	160, 179-180
2-methylhexane	C_7H_{16}	591-76-4	179-180
3-methylhexane	C_7H_{16}	589-34-4	181, 290
2-methylpentane	C_6H_{14}	107-83-5	179, 227
3-methylpentane	C_6H_{14}	96-14-0	161
4-methyl-2-pentanone	$C_6H_{12}O$	108-10-1	173
2-methylpropane	C_4H_{10}	75-28-5	91, 101, 104, 226, 264, 331
neon	Ne	7440-01-9	334
nitroethane	$C_2H_5NO_2$	79-24-3	114, 135-139, 174, 320, 442
nitrogen	N_2	7727-37-9	7, 252, 258-259, 264-265, 333-334
nitromethane	CH_3NO_2	75-52-5	114, 320
1-nitropropane	$C_3H_7NO_2$	108-03-2	114, 160, 320
2-nitropropane	$C_3H_7NO_2$	79-46-9	114, 320
n-nonane	C_9H_{20}	111-84-2	43-45, 89-92, 98-102, 110, 125, 128-133, 137, 141, 179-181, 307, 310, 316, 318, 325-330
n-octadecane	$C_{18}H_{38}$	593-45-3	335
2,2,3,3,4,4,5,5-octafluoro-1-pentanol	$C_5H_4F_8O$	355-80-6	439
octafluoropropane	C_3F_8	76-19-7	228
octafluorotoluene	C_7F_8	434-64-0	331
n-octane	C_8H_{18}	111-65-9	30-31, 86, 89-92, 98-104, 109-110, 113-117, 121-122, 125-126, 141, 179-181, 290-291, 307, 310, 316-320, 323, 327-330
1-octanol	$C_8H_{18}O$	111-87-5	128-133
1-octene	C_8H_{16}	111-66-0	90, 114-116, 320
oxygen	O_2	7782-44-7	333-334
pentafluoroethane	C_2HF_5	354-33-6	228
2,2,4,6,6-pentamethylheptane	$C_{12}H_{26}$	13475-82-6	291
n-pentane	C_5H_{12}	109-66-0	30-32, 39, 43, 110, 114-116, 127, 161, 179-181, 320, 331, 335
1-pentanol	$C_5H_{12}O$	71-41-0	128-133, 309, 324
2-pentanol	$C_5H_{12}O$	6032-29-7	128-133
2-pentanone	$C_5H_{10}O$	107-87-9	90, 115-116, 319, 323, 335
3-pentanone	$C_5H_{10}O$	96-22-0	114, 320

Name	Formula	CAS-RN	Page(s)
pentyl acetate	$C_7H_{14}O_2$	628-63-7	159, 182
phenol	C_6H_6O	108-95-2	319
propane	C_3H_8	74-98-6	195-196, 206, 225-227, 236, 253, 264-265
1-propanol	C_3H_8O	71-23-8	129-133, 265, 309, 322-324
2-propanol	C_3H_8O	67-63-0	112-115, 129-139, 161-162, 172, 320, 322
2-propanone	C_3H_6O	67-64-1	38-43, 49-50, 73-78, 81, 84, 89-92, 112-116, 125, 136-140, 150-151, 160, 162, 179, 264-265, 274, 303-304, 310, 320, 329, 436-438
propene	C_3H_6	115-07-1	203-206, 225-227, 265
propionic acid	$C_3H_6O_2$	79-09-4	161
propionitrile	C_3H_5N	107-12-0	114, 320
propyl acetate	$C_5H_{10}O_2$	109-60-4	56, 114, 138-140, 321
propylbenzene	C_9H_{12}	103-65-1	33, 94
styrene	C_8H_8	100-42-5	103, 137, 161, 173, 181, 333, 440-441
sulfur dioxide	SO_2	7446-09-5	112-115
sulfur hexafluoride	SF_6	2551-62-4	228, 265
1,1,2,2-tetrachloroethane	$C_2H_2Cl_4$	79-34-5	172, 275, 335-336, 437
1,1,2,2-tetrachloroethene	C_2Cl_4	127-18-4	174
tetrachloromethane	CCl_4	56-23-5	67-68, 90, 110, 113-114, 117, 121-122, 137, 141, 294-295, 314-315, 319-320, 323, 442
n-tetradecane	$C_{14}H_{30}$	629-59-4	129-133, 311
1,1,1,2-tetrafluoroethane	$C_2H_2F_4$	811-97-2	228
tetrafluoromethane	CF_4	75-73-0	228
tetrahydrofuran	C_4H_8O	109-99-9	89-90, 113-117, 136, 139, 141, 162, 172-174, 182, 293, 311, 321, 435-436, 439-440, 443-444
2,2,4,4-tetramethylpentane	C_9H_{20}	1070-87-7	179-181, 291-292
toluene	C_7H_8	108-88-3	32-41, 57-59, 63-70, 80-82, 89-94, 97-108, 110, 113-117, 125, 129-141, 145-146, 162, 173, 181, 307-313, 318, 320, 329, 335, 436-438, 441-442
1,1,2-trichloro-1,2,2-trifluoroethane	$C_2Cl_3F_3$	76-13-1	443
trichloromethane	$CHCl_3$	67-66-3	31, 39-40, 44, 59, 73, 81, 89-90, 94, 97, 105, 110, 114-116, 121-122, 137-138, 141, 171-174, 306, 311-315, 319-323, 437
2,2,2-trifluoroethanol	$C_2H_3F_3O$	75-89-8	114, 321
trifluoromethane	CHF_3	75-46-7	228, 265

Name	Formula	CAS-RN	Page(s)
2,2,3-trimethylbutane	C_7H_{16}	464-06-2	179-181
3,4,5-trimethylheptane	$C_{10}H_{22}$	20278-89-1	121-122, 323
2,3,4-trimethylhexane	C_9H_{20}	921-47-1	179
2,2,4-trimethylpentane	C_8H_{18}	540-84-1	179-181, 277-280, 292, 318
n-undecane	$C_{11}H_{24}$	1120-21-4	93, 125, 129, 131-132, 134, 310, 329
vinyl acetate	$C_4H_6O_2$	108-05-4	112-115, 138-139, 215, 244-263, 265, 336
vinyl chloride	C_2H_3Cl	75-01-4	138-140
1-vinyl-2-pyrrolidinone	C_6H_9NO	88-12-0	336, 443
water	H_2O	7732-18-5	87-88, 114, 147-148, 152, 154-156, 160-162, 164-170, 172, 174, 178, 181-182, 299-303, 321, 332-336, 438
o-xylene	C_8H_{10}	95-47-6	34, 59-60, 93-94, 125, 310, 329
m-xylene	C_8H_{10}	108-38-3	34, 94, 114, 321
p-xylene	C_8H_{10}	106-42-3	34, 41, 60-62, 81-82, 86, 94, 103, 105, 138, 313, 320

8.4. List of solvents in order of their molecular formulas

Formula	Name	CAS-RN	Page(s)
Ar	argon	7440-37-1	334
$CClF_3$	chlorotrifluoromethane	75-72-9	228
CCl_4	tetrachloromethane	56-23-5	67-68, 90, 110, 113-114, 117, 121-122, 137, 141, 294-295, 314-315, 319-320, 323, 442
CF_4	tetrafluoromethane	75-73-0	228
$CHCl_3$	trichloromethane	67-66-3	31, 39-40, 44, 59, 73, 81, 89-90, 94, 97, 105, 110, 114-116, 121-122, 137-138, 141, 171-174, 306, 311-315, 319-323, 437
$CHClF_2$	chlorodifluoromethane	75-45-6	226, 228, 264
CHF_3	trifluoromethane	75-46-7	228, 265
CH_2Cl_2	dichloromethane	75-09-2	110, 114, 116, 121-122, 311, 314-315, 321-323, 335
CH_2F_2	difluoromethane	75-10-5	228
CH_2O_2	formic acid	64-18-6	114, 321

Formula	Name	CAS-RN	Page(s)
CH_3Cl	methyl chloride	74-87-3	112-115
CH_3NO	formamide	75-12-7	161
CH_3NO_2	nitromethane	75-52-5	114, 320
CH_4	methane	74-82-8	235, 239-240, 265, 334
CH_4O	methanol	67-56-1	74-75, 78, 83, 86, 93, 114, 116, 125, 136, 138, 227, 265, 310, 320-324, 329
CO_2	carbon dioxide	124-38-9	112-115, 187, 223-228, 251, 255-256, 264-265, 333-334
$C_2Cl_3F_3$	1,1,2-trichloro-1,2,2-trifluoroethane	76-13-1	443
C_2Cl_4	1,1,2,2-tetrachloroethene	127-18-4	174
C_2F_6	hexafluoroethane	76-16-4	228
C_2HF_5	pentafluoroethane	354-33-6	228
$C_2H_2Cl_4$	1,1,2,2-tetrachloroethane	79-34-5	172, 275, 335-336, 437
$C_2H_2F_4$	1,1,1,2-tetrafluoroethane	811-97-2	228
C_2H_3Cl	vinyl chloride	75-01-4	138-140
$C_2H_3ClF_2$	1-chloro-1,1-difluoroethane	75-68-3	228
$C_2H_3F_3O$	2,2,2-trifluoroethanol	75-89-8	114, 321
C_2H_3N	acetonitrile	75-05-8	24, 29-31, 114, 136, 138, 320, 438-439
C_2H_4	ethene	74-85-1	112-115, 188-195, 197-199, 202, 205, 208-221, 225-227, 229-265
$C_2H_4Cl_2$	1,2-dichloroethane	107-06-2	35, 114, 135-139, 283-284
$C_2H_4F_2$	1,1-difluoroethane	75-37-6	228
C_2H_4O	acetaldehyde	75-07-0	114, 136, 138, 320
$C_2H_4O_2$	acetic acid	64-19-7	114, 136, 138, 161, 320
$C_2H_5NO_2$	nitroethane	79-24-3	114, 135-139, 174, 320, 442
C_2H_6	ethane	74-84-0	112-115, 205, 226, 234, 251
C_2H_6O	dimethyl ether	115-10-6	205, 225-226, 264
C_2H_6O	ethanol	64-17-5	92, 114, 125, 264-265, 303-304, 309-310, 320-324, 329
C_2H_6OS	dimethyl sulfoxide	67-68-5	174
C_3F_6	hexafluoropropene	116-15-4	228
C_3F_8	octafluoropropane	76-19-7	228
C_3H_3N	acrylonitrile	107-13-1	114, 137, 320, 333, 440-441
$C_3H_4O_2$	acrylic acid	79-10-7	229-232
C_3H_5N	propionitrile	107-12-0	114, 320
C_3H_6	propene	115-07-1	203-206, 225-227, 265
C_3H_6O	2-propanone	67-64-1	38-43, 49-50, 73-78, 81, 84, 89-92, 112-116, 125, 136-140, 150-151, 160, 162, 179, 264-265, 274, 303-304, 310, 320, 329, 436-438
$C_3H_6O_2$	methyl acetate	79-20-9	55-56, 81, 92, 124, 153, 181, 310, 329
$C_3H_6O_2$	propionic acid	79-09-4	161
C_3H_7NO	N,N-dimethylformamide	68-12-2	162, 173, 271-273, 435

Formula	Name	CAS-RN	Page(s)
$C_3H_7NO_2$	1-nitropropane	108-03-2	114, 160, 320
$C_3H_7NO_2$	2-nitropropane	79-46-9	114, 320
C_3H_8	propane	74-98-6	195-196, 206, 225-227, 236, 253, 264-265
C_3H_8O	1-propanol	71-23-8	129-133, 265, 309, 322-324
C_3H_8O	2-propanol	67-63-0	112-115, 129-139, 161-162, 172, 320, 322
C_4H_4O	furan	110-00-9	114, 321
C_4H_6	1,3-butadiene	106-99-0	91, 101-104
$C_4H_6O_2$	γ-butyrolactone	96-48-0	435
$C_4H_6O_2$	methyl acrylate	96-33-3	264
$C_4H_6O_2$	vinyl acetate	108-05-4	112-115, 138-139, 215, 244-263, 265, 336
C_4H_8	1-butene	106-98-9	201-202, 225-227, 241, 264-265
C_4H_8O	2-butanone	78-93-3	92, 105, 112-116, 125, 136-139,160-162, 173-174, 297-298, 310-313, 320, 329, 333, 439-442
C_4H_8O	tetrahydrofuran	109-99-9	89-90, 113-117, 136, 139, 141, 162, 172-174, 182, 293, 311, 321, 435-436, 439-440, 443-444
$C_4H_8O_2$	butanoic acid	107-92-6	161
$C_4H_8O_2$	1,4-dioxane	123-91-1	89-90, 113-114, 117, 126, 135-139, 174, 311, 321, 327, 330, 442
$C_4H_8O_2$	ethyl acetate	141-78-6	54-55, 89-90, 93-94, 113-117, 125, 161, 181-182, 303-304, 310, 320-321, 329, 439-441
C_4H_9Br	1-bromobutane	109-65-9	114, 320
C_4H_9Br	2-bromobutane	78-76-2	114, 320
C_4H_9Cl	1-chlorobutane	109-69-3	90, 115-116, 121-122, 319, 323
C_4H_9NO	N,N-dimethylacetamide	127-19-5	275, 435, 439-440
C_4H_{10}	2-methylpropane	75-28-5	91, 101, 104, 226, 264, 331
C_4H_{10}	n-butane	106-97-8	91, 101-104, 225-226, 233, 236-237, 264-265, 331
$C_4H_{10}O$	1-butanol	71-36-3	53, 85, 114-116, 126-133, 172, 264, 309, 311, 319-324, 327, 330
$C_4H_{10}O$	2-butanol	78-92-2	114, 128-133, 320, 322
$C_4H_{10}O$	*tert*-butanol	75-65-0	128-133
$C_4H_{10}O$	diethyl ether	60-29-7	114, 116, 321
$C_4H_{10}O_2$	2-ethoxyethanol	110-80-5	442-443
$C_5H_4F_8O$	2,2,3,3,4,4,5,5-octafluoro-1-pentanol	355-80-6	439
C_5H_{10}	cyclopentane	287-92-3	158, 179-182, 286
$C_5H_{10}O$	2-pentanone	107-87-9	90, 115-116, 319, 323, 335
$C_5H_{10}O$	3-pentanone	96-22-0	114, 320

Formula	Name	CAS-RN	Page(s)
$C_5H_{10}O_2$	propyl acetate	109-60-4	56, 114, 138-140, 321
C_5H_{12}	2-methylbutane	78-78-4	160, 179-180, 265
C_5H_{12}	n-pentane	109-66-0	30-32, 39, 43, 110, 114-116, 127, 161, 179-181, 320, 331, 335
$C_5H_{12}O$	1-pentanol	71-41-0	128-133, 309, 324
$C_5H_{12}O$	2-pentanol	6032-29-7	128-133
C_6F_6	hexafluorobenzene	392-56-3	331
C_6H_5Cl	chlorobenzene	108-90-7	71-72, 85, 106-108, 114-116, 137, 311, 319-320, 323, 335
C_6H_5F	fluorobenzene	462-06-6	138
C_6H_6	benzene	71-43-2	33, 40, 51-53, 79, 81, 85, 89-94, 97-105, 110, 113-124, 127-141, 173, 270, 276-280, 307-313, 318-320, 323, 329, 331, 333-335, 435-436
C_6H_6O	phenol	108-95-2	319
C_6H_9NO	1-vinyl-2-pyrrolidinone	88-12-0	336, 443
$C_6H_{10}O$	cyclohexanone	108-94-1	281-282, 296, 305, 319
$C_6H_{11}Cl$	chlorocyclohexane	542-18-7	319
C_6H_{12}	cyclohexane	110-82-7	30-31, 38-42, 46-47, 54, 74-77, 83, 90-91, 97-105, 110, 113-114, 117, 121-122, 135-141, 158, 161, 163, 173, 179-182, 227, 285, 308, 312-313, 318-320, 323, 333-335, 435-436
C_6H_{12}	1-hexene	592-41-6	202-203, 235-242, 264-265
C_6H_{12}	methylcyclopentane	96-37-7	160, 179-180
$C_6H_{12}O$	4-methyl-2-pentanone	108-10-1	173
$C_6H_{12}O$	cyclohexanol	108-93-0	128-133, 182, 311, 319, 323, 442-443
$C_6H_{12}O_2$	butyl acetate	123-86-4	53-54, 157, 182, 311, 320, 323
C_6H_{14}	2,2-dimethylbutane	75-83-2	91, 101, 104, 179-180
C_6H_{14}	2,3-dimethylbutane	79-29-8	179-180
C_6H_{14}	n-hexane	110-54-3	30-31, 39, 42, 89-91, 97-105, 110, 114-116, 127, 160, 179-180, 227, 265, 308, 312-313, 316, 318-320, 325-331, 335
C_6H_{14}	2-methylpentane	107-83-5	179, 227
C_6H_{14}	3-methylpentane	96-14-0	161
$C_6H_{14}O$	1-hexanol	111-27-3	128-134, 309
$C_6H_{14}O$	dipropyl ether	111-43-3	114, 321
C_7F_8	octafluorotoluene	434-64-0	331
C_7H_8	toluene	108-88-3	32-41, 57-59, 63-70, 80-82, 89-94, 97-108, 110, 113-117, 125, 129-141, 145-146, 162, 173, 181, 307-313, 318, 320, 329, 335, 436-438, 441-442

Formula	Name	CAS-RN	Page(s)
C_7H_8O	anisole	100-66-3	48-49, 149-150, 160, 179
C_7H_{14}	methylcyclohexane	108-87-2	114, 173, 179-180, 320, 323
$C_7H_{14}O_2$	pentyl acetate	628-63-7	159, 182
C_7H_{16}	2,2-dimethylpentane	590-35-2	179-180, 287
C_7H_{16}	2,3-dimethylpentane	565-59-3	179-180, 287-288
C_7H_{16}	2,4-dimethylpentane	108-08-7	179-180, 288
C_7H_{16}	3,3-dimethylpentane	562-49-2	288
C_7H_{16}	3-ethylpentane	617-78-7	179-180, 289
C_7H_{16}	n-heptane	142-82-5	89-90, 98-105, 110, 114, 126-127, 135-141, 179-180, 226-227, 307-308, 312-313, 316, 318, 321, 327-331, 335
C_7H_{16}	2-methylhexane	591-76-4	179-180
C_7H_{16}	3-methylhexane	589-34-4	181, 290
C_7H_{16}	2,2,3-trimethylbutane	464-06-2	179-181
$C_7H_{16}O$	1-heptanol	111-70-6	129-133
C_8H_8	styrene	100-42-5	103, 137, 161, 173, 181, 333, 440-441
C_8H_{10}	ethylbenzene	100-41-4	41-45, 67, 80-82, 85-86, 91, 101-105, 129-133, 137-138, 277-280, 313, 318, 323
C_8H_{10}	o-xylene	95-47-6	34, 59-60, 93-94, 125, 310, 329
C_8H_{10}	m-xylene	108-38-3	34, 94, 114, 321
C_8H_{10}	p-xylene	106-42-3	34, 41, 60-62, 81-82, 86, 94, 103, 105, 138, 313, 320
$C_8H_{10}O$	ethoxybenzene	103-73-1	160
C_8H_{16}	cyclooctane	292-64-8	285-286
C_8H_{16}	1-octene	111-66-0	90, 114-116, 320
$C_8H_{16}O_2$	hexyl acetate	142-92-7	159, 182
C_8H_{18}	2,4-dimethylhexane	589-43-5	180
C_8H_{18}	2,5-dimethylhexane	592-13-2	180
C_8H_{18}	3,4-dimethylhexane	583-48-2	179-180
C_8H_{18}	n-octane	111-65-9	30-31, 86, 89-92, 98-104, 109-110, 113-117, 121-122, 125-126, 141, 179-181, 290-291, 307, 310, 316-320, 323, 327-330
C_8H_{18}	2,2,4-trimethylpentane	540-84-1	179-181, 277-280, 292, 318
$C_8H_{18}O$	1-octanol	111-87-5	128-133
$C_8H_{18}O$	dibutyl ether	142-96-1	126, 327, 330
C_9H_{12}	mesitylene	108-67-8	41, 80, 82, 91, 101-104
C_9H_{12}	propylbenzene	103-65-1	33, 94
C_9H_{20}	2,2,4,4-tetramethylpentane	1070-87-7	179-181, 291-292
C_9H_{20}	2,3,4-trimethylhexane	921-47-1	179
C_9H_{20}	3,3-diethylpentane	4032-86-4	287
C_9H_{20}	n-nonane	111-84-2	43-45, 89-92, 98-102, 110, 125, 128-133, 137, 141, 179-181, 307, 310, 316, 318, 325-330

Formula	Name	CAS-RN	Page(s)
$C_{10}H_{10}O_4$	dimethyl phthalate	131-11-3	173
$C_{10}H_{14}$	butylbenzene	104-51-8	311, 323
$C_{10}H_{14}$	*tert*-butyl-benzene	98-06-6	121-122, 318, 323
$C_{10}H_{18}$	*cis*-decahydronahpthalene	493-01-6	286
$C_{10}H_{18}$	decahydronaphthalene	91-17-8	174
$C_{10}H_{18}$	*trans*-decahydronahpthalene	493-02-7	158, 286-287
$C_{10}H_{20}$	butylcyclohexane	1678-93-9	90, 323
$C_{10}H_{22}$	3,4,5-trimethylheptane	20278-89-1	121-122, 323
$C_{10}H_{22}$	n-decane	124-18-5	89-92, 98-108, 115-125, 128-133, 137, 141, 310, 318, 320, 323-330
$C_{10}H_{22}O$	1-decanol	112-30-1	129-133
$C_{11}H_{20}O_2$	2-ethylhexyl acrylate	103-11-7	264
$C_{11}H_{24}$	n-undecane	1120-21-4	93, 125, 129, 131-132, 134, 310, 329
$C_{12}H_{10}O$	diphenyl ether	101-84-8	161, 181
$C_{12}H_{26}$	2,2,4,6,6-pentamethylheptane	13475-82-6	291
$C_{12}H_{26}$	n-dodecane	112-40-3	92, 125, 129-133, 288-289, 310-311, 323, 329, 335
$C_{14}H_{26}O_4$	diisobutyl adipate	141-04-8	437
$C_{14}H_{30}$	n-tetradecane	629-59-4	129-133, 311
$C_{16}H_{34}$	2,2,4,4,6,8,8-heptamethylnonane	4390-04-9	289
$C_{16}H_{34}$	n-hexadecane	544-76-3	129-134, 290
$C_{18}H_{38}$	n-octadecane	593-45-3	335
$C_{21}H_{24}O_4$	bisphenol-A-diglycidyl ether	1675-54-3	160
H_2	hydrogen	1333-74-0	333-334
H_2O	water	7732-18-5	87-88, 114, 147-148, 152, 154-156, 160-162, 164-170, 172, 174, 178, 181-182, 299-303, 321, 332-336, 438
He	helium	7440-59-7	234, 238, 252, 257, 333
Kr	krypton	7439-90-9	334
N_2	nitrogen	7727-37-9	7, 252, 258-259, 264-265, 333-334
Ne	neon	7440-01-9	334
O_2	oxygen	7782-44-7	333-334
SF_6	sulfur hexafluoride	2551-62-4	228, 265
SO_2	sulfur dioxide	7446-09-5	112-115

Printed and bound by CPI Group (UK) Ltd, Croydon, CR0 4YY
17/10/2024
01775701-0004